Advanced Functional and Composite Materials

This book provides an in-depth understanding of the properties, characterization, and applications of advanced functional and composite materials for energy, electronics, biomedicine, and aerospace fields. Divided into four sections, it covers sustainable smart and composite materials, intelligent techniques for optimization of material properties, their characterization and processing, case studies based on modernization in processes and systems, sustainability-based approaches and novel strategies, and sustainability and Industry 4.0-based contents.

Key Highlights

- Provides smart materials and sustainability, novel product design, and materials implementation
- Reviews molecular dynamics simulation techniques
- Includes Industry 4.0 and sustainability interventions in functional composites
- Discusses life cycle engineering and analysis in composite processing
- Covers microwave-assisted processing of composites

This book is aimed at graduate students and researchers in materials sciences and metallurgy, mechanical and manufacturing engineering.

Advanced Materials Processing and Manufacturing

Series Editor: Kapil Gupta

The CRC Press Series in Advanced Materials Processing and Manufacturing covers the complete spectrum of materials and manufacturing technology, including fundamental principles, theoretical background, and advancements. Considering the accelerated importance of advances for producing quality products for a wide range of applications, the titles in this series reflect the state-of-the-art in understanding and engineering the materials processing and manufacturing operations. Technological advancements for enhancement of product quality, process productivity, and sustainability, are on special focus including processing for all materials and novel processes. This series aims to foster knowledge enrichment on conventional and modern machining processes. Micro-manufacturing technologies such as micro-machining, micro-forming, and micro-joining, and Hybrid manufacturing, additive manufacturing, near net shape manufacturing, and ultra-precision finishing techniques are also covered.

Advanced Materials Characterization: Basic Principles, Novel Applications, and Future Directions
Ch Sateesh Kumar, M. Muralidhar Singh and Ram Krishna

Thin-Films for Machining Difficult-to-Cut Materials: Challenges, Applications, and Future Prospects
Ch Sateesh Kumar and Filipe Daniel Fernandes

Advanced Materials Processing and Manufacturing: Research, Technology, and Applications
Amogelang Sylvester Bolokang and Maria Ntsoaki Mathabathe

Advanced Joining Technologies
Edited by Manjaiah M, Shivraman Thapliyal and Adepu Kumar

Nanofinishing of Materials for Advanced Industrial Applications
Edited by Faiz Iqbal, Dilshad Ahmad Khan and Zafar Alam

Environmentally Benign Machining
Edited by Şenol Bayraktar and Sunil Pathak

Post-Processing of Parts and Components Fabricated by Fused Deposition Modeling: Techniques and Advancements
Edited by Vinayak R. Malik, Vivek Tiwary and Arunkumar P.

Forming and Machining of Polymers, Ceramics, and Composites
Edited by Matruprasad Rout and Kishore Debnath

Fundamentals and Advances in Metal Matrix Composites
Edited by Tharmaraj Ramakrishnan and PM Gopal

Thermoelectric Materials and Devices for Clean Energy Harvesting: Properties, Fabrication, and Opportunities
Ram Krishna and A. D. Dhass

Novel Applications of Functionally Graded Materials
Edited by Neeraj Kumar Bhoi, Harpreet Singh and Himansu Sekhar Nanda

Advanced Functional and Composite Materials: Fabrication, Characterization, and Applications
Edited by Neeraj Kumar Bhoi, Harpreet Singh, and Himansu Sekhar Nanda

For more information about this series, please visit: www.routledge.com/Advanced-Materials-Processing-and-Manufacturing/book-series/CRCAMPM

Advanced Functional and Composite Materials

Fabrication, Characterization, and Applications

Edited by
Neeraj Kumar Bhoi, Harpreet Singh, and
Himansu Sekhar Nanda

CRC Press
Taylor & Francis Group
Boca Raton London New York

CRC Press is an imprint of the
Taylor & Francis Group, an informa business

First edition published 2026
by CRC Press
2385 NW Executive Center Drive, Suite 320, Boca Raton FL 33431

and by CRC Press
4 Park Square, Milton Park, Abingdon, Oxon, OX14 4RN

CRC Press is an imprint of Taylor & Francis Group, LLC

ISBN: 978-1-032-61934-7 (hbk)
ISBN: 978-1-041-12413-9 (pbk)
ISBN: 978-1-003-66461-1 (ebk)

DOI: 10.1201/9781003664611

Typeset in Times
by Newgen Publishing UK

Contents

Preface

Functional materials play a crucial role in defining our future in a rapidly transforming technological context. *Advanced Functional and Composite Materials: Fabrication, Characterization, and Applications* is a significant inclusion that captures the transformational capabilities and complex details of these state-of-the-art materials. This book is based on the acknowledgement of the crucial significance of materials science in propelling progress in several fields like electronics, energy, medicine, and aerospace.

The contributors accurately depict the research community's unwavering commitment to innovation, providing a comprehensive account of the many strategies used to improve material performance. This book explores the complicated process of material evolution by examining the synthesis of composite materials, the combining of different alloys, and the modification of microstructures. Their zeal and devotion are clearly visible, providing readers with a detailed understanding of the endeavour to expand the limits of what is achievable. This edited book offers a complete and holistic review of advanced functional and composite materials, filling a key gap in the current literature. This convergence of innovative research methodologies and practical applications ensures its relevance in both academic and industrial contexts. The incorporation of a wide range of developments, including metal and polymeric composites, biomaterials, smart structures, and nano-finishing technologies, highlights the extensive scope of the discipline.

The integration of contributions from prominent experts enhances the authenticity and depth of this resource, making it essential for scholars, students, and professionals worldwide. The book's thorough collection guarantees to be an indispensable reference, providing insights into current research trends and manufacturing tactics, with a focus on environmental sustainability and the evolving demands of clients. Although this first version aims to be accurate and complete, the authors freely welcome criticism and additions from readers. The purpose of this collaborative effort is to improve future editions, so guaranteeing that the book maintains its status as a reliable and influential source in the constantly evolving subject of materials science.

Advanced Functional and Composite Materials: Fabrication, Characterization, and Applications is a clear demonstration of the ingenuity and commitment of the individuals who contributed to it. This resource is positioned to motivate and educate the next cohort of materials scientists and engineers, providing them with guidance to drive their efforts towards innovation and excellence.

Chapter 1 explores functionally graded materials, focusing on their bioinspired design, fabrication, and applications in implants, drug delivery, and tissue engineering.

Chapter 2 focuses on active smart materials that respond to external stimuli like temperature, pH, and electric or magnetic fields, covering shape memory alloys (SMAs), chromogenic, magnetorheological (MR), thermoelectric, and piezoelectric materials, along with their synthesis, properties, and applications.

Chapter 3 presents a detailed overview of polymer composites, covering their types, properties, characterization techniques, and applications, with a special focus on sustainability, eco-friendly enhancements, and their role in smart material systems.

Chapter 4 investigates a polyvinylidene fluoride (PVDF) based hybrid composite with boron nitride nanosheets (BNNSs) and graphene nano platelets (GNPs) for flexible piezoelectric sensors, highlighting improved tensile strength and potential in wearable, self-powered sensing applications.

Chapter 5 focuses on carbon-based noble metal nanocomposites, highlighting their biocompatibility and multifunctionality in electrochemical sensing and anticancer applications, with detailed discussion on CNT, graphene, and carbon dot-based systems.

Chapter 6 examines the effect of varying jute and nettle fiber ratios on the water absorption and mechanical properties of natural fiber composites, highlighting the role of jute in enhancing strength and stiffness through improved interfacial bonding.

Chapter 7 highlights the integration of artificial intelligence (AI) in damage characterization of fiber-reinforced composites, focusing on advanced testing methods, microscopy techniques, and machine learning models to enhance microstructural analysis and damage assessment.

Chapter 8 presents molecular dynamics simulations of penta-graphene sheets, analyzing the effects of functionalization, temperature, strain rate, and length-to-width ratios on their mechanical properties.

Chapter 9 discuss numerical analysis comparing the blast performance of AL-6XN steel and carbon fiber reinforced polymer (CFRP) composite laminates, highlighting the superior energy absorption and lower deflection of CFRP, making it a promising material for protective military and naval applications.

Chapter 10 examines nanoparticle-enhanced green fiber composites, focusing on improved water resistance and sustainable applications in automotive, aerospace, biomedical, and construction fields.

Chapter 11 reviews severe plastic deformation (SPD) techniques for producing ultrafine-grained materials, highlighting processes like equal channel angular pressing (ECAP), accumulative roll bending (ARB), and high-pressure torsion (HPT) that enhance mechanical properties and enable novel functional materials.

Chapter 12 reviews Al–TiB2 metal matrix composites, emphasizing their mechanical strength, corrosion resistance, and challenges like brittleness, while exploring multicomponent alloy reinforcements to enhance performance for automotive and aerospace applications.

Chapter 13 discuss the synthesis, functionalization, and applications of carbon nanotubes (CNTs) in electronics, energy storage, medicine, and composites, highlighting their vast potential and encouraging further research to expand their use across diverse fields.

Chapter 14 examines functionally graded materials and graphene-based coatings for implants, enhancing biocompatibility, durability, and recovery outcomes in medical applications.

Acknowledgements

We would like to extend our heartfelt gratitude to all the contributors whose expertise and dedication have been instrumental in bringing this book to fruition. We would like to express our gratitude to Gagandeep Singh, our editorial office assistant, for his unwavering support throughout this journey, and to Prof. Kapil Gupta, our series editor, for his indispensable guidance.

About the Editors

Neeraj Kumar Bhoi is currently working as Assistant Professor in the Department of Mechanical Engineering at O. P. Jindal University, Raigarh, Chhattisgarh, India. Previously, he worked as Assistant Professor in the School of Automation (Mechatronics Engineering) at Banasthali Vidyapith, Banasthali, Rajasthan, India, from September 2022 to June 2025. He received his B.E. in Mechanical Engineering with distinction from Bhilai Institute of Technology, Durg, in 2014, and his M.Tech. in Production Engineering from Veer Surendra Sai University of Technology, Burla (formerly UCE Burla), in 2016. He earned his Ph.D. from the Department of Mechanical Engineering at the Indian Institute of Information Technology, Design and Manufacturing, Jabalpur, India. He serves as an academic editor for *Advances in Materials Science and Engineering* (John Wiley & Sons Ltd.) and as an editorial board member for *Scientific Reports* and *Discover Materials* (Springer Nature Publications). His research interests include microwave-assisted processing, electronic waste recycling, sustainable manufacturing, and 3D-printed soft actuation systems for medical rehabilitation. Dr. Bhoi is the recipient of the 35th Madhya Pradesh Young Scientist Award for his research titled "An Innovative and Sustainable Manufacturing Process for Uniform Sintering and Cladding in the Microwave Field", awarded by the Madhya Pradesh Council of Science and Technology, Bhopal, Madhya Pradesh, India. He has authored over 23 peer-reviewed articles in prestigious journals, edited two books, contributed to seven book chapters, and eight international conference articles. He also serves as a reviewer for several book proposals and international journals.

Harpreet Singh is currently serving as Assistant Professor in the Mechanical Engineering Department at the National Institute of Technology, Jalandhar, India. Before the current assignment, he was Assistant Professor in the Mechanical Engineering Department at IIIT Jabalpur during 2017–22. He received his Ph.D. degree from the Department of Mechanical and Industrial Engineering at the Indian Institute of Technology, Roorkee. His research interests are broadly related to the advancement of precision machining processes, nano-finishing science and technology, green manufacturing methods, microwave processing of metals and composites, polymer processing and PMCs, and material processing and characterizations. He is one of the first researchers in the world to propose and develop a pulse DC supply-controlled and ultrasonic vibration-assisted ECH machine. Presently, he, along with his research group, is working for the new era of nano-finishing science and technology, namely CMMRAF, for biomedical applications, which was first proposed for nano-finishing of complex shapes. His research group is also making tremendous contributions to studying green manufacturing processes such as microwave heating of metals, alloys, and composites and establishing near-to-net shape manufacturing possibilities. Similarly, studies on PMCs using graphite and aluminum oxide and natural composites provide remarkable data for genuine improvements in the defense and gear industries. He has over 14 years of research and teaching experience at UG and PG levels. He has published more than 70 research papers in international peer-reviewed journals of repute, national and international conferences in India and abroad, and 2 patents. He has also guided a number of Ph.D. (awarded: 03, ongoing: 07), M.Tech (awarded: 18), and B.Tech (6-month intern project: awarded: 37, ongoing: 21) students. He is a reviewer for various government-funded projects and leading international journals of repute. He has been a resource person and delivered invited lectures in several national and international conferences, STTPs, and QIP programs.

Himansu Sekhar Nanda received Ph.D. in Materials Science and Engineering, majoring in Biomaterials, from National Institute for Materials Science (NIMS), Japan. He was a Post-Doctoral Fellow at King Abdullah University of Science and Technology (KAUST), Saudi Arabia, and a

Post-Doctoral Research Fellow at Nanyang Technological University (NTU) Singapore. Currently, he is working as an Assistant Professor at the Indian Institute of Information Technology Design and Manufacturing (IIITDM) Jabalpur, India (an Institute of National Importance, Ministry of Education, Government of India) and heading Laboratory of Biomaterials and Biomanufacturing at IIITDM. From August 2022–Jan 2023, he was a Research Scientist at Terasaki Institute for Biomedical Innovation (TIBI), Los Angeles, USA. Since Feb 2023, he has been working as an Affiliate Faculty at TIBI and has collaborated with the core scientific team of Dr. Ali Khademhosseini. He holds a few visiting positions: as an adjunct Assistant professor at the Beijing University of Chemical Technology and a Visiting Scientist at the State Key Lab of Molecular Engineering of Polymers, Fudan University, China. Apart from these visiting positions, he also holds the position of Founding President for Sustainable Biomanufacturer and Research Society. He has edited the special issues "Biomaterials: Future of Biomaterials 2021" and Biomaterials: Intelligent Biomaterials 2023" in the journal *Current Opinion in Biomedical Engineering* (Elsevier, 2022 IF: 4.16), "Biodegradable Polymers for Biomedical Applications Vol-1 and Vol-2" in *Frontier in Materials*, (Frontiers, 2022 IF: 3.5) and "Advances in Metallic Biomaterials" in *Bioengineering* (MDPI, 2022 IF: 5.05) and "Recent Progress in Biosensor Technologies for Detection of Exosomes" in *Biosensors* (MDPI, 2022 IF: 5.74). His research interests include Biodesign and Biomimetics, Biomanufacturing for Tissue Engineering and Regenerative Medicine and Sustainable Science and Engineering for Better Human Health.

Contributors

R.G. Abaszade
Azerbaijan State Oil and Industry University
Baku, Azerbaijan
Azerbaijan University of Architecture and Construction
Baku, Azerbaijan

Shahram Ajori
Department of Mechanical Engineering
University of Maragheh
Maragheh, East Azerbaijan Province, Iran

Md. Irquam Alam
School of Mechanical Sciences
Indian Institute of Technology Bhubaneswar
Odisha, India

F.G. Aliyev
Azerbaijan University of Architecture and Construction
Baku, Azerbaijan.

Santhy Antherjanam
Department of Chemistry, School of Physical Sciences
Amrita Vishwa Vidyapeetham
Amritapuri, Kerala

Shahid Anwar
Materials Chemistry Department
CSIR-Institute of Minerals and Materials Technology
Bhubaneswar, India
Academy of Scientific and Innovative Research (AcSIR)
Ghaziabad, India

Sharmistha Anwar
Academy of Scientific and Innovative Research (AcSIR)
Ghaziabad, India
Advanced Materials Technology Department
CSIR-Institute of Minerals and Materials Technology
Bhubaneswar, India

I.A. Asgarli
Ministry of Science and Education, "Vatan" Republican Specialized Olympic Reserve Football School for Children and Youth
Baku, Azerbaijan

Alok Behera
Unitedworld Institute of Technology (UIT)
Karnavati University Uvarsad
Gandhinagar, Gujarat, India

Manoranjan Bera
Gustaf H. Carlson School of Chemistry
Clark University
Worcester, MA, United State of America

Aniruddha Dan
Biological Sciences and Engineering
Indian Institute of Technology Gandhinagar
Gujarat, India

Harshil Dave
Biological Sciences and Engineering
Indian Institute of Technology Gandhinagar
Gujarat, India

Janaki Dehury
Department of Mechanical Engineering
Veer Surendra Sai University of Technology
Burla, Odisha, India

Mukesh Dhanka
Biological Sciences and Engineering
Indian Institute of Technology Gandhinagar
Gandhinagar, Gujarat, India

P. Elorika
Materials Chemistry Department
CSIR-Institute of Minerals and Materials Technology
Bhubaneswar, India
Academy of Scientific and Innovative Research (AcSIR)
Ghaziabad, India

Moumita Gangopadhyay
Department of Chemistry
School of Physical Sciences
Amrita Vishwa Vidyapeetham
Amritapuri, Kerala

Debarupam Gogoi
School of Mechanical Sciences
Indian Institute of Technology Bhubaneswar
Bhubaneswar, Odisha, India

Pournami Hareesh
Department of Chemistry, School of Physical Sciences
Amrita Vishwa Vidyapeetham
Amritapuri, Kerala

Shabir Hassan
Department of Biological Sciences
Khalifa University, Main Campus
Abu Dhabi, United Arab Emirates
Functional Biomaterials Group
Khalifa University, SAN Campus
Abu Dhabi, United Arab Emirates,
Center for Biotechnology
Khalifa University, Main Campus
Abu Dhabi, United Arab Emirates

S. Jayasri
Materials Chemistry Department
CSIR-Institute of Minerals and Materials Technology
Bhubaneswar, India
Academy of Scientific and Innovative Research (AcSIR)
Ghaziabad, India

Sasmita Kar
Department of Mechanical Engineering
Veer Surendra Sai University of Technology
Burla, Odisha, India

E.A. Khanmamadova
Azerbaijan State Oil and Industry University
Baku, Azerbaijan

Avanish Kumar
Indian Institute of Technology (Indian School of Mines) Dhanbad
Dhanbad, Jharkhand, India

Deepak Kumar
Department of Mechanical Engineering,
PDPM Indian Institute of Information Technology, Design & Manufacturing, Jabalpur
Jabalpur, Madhya Pradesh, India

Hemant Kumar
Department of Mechanical Engineering
G.B. Pant Institute of Engineering and Technology
Pauri Garhwal, India

Ravi Kumar
Mechanical Engineering Department
National Institute of Technology Patna
Patna, India

Sanjeev Kumar
Department of Mechanical Engineering
Dr B R Ambedkar National Institute of Technology, Jalandhar
Jalandhar, Punjab, India

Vijyesh Kumar
Council of Scientific and Industrial Research (CSIR)
Chandigarh, India

Kishore Kumar Mahato
Department of Manufacturing Engineering, School of Mechanical Engineering
Vellore Institute of Technology
Vellore, India

Mihir Kumar Sutar
Department of Mechanical Engineering
Veer Surendra Sai University of Technology
Burla, Odisha, India

Haridev M
Department of Chemistry, School of Physical Sciences
Amrita Vishwa Vidyapeetham

Simi M
Department of Chemistry, School of Physical Sciences
Amrita Vishwa Vidyapeetham
Amritapuri, Kerala
Amritapuri, Kerala

T.O. Margitych
Kiev Institute for Nuclear Research
National Academy of Sciences of Ukraine
Kyiv, Ukraine

Lokesh Nair
Department of Mechanical Engineering
G.B. Pant Institute of Engineering and Technology
Pauri Garhwal, India

Bibekananda Nayak
Materials Chemistry Department
CSIR-Institute of Minerals and Materials Technology
Bhubaneswar, India
Academy of Scientific and Innovative Research (AcSIR)
Ghaziabad, India

Mihir Kumar Pandit
School of Mechanical Sciences
Indian Institute of Technology Bhubaneswar
Bhubaneswar, Odisha, India

Pramod Kumar Parida
School of Mechanical Sciences
Indian Institute of Technology Bhubaneswar
Bhubaneswar, Odisha, India

Murlidhar Patel
Department of Mechanical Engineering
PDPM Indian Institute of Information Technology, Design & Manufacturing, Jabalpur
Jabalpur, Madhya Pradesh, India

Shivdayal Patel
Department of Mechanical Engineering
PDPM Indian Institute of Information Technology, Design & Manufacturing, Jabalpur
Jabalpur, Madhya Pradesh, India

Sarojrani Pattnaik
Department of Mechanical Engineering
Veer Surendra Sai University of Technology
Burla, Odisha, India

K. Ponappa
Department of Mechanical Engineering
PDPM Indian Institute of Information Technology, Design & Manufacturing, Jabalpur
Jabalpur, Madhya Pradesh, India

Arun Kumar Pradhan
School of Mechanical Sciences
Indian Institute of Technology Bhubaneswar
Bhubaneswar, Odisha, India

Ravikant Ravi
Department of Mechanical Engineering
NIT Jalandhar
Jalandhar, India

Dimple Sagar
Department of Mechanical Engineering
G.B. Pant Institute of Engineering and Technology
Pauri Garhwal, India

Partha Sarathi De
Indian Institute of Technology Bhubaneswar
Bhubaneswar, Odisha, India

R. Seetharam
Department of Mechanical Engineering
Indian Institute of Information Technology, Design and Manufacturing
Kurnool, India

Chaitanya Sharma
Mechanical Engineering Department
BIT Sindri
Dhanbad, India

Ritukesh Sharma
Indian Institute of Technology Bhubaneswar
Odisha, India

Sumit Sharma
Department of Mechanical Engineering
Dr B R Ambedkar National Institute of Technology Jalandhar
Jalandhar, Punjab, India

Chandraveer Singh
Department of Mechanical Engineering
G.B. Pant Institute of Engineering and Technology
Pauri Garhwal, India

Hemant Singh
Department of Biological Sciences
Khalifa University, Main Campus
Abu Dhabi, United Arab Emirates,
Functional Biomaterials Group
Khalifa University, SAN Campus
Abu Dhabi, United Arab Emirates,
Center for Biotechnology
Khalifa University, Main Campus
Abu Dhabi, United Arab Emirates

M M Sreejaya
Department of Chemistry, School of Physical Sciences
Amrita Vishwa Vidyapeetham
Amritapuri, Kerala

M.O. Stetsenko
Key Laboratory of Optoelectronic Devices and Systems, College of Physics and Optoelectronic Engineering
Shenzhen University
Shenzhen, PR China

Lakshmi Krishnaa Suresh
Department of Chemistry, School of Physical Sciences
Amrita Vishwa Vidyapeetham,
Amritapuri, Kerala

Tiju Thomas
University School of Design and Innovation
Guru Gobind Singh Indraprastha University
New Delhi, India

Anu Tomy
Sri Sivasubramaniya Nadar College of Engineering
Rajiv Gandhi Salai (OMR)
Kalavakkam-603110
Tamil Nadu, India

Vikas Upadhyay
Mechanical Engineering Department
National Institute of Technology Patna
Patna, India

Hitasha Vithalani
Biological Sciences and Engineering
Indian Institute of Technology Gandhinagar
Gujarat, India

Indu Yadav
Biological Sciences and Engineering
Indian Institute of Technology Gandhinagar
Gujarat, India

Introduction

Neeraj Kumar Bhoi, Harpreet Singh, and Himansu Sekhar Nanda

Functional materials are at the forefront of modern technology, driving innovation and progress in fields ranging from electronics and energy to medicine and aerospace. These materials have the paradigm shift in the way we live and work due to their unique properties and competences. The enthusiasm to play with the materials' texture by the additions of different layers, combining different alloys and phases or by microstructure transformation had always been a quest for the research community. Crafting for the improved performance in materials chemistry is a key concern in the modern manufacturing industry. Advanced Functional and Composite Materials is one of the hottest areas of research and development in many industrial practices, educational institutions, and exploratory research work.

Despite the availability of a large choice of textbooks, reference resources, and literary content, a foundation book that may help in the absorption of essential information about modern materials and industrial processes is still lacking. As a result, a source of information that provides a complete overview of the latest materials and their uses in the context of modern production is required. The chapters in this book present prospective revolutionary research tactics conducted in advanced functional and composite materials, referencing quickly changing client requirements and environmental sustainability. The topics addressed in this book represent diverse advancements in metal and polymeric composites, biomaterials, smart structures, fast prototyping, nano-finishing technology, and a wide range of production strategies.

This book is a vital resource for academics, students, and professionals working in materials science, chemistry, physics, engineering, and related subjects, with contributions from prominent specialists in the field. It gives a thorough and up-to-date overview of the topic, making it an excellent resource for individuals who want to keep on top of functional materials research.

This initial version of the book was written with great care. However, there will be some inaccuracies, or certain topics may be left ignored or missed. As writers, we would welcome any such notes, contributions, and/or comments. They will be addressed appropriately in the subsequent version of this book.

DOI: 10.1201/9781003664611-1

1 Functionally Graded Materials in Medical Applications

Hemant Singh, Aniruddha Dan, Hitasha Vithalani, Harshil Dave, Indu Yadav, Shabir Hassan, and Mukesh Dhanka

1.1 INTRODUCTION

Traditional materials used in medical applications have proven effective and enhanced the life expectancy of patients. Still, they have common drawbacks like bio-mimicking, biocompatibility, mechanical mismatching, susceptibility to corrosion and wear over time, and limited customization for individual patient needs [1,2]. Further, traditional materials show stress concentrations at the interfaces of implants—natural tissue, inadequate drug release, unexpected degradation, high density, and many more drawbacks motivate the researchers and engineers to navigate the complex demands of medical applications to quest for advanced materials capable of seamlessly integrating with the human body's complexities has been a driving force behind transformative breakthroughs [1–4]. Among various innovations in material sciences, the idea of functionally graded materials (FGMs) was first time proposed in 1972 and has emerged as a cutting-edge paradigm, offering unparalleled versatility in tailoring mechanical, biological, and chemical properties across a gradient, addressing challenges ranging from biomechanical compatibility to long-term implant stability [1–4]. A brief timeline of the progress in FGMs' development [1,2,4] is portrayed in Figure 1.1(A). FGMs represent a departure from traditional homogeneous materials to featuring a gradual transition in composition, structure, or properties over a defined spatial domain. The idea of the FGMs comes from nature, like graded bone, teeth, and bamboo structures [1,2,4], as shown in Figure 1.1(B). This unique design enables FGMs to mimic the intricate and dynamic nature of biological tissues, opening doors to a myriad of possibilities in the medical engineering and healthcare sector [1,2,4]. Lin's research group [5] introduced a functionally graded hydroxyapatite and collagen (HAP/Col) hybrid to enhance bone remodeling, which suggests that an increase in the hydroxyapatite/collagen content in the FGM enhances bone remodeling. Powder sintering technologies emerge as a viable option for producing the appropriate difference in rigidity from the apical to the coronal ends, adapted to particular base materials with specified surface porosity [6]. Further, Parisi et al. [7] fabricated a gradient scaffold made of collagen matrix loaded with low-crystalline hydroxyapatite particles in a gradient manner for treating osteochondral defects. The outcomes of this research depict that the stiffness gradient, along with the mineral composition, favors the growth of both bone cells and cartilage cells. Another research group [8] reported a new FGM through a sequential electrospinning method for treating bone defects. In vitro studies in their research confirmed that the scaffolds promoted epithelial cell adhesion, proliferation, and osteodifferentiation, and in-vivo studies in rabbits indicated excellent bone regeneration capacity. Moreover, Zhu et al. [9] developed nerve conduits composed of aligned PCL fiber with nerve growth factor (NGF), facilitating axons' longitudinal and chemotactic growth in vitro and directing the regeneration of nerve cells through

 DOI: 10.1201/9781003664611-2

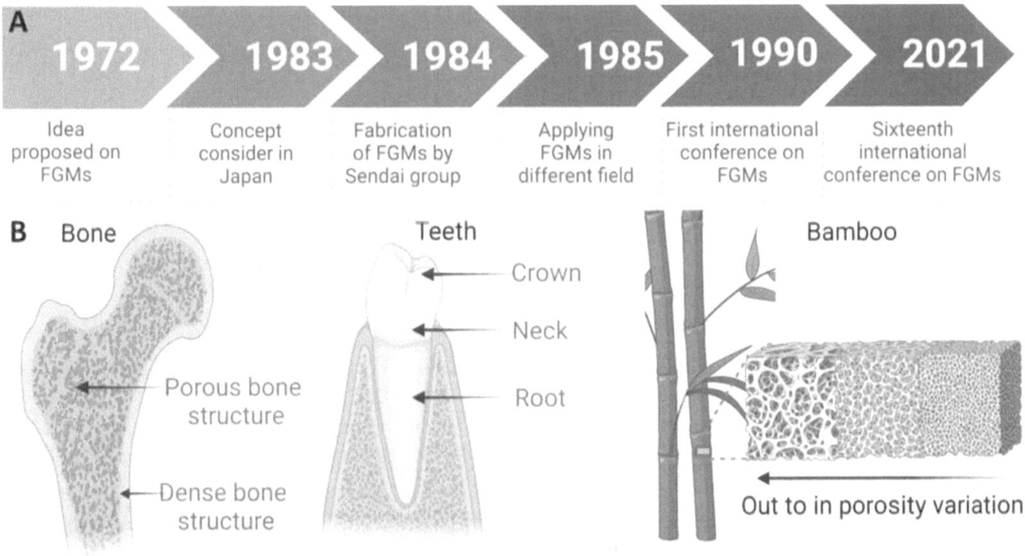

FIGURE 1.1 (A) A brief timeline of the FGMs' development [1,4] and (B) the idea of the FGMs comes from nature, like graded bone, teeth, and bamboo structures [1,4].

axonal communication. Many more researchers nowadays focus on FGM to explore different fields of biomedical application [1–4,10,11]. But this chapter unfolds against the backdrop of an escalating field where the intersection of materials science, engineering, and medicine paves the way for pioneering solutions, starting with the exploration of the FGMs and their pivotal role in revolutionizing implants, drug delivery, and tissue engineering (TE), elucidating their design principles, fabrication techniques, and the remarkable impact they hold across various medical disciplines.

1.2 FUNCTIONALLY GRADED MATERIALS IN IMPLANTS

A biomaterial is any natural or synthetic substance, surface, or assembly that has been developed to interact with biological systems to deliver therapeutic effect [12]. For instance, a material is designed to be introduced into the human body, either partially or entirely, through medical procedures and intended to remain there afterward to support or completely replace the biological tissue. In that case, it is termed an implant [13]. The demand for improved implants and the creation of artificial tissues is evident; a notable aspect of this research involves embracing biologically inspired approaches, where insights from biological structures guide the creation of novel materials. When examining natural materials, distinct design principles not commonly employed in traditional materials processing become apparent [14]. Biomaterials often exhibit hierarchical structures and intricate functionality in different tissues, continually presenting change from one framework or composition to another. For example, consider the clever architecture of bone, which transitions from a thick, hard outward structure (cortical bone) to a porous interior structure (cancellous bone), illustrating the use of functional gradation through biological adaptation [15–17]. This gradation improves the material's durability against external loads.

Consequently, an artificially created implant should ideally showcase similar gradation for optimized functionality [14]. The FGMs are specialized composite materials with different microstructures and compositions that are smooth and spatially arranged according to a predefined pattern. This gradual variation in composition also leads to corresponding variations in FGM properties, such as physical, mechanical, and biochemical characteristics, to meet specific requirements for a particular application. Biomedical applications, especially in the realm of implants, are showing an increased interest in FGMs due to their reported superior composition [15–17]. Functional gradation, a distinctive feature of living tissues, is employed in developing new implant materials. In recent decades, scientists have explored innovative materials for various reasons, including economic and environmental concerns. FGMs provide unique opportunities for developing advanced engineered implant materials, offering distinct advantages regarding their tailored properties and applications [18]. This section mainly focuses on the role of FGMs in dental implants, orthopedic implants, and some other advanced implants.

1.2.1 Dental Implants

Modern dentistry aims to restore patients to their normal function, wellness, speech, and appearance. Individuals with excellent oral hygiene who lost their teeth because of gum disease, an accident, or other reasons may consider dental implants as a primary repair option [19]. Developing novel materials for oral implants is a significant challenge in today's materials science. One composite with a uniform structure is insufficient for particular biomedical applications. Dental biomaterials must fulfill specified characteristics, such as body compatibility, desirable degradation rate, osteoconductivity, toughness, resistivity to corrosion, modulus of elastic behavior, fatigue endurance, and chemical resemblance to substances found in human hard tissues [20–23]. Various designs stemming from traditional and TE methods are standard in biomedical implants. Current implant research focuses on composites, TE, coatings of polymers on metallic implants, and FGMs [24]. Most oral implants are made of single materials; however, a coating layer is occasionally used. However, issues with recently coated metal implants include stress shielding of the neighboring bone and the coating's limited longevity, leading to significant biocompatibility concerns [24]. The concept of stress shielding suggests that bone deterioration surrounding dental and orthopedic implants occurs due to an implant's reduced normal stress on the bone [25] . Applying FGMs to enhance the acceptance of oral implants within living bones is considered a favorable approach [26,27], as demonstrated in Figure 1.2. Functionally graded oral implants aim to optimize mechanical behavior for improved compatibility and integration with surrounding bone [28]. Functionally graded oral implants (Figure 1.2(A)) utilize various materials like titanium (Ti), titanium nitride, different polymers, zirconia, and the design and optimization of these implants are also under consideration [28]. In oral implants, the elements are typically more minor and restore the chewing function of tooth roots when the root of a tooth is removed entirely or damaged. For instance, the implant is positioned in the jawbone to penetrate the bone. The implant's desired function varies on the jawbone's outer surface, within it, and at the boundary of the jawbone [29]. On the outer surface of the bone, the implant material must possess enough mechanical toughness to withstand the chewing load. Conversely, the portion inside the jawbone requires osteoconductivity and proper implant–bone contact to facilitate the swift creation of new bone that attaches directly to the implant. The primary benefits of using FGMs oral implants include improving biocompatibility, reducing the stress shielding effect seen with fully metallic oral implants [30–32], avoiding thermal and mechanical breakdown at the interface of metallic implants covered with HA [33], as well as improving biomechanical needs and control for graded bioreaction in each bone area [34].

1.2.1.1 Structural Consideration for Dental Implants

Environmental organs and tissues like snail shells, bamboo, bone, and skin have non-uniform morphologies with functional gradients (FGCs) [35]. When the various layers of a tissue or organ serve specific functions to meet local requirements, they are termed functionally graded. The implantation should mimic natural functionality for a successful dental implant. It is essential to incorporate FGM into the design [14,26,36,37]. Figure 1.2(A) illustrates a schematic view of an ideal FGM for oral implants. The implant should have a composition that varies along the axial direction to

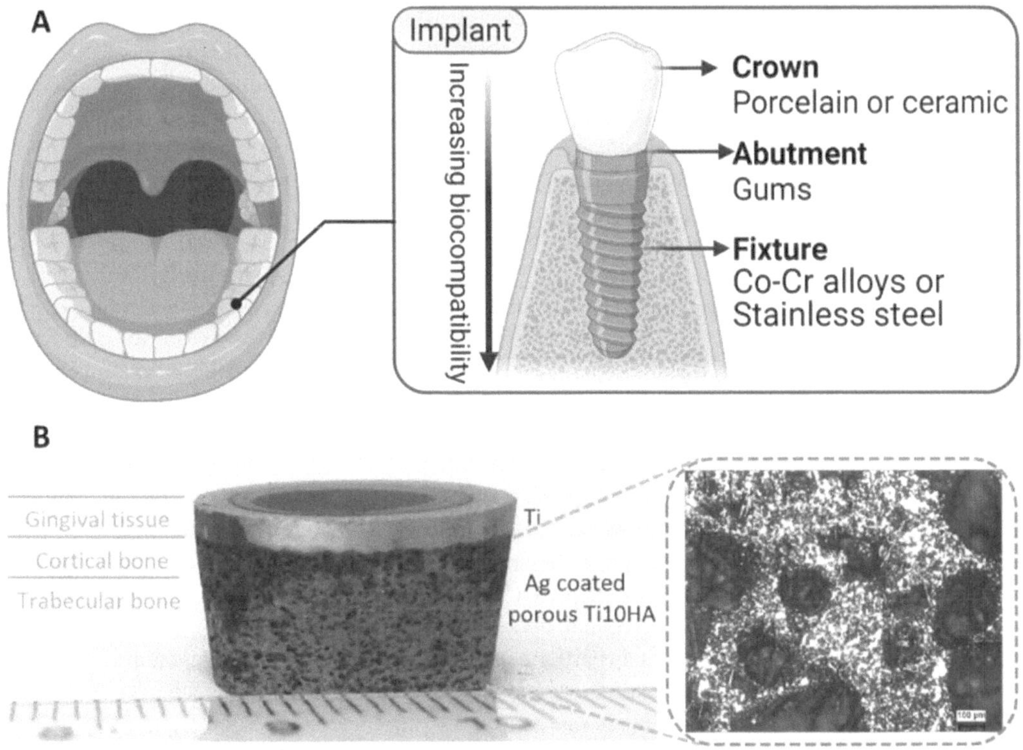

FIGURE 1.2 (A) Illustration of the need for the FGMs in teeth implants and (B) fabricated implants for dental applications using the FGM concept, adopted with Elsevier's permission [43].

optimize its functionality. The upper part bears direct occlusal force and needs enhanced strength and mechanical properties. This force is subsequently conveyed to the lower sections implanted into the trabecular bone, where it is advantageous to use more biocompatible materials [24,38,39]. Implant failures often stem from a lack of mechanical bonding with the surrounding jawbone. To counteract this, the lower parts of the implant should exhibit high osteoconductivity, promoting suitable implant-to-bone interaction and facilitating bone regeneration and rapid osseointegration [31,40,41]. Insufficient strength, mechanical properties, or overload can also lead to implant failure. A potential solution involves designing the upper parts of FGM oral implants with suitable mechanical properties [40–42]. Incorporating FGMs into oral implant design improves mechanical and biological performance, addresses critical issues, and improves overall success rates [4,43].

1.2.1.2 Advances in FGMs for Dental Implants

Functionally gradient oral implants aim to achieve the best oral-compatible features by imitating the natural configuration of human teeth. These implants can be designed with varying properties to optimize mechanical behavior and enhance biocompatibility and osseointegration [2,4,43]. Ti and its alloys are commonly used in oral implants; conversely, they are biologically inert and do not cause any adverse reaction . Ideally, an oral implant should have high mechanical strength to maintain integrity after its implantation [43]. To address this, Ti and Ti alloys are utilized along the longitudinal direction where occlusal force is directly applied, while other materials rich in ceramics or metals are employed in different parts implanted inside the jawbone [28].

Different researchers have suggested various materials for dental implantation, like titanium/hydroxyapatite (Ti/HA) [31], TiN/HA [44], and HA/ZrO_2 [6] . Watari et al. [44] studied titanium/

cobalt FGMs (Ti/Co FGMs), assessing their mechanical properties and implant biocompatibility. Their findings revealed that a thin fiber connecting tissue developed in the Ti region when Ti/Co implants were analyzed in soft tissue. However, when the Co concentration grew, the thickness of this fibrous tissue layer expanded, causing inflammation in the Co-rich region. It demonstrates that a change in Co concentration has a considerable influence on the biocompatibility of the implant material. Furthermore, the clinical studies supported these findings, indicating that epithelial cells and fibroblasts exhibited a more adverse effect to a Co-chrome alloy than titanium. Despite this, the Co demonstrated several mechanical characteristics, including wear and corrosion resistance advantages. Recently, Yılmaz and Caliskan [43] fabricated a teeth implant using the concept of FGMs, as demonstrated in Figure 1.2(B). In their study, the FGMs-based tooth implant depicts the desired compressive strength, , biocompatibility, and antibacterial activity. Further, Majzoobi et al. [45] explored the tribological characteristics of Ti/HA and titanium/silicon oxide (Ti-SiO_2) FGMs, which were manufactured using the powder metallurgy method. The results obtained from Vickers hardness tests and wear measurements indicated that the Ti-HA sample produced through the split-Hopkinson bar (SHB) method exhibited superior tribological properties. Remarkably, these properties were suitable for normal human teeth's natural behavior. One commonly employed material for coating of implants is bioactive glass-ceramic (BGC) coatings. *In-vitro* studies for composite biomaterials containing active glass-ceramic nanoparticles with less phosphorous and high silica content showed better bioactivity than BGC with greater phosphorous and less silica content [46,47]. Specifically, the silica particles play a vital role in improving the strength of HA coatings through a mechanical process known as particle-mediated reinforcement. This phenomenon results in crack deflection or arrest, contributing to the overall improvement in the mechanical properties of HA coatings. Amorphous SiO_2 stands out as a promising candidate for bolstering the strength of these coatings [48]. Matsuno et al. [49] found that a composite material consisting of laminated HA and partially stabilized zirconia (PSZ) with a gradient composition can be created by sintering. According to the study, due to its osteoconductivity and outstanding mechanical strength, this HA/PSZ gradient composition is expected to be widely accepted as a substitute for dense tissues for implant dentistry. Lin et al. [5] introduced a model involving a hybrid of HA and collagen (HA/Col) in a functionally graded manner. The implants created using these FGMs exhibited enhanced outcomes in bone remodeling. Oral implants with FGMs, in particular, were found to increase the remodeling of the bone's activity and minimize maximal pressure on the surrounding cancellous peri-implant tissues. These benefits lead to a better engagement with the oral structure, enabling recovery at the implant–bone interface and sustaining oral bone well-being. Furthermore, an increase in the hydroxyapatite/collagen content in the FGM was found to enhance bone remodeling. Powder sintering technologies emerge as a viable option for producing the appropriate difference in rigidity from the apical to the coronal ends, adapted to particular base materials with specified surface porosity. Guo et al. [6] utilized Y-TZP to create composites of HA and ZrO_2. The researchers noted a significant enhancement in mechanical characteristics for the HA/Y-TZP FGMs, increasing both microhardness and Young's modulus. In addition to oral implants with functionally graded structures, porous biomaterials featuring graded functionality, such as Ti and its alloys, have proven effective in biomedical applications. Notably, the prevalent use of Ti–6Al–4V (90% titanium, 6% aluminium, and 4% vanadium) underscores its favorable physicomechanical properties, resistance to corrosion, and biocompatibility in oral implants. Moayedee et al. [50] investigated the mechanical, microstructural, and biological characteristics of Yttria-stabilized zirconia nanoparticles combined with titanium, aluminium, and vanadium powder (3YTZP/Ti6Al4V) as FGMs produced through spark plasma sintering (SPS). Despite a metal coating showing a brittle fracture, the study revealed that higher bending strength and increased YSZ content correlated with decreased antimicrobial activity. Notably, the material exhibited high biocompatibility, as evidenced by MTT cytotoxicity tests conducted after one and seven days of culture. Dinu et al. researched that a CrSiN/CrSiON coating exhibited pore resistance owing to a more defect-free construction as an inhibitor

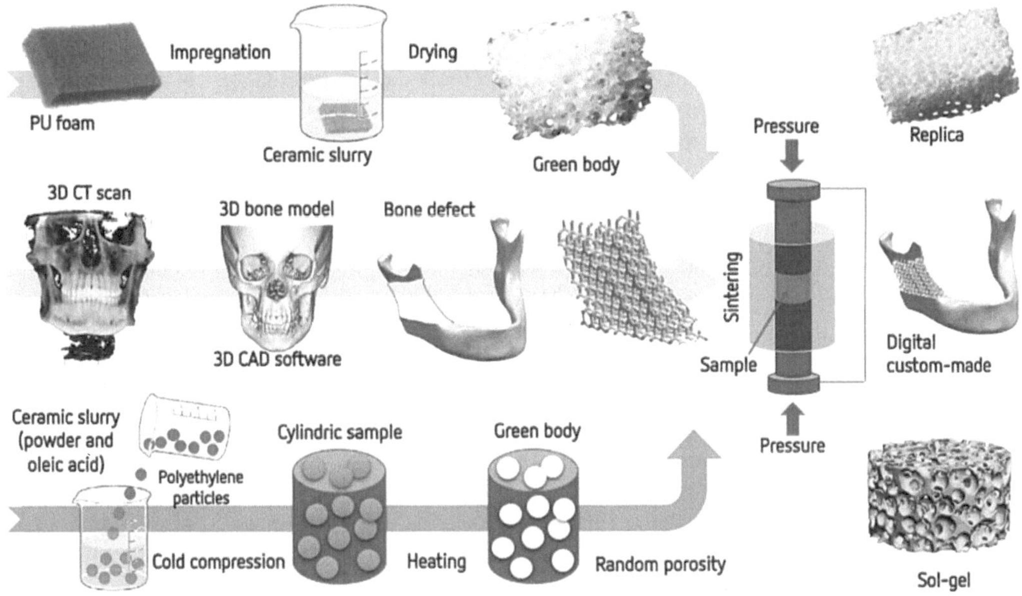

FIGURE 1.3 Diagram depicting the manufacturing process of porous zirconia. It is adapted from MDPI Ceramics [53].

against electrolyte infiltration and corrosion. Among the coatings studied, Cr/CrSi/CrSiN/CrSiON demonstrated the most favorable support for human gingival fibroblasts, displaying superior proliferation and metabolic activity. These findings were substantiated by comprehensive mechanical characterization [51]. Zhu et al. [52,53] utilized porous zirconia implants, as illustrated in Figure 1.3. These implants effectively restored the crown structure and demonstrated commendable biocompatibility. Using porous zirconia implants represents a promising avenue for dental applications, showcasing the continual evolution and advancement in biomaterial design for improved clinical outcomes.

Researchers have also looked at other materials, and their discovery of various materials for certain oral implants indicates that producing implants with FGM is necessary. Consequently, the fields of dentistry and bone TE both have considerable promise for using FGM. The concept of FGM is a valuable way to improve oral implants' acceptance into functioning bone.

1.2.2 Orthopedic Implants

In past few decades, the design of orthopedic implants has improved remarkably to achieve better long-term stability and fast integration with the bone. Despite technological progress in total hip arthroplasty, revision procedures are still performed on over 13% of hip prostheses because of problems, including implant loosening and bone resorption [54]. Current orthopedic implants are typically constructed from uniform-density materials including, 316L stainless steel, titanium, and cobalt-chromium alloys and titanium-based alloys. Besides ensuring biocompatibility, successful implants need to meet three critical criteria: (i) stable functioning over time, (ii) prevention of tissue resorption of the bone surrounding the implant, and (iii) excellent wear and corrosive resilience of the areas in contact [55]. One major challenge in orthopedics is stress shielding, an issue caused by a mismatch in mechanical characteristics between the bone and implant that could result in the implant malfunctioning. Traditional manufacturing processes contribute to this issue as properties like porosity, strength, and hardness cannot be customized to individual patients, potentially causing implant resorption. Additive manufacturing (AM) presents a solution by allowing precise control over implants' mechanical and structural properties, thereby reducing stress concentration and minimizing the risk of implant failure. Various orthopedic fixtures, such as bone screws, pins, wires-like structures, intramedullary rods, and spinal fixtures, are widely used during orthopedic surgeries [56]. Improper torque applied to these fixtures during implantation can result in implant failure. Addressing these challenges through advanced manufacturing techniques enhances orthopedic implants' overall success and longevity [57].

FGMs are becoming popular for creating orthopedic prostheses. This is because their unique property gradient can be adjusted to replicate the specific qualities of natural bones. Specifically, the quick and proper adaptation of implants help in minimizing the stress shielding effect and, simultaneously, reduce the shear tension between the implant and the surrounding bone tissue, which are considered the crucial factors for extending the duration of the implant [15]. FGMs can be utilized to craft implants for orthopedic purposes. When exploring the features of materials used in orthopedic implants, two key concerns with using FGMs are addressed: (i) designing materials and optimizing geometry using computer-based simulation and (ii) exploring the current methods of manufacture for grafts based on FGMs. Additionally, to address the manufacturing challenges of functionally graded coatings (FGCs), specific devices such as functionally graded scaffolds and three-dimensional functionally graded components with controlled porosity gradients have been developed. Unlike FGCs, FGMs do not rely on a substrate.. The significance of integrating design and manufacturing processes is strongly emphasized to capitalize fully on the benefits of FGMs [15].

One remarkable feature of biomaterials is their ability to form gradable structures. For instance, a robust material like alumina can be combined on one end of an implant and a bioactive material like bioglass on the other. Yet, the lack of any unforeseen interface considerably increases the system's durability [15]. Consider orthopedic implants used in shoulder and knee joint replacements, for instance. Improving the acceptability of engineered implants by living tissues is achievable by incorporating a biopolymer interlayer with FGCs. A good illustration of this concept involves titanium implants that have been improved by a graded coating that comprises three distinct biopolymers. This example underscores the importance of enhancing mechanical performance and optimizing interactions with cells in the surrounding tissue when designing the interlayer [14]. Hence, FGMs prove valuable since the composition of tissues naturally undergoes a continuous transformation from one form to another [58]. For instance, demonstrating the application of functional gradation through biological adaptation involves effectively designing porous bone. This design includes a porosity gradient that transitions from a dense, rigid outer structure (Mimicking to cortical bone) to a porous internal osseous tissue (cancellous bone) while ensuring a suitable level of interconnectivity [59–61].

1.2.2.1 Structural Consideration for Orthopedic Implants

Various materials employed in orthopedic implants, such as Ti and its alloys, stainless steel, alumina, zirconia, and calcium phosphates (CaPs), are typically more rigid and less flexible than the hard bone tissue, except for scaffolds, which are widely being used to support bone healing rather than for load-bearing applications. This difference in stiffness changes the load distribution compared to the usual physiological situation [62]. When an implant with differing mechanical qualities substitutes a portion of bone, the initial strain and stress patterns are unavoidably changed, even though the outside force stays constant. The inherent load transmission within the bone's cortex, trabecular structure, and cortex is redirected to the implant–bone interface. Functional materials with graded properties go beyond being two-phase materials with a gradual composition change that can be customized to meet specific optimized requirements, particularly in complex stress situations [63,64]. Advanced numerical optimization methods have been employed to determine the optimal stiffness distribution, addressing conflicting stress needs. This has made the steady advancement of FGMs and functionally graded coatings for orthopedic applications [55,65]. The difference in density, combined with variations in mechanical qualities, can enhance the functionality of implants compared to a single constitutive material. Implant parts with less porosity offer enhanced mechanical durability, while regions with a high degree of porosity promote bone formation and assist in the implant's anchoring [66]. Graded porosity can be created using any technique for fabricating lattice structures. AM is attractive because it produces precise geometries and controls unit cells. Two common AM technologies for printing functional gradations are powder bed fusion (PBF) and direct energy deposition (DED) [67]. Specifically, PBF techniques like selective laser melting (SLM) and electron beam melting (EBM) can produce precisely controlled pores following specific design rules in terms of size and allocation [68].

1.2.2.2 Advanced in FGMs for Orthopedic Implants

To avoid additional surgeries after implantation of biomaterial, choosing the right biodegradable material for developing the orthopedic implants is crucial. Researchers are consistently working on developing mechanically robust and biodegradable materials tailored to specific job requirements for individual patients. Various FGMs are employed in orthopedic implants like tantalum (Ta) coated Ti6Al4V. This bone plate has similar elastic properties to cortical bone, preventing stress shielding. The Ta coating promotes cell growth on the scaffolds, improving osseointegration and osteogenic qualities compared to regular bone plates [69]. Another example is the use of Agilus30 (FLX935), Tango (FLX930) rubber-like materials, and digital acrylonitrile butadiene styrene (ABS) (RGD5130) in knee joint replacements designed for educational purposes. These alternatives have mechanical properties robust enough to endure flexo-extension [70]. Figure 1.4 shows various file formats used in their study.

Looking ahead, fiber matrices, such as polylactic acid (PLA) coated with HA and PVA fibers, may replace other materials to mimic soft tissues. The scaffolds created are both safe and water-attracting (hydrophilic). The coating enhances mechanical properties like tensile strength, although

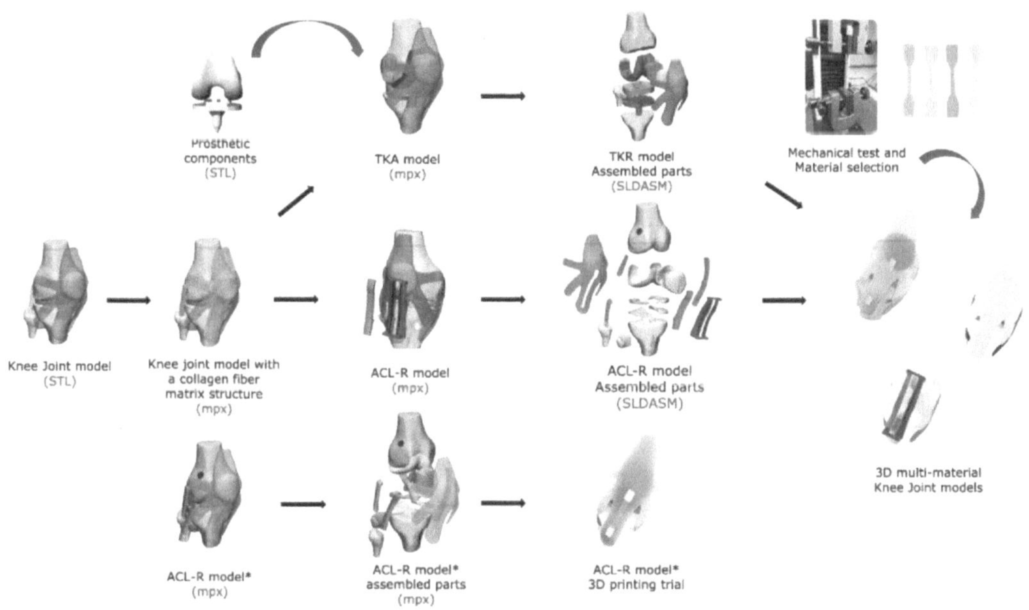

FIGURE 1.4 Diagram depicting study formats including a normal knee joint model with a designed fiber matrix, an ACL-R model without a fiber matrix, an improved ACL-R model with a custom surgical guide SG to prevent graft tunnel length mismatch, and a TKA model with custom-made implants. (Adapted from Grimaldo Ruiz and Dhaher [70]. Copyright the Author(s). 2021, 3D Printing in Medicine.)

it is vital to emphasize that the failure strain is reduced [71,72]. Sutou et al. [73] created shape memory alloys using polycrystalline Cu–Al–Mn, achieving remarkable ductility and significant super elasticity strain by carefully controlling the degree of order. This showed effective application in the guide wire, where the tip displays super elasticity while its end exhibits high stiffness. This functionally graded guide wire surpasses the Ni-Ti counterpart in terms of excellent torquability and easy to push. Rausch et al. [74] researched an annuloplasty device for repairing leaking valves. Their findings indicated that variable strain and curvature patterns highlight the necessity for diverse support, suggesting a more reasonable design approach for annuloplasty devices using concepts from FGMs. In another study, Xiong et al. [75] utilized SLM with Ti6Al4V as the building material for developing structurally four different types of FGM scaffolds. Similar to the graded-porous composition found in human bones, these scaffolds featured hollow cylindrical structures with various three-dimensional designs with gradient porosity. Among them, the FGM structures built with honeycomb-like unit cells and a supporting framework in the outer layer had the maximum yield strength, hardness, and stability in mechanical features. This makes them particularly suitable for constructing orthopedic scaffolds intended for load-bearing applications. Moreover, Taheri et al. [76] devised a design for tibial stems utilizing porous titanium with varied composition and directional properties. These stems included cellular structures with radially and axially FGC, meticulously customized to achieve specific stiffness characteristics. To predict nodal micro-motion at the implant–plate interface and assess the stress-shielding effect in the surrounding bone, a 3D voxel-based finite element model was employed. The model assumed frictional contact and a perfectly bonded connection. The implant was modeled with an exponential FGC material geometrically in radial and axial directions, with the gradient parameters treated as design variables. The research findings suggested that the biomechanical performance of the tibial stem would be most favorable with an ideal structure featuring bi-directional axial and radial gradients organized in a cylindrical lattice arrangement. Xie et al. [77] created bioinspired interpenetrating phase composites (IPCs) using Ti6Al4V and polyether ether ketone (PEEK). They achieved this by injecting melted PEEK with built Ti6Al4V scaffolds that were precisely engineered with minimum surface features. This study provides valuable insights into the mechanical characteristics and deformation mechanisms of Ti-PEEK IPCs when subjected to compression. These IPCs show the potential to improve the flexibility of orthopedic implants, and additional applications are outlined in Table 1.1. Lee et al. [78] developed a composite material using PLA that incorporates tricalcium phosphate (TCP) microspheres along with and Mg particles. The objectives behind designing this material were to boost physical strength, expedite pore formation, and enhance biological and antibacterial performance. The composite exhibited increased bioactivity with preosteoblast cells and demonstrated antibacterial properties. With enhanced mechanical qualities, heightened bioactivity, and a built-in antibacterial effect, these advanced biodegradable PLA/TCP-Mg composites have the potential to effectively address the challenges associated with implant-related infections, offering enhanced outcomes in orthopedic implant applications.

1.3 FUNCTIONALLY GRADED MATERIALS IN THERAPEUTICS DELIVERY

Several types of gradients are everywhere in naturally occurring tissues and organs [1,2,4]. These gradients are arranged or oriented in a specific manner, which plays the role of the structure and proper functioning of the tissues and organs [91]. Mainly, five gradients found in natural tissues are being looked at during the development of FGMs for biomedical applications. Gradients of cells are developed when several types of cells or densities of particular cell types are oriented explicitly according to the need [91–93]. Gradients of composition are created due to shifting in the components of certain organs or specific portions of a particular organ [91,94,95]. Gradients of the structure are shifting in orientation or organizations of different entities in a particular tissue [91,96,97]. Mechanical gradients are produced due to variations in the tissue composition,

TABLE 1.1
Recent Advances in FGMs as Implants

Composition	Applications	Ref.
Co-Cr-Mo (monoblock cobalt chrome molybdenum) alloy coated on porous Ti6Al4V.	Improve the lifespan of weight-bearing implants like total hip prostheses by enhancing implant-tissue interaction.	[79]
Co–Cr–Mo femoral stem	lightweight femoral stem to minimize implant loosening due to stress shielding while ensuring stability at the implant–bone interface in total hip arthroplasty.	[80]
Porous cobalt chromium (CoCr)	Show promising potential for implants with less stress-shielding and enhanced durability.	[81]
Metal–ceramic porous FGM using a 3D finite element model,	Bone stress shielding reduces wear on joint areas and tissue formation at the bone-prosthesis interface due to implant movement.	[82]
3D-printed fully porous Titanium femoral stem.	Highlight the value and potential of adjusting material architecture to significantly reduce stress shielding-induced bone resorption.	[83]
Hydroxyapatite, Bioglass, and collagen – 2D FGM model	Evenly distributing shear stresses at the bone-stem interface minimizes the risk of failure and stem loosening.	[84]
Carbon fiber–polysulfone composite	Suitable for intervertebral disc replacement as it met the fatigue strength requirements.	[85]
Biodegradable poly-L/DL-lactide [P(L/DL)LA 70/30] implants	Facilitate healing and regeneration of the bone at the inferior orbital wall defect site.	[86]
Stiffness-graded bone plates made of composite material	Improved bone healing, minimized stress shielding, and increased compressive stress at the fractured interface	[87]
Graded cellular materials	An uncemented stem in the proximal femur aims to reduce bone resorption and prevent implant failure simultaneously.	[55]
Ti6AL4V varied cell sizes.	The femoral implant enhances the load-sharing capability and reduces bone resorption.	[88,89]
Titanium alloy + layer of tantalum	Prostheses to enhance stability of implant–atbone interfaces.	[90]

ultimately leading to varying compressive, tensile, and shear properties [98]. *Morphogen gradients* rise due to transitions in the concentrations/diffusion pattern from a point source [99–102]. This relation between structure and function has inspired researchers to develop biomaterials that mimic the physiological gradients of actual tissue/organs [3,4,43].

1.3.1 Sequential Layering Approach in FGMs for Therapeutics Delivery

In the case of early approaches involved in AM, adhesives were used to clump together several layers of biomaterials. Using this approach, Niederauer et al. [103] reported the synthesis of autologous costochondral chondrocytes loaded multiphase biomaterial composites to treat osteochondral defect based on poly (D, L) lactide-co-glycolide using PGA nanofibers, bioglass, and medical grade calcium sulfate (MGCS) as additives for varying the stiffness along with composition according to need. *In-vivo* studies in Spanish goats have exhibited that these multiple-phase composites have successfully induced hyaline cartilage production and bone restoration. Also, a few years back, alginate-boronic acid-based glues were reported to bind several layers of precast hydrogels [104]. However, these adhesives or glues are not long-lasting. Gradually, their attractive interactions will lessen, resulting in the dissociation of the stacked layers. Recently, the research has focused on generating multiple layers without requiring an intermediate adhesive. Using this recent approach, Guo et al. [105] synthesized a biomimetic gradient composite based on silk fibro and R5 peptides for the osteochondral interface. These fabricated gradient composites have the desired stiffness due to the presence of mineralization domains (R5 peptide). These gradient systems showed cytocompatibility and biodegradability. In-vitro studies confirmed the fate of human mesenchymal stem cells (hMSCs) toward osteoblasts, osteoclasts corresponding to the gradient of R5 peptide.

Ko and researchers [106] reported a method of simple layer stacking to develop gradients inside photocrosslinkable hydrogel for 3D cell culture. The concentration gradient of biomolecules inside hydrogels composed of PEGDA and GelMA was confirmed using BSA as a model biomolecule. They have shown that by varying the UV intensity, the extent of crosslinking and the addition of several hydrogel layers can be tuned according to the need. Also, this technique has been used to create mechanical gradients inside the hydrogel by varying the concentrations of each layer. In-vitro cell culture studies have exhibited that fibroblast cells behave differently and migrate along the axis of the gradient. Recently, Parisi et al. [7] reported the synthesis of a novel gradient scaffold made of collagen matrix loaded with low-crystalline hydroxyapatite particles in a gradient manner for treating osteochondral defects. Mechanical studies proved that the stiffness gradient along the scaffold corresponded with the mineral composition. *In-vitro* studies confirmed that the fabricated gradient scaffolds aided the viability and proliferation of hBMSCs (human bone marrow-derived stem cells). Results showed that in both conditions favoring for bone cells and cartilage cells, the fabricated scaffold played its role in stimulating the hBMSCs towards specific fates (osteogenic or chondrogenic) according to the compositional gradient of the scaffolds. *In-vivo* studies in rats confirmed the biocompatibility of these gradient scaffolds. Furthermore, Wei et al. [107] reported the synthesis of a dual crosslinked self-healable novel modular gradient CEC-OAHA-MMP hydrogel composed of acryl-modified aldehyde hyaluronic acid (OAHA), N-carboxyethyl chitosan (CEC), dithiol matrix metalloproteinase sensitive crosslinker, and RGD. The dynamic Schiff-base interaction formed between the OAHA and CEC provided the self-healing behavior. Dual crosslinked gradient hydrogel exhibited excellent injectability. Also, the encapsulated KIA-GFP cells were behaving according to the concentration gradient of RGD peptide (elongated KIA-GFP cells were observed in the area containing high concentration of RGD compared to the area containing RGD in lower concentration and MMP-sensitive crosslinker in the CEC-OAHA-MMP hydrogel). They proposed that by exploring the fabricated self-healable modular gradient hydrogels, we might be able to recreate the microenvironments of complex tissues. After that, Adarsh and co-researchers [108] reported the fabrication of functionally graded bioactive composites made of thiol-modified poly(vinyl alcohol) (TPVA) and nano-hydroxyapatite (nano-HA), as shown in Figure 1.5(A). This bioactive composite was formed by TPVA layers containing different HA concentrations (Figure 1.5(B)). The polymer composite exhibited expected tensile strength and degradation behavior in in-vitro studies. Also, in-vitro bioactivity tests confirmed the formation of apatitic layers formation on the portion of the composite containing HA particles, and these bioactive composites were noncytotoxic when treated with human periodontal ligament cells (Figure 1.5).

Despite all the advantages, the main limitation of this approach is that, through this sequencing layering process, the fabricated biomaterial consists of discrete layers, leading to the formation of a stepped transition of biomolecules but not a smooth transition.

1.3.2 Bioprinting Approach in FGMs for Therapeutics Delivery

Rather than adding several layers of biomaterials to form a discrete gradient, it would be better to add the biologically mimicking structural complexity by using 3D bioprinting technique. Mechanically graded structures can be fabricated by adjusting the composition concentration of 3D printing ink [109,110]. Using the 3D printing method, Di Luca and co-researchers [111] fabricated a scaffold containing an in-built pore size gradient to treat articular cartilage lesions. These scaffolds were composed of poly(ethylene oxide terephthalate)/poly(butylene terephthalate) (PEOT/PBT). hMSCs were seeded within these gradient scaffolds. Glycosaminoglycan (GAG) was deposited in a larger amount than the non-gradient scaffolds. Also, the chondrogenic markers were expressed in

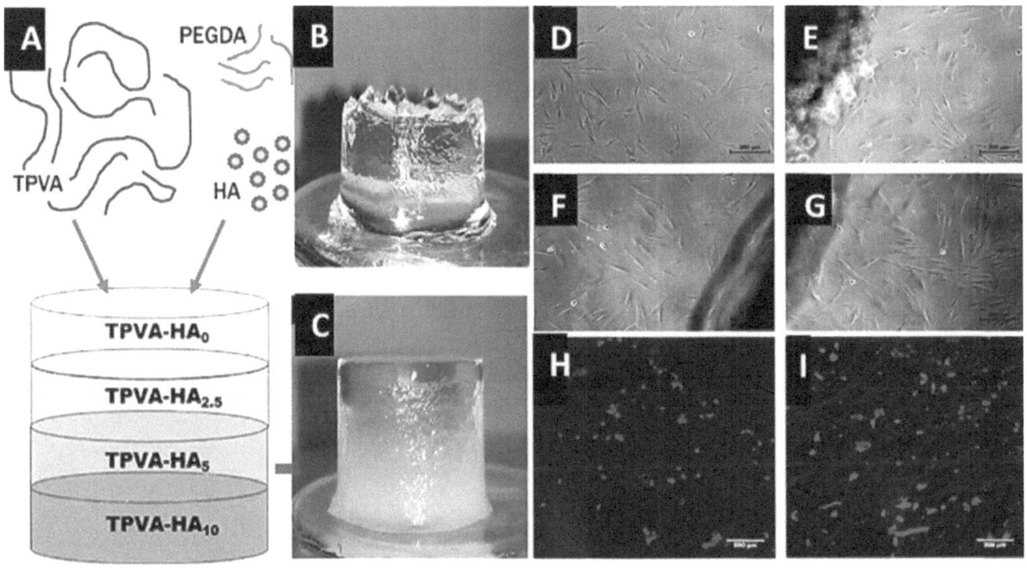

FIGURE 1.5 (A) Schematic depiction for fabricating layered bioactive composite. (B) and (C) Images of hydrogels of different compositions, TPVA and TPVA-HA hydrogels, respectively. (D–G) Microscopic images regarding direct contact assay of materials against hPDL cells (D) only cells, (E) PVA, (F) TPVA, and (G) TPVA-HA. (H–I) Confocal laser scanning microscopy images confirm actin staining on composites incubated with hPDL cells (H) TPVA and (I) TPVA-HA. Adapted from ACS Omega [108].

an increasing manner by several folds in scaffolds with gradients containing decreasing pore sizes compared to non-gradient scaffolds with more significant pore sizes. Di Luca and coworkers [103] reported synthesizing scaffolds with pore-size gradients along the Z axis to cure large bone defects in the same year. These scaffolds were composed of PCL and PEOT/PBT. ECM deposition was greater in the case of gradient scaffolds, and the seeded hMSCs exhibited a trend towards osteogenic differentiation according to the pore size gradients, possibly caused by better uptake of nutrients and surplus oxygen supply.

Further, Byambaa et al. [112] synthesized two different types of hydrogels based on GelMA for vasculogenesis and osteogenesis applications. These GelMA-based hydrogels with varying concentrations of VEGF were printed into 3D structures to create a gradient of VEGF. Outer layers were bio-printed with hMSCs along with silica nanoplatelets loaded in GelMA-VEGF hydrogels. The cocultured hMSCs differentiated into smooth muscle cells, which helped form and mature new blood vessels. VEGF released from the GelMA-VEGF hydrogel was almost four-fold reduced compared to the simple physical mixture of GelMA and VEGF. The printed hydrogel released the VEGF at a therapeutic level for more than seven days. It was confirmed through immune studies and RT-qPCR that the mature bone niche was formed after 21 days of culture due to the encapsulation of hMSCs. Further, Wang et al. [113] developed bi-layered scaffolds for osteochondral defects through an innovative approach. They developed double-layered scaffolds using a novel cryogenic 3D printing technique. In this research work, a composite material of water-in-oil emulsions containing osteogenic peptide/β-TCP and PLGA polymer was printed precisely into a highly porous subchondral layer. Further, on the top layer of subchondral tissue water-in-oil emulsions composed of poly(D,L-lactic acid-co-trimethylene carbonate) were printed into thermos-responsive cartilage frame. Further, the cartilage frame was filled with collagen hydrogel loaded with TGF-β1 (transforming growth factor). These two layers were clumped together and exhibited excellent shear strength. Both the TGF-β1 and osteogenic peptide were released from the subchondral area and the cartilage module over a span of 30 days. Also, through the CCK8 assay, it was confirmed that these osteochondral scaffolds provided a favorable environment for the proliferation of rBMSCs. TCP and osteogenic peptides aided the differentiation of rBMSCs at the subchondral layer towards osteoblast and osteoclast cells in this hierarchically porous bone microenvironment. Subsequently, the chondrogenic fate was improved due to the efficacy of TGF-β1.

Furthermore, Eryilidiz et al. [114] reported the synthesis of Ibuprofen-loaded PLA scaffolds observed with porosities in a radially gradient manner. The drug-loaded PLA scaffolds were fabricated with different infill zones mimicking the heterogenous bimodal structure of cortical and cancellous bone. The PLA scaffolds were fabricated through fused deposition modeling (FDM) with gyroid infill. In-vitro studies confirmed that the PLA scaffolds had released the drug for more than 120 hrs also, the scaffolds containing radial gradients aided in the highest amount of cell proliferation when seeded with L929 cells compared to the homogenous porosity scaffolds, which concluded that these scaffolds could be used to deliver antibiotics as they have desirable compressive strength alongside cell proliferation capacity.

1.3.3 Microfluidic Approach in FGMs for Drug Delivery

In the case of natural tissues/organs, along with the structural gradient, a biochemical gradient plays a major role in the proper functioning and development of specific tissues/organs. Therefore, graded materials with mimicking biochemical gradients might come in handy. Hubka et al. [115] reported, for the first time, creating heparan sulfate gradients inside 3D hydrogels to recreate the (heparin-binding growth factor) HBGF gradient, which can directly enhance the behavior of cells in the living systems. Here for the backbone of hydrogel, thiol-modified hyaluronic acid (HA-SH) was used, and acrylated poly(ethylene glycol) (PEGDA) was used as the crosslinker. A dynamic multichannel gradient maker device (MGMD) was designed, and it was further used to create the expected microparticle gradients in the hydrogel, followed by creating the gradients of PEGylated PlnD1 in hydrogels. Later, fibroblast growth factor-2 (FGF-2) was loaded inside the hydrogel. In-vitro studies confirmed that primary human salivary stem/progenitor cells were viable and growing without getting ruptured during the flow through MGMD. Also, to investigate the migratory behavior of cells in response to the HBGH gradients, two different cell types, including, preosteoblastic MC3T3-E1 and breast cancer cell line MDA-MB-231, were loaded inside hydrogels matrix. Both cells migrated towards the HBGFs bound to the PlnD1. Recently, Idaszek and co-researchers [116] developed biomaterials mimicking cartilage configuration and compositional distribution of extracellular matrix (ECM) using microfluidics. The system was devised in such a manner that different types of bioinks can be ejected either separately or simultaneously. Using this microfluidics system, cell-laden hydrogel composed of GelMA, HAMA, and CS-AEMA was constructed to recapitulate the exact layering of hyaline and calcified cartilage, respectively, to ensure the presence of proper biological cues hydrogels were loaded with human bone marrow-derived mesenchymal stem cells (hBMMSCs) and human articular chondrocytes (hACs). It was demonstrated that hMSCs and hACs loaded ALG+GelMA+CS-AEMA hydrogel have better hyaline phenotype than a monoculture of hMSCs. Also, once TCP microparticles and HAMA were introduced in ALG+GelMA+CS-AEMA hydrogel, it enhanced the development of the hypertrophic chondrocytes, which expresses ALPL, COL10A in an upregulated manner. It was proposed that microfluidics combined with extrusion-based bioprinting can enable the fabrication of several heterogeneous tissues like osteochondral interfaces with excellent viability. Xin et al. reported [117] PEG-based MAP (microporous annealed particle) hydrogels by using microfluidics to create physicochemical gradients. Microgels with different features were synthesized by maintaining the relative flow rates between the precursor solutions, which were later deposited in the syringe layer-by-layer. Further, these microgels were injected and annealed through thiol-ene click chemistry. The seeded hMSCs exhibited different spreading trends according to the stiffness gradient of the scaffolds.

1.3.4 Electrospinning Approach in FGMs for Therapeutics Delivery

Gupta et al. [118] reported a study demonstrating a novel approach for fabricating bilayered scaffolds using mulberry and non-mulberry silk for treating vascular tissue diseases. To form the first layer, they used BM or BM & AA protein; later, upon this wet lyophilized inner layer, the second layer was coated through electrospinning. The outer layer was composed of PCL and BM silk. It was observed that the bilayered silk grafts were biomimetic as their inner porous layer and outer nanofibrous layer replicated the tunica media and tunica adventitia. The porosity, pore size, and biodegradability can be regulated by varying the fabrication methodology and freezing temperature. These bi-layered scaffolds have desired physical, morphological, and mechanical properties. In-vivo studies in Lewis rat models confirmed the viability of the human adipose tissue-derived stromal vascular fraction (SVF) seeded on the scaffold. Furthermore, histological investigation confirmed the formation of new tissue, host cell infiltration, and ECM protein (collagen and elastin) production. Altogether, these bilayered silk-based scaffolds might be promising for treating vascular tissue-related diseases. Further, Tanes and colleagues [119] reported a simple yet facile process for creating bioactive protein gradients on nanofibrous mats. In this study, the BSA (bovine serum albumin) was physically adsorbed on the nanofibers in a time-dependent manner. However, during the processes it was observed that increasing the volume of the BSA solution into the system created a vertical BSA gradient on the PCL nanofiber strip. Furthermore, to investigate the outgrowths of neurites from dorsal root ganglion (DRG), neurite cells were seeded on PCL nanofibers coated with NGF uniformly or in a concentration gradient. Astonishingly, when the nanofibers were uniformly covered by NGF, the extension of neurites from the DRG was similar in length. Conversely, the sample with a gradient with NGF promoted the extension of neurites along with the gradient. Furthermore, He et al. [8] reported a novel FGM nanofibrous membrane through a sequential electrospinning method for bone tissue defects, as shown in Figure 1.6. This FGM comprises nanofiber surface layers covering the core of PCL nanofiber layering. Pro-metronidazole was encapsulated into the gelatin nanofibrous layer to eliminate any type of bacterial infection. The outer shell layer was to promote the compatibility and regeneration of bone, and core PCL nanofibrous layering provided mechanical strength. The loaded pro-metronidazole was released

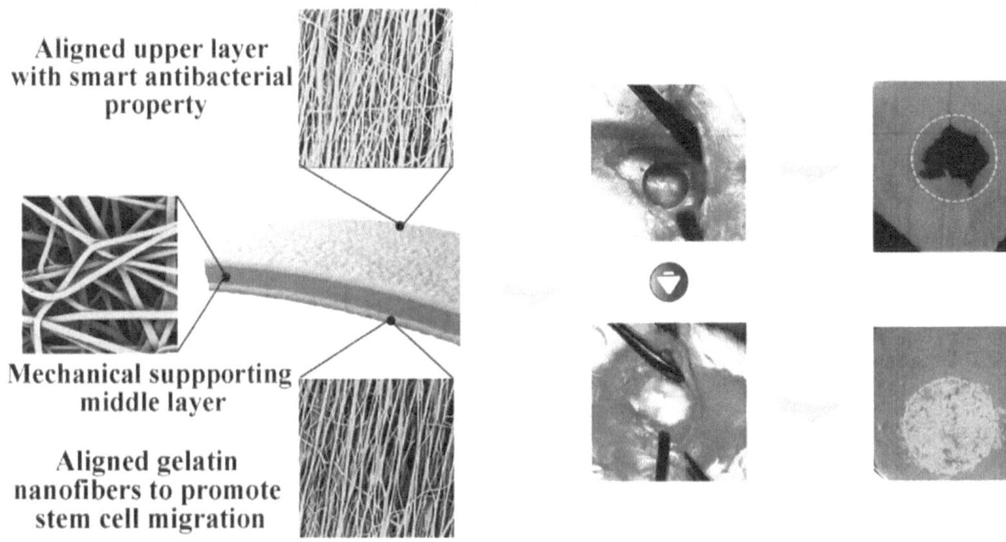

FIGURE 1.6 A multi-functionalized graded membrane was fabricated by electrospinning with enhanced bone regeneration. (Adopted with the permission from the Elsevier [8].)

in an infection-responsive manner and exhibited excellent antibacterial efficacy. Further, in vitro studies confirmed that the scaffolds promoted epithelial cell adhesion proliferation. Excellent osteoblast proliferation and osteodifferentiation confirmed the bio-compatibility of these FGMs. Also, in-vivo studies in rabbits indicated that FGMs containing both aligned nanofiber surface layers have exhibited excellent bone regeneration capacity [8]. Zhu et al. [9] developed conduits of aligned PCL fiber with NGF-gradient to treat peripheral nerve defects. In in-vitro conditions, these conduits promoted longitudinal and chemotactic growth of axons. These conduits directed the regeneration of nerve cells through several (19) signaling cascades related to axonal communication. These conduits also promoted regeneration of axons and functional recovery superior to the conduits with uniform NGF.

Xue et al. [120] reported creating unidirectional and bidirectional gradients based on the density of bio-macromolecular particles on nanofibers using a novel approach based on masked electrospray. In this work, PCL nanofibers were prepared initially through electrospinning, and then collagen particles were directly electro-sprayed onto the nanofibers. This method was applied to many types of biomacromolecules. Particles of collagen arranged in a unidirectional gradient aided in the motility of BMSCs towards the gradually rising particle density, applicable for recruiting BMSCs to the position of infection/ disease. Another investigation of bidirectional gradient enhanced the movement of NIH-3T3 fibroblasts from opposite sides towards the center, which can be applicable for treating a wound. For treating peripheral nerve repair, the particles containing collagen–fibronectin mixture were electro-sprayed to create a bidirectional gradient, which promoted the movement of Schwann cells from the periphery toward the center. In the context of neurite extension, the particles having a collagen–laminin mixture were electro-sprayed to create a unidirectional gradient, which promoted the extension of neurites from DRG body toward the direction of gradually increasing density. They proposed that this method can be applied for biomaterial preparation for several different applications, including wound healing, bone regeneration, and peripheral nerve repair, by varying the composition of particles and gradients of the particles.

1.3.5 Component Redistribution Approach in FGMs for Therapeutic Delivery

All of the previously mentioned works have involved directly depositing the selected materials at various spatial coordinates according to the gradient axis. Alternatively, it can be started from a homogenous/uniform distribution and then redistributing the components into the desired gradient by applying an external force /influence. Li and coworkers [121] have reported an innovative approach using the natural buoyancy mechanism for patterning materials with changeable gradients in this context. This method was applied to different types of drug delivery carriers (liposomes, inorganic nanoparticles, hydrogels). Also, this work demonstrated the capability of this method by forming a BMP-2 gradient in HepMA-GelMA hydrogel for osteochondral tissue regeneration. The BMP-2 from the hydrogel was released sustainably for 28 days. The hydrogel exhibited no toxicity toward the encapsulated hMSCs. Furthermore, the fabricated tissue structures were composed of distinct bony and cartilage regions, also having a transitional phase resembling the native osteochondral interface. Including buoyancy gravity-based approaches, there are other approaches where the individual components can be selectively manipulated by applying some type of force field [121]. For example, Li et al. [122] developed a glycosylated superparamagnetic nanoparticles (SPIONs) gradient for osteochondral defects. This work used a magnetic field to align the glycosyl-modified SPIONs already loaded with desired biomolecules (growth factors) to create gradients different from biomolecules. Also, they created bone morphogenic protein-2 gradient in hydrogels composed of agarose to spatially stimulate the differentiation of human mesenchymal stem cells toward bone cells. To investigate if that arrangement can be used for drug delivery, they incorporated BMP-2-loaded glycosylated SPIONs inside agarose hydrogel. The data showed that BMP-2 was released sustainably over 28 days through diffusion.

Furthermore, the osteochondral tissue constructs exhibited a distinct transition of minerals moving from bone to cartilage. Zwi-Dantsis and coworkers [123] reported work regarding manipulating cellular orientation within 3D collagen hydrogel to develop microstructures of cardiac tissue. In this research, iron oxide nanoparticles were incorporated inside cardiomyocytes, and the cells were oriented in a short-term external field. The main advantage is that during the gelation period, MNP-labeled cells loaded hydrogels can be injected into the infected site or injured site, and then, by applying an external magnetic field, the shape can be tailored according to the pathology needed. However, another limitation is that the cells must be mixed in the hydrogel in the fluid state to get the desired arrangement and shape. Still, once the gel solidifies, the orientation can't be changed by applying another magnetic field.

Also, there are other approaches where gradients were formed through spatially controlled crosslinking instead of controlling the distribution of bulk materials or material components. Alike Mirab and coworkers [124] developed a highly porous nanocomposite scaffold composed of cellulose nanofibers and hydroxyapatite nanoparticles, freezing unidirectionally and further by freeze drying to get the scaffolds. Starch was blended with PVA to impart stability, further crosslinking using citric acid. The porous structure of the scaffold is favorable for bone tissue regeneration. Production of the apatite phase through the mineralization process confirmed the bioactivity of the scaffold. The nanocomposite scaffold was cytocompatible with MG-63 human osteoblast cells. The previously mentioned research was regarding gradient formation during the synthesis step. But, several reports focus on post-modifying the preformed solids or readily-armed hydrogel materials. Fisher et al. [125] reported a hydrogel based on HA. Hydrogels composed of HA were crosslinked with MMPs sensitive peptides and modified by multiphoton labile nitro benzofuran (NDBF). The results showed that the gradient of EGF influences breast cancer cell movement.

1.4 FUNCTIONALLY GRADED MATERIALS IN TISSUE ENGINEERING

TE has quickly become a recognized field in biotechnology, employing transdisciplinary and interdisciplinary approaches. Its primary goal is substituting lost, damaged, or injured body parts with physiologically acceptable alternatives. The co-cultivation of cells and biodegradable scaffolds in vitro are the alternative approaches [126]. Artificial ECM scaffolds play a significant role in various applications for TE, serving as a supportive environment for cells [127]. Biomaterials used in TE must satisfy biodegradable and biocompatible standards, offering a structural framework with an appropriate three-dimensional architecture for tissue reconstruction [128]. Despite significant strides in TE, insufficient understanding regarding dynamic interface interaction poses obstacles to creating efficient biomaterials. Numerous parameters, including surface topography, roughness, and chemical makeup direct contact with body fluids and cells, play a complex role in the early interaction between biomaterials and cells [129]. Advancements in technology, stem cell programming, three-dimensional bioprinters, and dynamic biomaterials have propelled the TE field forward. One notable advancement is the introduction of FGMs, which mimic the inherent diversity found in the body's tissues. FGMs consist of two-component structures that vary from each other in 3D space. Initially developed sought materials for spacecraft that could withstand heat in the 1980s, FGMs have since been leveraged to design materials with specific physical, chemical, and morphological traits that affect cell signaling and mechanical properties [130]. Harnessing the benefits of non-uniform substances, the ability of FGMs to mimic the physiological function of varied tissues might advance our knowledge of the integration of biomaterials in vivo. Various methods, such as composite materials, copolymers, or homopolymers, are used to produce FGMs with various characteristics [131]. This section highlights recent advances in FGMs, examining their applicability in novel tissue growth, substitution, and regeneration models. Examining gradients in composition, porosity, stiffness, and biochemical concentration is common due to their well-established influence on cellular behavior. These gradients can be continuous or non-continuous variations, either linear or non-linear. In softer tissues, scaffolds often utilize more flexible biomaterials, like hydrogels. Conversely, substitutes for hard tissues opt for rigid, ceramics, and slowly resorbing polymers, chosen for their ability to withstand increased mechanical stress [132]. This section will explore various fabrication techniques used to create functionally graded materials.

1.4.1 Microfluidics Approach in FGMs for Tissue Engineering

Microfluidic systems excel in generating gradients of bioactive compounds and shear stress by precisely controlling fluid flow, providing a miniature-scale environment for *in-vitro* model system studies. This technical capability complements advanced methods and techniques like electrospinning and 3D printing, which can fabricate highly personalized gradient scaffolds [133]. Bilayered microfluidic devices have become essential tools in TE for controlling stem and progenitor cell development *in vitro*. In a creative arrangement, mesenchymal stem cells were dispersed in the agarose in the lower section. The upper layer, differentiated by a microporous membrane, contains three-media channels for osteogenic, chondrogenic, and growth media. This strategic structure harmonizes the diffusion of three media channels, subjecting cells to a dissolved stimulus gradient [134]. The outcome is significant: cells primarily exposed to osteogenic media undergo osteogenesis, while cells exposed to chondrogenic media perform chondrogenesis. Moreover, the combination of mechanical and biochemical gradients proves to be a powerful method for examining the combined impact of factors influencing the behavior and functionality of cells. In a microfluidic system with nanofibers at different spacings, fibroblasts exhibit sensitivity to the dosage of the platelet-derived growth factor (PDGF-BB) [135]. At lower PDGF dosages and tighter nanofiber spacing, fibroblasts in this environment demonstrate excellent chemotaxis, enabling cells to move across two fibers and form punctate adhesion sites. These findings underscore the intricate relationship between microfluidic technology and biological reactions, offering valuable insights for developing functionally graded biomaterials.

1.4.2 Electrospinning Approach in FGMs for Tissue Engineering

The technique known as electrospinning constantly extracts nanofibers from a viscoelastic fluid by use of the repulsive force generated by electrostatic interaction between charges on the surface. The resulting fibers, comprising polymers, ceramics, and small molecules, have nanometer-scale sizes mimicking ECM found in natural systems [136]. The appeal of electrospinning lies in its ability to functionalize and align fibers, making it an attractive approach for developing FGMs for TE [137]. In TE, electrospinning has become an essential technique for producing substrates with diverse morphological and mechanical qualities. This is achieved by layering materials successively: one layer is spread, the material in the syringe is changed, and the next layer is added on top of the previous one.

A notable example is the creation of a trilayered scaffold composed of PLA, PCL, and collagen designed for dural replacement. The collagen exterior promotes cell attachment and proliferation, the PCL mid-layer forms a waterproof seal, and the PLA interior layer prevents tissue adhesion. Interestingly, this substrate made up of polymeric gradients exhibits significantly lower absorption of water in comparison to a control using a collagen matrix. When surgically positioned in a rabbit model with a defect in dural membranes, it demonstrates improved cell proliferation and biocompatibility similar to autologous fascial tissue control [138]. A novel method in electrospinning is bidirectional gradient electrospinning, which replaces conventional material stacking in constructing scaffolds with gradient. Briefly in this method, , two solutions were electrospun onto a collector simultaneously at inversely proportional flow rates [139]. Alternatively, syringe pumps could combine two solutions at opposite speeds before spinning, allowing for single fiber deposition with a variable composition changing with time [140]. Despite its cost-effectiveness, scalability, and versatility, electrospinning faces challenges such as solvent toxicity and shear stress during extrusion, necessitating careful process optimization [141]. This comprehensive exploration of electrospinning methods underscores the significance of developing FGMs for TE, offering customized solutions for various applications.

1.4.3 Light-Assisted Approach in FGMs for Tissue Engineering

These techniques in biomaterial synthesis involve introducing a photosensitive element that, when stimulated with a specific wavelength, triggers a chemical reaction leading to the creation or breakdown of crosslinking bonds [142]. An example is photolithography, which effectively controls the light reaching a photosensitive material to generate gradients. Using a photomask helps regulate crosslinking along the axis by managing light distribution based on translucency zones [143].

To explore applying these light-based techniques in TE, adipose-derived stem cells (ASCs) were enclosed within a GelMA hydrogel with a constant toughness gradient. In areas with low stiffness (~8 kPa), ASCs exhibited enhanced distribution of mechanosensitive proteins in the nucleus and increased cellular and nuclear volume. Conversely, regions with high stiffness (~30 kPa) displayed the opposite pattern. Similarly, ASCs placed over a gradient PA hydrogel for seeding at higher stiffness levels showed improved nucleus localization of YAP and Lamin A, along with increased cell spreading [144]. MSCs were introduced into methacrylated alginate hydrogels featuring stiffness gradients and adhesive peptide concentration (RGD). This approach aimed to study the impact of various interactions on cell proliferation. Higher concentrations of MSCs were observed in regions with increased RGD concentration and lower stiffness, emphasizing the significance of these gradients. The platform's ability to spatially regulate gene expression is further demonstrated by creating gradients of short interfering RNA (siRNA) [145,146]. While light-based manufacturing techniques are rapid and cost-effective, knowing their cytotoxic nature and the potential for significant DNA damage is crucial. This section underscores the dual nature of light-based TE techniques, emphasizing their advantages and associated challenges [131].

1.4.4 3D Printing Approach in FGMs for Tissue Engineering

It presents an attractive method for crafting intricate structures using materials highly compatible with the body and incorporating cells. Initially relying on micromachining or sacrificial polymers, it achieves precision in fabricating FGMs by accurately depositing materials using pre-established models [147]. Collagen, known for its abundance in native cartilage, excellent biocompatibility, and T-cell adhesiveness, is popular in cartilage TE. Printing with different collagen concentrations changed the compressive modulus from 1 kPa to 30 kPa without impacting the cell's survival. By printing with varying collagen concentrations, the compressive modulus can be adjusted from 1 kPa to 30 kPa without causing any adverse effect on cell survival. The distribution of cartilage in vivo, collagen type II is successively printed with a rising chondrocyte concentration [148]. For the osteochondral interface, 3D printing incorporates HA, a ceramic based on CaP. Combining HA with PCL facilitates the creation of structures with diverse compositions and porosity. These scaffolds exhibit an inverse correlation between porous nature and compressive modulus regardless of ceramic content. Including the ceramic component results in a compressive modulus of 86 MPa –220 MPa, aligning with the natural structure of trabecular bone [149]. Bioprinting introduces a quick prototyping of advanced structural and chemical signals. Operating in 3D allows for developing increasingly intricate gradients by integrating multiple materials using several printing heads. Nevertheless, the materials utilized in 3D printing must find a balance between viscosity to sustain structure after printing and shear thinning properties for extrusion from the head of the printer [150]. This exploration underscores the transformative potential of 3D printing in producing functionally graded biomaterials for TE applications.

1.4.5 Freeze Drying Approach in FGMs for Tissue Engineering

The technique in which the materials are frozen to form crystals and are eliminated by sublimation is a flexible technology capable of producing various gradient materials [151]. This process is helpful for creating porous materials, with the resultant pores mirroring the inverted dimensions of the frozen solvent crystals. Depending on the different solvents and solutes used, a material's mechanical characteristics can be uniquely modified [152]. In TE, a tailored layered scaffold for articular healing is crafted by combining collagen–chitosan–polycaprolactone (CH-PCL) copolymer and chondroitin sulfate (CS) in different ratios. The scaffold undergoes freeze-drying after precisely building each layer to introduce porosity and composition variations between layers. Widely applied with diverse biomaterials, this approach has proven beneficial for addressing osteochondral defects [153]. Another application involves freeze-casting hydroxyapatite particles around a frozen copper rod. This action prompts the evacuation of HA particles, forming a lamellar structure that runs parallel to the freezing direction. Ice's radial gradient production increases lamellar spacing toward the outside, facilitating cell self-seeding through capillary action [154].

The versatility of freeze-casting and other processes, like electrospinning or solvent-casting, to create FGMs underscores its flexibility. While it offers a cost-effective approach for generating porosity gradients, it's essential to highlight its limitation in simultaneous cell seeding [155]. This exploration delves into the numerous applications of freeze-drying in developing functionally graded biomaterials for TE, providing valuable insights into the potential contributions of this method to the field.

1.5 FUNCTIONALLY GRADED MATERIALS IN MODEL SYSTEMS

1.5.1 Tumor Models

FGMs find diverse applications in cancer modeling, capitalizing on the intricate signaling within the tumor microenvironment. Tumor cells continuously modify their surroundings, creating gradients concerning solubility signals and rigidity enabling interactions with neighboring cells [131]. In cancer modeling, establishing a gradient of oxygen to replicate hypoxia becomes crucial due to its significant impact on cancer growth and metastasis. Hydrogels with varying oxygen levels were utilized to encapsulate primary mouse sarcoma cells in vitro, mimicking pathophysiological oxygen conditions. Cell infiltration occurred more quickly and spread farther in the direction of rising oxygen tension than it did in control gels. This application of FGMs provides valuable insights into the mechanisms influencing cell migration and metastasis. Microfluidics, with its precise control capabilities, is emerging as a popular alternative for modeling the cancer cell environment. It allows the analysis of cell responses to specific factors like protein concentration or oxygen tension [156].

Haptotaxis, characterized as cell movement in reaction to the gradient of surface-bound chemicals, plays a pivotal role in cancer signaling [157]. The variation in fibronectin (FN) concentrations across tumors led to the creation of FN gradients in a microfluidic device. This investigation aimed to investigate the impact effects of matricellular proteins on the migration of MDA-MB-231 breast cancer cells. The research unveiled a novel haptotaxis mechanism mediated by the actin-regulating protein Mena, promoting ECM remodeling. Cell migration responses to ECM fiber alignment were also observed, with MDA-MB-231 cells migrating more quickly on oriented collagen fibers than random fibers [158].

1.5.2 Enthesis Models

The enthesis, which refers to the junction between tendons or ligaments and bones, is a most frequent site for regeneration in conditions like tennis or golfer's elbow, jumper's knee, or Achilles insertional tendinopathies [159]. Treatment options vary from steroidal injections to surgical transplants, depending on the extent of damage [160]. However, allogeneic grafts face challenges such as tissue responses and the risk of disease spread. In contrast, autologous tissue graftings are less affected by the accessibility of tissue and donor site morbidity [161]. To address these issues, ongoing research explores the instructional possibility of fiber orientation FGMs to replicate and exchange patellar–bone bone–tendon grafts, which are considered the best treatment for tears in the anterior cruciate ligament. The electrospinning technique involves a carefully staged setup to create consistent FGCs in biomaterials for TE. The solution filled the syringe and connected it with a regulated pump, and a needle at the end of the syringe was electrospun with a positive voltage. The target comprises parallel aluminium bars spaced 3.5 cm apart and linked to a negative voltage source. Squares of glass positioned strategically on the bars aid sample processing and prevent material folding. After carefully calibrating the needle tip to the target top distance at 15 cm and 20.5 cm, the target positioning was changed to center the fibers. The material is produced by electrospinning each sample for 5–10 minutes. It is cut along the glass square edge for cellular growth or imaging applications.

Another variation involves adding a cover halfway through to minimize fiber density in the random region (see Figure 1.1(B)). This novel procedure includes cutting the fibers along the edge of the cover before removing it, ensuring a more consistent thickness of fibers across the random and aligned portions, and minimizing potential disparities during ongoing electrospinning [137]. To simulate the dimensional characteristics of patellar–bone bone–tendon grafts, tissue-manufactured enthesis models were developed, in which randomly aligned PLGA fibers transitioned to aligned PCL and back to randomly aligned PLGA fibers. Surprising results were obtained using a random-to-aligned PCL scaffold populated with fibroblasts in the aligned portion and osteosarcoma cells in the random section. Osteosarcoma cells proliferated randomly along the fibers, whereas fibroblasts expanded in a straight line [162]. Both types of cells migrated toward the transition zone, pointing to a distinctive model for the transitory interface between soft tissue and bone. This innovative technique can potentially advance TE solutions in FGMs for improved enthesis repair and regeneration [163].

1.5.3 Osteochondral Model

Utilizing FGMs to replicate the bone–cartilage interface follows a similar approach to enthesis models [163]. Ongoing research focuses on gradients in inductive cues, particularly growth factors, to mimic osteochondral tissue. An exemplary instance involves a chitosan–gelatin hydrogel/PLGA scaffold designed for simultaneous TGF-1 and BMP-2 administration, aiming to facilitate the development of mesenchymal stem cells differentiated into chondrocytes and osteoblasts. *In-vitro* experiments revealed that the hydrogel and PLGA phases induced differentiation of chondrocytes and osteocytes, respectively. This method capitalizes on the influence of the sturdiness of the substrate to guide cells toward an osteogenic phenotype, demonstrating the harmonious fusion of calcified tissue and hyaline-like cartilage within two months in rabbit osteochondral lesions [164]. Considering cartilage degeneration and the underlying healthy bone requires focusing on spatial gradients—unidirectionally lyophilized alginate scaffolds supplemented with collagen types I and II produced geometries of anisotropic pores. Compared to isotropic pore geometries, this technique enhanced the mechanical characteristics of the scaffold and enabled a more substantial deposition of collagen and sGAG when seeded with stromal cells obtained from a human infrapatellar fat pad (IFP). Another strategy was to make silk fibroin–chitosan-nano-HAp scaffolds with a porosity gradient through lyophilization and centrifugation. This promoted MSC proliferation with 82–91% porosities, showcasing the ability to mimic the increasing osteochondral porosity structure observed in vivo [165]. This innovative use of FGMs in modeling the bone–cartilage interface holds the potential to advance TE solutions tailored to the complex demands of osteochondral tissues.

1.5.4 Drug Screening Platform

The versatility of FGMs extends to diagnostic devices, enabling the exposure of cells to diverse conditions, determining optimal and cytotoxic drug doses, and exploring cellular responses to change in the ECM. A revolutionary high-throughput drug system was fabricated using combinatorial drug concentrations produced by two microfluidic solutions to yield different combinations [166]. In this system, adenocarcinoma (A549) cells cultured with tirapazamine (TPZ) under different oxygen concentrations demonstrated that TPZ exhibited higher potency under hypoxia compared to traditional chemotherapeutic drugs like cisplatin [167]. A 3D ECM gradient array was created via bioprinting by adjusting the ratio of PEGDA hydrogel, which is bioinert and devoid of intrinsic adhesion patterns, to GelMA hydrogel, which is abundant in native adhesion ligands. This array facilitated the study of the human periodontal ligament stem cell's reactions to varied ECM composition, revealing cell survival and decreased spreading as PEG ratios increased. These platforms hold significant potential for advancing personalized medicine, identifying effective medications, and reducing healthcare costs [168].

This dynamic integration of FGMs into diagnostic devices underscores their crucial role in enhancing our comprehension of cellular behaviors and pharmacological responses in complex microenvironments. This contribution contributes to the development of state-of-the-art diagnostic and therapeutic techniques.

1.6 CONCLUSION

FGMs have emerged as a cutting-edge bio-mimicking paradigm, offering unparalleled versatility in tailoring mechanical, biological, and chemical properties across a gradient, addressing challenges ranging from biomechanical compatibility to long-term implant stability. FGMs represent a departure from traditional homogeneous materials to featuring a gradual transition in composition, structure, or properties over a defined spatial domain. This unique design enables FGMs to mimic biological tissues' intricate and dynamic nature, opening doors to many possibilities in the medical engineering and healthcare sector. Therefore, this chapter explores the FGMs and their pivotal role in revolutionizing implants, drug delivery, and TE, elucidating their design principles, fabrication techniques, and remarkable impacts. Still, FGMs face many drawbacks and challenges in their fabrication to achieve the desired properties, including the need for advanced fabrication tools, design complexity, uniformity, and quality and interface issues in FGMs.

1.7 FUTURE PROSPECTIVE

High-performance FGMs with gradient topologies have been achieved significantly in clinical applications. However, FGMs face numerous challenges and drawbacks in their fabrication to achieve the desired properties, including the need for advanced fabrication tools, design complexity, uniformity, enhanced quality, and interface issues in FGMs. In this regard, researchers should focus on designing and developing advanced fabrication techniques and standardizing the protocols for exploring the FGMs in progressive ways in the biomedical field. Further, advanced characterization techniques are needed to characterize the properties of the FGMs in a way that is dedicated to progressive exploration in the biomedical field. Like, an instrument must be able to determine the mechanical behavior of FGMs to match specific tissues or physiological conditions, considering factors such as stiffness, flexibility, and fatigue resistance. Further, the long-lasting cytotoxicity of FGMs in different medical contexts, paying particular attention to the cellular reactivity, host response, inflammatory responses, biomechanical stability, and long-term tissue integration, must be required for clinical application. Furthermore, a huge research gap between the FGMs and patient-specific design needed innovation in advanced imaging and modeling procedures dedicated to patient-specific design and fabrication of FGM-based implants or prosthetics utilized for personalized medical interventions.

ACKNOWLEDGMENTS

The authors acknowledge Khalifa University, Abu Dhabi, United Arab Emirates, Indian Institute of Technology Gandhinagar, Gujarat, India, and Biorender Software.

DISCLOSURE STATEMENT

No competing financial interests exist.

FUNDING INFORMATION

S.H. and H.S. thanks Khalifa University for funding through Grant No. 8474000442/FSU-2022-022. M.D. I.Y, H.V, H.D, and A.D thanks Project No. IP/IITGN/BE/MD/2223/13 and IP/IITGN/BE/MD/2223/10 from the Indian Institute of Technology Gandhinagar, India.

REFERENCES

[1] B. Saleh, J. Jiang, R. Fathi, T. Al-hababi, Q. Xu, and L. Wang et al., 30 years of functionally graded materials: An overview of manufacturing methods, applications and future challenges, *Composites Part B: Engineering*, 201 (2020), p. 108376.

[2] M. Sam, R. Jojith and N. Radhika, Progression in manufacturing of functionally graded materials and impact of thermal treatment—A critical review, *Journal of Manufacturing Processes*, 68 (2021), pp. 1339–1377.

[3] B. Saleh, J. Jiang, A. Ma, D. Song, D. Yang and Q. Xu, Review on the influence of different reinforcements on the microstructure and wear behavior of functionally graded aluminum matrix composites by centrifugal casting, *Metals and Materials International*, 26 (2020), pp. 933–960.

[4] M. Teacher and R. Velu, Additive manufacturing of functionally graded materials: A comprehensive review, *International Journal of Precision Engineering and Manufacturing*, 25 (2024), pp. 165–197.

[5] D. Lin, Q. Li, W. Li and M. Swain, Bone remodeling induced by dental implants of functionally graded materials, *Journal of Biomedical Materials Research - Part B Applied Biomaterials*, 92 (2010), pp. 430–438.

[6] H. Guo, K. A. Khor, Y. Chiang Boey and X. Miao, Laminated and functionally graded hydroxyapatite/yttria stabilized tetragonal zirconia composites fabricated by spark plasma sintering, *Biomaterials*, 24 (2003), pp. 667–675.

[7] C. Parisi, L. Salvatore, L. Veschini, M.P. Serra, C. Hobbs, M. Madaghiele et al., Biomimetic gradient scaffold of collagen–hydroxyapatite for osteochondral regeneration, *Journal of Tissue Engineering*, 11 (2020), p. 2041731419896068.

[8] M. He, Q. Wang, L. Xie, H. Wu, W. Zhao and W. Tian, Hierarchically multi-functionalized graded membrane with enhanced bone regeneration and self-defensive antibacterial characteristics for guided bone regeneration, *Chemical Engineering Journal*, 398 (2020), p. 125542.

[9] L. Zhu, S. Jia, T. Liu, L. Yan, D. Huang, Z. Wang et al., Aligned PCL fiber conduits immobilized with nerve growth factor gradients enhance and direct sciatic nerve regeneration, *Advanced Functional Materials*, 30 (2020), p. 2002610.

[10] H. Singh, R. Singh, A. Dan, H. Vithalani, S.S. Das, A.V. Fernando et al., Chapter two – Bioconjugated materials as potential vehicles for delivery of antibiotics/drugs, in *Comprehensive Analytical Chemistry*, S.K. Verma and A.K. Das, eds., Elsevier (2023), pp. 29–58.

[11] H. Singh, A. Dan, M.K. Kumawat, V. Pawar, D.S. Chauhan, A. Kaushik et al., Pathophysiology to advanced intra-articular drug delivery strategies: Unravelling rheumatoid arthritis, *Biomaterials*, 303 (2023), p. 122390.

[12] M. Pavlovic, What are biomaterials?, Bioengineering*: A Conceptual Approach*, (2015), pp. 229–244.

[13] Y.H. Joung, Development of implantable medical devices: From an engineering perspective, *International Neurourology Journal*, 17 (2013), pp. 98–106.

[14] W. Pompe, H. Worch, M. Epple, W. Friess, M. Gelinsky, P. Greil et al., Functionally graded materials for biomedical applications, *Materials Science and Engineering: A*, 362 (2003), pp. 40–60.

[15] A. Sola, D. Bellucci and V. Cannillo, Functionally graded materials for orthopedic applications – An update on design and manufacturing, *Biotechnology Advances*, 34 (2016), pp. 504–531.

[16] V. Birman and L.W. Byrd, Modeling and analysis of functionally graded materials and structures *Applied Mechanics Reviews,* 60(5) (2007), pp. 195–216.

[17] A. Kawasaki and R. Watanabe, Concept and P/M fabrication of functionally gradient materials, *Ceramics International*, 23 (1997), pp. 73–83.

[18] A. Kawasaki and R. Watanabe, Concept and P/M fabrication of functionally gradient materials, *Ceramics International*, 23 (1997), pp. 73–83.

[19] E.M. Christenson, K.S. Anseth, J.J.J.P. Van den Beucken, C.K. Chan, B. Ercan, J.A. Jansen et al., Nanobiomaterial applications in orthopedics, *Journal of Orthopaedic Research*, 25 (2007), pp. 11–22.

[20] E. Askari, M. Mehrali, I.H.S.C. Metselaar, N.A. Kadri and M. Rahman, Fabrication and mechanical properties of Al_2O_3 /SiC/ZrO_2 functionally graded material by electrophoretic deposition, *Journal of the Mechanical Behavior of Biomedical Materials*, 12 (2012), pp. 144–150.

[21] K.F. Leong, C.K. Chua, N. Sudarmadji and W.Y. Yeong, Engineering functionally graded tissue engineering scaffolds, *Journal of the Mechanical Behavior of Biomedical Materials*, 1 (2008), pp. 140–152.

[22] B. An, R. Wang, D. Arola and D. Zhang, The role of property gradients on the mechanical behavior of human enamel, *Journal of the Mechanical Behavior of Biomedical Materials*, 9 (2012), pp. 63–72.

[23] W.-J. Li, C.T. Laurencin, E.J. Caterson, R.S. Tuan and F.K. Ko, Electrospun nanofibrous structure: A novel scaffold for tissue engineering, *Journal of Biomedical Materials Research: An Official Journal of The Society for Biomaterials, The Japanese Society for Biomaterials, and The Australian Society for Biomaterials and the Korean Society for Biomaterials*, 60 (2002), pp. 613–621.

[24] H.S. Hedia, Design of functionally graded dental implant in the presence of cancellous bone, *Journal of Biomedical Materials Research - Part B Applied Biomaterials*, 75 (2005), pp. 74–80.

[25] A. M. O'Mahony, J.L. Williams and P. Spencer Anisotropic elasticity of cortical and cancellous bone in the posterior mandible increases peri-implant stress and strain under oblique loading, *Clinical Oral Implants Research*, 12 (2001), pp. 648–657.

[26] F. Watari, A. Yokoyama, F. Saso, M. Uo and T. Kawasaki, Fabrication and properties of functionally graded dental implant, *Composites Part B: Engineering*, 28 (1997), pp. 5–11.

[27] F. Watari, A. Yokoyama, M. Omori, T. Hirai, H. Kondo, M. Uo et al., Biocompatibility of materials and development to functionally graded implant for bio-medical application, *Composites Science and Technology*, 64 (2004), pp. 893–908.

[28] M. Mehrali, F.S. Shirazi, M. Mehrali, H.S.C. Metselaar, N.A.B. Kadri and N.A.A. Osman, Dental implants from functionally graded materials. *Journal of Biomedical Materials Research Part A: An Official Journal of the Society for Biomaterials, the Japanese Society for Biomaterials, and the Australian Society for Biomaterials and the Korean Society for Biomaterials*, 101 (2013), pp. 3046–3057.

[29] A. Sadollah and A. Bahreininejad, Optimum gradient material for a functionally graded dental implant using metaheuristic algorithms, *Journal of the Mechanical Behavior of Biomedical Materials*, 4 (2011), pp. 1384–1395.

[30] Y. Zhang, M.-J. Sun and D. Zhang, Designing functionally graded materials with superior load-bearing properties, *Acta Biomaterialia*, 8 (2012), pp. 1101–1108.

[31] J. Yang and H.-J. Xiang, A three-dimensional finite element study on the biomechanical behavior of an FGBM dental implant in surrounding bone, *Journal of Biomechanics*, 40 (2007), pp. 2377–2385.

[32] H. Sayed Hedia, M. Nemat-Alla, H.S. Hedia, A. Mahmoud and P. Eng, Design optimization of functionally graded dental implant, *Bio-Medical Materials and Engineering*, 14 (2004), pp. 133–143.

[33] F. Wang, H.P. Lee and C. Lu, Thermal-mechanical study of functionally graded dental implants with the finite element method, *Journal of Biomedical Materials Research Part A*, 80 (2007), pp. 146–158.

[34] D. Lin, Q. Li, W. Li, S. Zhou and M.V Swain, Design optimization of functionally graded dental implant for bone remodeling, *Composites Part B: Engineering*, 40 (2009), pp. 668–675.

[35] E.C.N. Silva, M.C. Walters and G.H. Paulino, Modeling bamboo as a functionally graded material: Lessons for the analysis of affordable materials, *Journal of Materials Science*, 41 (2006), pp. 6991–7004.

[36] M.C. Bottino, V. Thomas and G.M. Janowski, A novel spatially designed and functionally graded electrospun membrane for periodontal regeneration, *Acta Biomaterialia*, 7 (2010), pp. 216–224.

[37] E. Askari, M. Mehrali, I.H.S.C. Metselaar, N.A. Kadri and M. Rahman, Fabrication and mechanical properties of Al_2O_3/SiC/ZrO_2 functionally graded material by electrophoretic deposition, *Journal of the Mechanical Behavior of Biomedical Materials*, 12 (2012), pp. 144–150.

[38] C.-L. Lin, J.-C. Wang and Y.-C. Kuo, Numerical simulation on the biomechanical interactions of tooth/implant-supported system under various occlusal forces with rigid/non-rigid connections, *Journal of Biomechanics*, 39 (2006), pp. 453–463.

[39] H.J. Chun, S.Y. Cheong, J.H. Han, S.J. Heo, J.P. Chung, I.C. Rhyu, Y.C. Choi, H.K. Baik, Y. Ku and M.H. Kim, Evaluation of design parameters of osseointegrated dental implants using finite element analysis, *Journal of Oral Rehabilitation*, 29 (2002), pp. 565–574.

[40] H. Van Oosterwyck, J. Duyck, J. Vander Sloten, G. Van der Perre, M. De Coomans, S. Lieven, R. Puers and L. Naert, The influence of bone mechanical properties and implant fixation upon bone loading around oral implants. *Clinical Oral Implants Research*, 9 (1998), pp. 407–418.

[41] O. Kayabaşı, E. Yüzbasıoğlu and F. Erzincanlı, Static, dynamic and fatigue behaviors of dental implant using finite element method, *Advances in Engineering Software*, 37 (2006), pp. 649–658.

[42] S.J. Ding, Y.H. Chu and D.Y. Wang, Enhanced properties of novel zirconia-based osteo-implant systems. *Applied Materials Today*, 9 (2007), pp. 622–632.

[43] E. Yılmaz and F. Çalışkan, A new functional graded dental implant design with biocompatible and antibacterial properties, *Materials Chemistry and Physics*, 277 (2022), p. 125481.

[44] F. Watari, A. Yokoyama, H. Matsuno, F. Saso, M. Uo and T. Kawasaki, Biocompatibility of titanium/hydroxyapatite and titanium/cobalt functionally graded implants, *Materials Science Forum*, 308–311 (1999), pp. 356–361.

[45] G.H. Majzoobi, K. Rahmani, M. Mohammadi, H. Bakhtiari and R. Das, Tribological behaviour of Ti/HA and Ti/SiO_2 functionally graded materials fabricated at different strain rates, *Biotribology*, 35 (2023), p. 100233.

[46] Z. Hong, G.M. Luz, P.J. Hampel, M. Jin, A. Liu, X. Chen et al., Mono-dispersed bioactive glass nanospheres: Preparation and effects on biomechanics of mammalian cells, *Journal of Biomedical Materials Research Part A*, 95 (2010), pp. 747–754.

[47] Z. Hong, R.L. Reis and J.F. Mano, Preparation and in vitro characterization of novel bioactive glass ceramic nanoparticles, *Journal of Biomedical Materials Research Part A*, 88 (2009), pp. 304–313.

[48] M.F. Morks, Fabrication and characterization of plasma-sprayed HA/SiO_2 coatings for biomedical application, *Journal of the Mechanical Behavior of Biomedical Materials*, 1 (2008), pp. 105–111.

[49] T. Matsuno and M. Koishi, Preparation of laminated hydroxyapatiteazirconia sintered composite with the gradient composition, *Journal of Materials Science Letters*, 17 (1998), pp. 1349–1351.

[50] Y. Moayedee, L. Nikzad and H. Majidian, Exploration into the microstructural, mechanical, and biological characteristics of the functionally graded 3Y-TZP/Ti6Al4V system as a potential material for dental implants, *Journal of the Mechanical Behavior of Biomedical Materials*, 151 (2024), p. 106380.

[51] M. Dinu, C. Mihai Cotrut, T. Petreus, T. Hauffman, F. Baciu and A. Vladescu, Structural, electrochemical, biological, and mechanical assessment of functionally graded Cr-based multilayers for enhanced metal-ceramic bond strength in dental restorations, *Materials Chemistry and Physics*, 312 (2024), p. 128655.

[52] Y. Zhu, R. Zhu, J. Ma, Z. Weng, Y. Wang, X. Shi et al., In vitro cell proliferation evaluation of porous nano-zirconia scaffolds with different porosity for bone tissue engineering, *Biomedical Materials (Bristol)*, 10 (2015), p. 055009.

[53] C.I. Resende-Gonçalves, N. Sampaio, J. Moreira, O. Carvalho, J. Caramês, M.C. Manzanares-Céspedes et al., Porous zirconia blocks for bone repair: An integrative review on biological and mechanical outcomes, *Ceramics*, 5 (2022), pp. 161–172.

[54] S. Kurtz, K. Ong, E. Lau, F. Mowat and M. Halpern, Projections of primary and revision hip and knee arthroplasty in the united states from 2005 to 2030, *The Journal of Bone and Joint Surgery*, 89 (2007), pp. 780–785.

[55] D. Pasini and S. Arabnejad Khanoki, Multiscale design and multiobjective optimization of orthopedic hip implants with functionally graded cellular material 134(3) (2012), 031004(10 pages).

[56] J.W. Sharp, K.K. Kani, A. Gee, H. Mulcahy, F.S. Chew and J. Porrino, Anterior cruciate ligament fixation devices: Expected imaging appearance and common complications, *European Journal of Radiology*, 99 (2018), pp. 17–27.

[57] M. Jamil, S. Rafique, A.M. Khan, H. Hegab, M. Mia, M.K. Gupta et al., Comprehensive analysis on orthopedic drilling: A state-of-the-art review, *Proceedings of the Institution of Mechanical Engineers, Part H*, 234 (2020), pp. 537–561.

[58] B. Henriques, S. Gonçalves, D. Soares and F.S. Silva, Shear bond strength comparison between conventional porcelain fused to metal and new functionally graded dental restorations after thermal – Mechanical cycling, *Journal of the Mechanical Behavior of Biomedical Materials*, 13 (2012), pp. 194–205.

[59] K.A. Hing, S.M. Best and W. Bonfield, Characterization of porous hydroxyapatite, *Journal of Materials Science: Materials in Medicine*, 10 (2003), pp. 135–145.

[60] X. Miao and D. Sun, Graded/gradient porous biomaterials, *Materials*, 3 (2010), pp. 26–47.

[61] A. Tampieri, G. Celotti, S. Sprio, A. Delcogliano and S. Franzese, Porosity-graded hydroxyapatite ceramics to replace natural bone, *Biomaterials*, 22 (2001), pp. 1365–1370.

[62] M.P. Staiger, A.M. Pietak, J. Huadmai and G. Dias, Magnesium and its alloys as orthopedic biomaterials: A review, *Biomaterials*, 27 (2006), pp. 1728–1734.

[63] J. Jitcharoen, N.P. Padture, A.E. Giannakopoulos and S. Suresh, Hertzian-crack suppression in ceramics with elastic-modulus-graded surfaces, *Journal of the American Ceramic Society*, 81 (1998), pp. 2301–2308.

[64] Y. Miyamoto, *Functionally Graded Materials: Design, Processing, and Applications*, vol. 5. Springer Science & Business Media (2013), p. 330.

[65] J.H. Kuiper and R. Huiskes, Mathematical optimization of elastic properties: Application to cementless hip stem design 119(2) (1997): 166–174.

[66] A. Boccaccio, A.E. Uva, M. Fiorentino, G. Mori and G. Monno, Geometry design optimization of functionally graded scaffolds for bone tissue engineering: A mechanobiological approach, *PLoS One*, 11 (2016), p. e0146935.

[67] S.L. Sing, J. An, W.Y. Yeong and F.E. Wiria, Laser and electron-beam powder-bed additive manufacturing of metallic implants: A review on processes, materials and designs, *Journal of Orthopaedic Research*, 34 (2016), pp. 369–385.

[68] S.Y. Choy, C.N. Sun, K.F. Leong, K.E. Tan and J. Wei, Functionally graded material by additive manufacturing, *Proceedings of the 2nd International Conference on Progress in Additive Manufacturing* (2016), pp. 206–211, ISSN: 2424-8967. https://doi.org/10.1016/j.msea.2019.138209

[69] B. Liu, Z. Ma, J. Li, H. Xie, X. Wei, B. Wang et al., Experimental study of a 3D printed permanent implantable porous Ta-coated bone plate for fracture fixation, *Bioactive Materials*, 10 (2022), pp. 269–280.

[70] O. Grimaldo Ruiz and Y. Dhaher, Multi-color and Multi-Material 3D Printing of Knee Joint models, *3D Printing in Medicine*, 7 (2021), p. 12.

[71] M. Farsi, A. Asefnejad and H. Baharifar, A hyaluronic acid/PVA electrospun coating on 3D printed PLA scaffold for orthopedic application, *Progress in Biomaterials*, 11 (2022), pp. 67–77.

[72] S. Rouf, A. Malik, A. Raina, M. Irfan, U. Haq, N. Naveed et al., Functionally graded additive manufacturing for orthopedic applications, *Journal of Orthopaedics*, 33 (2022), pp. 70–80.

[73] Y. Sutou, T. Omori, A. Furukawa, Y. Takahashi, R. Kainuma, K. Yamauchi et al., Development of medical guide wire of Cu-Al-Mn-base superelastic alloy with functionally graded characteristics, *Journal of Biomedical Materials Research Part B: Applied Biomaterials: An Official Journal of the Society for Biomaterials, the Japanese Society for Biomaterials, and the Australian Society for Biomaterials and the Korean Society for Biomaterials*, 69 (2004), pp. 64–69.

[74] M.K. Rausch, W. Bothe, J.-P.E. Kvitting, J.C. Swanson, D. Craig Miller and E. Kuhl, Mitral valve annuloplasty a quantitative clinical and mechanical comparison of different annuloplasty devices, *Annals of Biomedical Engineering*, 40 (2012), pp. 750–761.

[75] Y.-Z. Xiong, R.-N. Gao, H. Zhang, L.-L. Dong, J.-T. Li and X. Li, Rationally designed functionally graded porous Ti6Al4V scaffolds with high strength and toughness built via selective laser

melting for load-bearing orthopedic applications, *Journal of the Mechanical Behavior of Biomedical Materials*, 104 (2020), p. 103673.

[76] A. Taheri, F. Farahmand and M. Bahraminasab, Radially and axially graded cellular tibial stems for total knee replacement, *International Journal of Mechanical Sciences*, 263 (2024), pp. 108772.

[77] H. Xie, J. Chen, F. Liu, R. Wang, Y. Tang, Y. Wang et al., Ti-PEEK interpenetrating phase composites with minimal surface for property enhancement of orthopedic implants, *Composite Structures*, 327 (2024), p. 117689.

[78] H. Lee, D.Y. Shin, S.-J. Bang, G. Han, Y. Na, H.S. Kang et al., A strategy for enhancing bioactivity and osseointegration with antibacterial effect by incorporating magnesium in polylactic acid based biodegradable orthopedic implant, *International Journal of Biological Macromolecules*, 254 (2024), p. 127797.

[79] F.A. España, K. Balla, S. Bose, A. Bandyopadhyay and W. M. Keck, Design and fabrication of CoCrMo alloy based novel structures for load bearing implants using laser engineered net shaping, *Materials Science and Engineering: C*, 30 (2010), pp. 50–57.

[80] K.B. Hazlehurst, C.J. Wang and M. Stanford, The potential application of a cobalt chrome molybdenum femoral stem with functionally graded orthotropic structures manufactured using laser melting technologies, *Medical Hypotheses*, 81 (2013), pp. 1096–1099.

[81] S. Limmahakhun, A. Oloyede, N. Chantarapanich, P. Jiamwatthanachai, K. Sitthiseripratip, Y. Xiao et al., Alternative designs of load–sharing cobalt chromium graded femoral stems, *Materials Today Communications*, 12 (2017), pp. 1–10.

[82] M. Bahraminasab, B. Sahari, K. Edwards, F. Farahmand, T. Sai Hong and H. Naghibi, Material tailoring of the femoral component in a total knee replacement to reduce the problem of aseptic loosening, *Materials & Design*, 52 (1980–2015), pp. 441–451.

[83] S. Arabnejad, B. Johnston, M. Tanzer and D. Pasini, Fully porous 3D printed titanium femoral stem to reduce stress-shielding following total hip arthroplasty, *Journal of Orthopaedic Research*, 35 (2017), pp. 1774–1783.

[84] H.S. Hedia, M.A.N. Shabara, T.T. El-Midany and N. Fouda, A method of material optimization of cementless stem through functionally graded material, *International Journal of Mechanics and Materials in Design*, 1 (2004), pp. 329–346.

[85] K. Migacz, J. Chłopek, A. Morawska-Chochół and M. Ambroziak, Gradient composite materials for artificial intervertebral discs, *Acta of Bioengineering and Biomechanics*, 16 (2014), pp. 3–12.

[86] J. Al-Sukhun, H. Penttilä and N. Ashammakhi, Orbital stress analysis, part IV: Use of a "stiffness Graded" biodegradable implants to repair orbital blow-out fracture, *Journal of Craniofacial Surgery*, 23 (2012), pp. 126–130.

[87] V.K. Ganesh, K. Ramakrishna and D. N Ghista, Biomechanics of bone-fracture fixation by stiffness-graded plates in comparison with stainless-steel plates, *Biomedical Engineering Online*, 4 (2005), pp. 1–15.

[88] S.A. Khanoki and D. Pasini, Fatigue design of a mechanically biocompatible lattice for a proof-of-concept femoral stem, *Journal of the Mechanical Behavior of Biomedical Materials*, 22 (2013), pp. 65–83.

[89] S. Arabnejad Khanoki Damiano Pasini, The fatigue design of a bone preserving hip implant with functionally graded cellular material, *Journal of Medical Devices*, 7 (2013), p. 020907.

[90] H.V. Wang, S.R. Johnston, P. Fellow and D.W. Rosen, Design of a graded cellular structure for an acetabular hip replacement component (2006–2008). https://doi.org/10.1016/j.surfcoat.2024.131593

[91] C. Li, L. Ouyang, J.P. Armstrong and M.M., Stevens, Advances in the fabrication of biomaterials for gradient tissue engineering. *Trends in Biotechnology*, 39 (2021), pp. 150–164.

[92] A.J. Sophia Fox, A. Bedi and S.A. Rodeo, The basic science of articular cartilage: Structure, composition, and function, *Sports Health*, 1 (2009), p. 461.

[93] X. Ren, F. Wang, C. Chen, X. Gong, L. Yin and L. Yang, Engineering zonal cartilage through bioprinting collagen type II hydrogel constructs with biomimetic chondrocyte density gradient, *BMC Musculoskelet Disord*, 17 (2016), pp. 1–10.

[94] M.S. Bergholt, J.-P. St-Pierre, G.S. Offeddu, P.A. Parmar, M.B. Albro, J.L. Puetzer et al., Raman spectroscopy reveals new insights into the zonal organization of native and tissue-engineered articular cartilage, *ACS Central Science*, 2 (2016), pp. 885–895.

[95] S.I. Djomehri, S. Candell, T. Case, A. Browning, G.W. Marshall, W. Yun et al., Mineral density volume gradients in normal and diseased human tissues, *PloS One*, 10 (2015), p. e0121611.
[96] C. Tobón, C.A. Ruiz-Villa, E. Heidenreich, L. Romero, F. Hornero and J. Saiz, A three-dimensional human atrial model with fiber orientation. Electrograms and arrhythmic activation patterns relationship, *PLoS One*, 8 (2013), p. e50883.
[97] C. Baron, Using the gradient of human cortical bone properties to determine age-related bone changes via ultrasonic guided waves, *Ultrasound in Medicine and Biology*, 38 (2012), pp. 972–981.
[98] S. Thomopoulos, G.R. Williams, J.A. Gimbel, M. Favata and L.J. Soslowsky, Variation of biomechanical, structural, and compositional properties along the tendon to bone insertion site, *Journal of Orthopaedic Research*, 21 (2003), pp. 413–419.
[99] H.M. Kronenberg, Developmental regulation of the growth plate, *Nature*, 423(6937) (2003), pp. 332–336.
[100] J.L. Christian, Morphogen gradients in development: From form to function, *Wiley Interdisciplinary Reviews: Developmental Biology*, 1 (2012), pp. 3–15.
[101] A.J. Zhu and M.P. Scott, Incredible journey: How do developmental signals travel through tissue?, *Genes & Development*, 18 (2004), pp. 2985–2997.
[102] M. Zhu, H. Tao, M. Samani, M. Luo, X. Wang, S. Hopyan et al., Spatial mapping of tissue properties in vivo reveals a 3D stiffness gradient in the mouse limb bud, *Proceedings of the National Academy of Sciences*, 117 (2020), pp. 4781–4791.
[103] G.G. Niederauer, M. A. Slivka, N.C. Leatherbury, D.L. Korvick, H.H. Harroff, W.C. Ehler et al., Evaluation of multiphase implants for repair of focal osteochondral defects in goats, *Biomaterials*, 21 (2000), pp. 2561–2574.
[104] S.H. Hong, M. Shin, E. Park, J.H. Ryu, J.A. Burdick and H. Lee, Alginate-boronic acid: pH-triggered bioinspired glue for hydrogel assembly, *Advanced Functional Materials*, 30 (2020), p. 1908497.
[105] J. Guo, C. Li, S. Ling, W. Huang, Y. Chen and D.L. Kaplan, Multiscale design and synthesis of biomimetic gradient protein/biosilica composites for interfacial tissue engineering, *Biomaterials*, 145 (2017), pp. 44–55.
[106] H. Ko, K. Suthiwanich, H. Mary, S. Zanganeh, S.K. Hu, S. Ahadian et al., A simple layer-stacking technique to generate biomolecular and mechanical gradients in photocrosslinkable hydrogels, *Biofabrication*, 11 (2019), p. 025014.
[107] Z. Wei, D.M. Lewis, Y. Xu and S. Gerecht, Dual cross-linked biofunctional and self-healing networks to generate user-defined modular gradient hydrogel constructs, *Advanced Healthcare Materials*, 6 (2017), p. 1700523.
[108] R.K. Adarsh, E.C. Das, G. V. Gopan, S. Selvam and M. Komath, Functionally graded bioactive composites based on poly(vinyl alcohol) made through thiol-ene click reaction, *ACS Omega*, 7 (2022), pp. 29246–29255.
[109] L. Ouyang, J.P.K. Armstrong, M. Salmeron-Sanchez and M.M. Stevens, Assembling living building blocks to engineer complex tissues, *Advanced Functional Materials*, 30 (2020), p. 1909009.
[110] L.G. Bracaglia, B.T. Smith, E. Watson, N. Arumugasaamy, A.G. Mikos and J.P. Fisher, 3D printing for the design and fabrication of polymer-based gradient scaffolds, *Acta Biomaterialia*, 56 (2017), pp. 3–13.
[111] A. Di Luca, K. Szlazak, I. Lorenzo-Moldero, C.A. Ghebes, A. Lepedda, W. Swieszkowski et al., Influencing chondrogenic differentiation of human mesenchymal stromal cells in scaffolds displaying a structural gradient in pore size, *Acta Biomaterialia*, 36 (2016), pp. 210–219.
[112] B. Byambaa, N. Annabi, K. Yue, G. Trujillo-de Santiago, M.M. Alvarez, W. Jia et al., Bioprinted osteogenic and vasculogenic patterns for engineering 3D bone tissue, *Advanced Healthcare Materials*, 6 (2017), p. 1700015.
[113] C. Wang, H. Yue, W. Huang, X. Lin, X. Xie, Z. He et al., Cryogenic 3D printing of heterogeneous scaffolds with gradient mechanical strengths and spatial delivery of osteogenic peptide/TGF-β1 for osteochondral tissue regeneration, *Biofabrication*, 12 (2020), p. 025030.
[114] M. Eryildiz, Fabrication of drug-loaded 3D-printed bone scaffolds with radial gradient porosity, *Journal of Materials Engineering and Performance*, 32 (2023), pp. 4249–4257.
[115] K.M. Hubka, D.D. Carson, D.A. Harrington and M.C. Farach-Carson, Perlecan domain I gradients establish stable biomimetic heparin binding growth factor gradients for cell migration in hydrogels, *Acta Biomaterialia*, 97 (2019), pp. 385–398.

[116] J. Idaszek, M. Costantini, T.A. Karlsen, J. Jaroszewicz, C. Colosi, S. Testa et al., 3D bioprinting of hydrogel constructs with cell and material gradients for the regeneration of full-thickness chondral defect using a microfluidic printing head, *Biofabrication*, 11 (2019), p. 044101.
[117] S. Xin, J. Dai, C.A. Gregory, A. Han and D.L. Alge, Creating physicochemical gradients in modular microporous annealed particle hydrogels via a microfluidic method, *Advanced Functional Materials*, 30 (2020), p. 1907102.
[118] P. Gupta, K.L. Lorentz, D.G. Haskett, E.M. Cunnane, A.K. Ramaswamy, J.S. Weinbaum et al., Bioresorbable silk grafts for small diameter vascular tissue engineering applications: In vitro and in vivo functional analysis, *Acta Biomaterialia*, 105 (2020), p. 146.
[119] M.L. Tanes, J. Xue and Y. Xia, A general strategy for generating gradients of bioactive proteins on electrospun nanofiber mats by masking with bovine serum albumin, *Journal of Materials Chemistry B*, 5580 (2017), p. 5580.
[120] J. Xue, T. Wu, J. Qiu, S. Rutledge, M.L. Tanes and Y. Xia, Promoting cell migration and neurite extension along uniaxially aligned nanofibers with biomacromolecular particles in a density gradient, *Advanced Functional Materials*, 30 (2020), p. 2002031.
[121] C. Li, L. Ouyang, I.J. Pence, A.C. Moore, Y. Lin, C.W. Winter et al., Buoyancy-driven gradients for biomaterial fabrication and tissue engineering, *Advanced Materials*, 31 (2019), pp. 1900291.
[122] C. Li, J.P. Armstrong, I.J. Pence, W. Kit-Anan, J.L. Puetzer, S. Correia Carreira et al., Glycosylated superparamagnetic nanoparticle gradients for osteochondral tissue engineering *Biomaterials*, 176 (2018), pp. 24–33.
[123] L. Zwi-Dantsis, B. Wang, C. Marijon, S. Zonetti, A. Ferrini, L. Massi et al., Remote magnetic nanoparticle manipulation enables the dynamic patterning of cardiac tissues, *Advanced Materials* 32(6) (2020), p. 1904598.
[124] F. Mirab, M. Eslamian and R. Bagheri, Fabrication and characterization of a starch-based nanocomposite scaffold with highly porous and gradient structure for bone tissue engineering, *Biomedical Physics and Engineering Express*, 4 (2018), p. 055021.
[125] S.A. Fisher, R.Y. Tam, A. Fokina, M.M. Mahmoodi, M.D. Distefano and M.S. Shoichet, Photo-immobilized EGF chemical gradients differentially impact breast cancer cell invasion and drug response in defined 3D hydrogels, *Biomaterials*, 178 (2018), pp. 751–766.
[126] Y. Tabata, Recent progress in tissue engineering, *Drug Discovery Today*, 6 (2001), pp. 483–487.
[127] J.A., Hubbell, Biomaterials in tissue engineering. *Biotechnology*, 13 (1995), pp. 565–576.
[128] A.E. Eldeeb, S. Salah and N.A. Elkasabgy, Biomaterials for tissue engineering applications and current updates in the field: A comprehensive review, *AAPS PharmSciTech*, 23 (2022), p. 3.
[129] V.I. Kodolov, A.P. Kuznetsov, O.A. Nicolaeva, E. Sh Shayakhmetova, L.G. Makarova, I.N. Shabanova et al., Bioengineered material surfaces for medical applications, *Surface and Interface Analysis*, 32 (2001), pp. 3–9.
[130] Y. Miyamoto, W. Kaysser, B. Rabin and A. Kawasaki, Functionally Graded Materials: Design, Processing and Applications, Edition 1, Springer New York, (2013), XV-330.
[131] J.M. Lowen and J.K. Leach, Functionally graded biomaterials for use as model systems and replacement tissues, *Advanced Functional Materials*, 30 (2020)1909089.
[132] A. Pattnaik, A.S. Sanket, S. Pradhan, R. Sahoo, S. Das, S. Pany et al., Designing of gradient scaffolds and their applications in tissue regeneration, *Biomaterials*, 296 (2023), p. 122078.
[133] W. Xiang, Z. Liu, and Y. Pang, Concentration Gradient Generation Methods Based on Microfluidic Systems, *Royal Society of Chemistry Advances* 7(48) (2017), pp. 29966–29984.
[134] X. Shi, J. Zhou, Y. Zhao, L. Li and H. Wu, Gradient-regulated hydrogel for interface tissue engineering: Steering simultaneous osteo/chondrogenesis of stem cells on a chip, *Advanced Healthcare Materials*, 2 (2014), pp. 846–853.
[135] C.M. Morrow, A. Mukherjee, M.A. Traore, E.J. Leaman, A. Kim, E.M. Smith et al., Integrating nanofibers with biochemical gradients to investigate physiologically-relevant fibroblast chemotaxis, *Lab on a Chip* 19 (2019), pp. 3641–3651.
[136] G. Ingavle and J. Kent Leach. Reviews and undefined 2014, Advancements in electrospinning of polymeric nanofibrous scaffolds for tissue engineering, liebertpub.comGC Ingavle, JK Leach *Tissue Engineering Part B: Reviews*, 2014•liebertpub.com 20 (2014), pp. 277–293.
[137] D.I. Braghirolli, D. Steffens and P. Pranke, Electrospinning for regenerative medicine: A review of the main topics, *Drug Discovery Today*, 19 (2014), pp. 743–753.

[138] Y.F. Wang, H.F. Guo and D.J. Ying, Multilayer scaffold of electrospun PLA–PCL–collagen nanofibers as a dural substitute, *Journal of Biomedical Materials Research – Part B Applied Biomaterials*, 101 (2013), pp. 1359–1366.

[139] T. Wu, C. Huang, D. Li, A. Yin, W. Liu, …, J. Wang et al., A multi-layered vascular scaffold with symmetrical structure by bi-directional gradient electrospinning, *Colloids and Surfaces B: Biointerfaces*, 133 (2015), pp. 179–188.

[140] A.P. Kishan, A.B. Robbins, S.F. Mohiuddin, M. Jiang, M.R. Moreno and E.M. Cosgriff-Hernandez, Fabrication of macromolecular gradients in aligned fiber scaffolds using a combination of in-line blending and air-gap electrospinning, *Acta Biomaterialia*, 56 (2017), pp. 118–128.

[141] H.Y. Mi, X. Jing, E. Yu, J. McNulty, X.F. Peng and L.S. Turng, Fabrication of triple-layered vascular scaffolds by combining electrospinning, braiding, and thermally induced phase separation, *Materials Letters*, 161 (2015), pp. 305–308.

[142] I. Gadjanski, Recent advances on gradient hydrogels in biomimetic cartilage tissue engineering, *F1000Research*, 6 (2017), 2158.

[143] M.J. Madou, B. Raton, L. New and Y. Washington, *Fundamentals of Microfabrication: The Science of Miniaturization*. CRC Press (2002).

[144] L.G. Major, A.W. Holle, J.L. Young, M.S. Hepburn, K. Jeong, I.L. Chin et al., Volume adaptation controls stem cell mechanotransduction, *ACS Publications*, 11 (2019), p. 45530.

[145] M.C. Hill, M.K. Nguyen, O. Jeon and E. Alsberg, Spatial control of cell gene expression by siRNA gradients in biodegradable hydrogels, *Advanced Healthcare Materials*, 4 (2015), pp. 714–722.

[146] O. Jeon, D.S. Alt, S.W. Linderman and E. Alsberg, Biochemical and physical signal gradients in hydrogels to control stem cell behavior, *Advanced Materials*, 25 (2013), pp. 6366–6372.

[147] S. Khorshidi and A. Karkhaneh, A review on gradient hydrogel/fiber scaffolds for osteochondral regeneration, *Journal of Tissue Engineering and Regenerative Medicine*, 12 (2018), pp. e1974–e1990.

[148] X. Ren, F. Wang, C. Chen, X. Gong, L. Yin and L. Yang, Engineering zonal cartilage through bioprinting collagen type II hydrogel constructs with biomimetic chondrocyte density gradient, *BMC Musculoskeletal Disorders*, 17 (2016), pp. 1–10.

[149] S. Bittner, B. Smith and L. Diaz-Gomez, Fabrication and mechanical characterization of 3D printed vertical uniform and gradient scaffolds for bone and osteochondral tissue engineering, *Acta Biomaterialia*, 90 (2014), pp. 37–48.

[150] C. Mandrycky, Z. Wang, K. Kim and D.K.-B. 3D bioprinting for engineering complex tissues, *Biotechnology Advances*, 34 (2016), pp. 422–434.

[151] A. do V. Morais, É., do Nascimento Alencar, F.H.X., Júnior, C.M., de Oliveira, H.R., Marcelino, G., Barratt, H., Fessi, E.S.T. do Egito and A. Elaissari, Freeze-drying of emulsified systems: A review, *International Journal of Pharmaceutics*, 503 (2016), pp. 102–114.

[152] S. Deville, Freeze-casting of porous biomaterials: Structure, properties and opportunities, *Materials*, 3 (2010), pp. 1913–1927.

[153] Y. Zhu, H. Wu, S. Sun, T. Zhou, J. Wu et al., Designed composites for mimicking compressive mechanical properties of articular cartilage matrix, *Journal of the Mechanical Behavior of Biomedical Materials*, 36 (2014), pp. 32–46.

[154] H. Bai, D. Wang, B. Delattre, W. Gao, J. De Coninck et al., Biomimetic gradient scaffold from ice-templating for self-seeding of cells with capillary effect, *Acta Biomaterialia*, 20 (2015), pp. 113–119.

[155] S. Zhang, L. Chen, Y. Jiang, Y. Cai, G. Xu et al., Bi-layer collagen/microporous electrospun nanofiber scaffold improves the osteochondral regeneration, *Acta Biomaterialia*, 9 (2013), pp. 7236–7247.

[156] D.M. Lewis, K.M. Park, V. Tang, Y. Xu, K. Pak, T.S.K. Eisinger-Mathason et al., Intratumoral oxygen gradients mediate sarcoma cell invasion, *Proceedings of the National Academy of Sciences of the United States of America*, 113 (2016), pp. 9292–9297.

[157] M. Oudin, O. Jonas, T. Kosciuk, L. Broye et al., Tumor cell–driven extracellular matrix remodeling drives haptotaxis during metastatic progression, *Cancer Discovery*, 6 (2016), pp. 516–531.

[158] S. Carey, Z. Goldblatt, K. Martin, … et al., Local extracellular matrix alignment directs cellular protrusion dynamics and migration through Rac1 and FAK, *Integrative Biology*, 8 (2016), pp. 821–835.

[159] M. Benjamin, H. Toumi, J.R. Ralphs, G. Bydder, T.M. Best and S. Milz, Where tendons and ligaments meet bone: attachment sites ('entheses') in relation to exercise and/or mechanical load, Wiley Online LibraryM Benjamin, H Toumi, JR Ralphs, G Bydder, TM Best, S Milz *Journal of Anatomy*, 2006•Wiley Online Library 208 (2006), pp. 471–490.

[160] I. Calejo, R. Costa-Almeida, ..., R. L Reis, Enthesis tissue engineering: Biological requirements meet at the interface, liebertpub.comI Calejo, R Costa-Almeida, RL Reis, ME Gomes *Tissue Engineering Part B: Reviews*, 2019•liebertpub.com 25 (2019), pp. 330–356.

[161] C.C. Kaeding, B. Aros, A. Pedroza, E. Pifel, A. Amendola, J.T. Andrish et al., Allograft versus autograft anterior cruciate ligament reconstruction: Predictors of failure from a moon prospective longitudinal cohort, *Sports Health*, 3 (2011), pp. 73–81.

[162] S. Samavedi, P. Vaidya, P. Gaddam, A.R. Whittington and A.S. Goldstein, Electrospun meshes possessing region-wise differences in fiber orientation, diameter, chemistry and mechanical properties for engineering bone-ligament-bone tissues, *Biotechnology and Bioengineering*, 111 (2014), pp. 2549–2559.

[163] J. Nowlin, M.A. Bismi, B. Delpech, P. Dumas, Y. Zhou and G.Z. Tan, Engineering the hard–soft tissue interface with random-to-aligned nanofiber scaffolds, *Nanobiomedicine (Rij)*, 5 (2018), p. 184954351880353.

[164] F. Han, F. Zhou, X. Yang, J. Zhao, Y. Zhao and X. Yuan, A pilot study of conically graded chitosan-gelatin hydrogel/PLGA scaffold with dual-delivery of TGF-β1 and BMP-2 for regeneration of cartilage-bone interface, *Journal of Biomedical Materials Research – Part B Applied Biomaterials*, 103 (2015), pp. 1344–1353.

[165] H. Xiao, W. Huang, K. Xiong, S. Ruan, C. Yuan, G. Mo et al., Osteochondral repair using scaffolds with gradient pore sizes constructed with silk fibroin, chitosan, and nano-hydroxyapatite, *International Journal of Nanomedicine*, 14 (2019), pp. 2011–2027.

[166] D. An, K. Kim and J. Kim, Microfluidic system based high throughput drug screening system for curcumin/TRAIL combinational chemotherapy in human prostate cancer PC3 cells, *Biomolecules & Therapeutics*, 22 (2014), p. 355.

[167] T. Mulholland, M. McAllister and S. Patek, Drug screening of biopsy-derived spheroids using a self-generated microfluidic concentration gradient, *Scientific Reports*, 8 (2018),pp. 14672.

[168] Y. Ma, Y. Ji, G. Huang, K. Ling, X. Zhang and F. Xu, Bioprinting 3D cell-laden hydrogel microarray for screening human periodontal ligament stem cell response to extracellular matrix, *Biofabrication*, 7 (2015),pp. 044105.

2 Sustainable Smart Materials and Composite Structures

Bibekananda Nayak, S. Jayasri, P. Elorika, Sharmistha Anwar, and Shahid Anwar

2.1 INTRODUCTION

In the wide domain of materials science, the exploration and application of advanced functional polymer and composite materials have paved the way for revolutionary progress across diverse industries. These materials go beyond enhancing mechanical aspects and act as catalysts for cutting-edge breakthroughs in electronics and healthcare, influencing the evolving technological landscape significantly. This chapter embarks on an in-depth exploration of these advanced functional materials, focusing on their inherent properties, characterization methods, and a broad spectrum of applications [1]. As we delve into the intricacies of these materials, we uncover the fascinating interplay between molecular design, structural composition, and the unique performance attributes that define them. The rise of smart materials introduces a transformative era where materials respond to environmental stimuli, showcasing properties beyond conventional expectations. However, in the pursuit of progress, a new imperative has emerged – sustainability. This book chapter explores the dual narrative of intelligence and eco-responsibility, examining the defining properties of smart materials and their role in sustainable practices [2].

Over the years, smart functional materials have found applications in small- and microscale transducers, precision mechatronic control systems, and large-scale structures as in situ sensors and actuators since the mid-1980s, giving rise to the concept of smart materials, smart structures, and smart systems/applications [3]. The chapter provides a comprehensive overview of current smart materials and their applications in sensors, actuators, and structures. The synthesis and characterization of advanced materials have played a crucial role in the development of smart materials, also known as intelligent materials. These materials, gaining popularity in the contemporary era, possess the unique ability to react to their surroundings and sense stimuli, enhancing overall material performance. Smart materials are categorized into metals, ceramics, polymers, and other innovative materials that exhibit extraordinary properties, such as self-healing and self-destructing. The "Smart Materials Era" in materials science and engineering signifies a transformative period where innovative and responsive materials drive technological advancements across diverse applications. Smart materials play a pivotal role in industries like aerospace, healthcare, robotics, and more, offering opportunities to enhance efficiency, functionality, and adaptability. The ongoing exploration of these materials shapes the landscape of modern technology and promises unprecedented opportunities. Certain materials in the Smart Materials Era will possess the remarkable ability to autonomously perform specific functions in response to changing environmental conditions. The historical epochs representing distinct eras, like the Stone Age, Bronze Age, and Iron Age, are paralleled by the present era of synthetic materials, leading to the upcoming Smart Materials Era [4,5]. This forthcoming era is poised to leverage synthetic materials, using them to harness emerging technologies for the synthesis of intelligent materials with nervous systems, cognitive capacities, and muscular

DOI: 10.1201/9781003664611-3

functionalities. This evolution represents a continuous trajectory of human innovation, where each age builds upon the achievements of its predecessors. The future promises a revolution, where smart materials impact industries and technologies in various domains. Some materials will have embedded sensory capabilities, ensuring structural components meet quality control specifications. Anticipated features of advanced smart materials include self-repair, self-diagnosis, self-multiplication, and self-degradation, particularly in engineering applications. These supreme classes of intelligent materials are expected to significantly impact various facets of civilization, with designers harnessing their unique capabilities across diverse industries such as aerospace, manufacturing, automotive, sports, medicine, semi-conductive technology, and civil engineering. The integration of these innovative materials is set to revolutionize the way we approach and solve challenges in numerous fields [6].

The advancement of science and technology has led to the widespread use of lithography-based strategies, particularly in the micro/nanofabrication field. This method is commonly employed for creating devices based on smart materials, particularly in areas like optics, electronics, and biology. However, the direct patterning of smart materials using lithography-based approaches poses challenges. In this process, a photoresist, acting as a graphic transfer medium, is initially patterned onto the target substrate. Subsequently, smart materials are applied to the substrate using techniques like magnetron sputtering, vacuum evaporation, or atomic deposition. The final step involves achieving a patterned functional layer in devices by removing the photoresist. Unfortunately, this method limits the selection of smart materials that can be effectively patterned through lithography. Despite the advantages offered by lithography-based technologies, there is a growing need to explore post-lithography techniques. Researchers are actively investigating alternative manufacturing methods to fulfill the requirements of smart material-based devices. Among these approaches, printing stands out as an additive manufacturing technology, showcasing exceptional universality, diversity, and stability in patterning smart materials. This capability is expected to significantly impact micro/nanomanufacturing and device integration technologies. In recent years, significant progress has been made in the design and construction of printable smart materials and devices. These printed materials and micro/nanostructures find applications in diverse fields such as communication, sensing, actuating, and imaging. To enhance the practical application of printed smart devices in engineering fields like aerospace, biomedicine, intelligent robotics, and smart electronics, extensive interdisciplinary studies will be essential in the future [7,8].

2.2 SMART MATERIALS CLASSIFICATIONS AND STRUCTURES

Defining smart materials poses a challenge due to the variety of definitions found in the literature, often characterized by ambiguity and contradictions. While various interpretations exist, there is a consensus within the scientific community that smart materials encompass substances or derived products capable of reversibly altering their physical or chemical properties in response to an external stimulus [9]. However, an issue with this definition arises as it could potentially be applied to nearly every existing material; for instance, steel undergoes reversible changes in dimension, and water reversibly changes its density with external temperature variations, yet these materials are not considered smart materials. Therefore, a more precise definition of smart materials should exclusively encompass non-conventional programmable materials distinguished by exceptional material properties. Smart materials encompass various types that can be categorized based on their compound compositions like organic or inorganic, material nature like metallic or polymeric, and physical state of materials like solids, gels, or fluids. Regardless of whether they are organic, metallic, or gel-based, the classification of smart materials typically revolves around their behavior, specifically how they respond to diverse external conditions [3,7].

Apart from that the smart materials are basically classified into two classes; one is passive smart materials and the other one is active smart materials as depicted in Figure 2.1 [3]. Passive smart materials and Active smart materials differ in their response to external stimuli and their energy

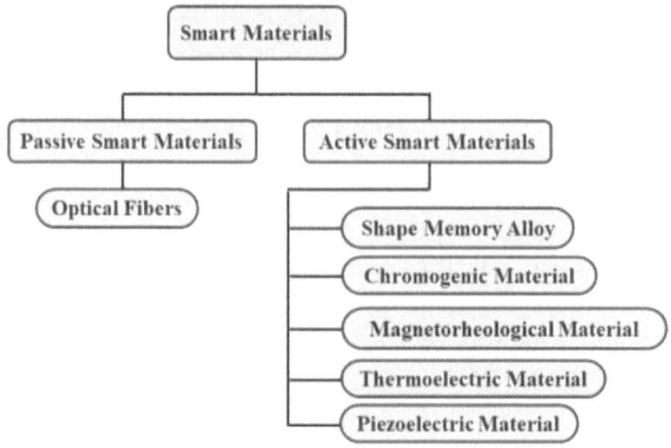

FIGURE 2.1 Classification of smart materials.

transduction capabilities [3]. Passive smart materials respond to stimuli but transfer energy without transducing it, meaning they undergo minimal changes in their characteristics. These materials do not actively convert energy and typically do not require an external power supply for their responses. Optical fibers, which can transmit signals without substantial alterations, are considered examples of passive smart materials. The key distinction lies in the active response and energy transduction capabilities of active smart materials, while passive smart materials transfer energy without significant changes in their properties. On the other hand, active smart materials actively react to external stimuli by changing their properties, and they possess the ability to transduce or convert energy from one form to another. This dynamic behavior often requires an external power source to facilitate their responses. Examples of active smart materials include shape-memory alloys, chromogenic material, magneto rheological material, thermoelectric materials, piezoelectric materials. Here in this book chapter, we mainly focus on active smart materials. The classification of active smart materials is shown in Figure 2.1.

2.2.1 Shape Memory Alloys

Shape memory alloys (SMAs) can remember their shape and attain their previous shapes when undergoing some deformations and this property is called the shape memory effect (SME). External stimuli such as changes in temperature, stress, and magnetic field can be used to control their behavior [10]. For example, the temperature variation can be used to harden or soften the material. These materials are generally a combination of martensite and austenite. SMAs follow a hysteresis curve which involves the transformation of martensite and austenite state. M_s, M_f, A_s, and A_f are the four characteristic temperatures associated with this hysteresis curve. M_s and M_f are the start and end points of martensite transformation. Similarly, A_s and A_f are the starting and end points of austenite transformation [11]. At lower temperatures, a martensite state is observed, and at higher temperatures, austenite state is present.

2.2.1.1 Classification of Shape Memory Alloy

Based on how SMAs regain their original configuration after deformation, they are categorized into three groups [12].

i. One-way shape memory: SMAs belonging to this group directly regain their original structure when the deformation is removed. Temperature is used here as an external stimulus.

ii. Two-way shape memory: SMAs of this group have the potential to memorize their shape at both higher and lower temperatures. But sometimes because of strain, their properties degrade at higher temperatures.
iii.Super elasticity: SMAs in this group regain their shapes by applying mechanical force in a particular temperature range. Here, temperature is not necessary to bring back the SMAs to their previous form.

Figure 2.2 (a) and (b) show the SME and super elasticity in SMAs. In case of (a), a thermal external stimulus is provided to the SMA to bring it back to an undisturbed state. However, because of super elasticity, no external stimulus is provided to SMA in case (b).

SMAs possess super elasticity and their behavior can be studied by focusing on the local microstructure. Arne Ölander was the first to discover SMA in 1932 in Au-Cd alloy [12]. This phenomenon was first observed in Au-Cd alloy in 1951, and later in Ti-Ni alloy in 1962 [13]. Even some ceramics and polymers belong to the shape memory category. However, the strain associated with them is lower in comparison to alloys. In the case of ceramics, the SME is associated with phase transition which may be any of the two: (i) paraelectric to ferroelectric and (ii) anti-ferroelectric to ferroelectric. In the latter case of polymers, this effect is observed because of glass transition temperature [13]. Composite of ceramic and polymers based SMA are done to gain elasticity and strength [14]. Shape memory materials also show responses toward mechanical stress, thermal factors, photons, and chemical environments. They are even used as actuators because of their super elasticity or rubber-like behavior. Because of biocompatibility, they are also used in medical and dental implants. Nickel-titanium generally known as nitinol is used for actuation in soft robotics, as an implant in the human body. In robotics, the motion of limbs, and movable parts, often needs soft materials. Also, in the case of some patients whose body parts are immovable, SMAs are the best option for providing flexibility. Sometimes after loading and unloading or physical movement of robotic parts, SMAs shape enters a constant shape. To tackle this issue temperature, heat reservoirs, oil tanks, etc. are used to bring their shape back to previous conditions. Actuation in SMAs is controlled via temperature which converts the martensite state (malleable) to austenite phase. They need to be protected from overheating; otherwise, there is a chance for them to enter into a permanent metallic state. NiTi is one of the most studied SMAs because of its higher corrosion resistance in

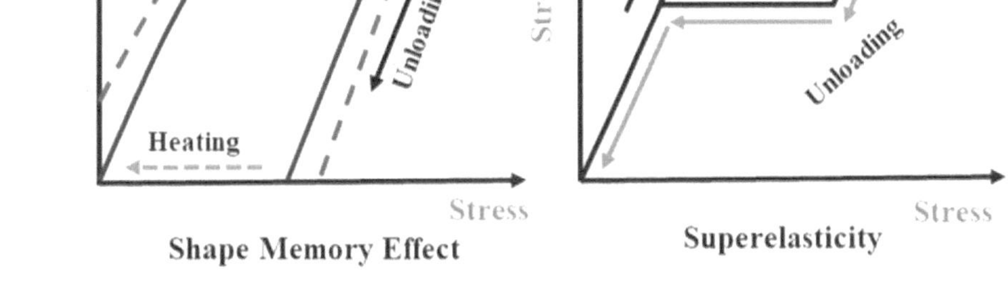

FIGURE 2.2 Stress vs. strain curve representing (a) shape memory effect and (b) super elasticity.

comparison to steel, better wear resistance, and excellent biocompatibility [15]. The recoverable strain of NiTi is four times greater than that of bones with 2%. Its elastic modulus is 48 GPa which is two times greater than cortical bone. Because of its mechanical behavior similar to bones, it is a better human implantable material [16]. Cu and Fe-based alloys are also there and they are more cost-effective in comparison to NiTi-based alloys [17]. Fe-based SMAs are also referred to as smart memory irons because of their superior strength, stiffness, and shape recovery stress.

2.2.1.2 Synthesis and Characterization of Shape Memory Alloy and Composite

The grain size in the case of SMAs plays a crucial role in defining their properties. The grain coarsening deteriorates the performance of SMAs. The SMAs are generally prepared by the mechanical alloying process. The elemental metal powders are taken in stoichiometric weight and milled for a different time duration in planetary mills. Steel balls are used and a definite power-to-ball ratio is maintained for the whole milling process. Mechanical alloying before heat treatment of SMAs helps to achieve a desirable morphology [18]. Combustion synthesis or self-propagating high-temperature synthesis is another method used to synthesize SMAs. Here, the metal powders are taken in stoichiometric weight and ball-milled for homogeneous mixing. This mixture is hydraulic pressed to obtain a desirous shape and then put into a furnace for pre-heating. After this, in the self-propagating high-temperature equipment, with the required pressure and flow of gases like argon, a combustion reaction is carried out [15]. The synthesis methods used to prepare SMA thin films are chemical vapor deposition, physical vapor deposition, and magnetron sputtering. The SMA polymers are prepared by molding, spin-coating techniques, or 3D printing techniques [19].

SMAs and their work are based on mechanical properties associated with temperature. Smart materials alloys are formed by mixing two or more alloys. Occasionally some intermetallic phases of alloys are also formed at the interfaces of the grains in the material. For example, in the case of NiTi alloy, $NiTi_2$ and Ni_4Ti_3 are intermetallic alloys. These unwanted phases damage the mechanical properties as they are hard with less toughness and make the material brittle [20]. Confirmation of the formation of pure and intermetallic alloys is done using X-ray diffraction (XRD) and scanning electron microscopy (SEM). The morphology of SMAs is determined by SEM. Sometimes some impure phases are also formed and remain at the grain boundaries of SMAs due to either precipitation or chemical etching done on the alloy, respectively [21]. Additionally, coarse grains in SMAs degrade the mechanical properties. To confirm the formation of grains of smaller grain size, also SEM is utilized. The porosity % of SMAs is determined using Equation (2.1).

$$\rho\% = \left(1 - \frac{\rho_s}{\rho_0}\right) \times 100 \tag{2.1}$$

The notation in the equation represents the porosity percentage (ρ%), where ρ_s and ρ_0 are the material density and theoretical density, respectively. The density of SMA material under investigation is determined by measuring its weight and dimensions. Porosity % is usually calculated using the above equation. Porosity in NiTi SMA makes it an excellent material for biomechanical implantable material [18]. However, in other mechanical-based applications, the same porosity results in the breakdown of this material.

2.2.1.3 Materials in Sensing and Actuation Applications and Process Control Study

Figure 2.3 shows the actuation mechanism of SMAs by flowing warm and cold water through tubes [20]. The tube gets bent by acquiring heat from hot water and returns to its previous state through the flow of cold water. This is an example of actuation in SMA tubes through fluid flow. Similarly, in robotic arms, the movement of the hand and arm actuated with the SMA device is shown in

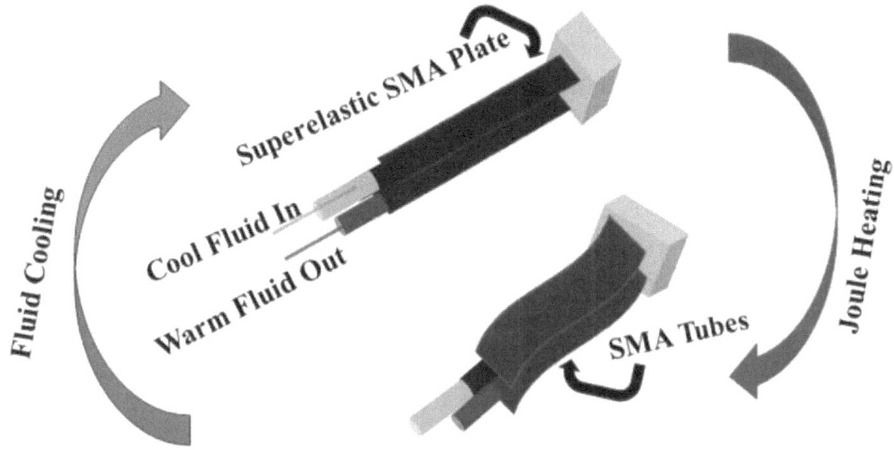

FIGURE 2.3 Schematic representation of SMA tube actuation.

Figure 2.4 (a, b, c, and d) [20]. The dislocation of SMA wires is prevented by mounting them on a glove. In this case, SMA wire coils are used for performing the actuation mechanism. A person with hand immovability can wear this arm for doing daily tasks like grasping an object, hand motion in different directions denoted as flexion, extension, radial deviation, ulnar deviation.

2.2.2 Chromogenic Material

Chromogenic systems possess the property to change color when applied with a variation in voltage/charge (electrochromic (EC)), photon (photochromic (PC)), or thermal (thermochromic (TC)) energy. PC devices change their color under UV light exposure. TC devices change color when the temperature is changed and EC devices change color under the influence of variation in voltage or charge. Liquid crystals and gasochromics (oxidation or reduction on exposure to gases) also belong to chromogenic systems. Chromic materials because of their color-changing behavior are used in smart windows to block electromagnetic (EM) radiation keeping cars and buildings cool in summer [22]. The type of chromogenic system is represented in Figure 2.5.

2.2.2.1 Classification of Chromogenic Materials

Chromogenic systems are classified into three categories: EC materials, TC materials, and photo chromic materials, the schematic diagram represented in Figure 2.5.

2.2.2.1.1 Electrochromic (EC) Materials

These materials change their color either by transfer of electrons or by electrochemical potential change. Schematic representation of an EC device used in a smart window is represented in Figure 2.6. There is historical evidence of EC from the nineteenth century. EC materials are used in glazing. They prevent some of the EM radiation from passing through them thus being able to cool the building in summer and hot in winter. An EC device generally consists of two transparent substrates that act as an anode and cathode and are separated by an electrolyte. The transparent substrates undergo coating in two steps first with a conducting material and second with EC material. The electrolytes consist of a small number of cations (H^+, Li^+, Na^+, etc.). Depending on the amount of charge transfer in the electrolyte the color/transparency of EC devices changes [23]. WO_3 is a cathodic EC material, while NiO is an anodic EC material. Some examples of EC materials are tungsten oxide (WO_3), titanium dioxide (TiO_2), and molybdenum oxide (MoO_3). The insertion and

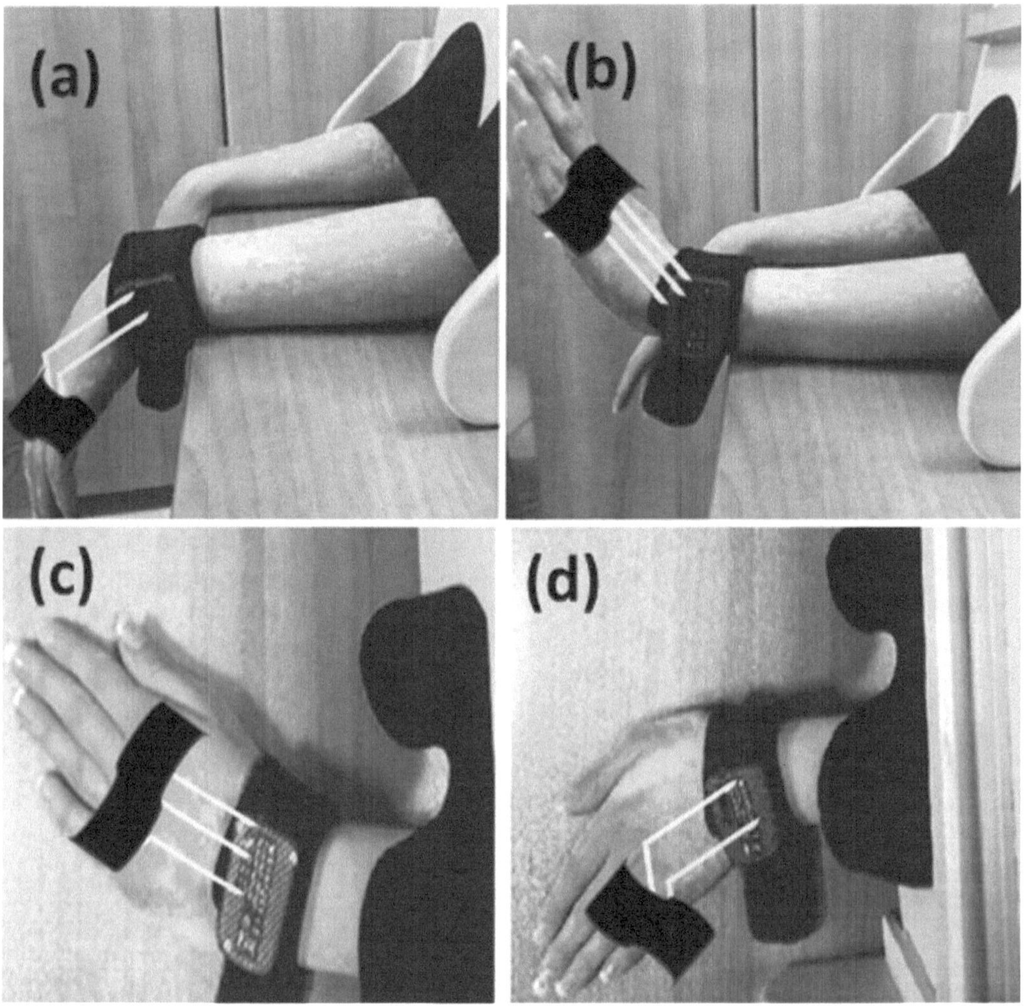

FIGURE 2.4 (a) Flexion, (b) extension, (c) radial deviation, (d) ulnar deviation activities by using a SMA device.

removal of any of H^+, Li^+, Na^+, etc. into and from the EC layer via electrolyte result in coloration and decoloration/bleaching of the device. The chemical reaction for H^+ ion in MoO_3-based EC device is as follows as per Equation (2.2).

$$MoO_3 + x(H^+ + e) = H_x Mo^{VI}_{1-x} Mo^{V}_{x} O_3 \tag{2.2}$$

where MoO_3 is colorless, and after the insertion of H^+, it becomes deep color [24]. Volatilization of liquid electrolytes is a major limitation in EC devices. The use of solid electrolytes such as tantalum oxide (Ta_2O_5) is a better option to solve this issue. Polymers and gels are also used as an electrolyte in EC devices. Polymethyl methacrylate (PMMA) is an example of a polymer electrolyte.

2.2.2.1.2 Thermochromic (TC) Materials

TC materials are capable of changing color with the variation in surrounding temperature. This happens due to a change of phase at the transition temperature. Because of their significance in a

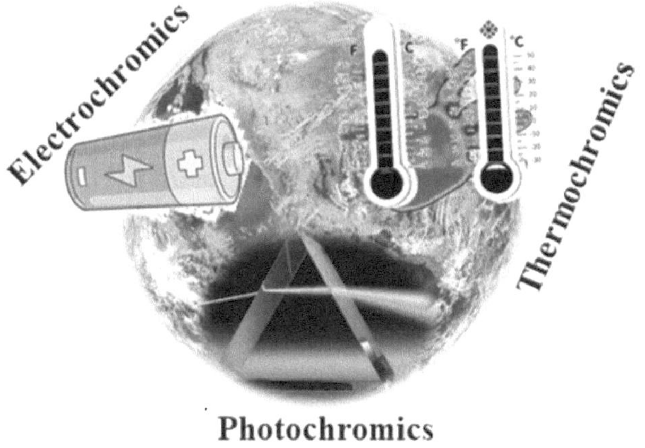

FIGURE 2.5 Types of chromogenic systems.

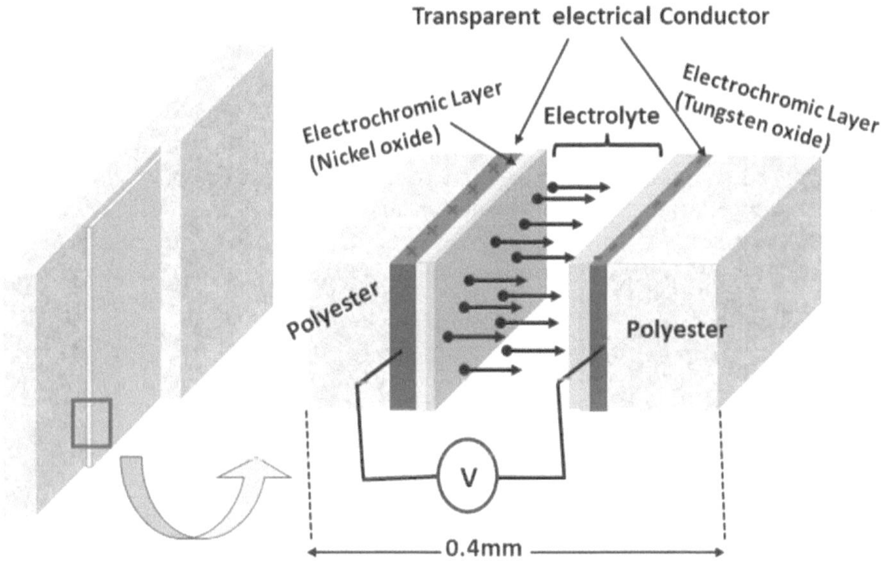

FIGURE 2.6 Schematic representation of an EC device used in a smart window.

variety of applications related to energy-efficient building constructions, thermal or heat storage, vintage maintenance processing, textile industries, and sensors, TC materials research and development are of tremendous interest. VO_2 and TiO_2 are examples of TC materials. VO_2 behaves differently at lower and higher temperatures. They are metallic at higher temperatures and semiconducting at lower temperatures. Its phase transition temperature is at 68 where it changes from monoclinic to rutile phase [25]. The phase transition temperature of VO_2 is reduced by doping it with molybdenum, tungsten, and niobium. Ruby is a naturally occurring TC material; its color is red at room temperature which changes to shiny green at higher temperatures. On cooling, it again regains the red color [25]. Schematic color change in TC dye encapsulated with metal oxide (e.g., SiO_2) over temperature variation is depicted in Figure 2.7.

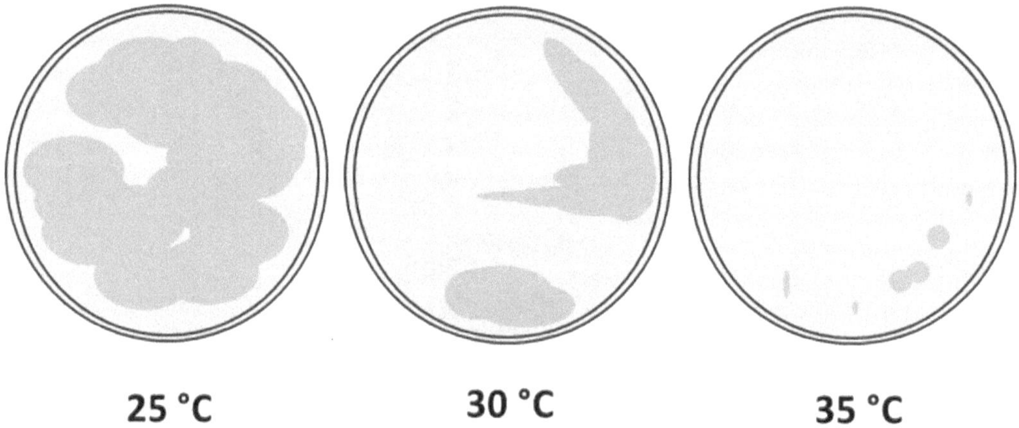

FIGURE 2.7 Color change in TC dye encapsulated with SiO_2 over temperature variation.

2.2.2.1.3 Photochromic (PC) Materials

PC materials change their color with exposure to photons. Since their behavior is photon or light-driven, they are used in sensors, bio-imaging, ultrahigh-density optical data storage devices, logic gates, etc. For this class of materials, light functions as an external stimulus; therefore, variations in light wavelength also affect their electrical properties and optical properties like dielectric constant, dipole interaction, refractive index, and absorption spectra [26]. Organic and inorganic and their hybrid, i.e., organic–inorganic composite materials belong to this class. Azobenzene and diarylethene are examples of organic PC materials and WO_3, TiO_2, and ZnO are examples of inorganic PC materials. Based on cost-effectiveness and stability, inorganic transition metal oxide PC materials are widely studied. The PC mechanism in WO_3 is understood by following the steps of Equations (2.3–2.5),

$$WO_3 + h\nu \rightarrow WO_3^* + e^- + h^+ \ldots\ldots\ldots \tag{2.3}$$

$$h^+ + \frac{1}{2}H_2O \rightarrow H^+ + \frac{1}{4}O_2 \tag{2.4}$$

$$WO_3 + x\,e^- + xH^+ \rightarrow H_xWO_3 \ldots\ldots\ldots \tag{2.5}$$

where e^- and h^+ are electrons and holes, respectively. The presence of defect states and impurities helps in the PC mechanism. When light falls on WO_3 film, electrons and holes are created. Electrons enter into the conduction band and holes react with the surface adsorbed water to form H^+ ions. The diffusion of electrons and H^+ ions in WO_3 results in its color change [26].

2.2.2.2 Synthesis and Characterization of Chromogenic Materials and Composite

Generally, EC coating is done by using RF magnetron sputtering under high pressure by using a mixture of O_2 and Ar. However, the electrolytes are prepared by chemical or physical vapor deposition [29,30]. Sol–gel method and hydrothermal, physical, and chemical deposition method are some of the techniques used for synthesis of TC and PC materials.

The general characterization of chromogenic systems includes XRD, Fourier transform infrared (FTIR) spectroscopy, and SEM. Apart from these basic characterizations, the performance of chromogenic systems is investigated through the following characterizations.

The decrease in the transmission of EM radiation in the UV-visible range results in EM shielding. This will forbid the transmission of EM light through EC, TD, and PC devices. The change in reflectance, absorbance, and transmittance under the application of an electric field, temperature change, and photon wavelength is conducted by UV-VIS-NIR spectroscopy. The difference between the transmittance percentage (T%) of color and bleached samples at a fixed wavelength provides color modulation (ΔQ) for chromogenic systems [27].

An electrochemical analyzer and three electrode cells with an electrolyte solution are used to investigate the EC behavior of the device. The reference electrode and counter electrode used for measurement are Ag/AgCl and Pt, respectively. A voltage window is selected for a fixed time interval and the amount of current generated is measured. The EC device is colored and bleached in cycles by supplying and removing the voltage, respectively [28]. In this way, the stability and reversibility of EC devices are confirmed. Chronoamperometry is also conducted using the Electrochemical workstation (CHI-600E, USA) . The ionic conductivity describes the transfer mechanism of H^+, Li^+, etc. in the chromogenic systems. Indium-Tin Oxide (ITO)-coated glass substrates are used as transparent conductive electrode in the device configuration.

2.2.2.3 Chromogenic Polymers and Composite Applications

Chromogenic polymers serve as a key element in these systems, allowing for optical detection of stimuli within packaged products. Smart packaging systems in the industry can utilize chromogenic polymers to visually alert consumers to potential hazards. This detection is achieved by synchronizing optical changes in the chromogenic polymer with variations in the quality parameters of the goods. By mimicking human senses and perception, chromogenic materials play a crucial role in enhancing the visual assessment of packaged product quality. Numerous investigations, conducted in both industry and academia, have focused on chromogenic materials, particularly in relation to chromogenic polymers. Smart materials are applicable in various packaging applications like EC polymer, TC polymer, and PC [29].

The utilization of EC polymers in packaging represents a burgeoning and actively researched method. Conjugated polymers, due to their high optical contrast and adaptable color characteristics, are extensively employed as EC materials in the field of packaging science. An EC device, employing poly[oxyethylene-oligo(oxyethylene)], has been developed based on the electro chromogenic concept. This device serves as a compact smart label designed for frozen goods, integrating time-temperature functionality. The system involves incorporating salt into the polymer, utilizing a metallic foil for the anode, and employing conduction ITO for circuit formation. Initially, the color-changing process occurs near the anode, saturating before transitioning within the device based on electrolyte conductivity and temperature variation rates. This thin EC cell functions as a frozen label attached to food products. A dual-layer conductive electrode, featuring a metal segment and a transparent system with an electrochrome and ion-bearing polymer electrolyte ($LiClO_4$ or ZnI_2) in oxomethylene-linked polyethylene glycol, ensures that the cell remains colorless at freezing temperatures due to the non-conductive nature of the polymer electrolyte. Proper storage and transportation conditions are essential for the effective application of this system. Chen et al. employed Ta_2O_5 films, generated via cathodic arc plasma, as solid electrolytes in an all-solid-state EC device. The device structure included a 170 nm-thick ITO film on glass, a 60 nm NiO film deposited by DC pulsed sputtering, and a subsequent 130 nm Ta_2O_5 film. An electrolyte with PC, $LiClO_4$, and deionized water was introduced into the NiO film at –2.5 V. The final ITO/WO_3/Ta_2O_5/NiO/ITO device demonstrated a notable 50% optical modulation, rapid 13 s coloring, and 5 s bleaching times. The swift response was attributed to the porous Ta_2O_5 layer's voids, enhancing ion transport, supported by lower charge-transfer resistance revealed in electrochemical impedance spectroscopy the layering structure Ti_2O_5 solid electrolyte film depicted in Figure 2.8 [30].

Numerous TC polymers are employed in commercial time-temperature indicator (TTI) products, particularly those containing the acetylene group ($R_1C \equiv C–C \equiv R_2C$). Upon exposure to

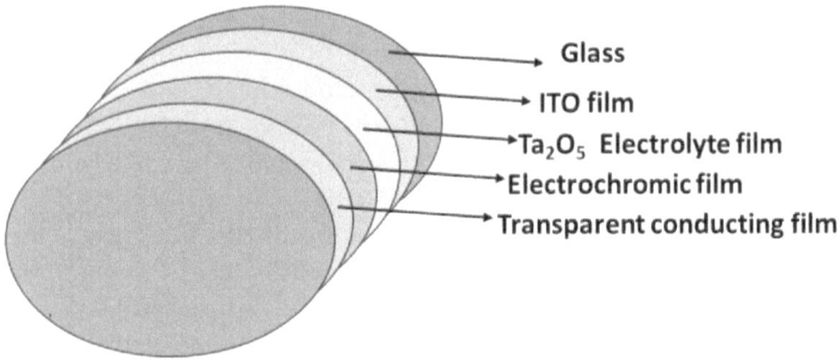

FIGURE 2.8 Electrochromic electrolyte film.

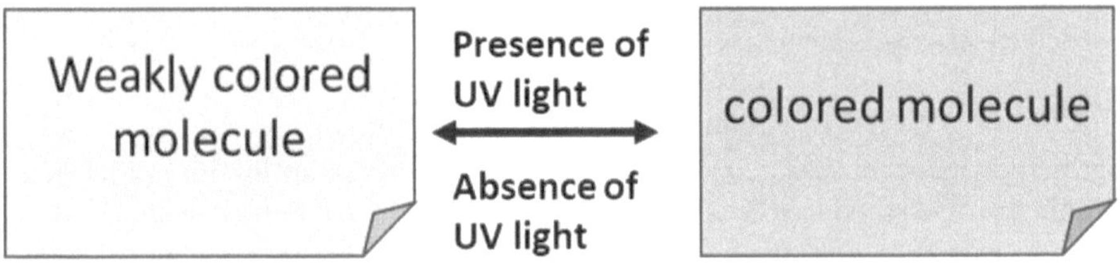

FIGURE 2.9 Behavior photochromic colorant.

environmental stimuli such as high temperature and radiation, these compounds undergo solid-state polymerization, forming polydiacetylene (PDA) and inducing a visible color change. PDA serves as a TTI system, offering optical representation for color changes, quality factors, reaction rates, and temperature variations. Temperature alterations influence polymerization rates in PDA, leading to observable color shifts. Ongoing research explores modifications to PDA side chains, end groups, or both, enhancing the versatility of color-changing systems.

PC materials find application as anti-counterfeit markers, particularly in security measures. These materials can also serve as packaging components or be integrated into packing systems. TTIs act as smart devices in food packaging, detecting key parameters related to food quality by monitoring time and temperature. Easily fabricated into small label shapes for embedding into packaging, TTI systems have broad applications in industries such as pharmaceuticals, food packaging, and floral products, based on Figure 2.9. The underlying mechanism categorizes TTIs into biological, enzymatic, chemical, and physical systems, with diaryl vinyl and spiro-aromatic hydrocarbon compounds being prevalent in current photochromatic indicators for intelligent packaging.

2.2.3 Magnetorheological Materials

Magnetorheological (MR) materials change their properties when they are exposed to a magnetic field. They consist of iron or ferrous materials dispersed in different mediums.

2.2.3.1 Classification of Magnetorheological Materials

Based on the type of dispersion medium, they are classified into MR fluid (MRF), MR grease (MRG), MR plastomer (MRP), MR elastomer (MRE), etc. Since all these materials are dispersed in fluids, they are commonly known as MRFs and change their rheological properties like stress and viscosity on the application of the magnetic field. In the year 1948, Rabinow was the first to discover a mixture of finely magnetic particles in liquid such as kerosine, petroleum, and silicon oil showing response to magnetic field. Iron-cobalt alloy, nickel zinc ferrite, manganese zinc ferrite, etc. are some of the examples of magnetic materials used in fluids [31]. MRF has gained a lot of attention because of its fast response, negligible or low rate of contamination, excellent magnetic response, and easy synthesis methods. Surfactants such as oleic acid, citric acid, etc. are sometimes used to prevent sedimentation of magnetic particles in MRFs [32]. MREs are rubber-like soft particles. MREs contain basic components which include magnetic particles, nonmagnetic elastic matrices, and additives. The viscous properties of these materials can be altered by changing the magnetic field.

2.2.3.2 Synthesis and Characterization of Magneto Rheological Materials and Composite

Magnetic particles are normally synthesized by solvothermal method. The magnetic particles are mixed in fluids and are ball-milled for homogeneous distribution [33]. MR composites are prepared by solvent evaporation, suspension polymerization, and emulsion polymerization [34].

XRD is performed to confirm the phase formation of MR materials. SEM demonstrates the particle size. To improve the yield strength and to lower sedimentation of magnetic particle, sometimes nano or micro wires are used in place of spherical powders. The particle size distribution analysis helps to investigate how MR particles are dispersed in different fluids. Density also plays a significant role toward rheological properties. The ON and OFF state rheological measurement demonstrate the viscosity of MRF. Based on the required viscosity of operation, the magnetic nanoparticle/microparticle concentrations are changed. The temperature stability test demonstrates the working temperature range of MR materials. Beyond the optimum working temperature, magnetic particles oxidized which degrade their performance in practical applications [35,33].

2.2.3.3 Magnetorheological Materials Applications

MRFs are used in dampers, clutches buffers, vibration absorbers, hydraulic systems, MR polishing, etc. [36]. Introduction of composites involving magnetite nanoparticles and diverse polymers, encompassing poly (vinyl butyral) (PVB), polystyrene (PS), as well as conducting polymers (CPs) like polypyrrole (PPy) and PANI, has been conducted [37]. The utilization of MR fluids in engineering applications is on the rise, particularly in areas requiring active vibration control or the transmission of torque and force. Exposure of the rheological fluid to a magnetic field alters its stiffness and damping coefficients, rendering MR semi-active dampers appealing devices for implementing semi-active control in civil, aerospace, and automotive applications. Consequently, numerous researchers have concentrated on scrutinizing motion analysis, damping control, and the design of devices suitable for a broad spectrum of industries, including automobile, construction, and aviation, where substantial yield stress for damping is imperative [37,38]. MRF brakes [38] are highly used in automotive industries. MR fluids act as a working medium in dampers, when a current passes through the magnetic coil, its resistance changes, which drops the pressure in the cylinder causing the movement of piston as shown in Figure 2.10. The same principle is also used in MRF brakes, MRF vehicle seats, etc., with some additional magnetic circuits (rotor and stator). Some companies also use MRF in washing machines to reduce noise and vibration during their operation [38]. The MR composites are prepared by mixing magnetic particles in polymers like PVB, polypyrrol, and PANI.

2.2.4 Thermoelectric Materials

Thermoelectric materials have drawn immense attention in past few years. They can enable direct conversion between electrical and thermal energy, hence providing an alternative for refrigeration and power generation. Today, more than 60% of the produced energy is lost in vain worldwide, predominantly in the form of waste heat. High-performance thermoelectric materials

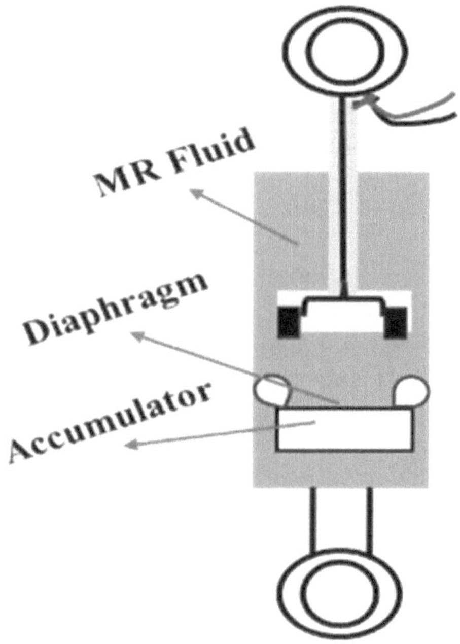

FIGURE 2.10 Schematic illustration of MR damper.

are an environment-friendly energy conversion technology with the advantages of small size, high reliability, can work in a wide temperature range, and have no pollutants. Nonetheless, the efficiency of thermoelectric devices is not yet high enough to be an alternative to the Carnot cycle [39].

2.2.4.1 Thermoelectric Effect

The thermoelectric effect, otherwise known as "Seebeck effect," it is the direct conversion of a temperature difference between two dissimilar electrical conductors or semiconductors to an electrical voltage. When the sides of TE materials are exposed to different temperatures, a voltage is created across the two sides of the material. Conversely, when a voltage is applied, a temperature difference can be created known as "Peltier effect" [39]. Figure 2.11 shows that how a thermoelectric material can convert heat directly to electricity (Seebeck effect) and also how the material can heat up or cool down at an electrified junction (Peltier effect). At the atomic scale, when a temperature gradient is applied at the two end sides of a thermocouple, the electrons and holes move faster, and they have a lower density at the hot side, resulting in diffusion of electrons/holes toward the cold side as schematically demonstrated in Figure 2.12. This movement of carriers (electrons for n-type and holes for p-type materials) is translated into the generation of an electric field across the thermocouple [40]. This is called the "Seebeck effect," and the voltage created for a temperature difference, ΔT, under thermodynamic equilibrium is $S = \Delta T$, where S is the Seebeck coefficient [39,40].

TE materials are one potential candidate for harvesting waste thermal energy, due to their ability to convert it into electricity, even under very low-temperature gradients relative to the environmental temperature. This technology exhibits distinct advantages over other energy-harvesting technologies: (i) TE conversion is reliable and operates in silence as it works without mechanical movement, (ii) it is an environmentally friendly green technology, since no heat and no gaseous or chemical wastes are produced during operation, and (iii) it can be widely used in places where other energy conversion technologies are unavailable, such as in the remote outer space [41].

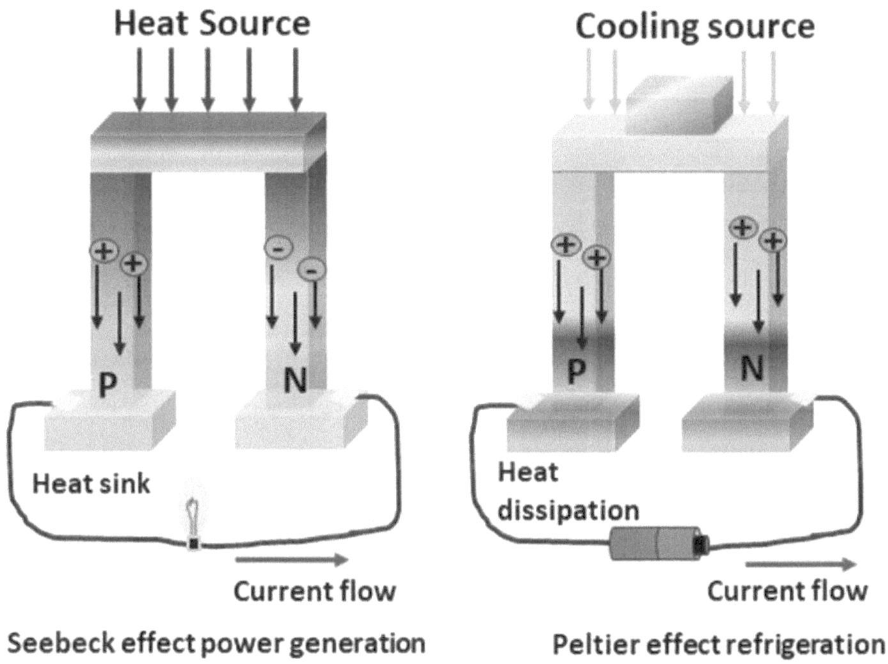

FIGURE 2.11 Illustration of how a thermoelectric material can convert heat directly to electricity (Seebeck effect) and also how the material can heat up or cool down at an electrified junction (Peltier effect).

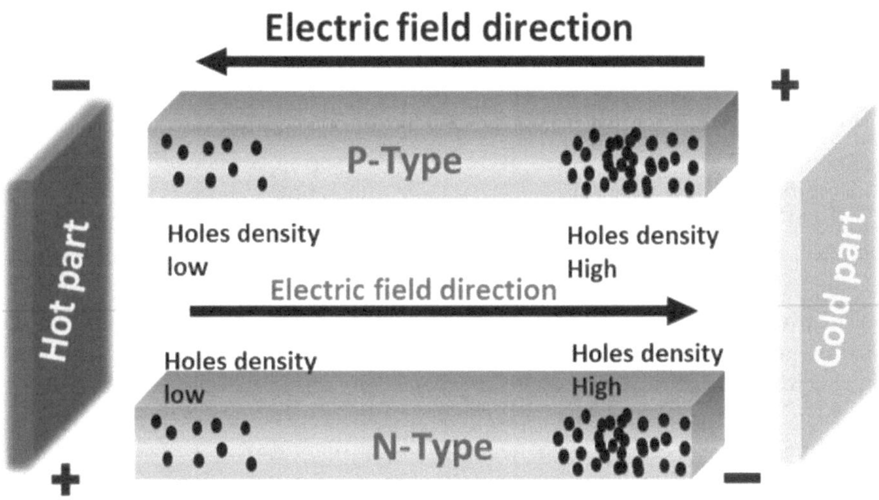

FIGURE 2.12 A schematic representation of the TE effect showing the charge carriers of a p- and n-type material to diffuse from the hot side to the cold, when a temperature gradient is applied.

2.2.4.2 Conducting Polymer

Most polymers are generally good insulating materials. Nevertheless, the emergence of CPs in 1976 has developed enormous opportunities for the development of polymers with unique properties [41,42]. CPs are organic materials that exhibit mechanical and physical properties related to conventional polymers and the unique electrical properties of metals. CPs belong to a class of polymers that

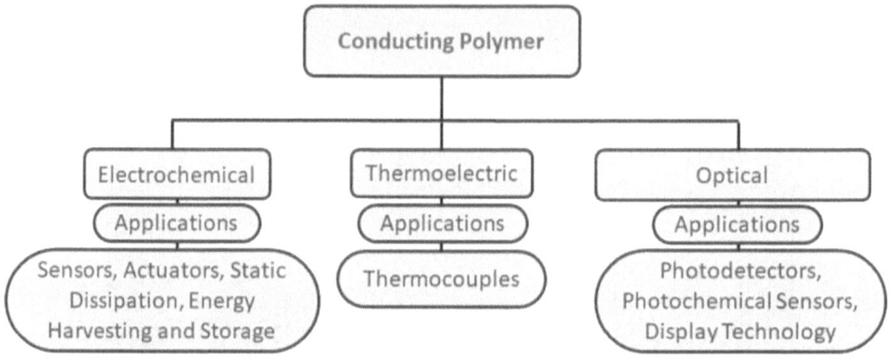

FIGURE 2.13 Conducting polymers along with their respective applications.

can inherently conduct electricity attributed to conjugation structures (i.e., having π-electron delocalization) in their polymeric backbone. CPs possess several advantages, such as corrosion resistance, easy synthesis, good mechanical and optical properties, high flexibility in preparation, and tuneable conductivity. Moreover, the light weight of CPs enables them to be used as replacement for metals, particularly in weight-sensitive applications. Examples of CP are polyaniline (PANI), PPy, poly (3,4-ethylene dioxythiophene) (PEDOT), polythiophene, and polyacetylene polyphenylene. Out of this PPy is one of the most commonly used CPs because of its easy synthesis, unique electrical conductivity, and environmental stability. Numerous comprehensive reviews have reported on the optical properties, thermoelectric properties, and applications in energy storage, light-emitting devices, photovoltaics, actuators, and sensor applications on using CPs [43]. The distinctive characteristics of CP have attracted research across various domains on exploring both fundamental properties and applications. Figure 2.13 depicts the applications of conductive polymer in various domains of research.

The use of thermoelectric modules combined with photovoltaics and energy harvesting, i.e., the conversion of waste heat into electricity (energy recovery), is one of the new interesting applications. PPy is a heterocyclic and positively charged CP that contains nitrogen in its oxidized form, and it loses its conductivity and charge when overoxidation occurred. Besides, PPy is electroactive in organic electrolyte and aqueous solutions. PPy is one of the most important CPs with a wide range of applications in anti-electrostatic coatings, biosensors, solid electrolytic capacitors, polymer batteries, etc. [44,45]. Because of its diverse advantages such as ease of synthesis, good electrical conductivity, low thermal conductivity, good environmental stability, and low toxicity, PPy has been believed to be a promising candidate for thermoelectric materials. Furthermore, PPy is known for its nontoxic and biocompatibility, and it is utilized in wide-ranging applications (for example, supercapacitors/electrodes, nanocomposites, gas sensors, biosensors, drug delivery, protective clothing, anticorrosion coatings, actuators, and adsorbents for the removal of heavy metals and dye). Figure 2.14 illustrates the structure of PPy [46].

The monomer units in PPy chains are primarily bonded at α–α positions, a minimum amount of pyrrole (Py) monomer is bonded at the α–β and β–β positions as shown in Figure 2.15. PPy is conductive due to alternating single and double bonds, which lead to a certain delocalization of electron density in the molecule. However, in its undoped state (neutral), PPy is an insulator with a large band gap of approximately 3.16 eV. Chemical or electrochemical doping can be used to enhance PPy conductivity.

2.2.4.3 Synthesis and Characterization of Thermoelectric Composite Materials

The preparation of thermoelectric materials can be synthesized by both bulk materials and thin films depending upon their application. Bulk materials are made using methods like vacuum

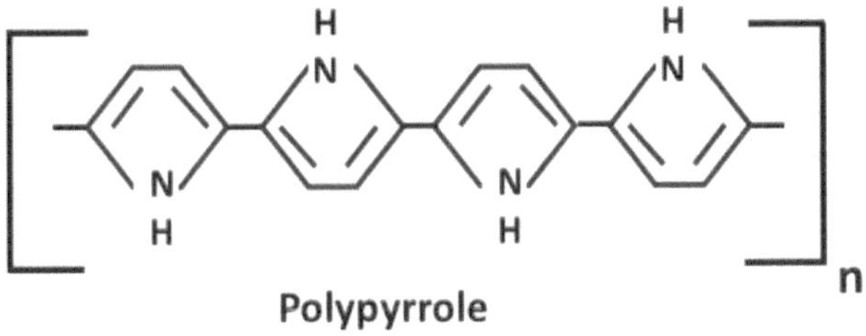

FIGURE 2.14 The chemical structure of polypyrrole.

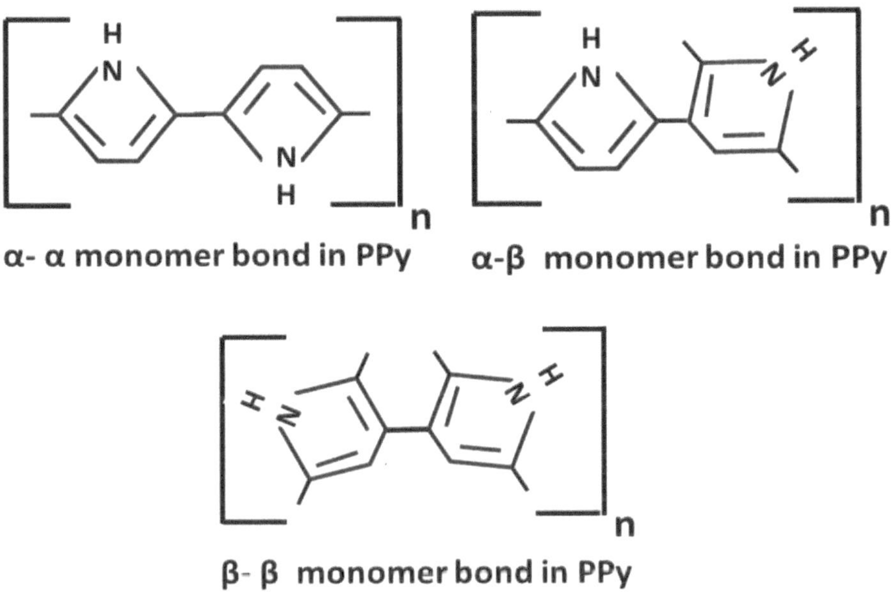

FIGURE 2.15 The possible monomer unit bonding in polypyrrole (PPy).

melting, solvothermal, and co-precipitation method. On the other hand, thin films are prepared by using techniques such as magnetron sputtering, molecular beam epitaxy (MBE), ink printing, chemical vapor deposition, atomic layer deposition, and electrodeposition. Some advance techniques that have developed recently for bulk synthesis of thermoelectric materials are as follows: solvothermal method, hot-pressing method, spark plasma sintering etc.; the solvothermal method is a widely used technique because it's energy-efficient and easy to operate. However, it has a drawback – controlling the purity and shape of the product is challenging, and the reaction process can't be closely monitored. The hot-pressing method involves melding, heating, and sintering under pressure simultaneously. This technique has its merits: it lowers sintering temperature, reduces sintering time, and makes it easier to create a sintered block with nearly ideal density and minimal pore formation. To enhance both mechanical and thermoelectric properties, researchers used a YbFe-doped $CoSb_3$ skutterudite. This material was shown an impressive ZT value of 0.48, through temperature-gradient zone melting process with a hot press process [47]. For preparation of thermoelectric thin film, 3D printing technique, ink printing, atomic layer deposition are newly employed for easy synthesis of thin film thermoelectric materials. 3D printing is a fast and affordable manufacturing method. This technology allows the creation of objects by layering molten materials according to digital models. As an illustration, Bian et al.

utilized 3D printing to produce thermoelectric legs with distinctive helical shapes and supporting structures, examining their performance [48].

2.2.4.4 Thermoelectric Application and Wearable Sensors in IoT Device

Currently, around 85% of the world's energy comes from non-renewable fossil fuels and has unexpected problems that may arise after a century of industrialization. The challenge is that suddenly cutting down on fossil fuels would seriously downgrade our living standards, but continuing to use them massively would make environmental problems and climate change and environmental pollution. That's why researchers are on the lookout for clean energy sources nowadays. Options like wind power, solar power, and nuclear fission thermal energy have been developed, but they have their limitations. Solar and wind energy, for example, depend on specific conditions. That's where thermal energy comes in – it's seen as a more stable energy source, making it crucial for addressing the global energy crisis.

Compared with thermoelectric generators (TEGs) to regular heat engines like steam turbines, TEGs are much smaller and work differently. This means that they can only produce a small amount of energy, in the range of milliwatts (mW) to microwatts (μW). This might not be enough for everyday energy needs. However, it fits perfectly for certain small devices like sensors used in the Internet of Things. To explore the TEG powered sensors can be used inside the human body, such a way that wearable and implantable medical devices etc. In pursuit of highly efficient TEGs capable of producing a hundred μW in the presence of temperature variations, several critical considerations must be addressed. These considerations, outlined for wearer comfort and mass production feasibility, encompass the necessity for TEGs to intricately integrate with their environment, ensuring thermal harmony between the heat source and the ambient air radiator. Furthermore, optimal performance hinges on minimizing thermal resistance, exact matching of the thermal reactor and heat source, and generating sufficient voltage to power electronic devices effectively. Balancing functionality with user comfort is paramount, and achieving a compact size remains a key goal, enhancing the practicality and broad applicability of TEGs.

The rapid evolution of internet has paved the way for the development of advanced power sources for wireless connections. In an innovative application, Rosch A.G et al. [45] integrated TEGs with Bluetooth systems in mobile phones, enabling users to receive real-time weather updates. This integration significantly reduced the delay in transmitting current weather forecasts on mobile phones, enhancing the overall user experience. In aviation, wireless sensor networks are gaining traction for diverse applications in sensing, data processing, and wireless information transmission [49,50]. Detailed TEG installation positions suggest their inclusion in various hotspots such as the cabin, outer surface, and rear hanger fairing of aircraft. TEG-powered sensors on the fuselage offer advantages, monitoring mechanical loads during operation to detect material fatigue early on. This approach not only resolves wiring complexities but also addresses battery instability and high replacement costs in extreme environments [50]. The schematic diagram of TEG wearable device is depicted in Figure 2.16.

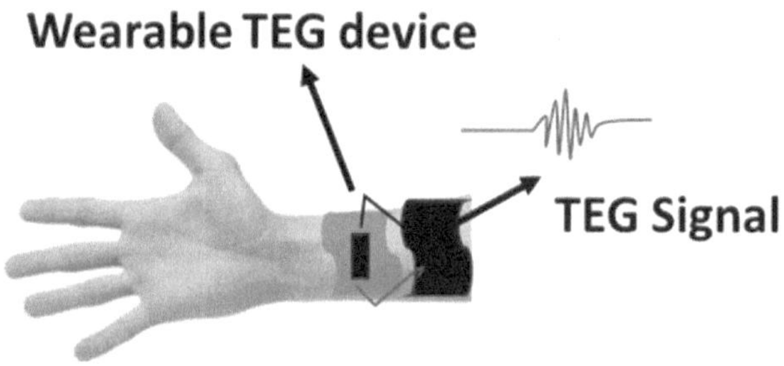

FIGURE 2.16 Schematic diagram of wrist band TEG.

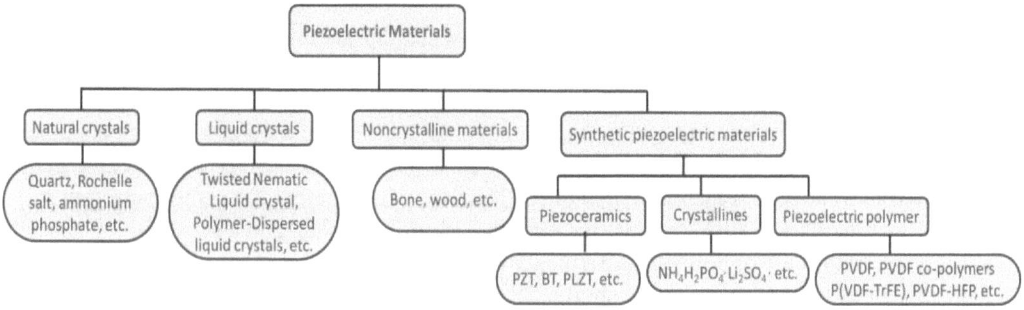

FIGURE 2.17 Common piezoelectric materials classifications.

2.2.5 Piezoelectric Materials

Piezoelectricity emerges in specific crystalline materials, showing the capacity to generate an electric charge in response to mechanical stress. Numerous materials, both natural and synthetic, display piezoelectric characteristics, broadly categorized as natural crystals, liquid crystals, nanocrystalline materials, textures, and synthetic piezoelectric materials. Common piezoelectric materials were categorized and depicted in Figure 2.17.

It became evident that materials demonstrating piezoelectricity also exhibit the converse effect of geometric strain or deformation proportional to an applied voltage. Both natural and synthetic materials display a spectrum of piezoelectric effects. While piezoceramics stand out as highly efficient in this regard, polymers possessing transducer characteristics offer distinct advantages over ceramics due to their less brittle nature and greater flexibility [51]. A diverse range of ceramic materials, including lead zirconate titanate (PZT), zinc oxide (ZnO), barium titanate (BT), potassium sodium niobate (KNN), etc. However, piezo ceramics are polycrystalline materials of perovskite crystal structure composed of numerous single crystal "grains" possessing identical chemical composition [52,53]. However, the orientation in the single grain is different from one another; there exists slightly different space between the ions. These materials are characterized by superior piezoelectric properties, and lead-zirconate-titanate (PZT) is considered an important piezoceramic that is widely used for energy harvesting and storage applications. Other examples of ceramics include barium titanate ($BaTiO_3$), potassium niobate ($KNbO_3$), high piezoelectric constant (d_{33}=160 pC/N), and NBT piezoelectric coefficient (d_{33}=80 pC/N) [54].

Whereas polymers such as polyvinylidene fluoride (PVDF) and its copolymers like hexafluoro propylene P(VDF-HFP) and tetrafluoroethylene P(VDF-TrFE), alongside polyvinyl acetate (PVAc), find application as piezoelectric materials. This class of material belongs to carbon-based materials with long polymer chains and is specified by their excellent flexibility compared to single crystals and ceramics. They have the capability to withstand much higher strain due to their inherent flexibility, making them more suitable for applications demanding a large amount of bending or twisting. This material is found suitable for a variety of energy-related applications such as ultrasonic transducers, audio transducers, medical transducers, display devices, vibrometers, shock sensors, and pressure sensors [55]. The major limitation of these ceramic materials is brittle, toxicity, non-biodegradability, and intricate fabrication processes. On the contrary, polymers offer notable advantages such as toughness, non-toxicity, and biocompatibility. These attributes render polymers suitable for applications demanding high flexibility, bending, and twisting, particularly in the realm of biomedical devices. Piezoelectric polymers come under categories of amorphous, crystalline, or semicrystalline structures. Among the semicrystalline piezopolymers is PVDF. The level of crystallinity is dictated by the molecular structure of the polymer and is influenced by the ratio of semicrystalline to amorphous regions. This ratio undergoes variations due to fabrication methods and thermal history. The piezoelectric effect is mainly attributed to molecular dipole orientation. Through thermal annealing and high voltage application, randomly oriented dipoles undergo arrangement, and the process is expedited by stretching and hot-pressing the polymer. In semi-crystalline polymers, the piezoelectricity is heightened through the influence of electric voltage. Conversely, in amorphous materials, a process called polarization near the glass transition temperature (T_g) is necessary to align and freeze the dipoles. Polarization increases the piezoelectric coefficient of amorphous polymers by four to five times. Notably, when compared to other piezopolymers, PVDF-TrFE, particularly PVDF, exhibits the highest dielectric constant and a substantial piezoelectric coefficient. In the case of polymer–ceramic composites, the matrix and reinforcement dispersion geometry also alter their piezoelectric properties. Piezoelectric composites with 0–3 geometry (fillers in the form of particles in polymer matrix) show poor piezoelectric properties, while composites with 1–3 (fillers in the form of fiber) or 2–2 (fillers in the form of geometry) exhibit superior piezoelectric properties, when the longitudinal direction of the fillers remains parallel to the direction of electric field.

2.2.5.1 Principle of Piezoelectricity

The phenomenon of piezoelectricity manifests as the generation of an electric charge or voltage when subjected to mechanical loads or, conversely, as a mechanical deformation resulting from the application of an electric field. Pierre and Jacques Curie, in their work between 1880 and 1881, unearthed these piezoelectric effects, subsequently categorized as the "direct piezoelectric effect" and the "converse piezoelectric effect." The word "piezoelectricity" comes from the Greek word "pie-zein," meaning "to press" or "to squeeze." It refers to generating electricity when pressure is applied. This phenomenon is usually seen in insulators or materials with low electrical conductivity. Piezoelectric materials show changes in electric charges or molecular dipoles when stressed, creating electricity (called the direct piezoelectric effect) [55,56]. Conversely, when an electric field is applied, they experience mechanical strain (known as the converse piezoelectric effect). Figure 2.18 depicted the direct and inverse piezoelectric effect.

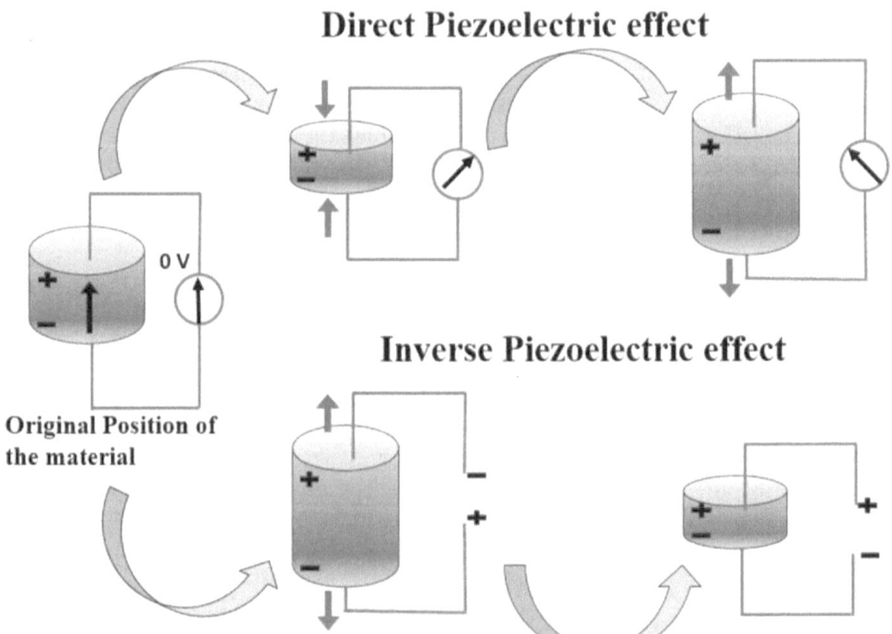

FIGURE 2.18 Phenomenon of direct and inverse piezoelectric effect of a piezoelectric materials.

The Curie brothers meticulously explored the impact of mechanical pressure on the induction of electricity in specific crystals like quartz, Rochelle salt, tourmaline, and cane sugar. The polarization of the piezoelectric material is contingent upon the direction of the applied mechanical load, while, conversely, the application of an electric field endows the material with the capacity to alter its shape, leading to the release of energy in the form of mechanical strain. This foundational research has paved the way for understanding and harnessing the unique capabilities of piezoelectric materials in various applications. The direct piezoelectric effect is crucial for nanogenerators (NGs), sensors, and biomedical applications. Here, mechanical stresses are used to create electrical charges in piezoelectric materials. On the other hand, the converse piezoelectric effect is more suitable for applications like actuators, vibration damping, and acoustic emitters. Interestingly, piezoelectric materials were first used during World War I, but it wasn't until the 1960s to 1970s that polymers with these properties found practical applications [57]. It's important to note that the properties of piezoelectric materials depend not only on their chemical structure but also on how they are made and the conditions of their manufacturing.

The theory of piezoelectricity of piezoelectric materials has constitutive equations; governing piezoelectric materials finds expression in multiple forms, with the two most commonly employed formulations being:

$$\varepsilon_{ij} = C_{ijkl}\sigma_{kl} + d_{kij}E_k \tag{2.6}$$

$$D_i = d_{ikl}\sigma_{kl} + e_{ij}E_k \tag{2.7}$$

Here, it indices significant directions different within the material coordinate system. Where ε_{ij} represents the strain tensor, C_{ijkl} denotes the elastic compliance matrix, σ_{kl} corresponds to the stress tensor, d_{kij} represents the matrix of piezoelectric coefficients, and E_k stands for the electric field. Similarly, D_i represents the electric flux density, while e_{ij} signifies the dielectric permittivity matrix. Equations (2.6 and 2.7) provide a comprehensive representation of the relationships governing the behavior of piezoelectric materials, offering insights into their response to mechanical and electrical stimuli [55]. The coordinate axes employed to define the orientation of a piezoelectric material shown in Figure 2.19. Typically, axis-3, as shown in Figure 2.19, is oriented through the thickness of the sample, while axis-1 is positioned within the plane along its longest side. Various mechanical and electrical parameters are expressed as tensors of distinct orders. Piezoelectric coefficients, specifically d_{33} and d_{31} mode, play a vital role in characterizing the piezoelectric behavior of materials along different coordinate axes. d_{33} refers to the piezoelectric charge output per unit applied stress in the direction perpendicular to the sample's surface or through its thickness. This coefficient is associated with the third axis in the coordinate system, reflecting the material's response when stress is applied through its depth. On the other hand, d_{31} represents the piezoelectric charge output per unit applied stress along the axis parallel to the sample's surface or in-plane. This coefficient is associated with the first axis in the coordinate system, typically aligned with the longest side of the sample. Together, d_{33} and d_{31} provide insights into the electromechanical characteristics of piezoelectric materials, particularly their response to stress in different directions [56,57].

2.2.5.2 Piezoelectric Polymer

Piezoelectricity in polymers is a result of their high polarization, determined by the net molecular dipole moments within the structure. The molecular dipoles in these materials can be aligned through the application of an external electric field, mechanical stretching, or the inclusion of external fillers. This alignment induces a piezoelectric response in the polymers. The examples of piezoelectric polymers include vinylidene fluoride (VDF) based fluoropolymers, odd-numbered

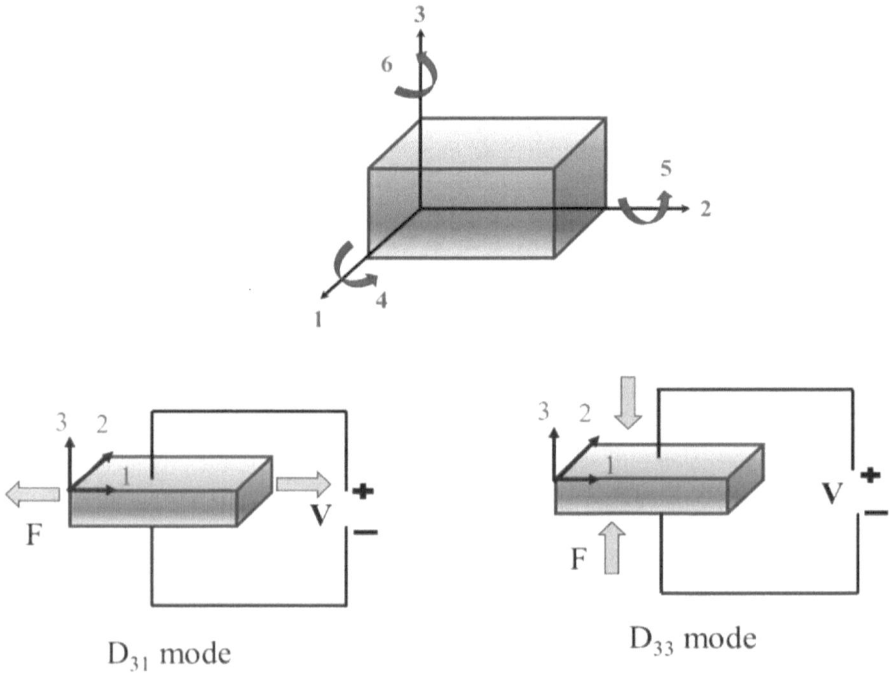

FIGURE 2.19 The coordinate axes employed to define the orientation of a piezoelectric material and illustration of D_{31} and D_{33} modes.

nylon, poly-L-lactide, and polyimide. These polymers, known for their piezoelectric properties, are fevered materials for multifunctional applications [57].

Piezoelectric polymers are a type of polymers with a semi-crystalline structure. Their ability to generate an electric charge depends on the arrangement of dipoles in crystalline regions when exposed to an electric field, as well as factors like chain conformation and overall macromolecular structure. However, these polymer materials face challenges as they tend to have lower piezoelectric coefficients and dielectric permittivity compared to ceramic materials. To overcome these limitations, various approaches are employed to enhance the piezoelectric properties. This includes modifying the molecular structure, creating composites, adjusting processing conditions, and using post-treatment methods like mechanical stretching and electrical poling. Among the piezoelectric polymers, PVDF and its copolymers stand out as excellent choices for energy harvesting and other multifunctional applications. They are preferred due to their higher piezoelectric coefficients and ease of fabrication when compared to other piezoelectric polymers. The piezoelectric properties of few polymers are reported in Table 2.1.

PVDF stands out as one of the most extensively researched piezoelectric polymers, finding versatile applications in recent years. Its structure is semi-crystalline, characterized by a straightforward chemical composition: two fluorine atoms alternate with each carbon atom in the hydrocarbon backbone, forming $[-CH_2-CF_2-]_n$. The polymer exhibits around 50% lamellar crystals within an amorphous region, contributing to its relatively robust piezoelectric properties compared to other polymers. In the crystalline regions, PVDF displays five distinct phases, denoted as α, β, γ, and δ, each having a complex structure. Among these phases, α, β, and γ are the most extensively studied and commonly utilized polymorphs of PVDF. Specifically, the α- phase is electrically nonactive due to the anti-parallel arrangement of dipoles (trans-gauche trans-gauche, i.e., TGTG/conformation) [55,56]. On the other hand, the β-phase (all-trans TTTT chain conformations) and γ-(T_3GT_3G' /

TABLE 2.1
Piezoelectric Properties of Polymers

Polymer	Relative Permittivity (ε_r)	Piezoelectric Coefficient (D_{33} or D_{31})
PVDF	7–12	D_{31} =8–22 D_{33} =-24 to –34
P(VDF-TrFE)	18–20	D_{31} = 12–25 D_{33} = -25 to –40
P(VDF-HFP)	11	D_{31} = 30 D_{33} = –24
P(VDF-CTFE)	13	D_{33} = –140

conformations) phases are considered electrically active. In these phases, the dipoles of individual molecules align parallel to each other, resulting in a nonzero dipole moment. Importantly, the be phase exhibits the highest polarization, showcasing superior properties such as piezoelectricity, pyroelectricity, and ferroelectricity of the material. Consequently, PVDF is more advantageous than other piezoelectric polymers, boasting a high piezoelectric coefficient (D_{33} = 20–to 30 pC/N; D_{31} = 8– to 22 pC/N).

Copolymerization is an effective process for adjusting the polymorph structure and electroactive phase nucleation of PVDF. The choice of comonomer and its composition ratio plays a crucial role in influencing the molecular structure as well as the ferroelectric, piezoelectric, and pyroelectric properties of the polymer. One such copolymer is P(VDF-TrFE), where the introduction of a third fluoride atom in the TrFE, unit helps induce a polar β-phase by arranging the molecular chain in an all-trans conformation. The nucleation of the β-phase is observed to depend on the TrFE content, typically exceeding 11 mol%. Importantly, this nucleation is independent of processing conditions or electrical poling when the TrFE content is over 20 mol%. P(VDF-TrFE) has found multifunctional applications in the form of thin films, free-standing films, or nanofibers (NFs). For instance, researchers have fabricated flexible NGs and nano-pressure sensors using electrospun NFs of P(VDF-TrFE), demonstrating enhanced polar phase and piezoelectric performance, especially with aligned NFs. Another copolymer, P(VDF-HFP), introduces bulky $-CF_3$ groups, providing more space for molecular dipoles to rearrange in the polymer chain. Solvent-casted films with 5 mol% HFP content exhibit high remnant polarization around 80 mC/m^2 and a high piezoelectric coefficient (30 pC/N). This makes P(VDF-HFP) promising for applications in piezoelectric and ferroelectric areas, such as the development of magneto-electric sensors and actuators [58].

2.2.5.3 Piezoelectric Polymer Composite

We have explored various methods to induce a polar phase in piezoelectric polymers, but these approaches come with drawbacks such as structural deformations or microstructural limitations, restricting applications like electro-optical sensors and non-volatile memories. In this response, researchers have sought alternative strategies by incorporating different fillers into the polymer matrix. These fillers include ceramic materials (such as $BaTiO_3$ and PZT), ionic liquids, carbon materials (like carbon nanotubes and graphene), metal nanoparticles (Pd, Ag, Pt), metal salts, clay, hydrated ionic salts, and PMMA. Importantly, the addition of fillers serves to induce or enhance the polar content in polymer materials through a nucleation mechanism. Intriguingly, polymer composites exhibit distinct properties based on the functional characteristics of the added fillers. Consequently, polymer composite materials have become a focal point of interest due to their multifunctional applications, which we will delve into in the following section.

In addition to PVDF and its copolymer composites, researchers have extensively investigated other polymer composites with piezoelectric characteristics. These composites involve the incorporation of various materials, including piezoelectric ceramics (such as PZT and $BaTiO_3$), metal/semiconductor nanoparticles (Ag, Au, ZnS), and natural crystals into the polymer matrix. Notably, these polymer composites showcase multifunctional properties and compatibility, serving as alternatives to inorganic and brittle piezoelectric materials, which are often impractical, especially in biomedical applications. For instance, Sakamoto et al. [57] incorporated ceramic filler PZT into polyurethane (PU), resulting in a PU/PZT composite with a piezoelectric value (D_{33}) of approximately 24 pC/N. These composites hold promise as a functional material for sensor and actuator applications. David et al. [58] reported on the preparation of a composite film using polyamide 11 (PA 11) and sodium niobate nanowires (NW). This composite film exhibited piezoelectric properties, making it suitable for piezoelectric sensor applications. Khanbareh et al. [59] introduced lead titanate (PT) ceramic materials into polyethylene oxide (PEO), creating a sensitive and flexible composite. The PT-PEO composite not only demonstrated enhanced electrical conductivity but also exhibited piezoelectric and pyroelectric properties. Sultana et al. reported on the preparation of an electrospun NF composite by incorporating piezoelectric semiconductor, oriented zinc sulfide (ZnS) nanorods (NRs), into a semi-crystalline poly (vinyl alcohol) (PVA) matrix [60]. This composite displayed piezoelectric properties due to the non-centrosymmetric nature of ZnS, making it suitable for mechanical energy harvesting applications on a large-area mat of polymer materials. These examples highlight the diverse and innovative applications of polymer composites with piezoelectric properties in various fields. A novel piezoelectric polymer ceramic composite, composed of PVDF as the polymer host and barium calcium zirconium titanium (BCZT) oxide ceramics as the filler, has been synthesized. The BCZT filler, derived through a solid-state reaction route, underwent processing to reduce particle size [61]. Recently, flexible piezoelectric generators have gained attention for potential applications in mechanical energy harvesting and portable electronic devices. Inspired by the notable properties of 2D MXene ($Ti_3C_2T_X$), we developed a nanocomposite with PVDF using varied $Ti_3C_2T_X$ filler via tape-casting [62].

2.2.5.4 Synthesis and Characterization of Piezoelectric Polymer and Polymer Composite

The production of piezoelectric PVDF-based materials offers a comprehensive discussion and comparison of the synthesis technique. The methods explored are MEMS (micro-electromechanical systems) techniques, melt blending, 3D printing, self-poling, electrospinning, soft lithography, drop-casting, solvent casting, as well as the utilization of composites and nanohybrid composites. Each of these approaches brings its unique attributes and considerations to the development of PVDF-based materials with piezoelectric properties for improvement of electroactive phase. 3D printing and self-poling electrospinning methods are acquired recently for the development of electroactive phase in PVDF. In recent years, the application of three-dimensional printing (3D printing) has extended to the assembly of electroactive fluoropolymer PEGs. Specifically, a 3D-printed PVDF homopolymer can be deposited and poled using an electric-poling-assisted method. Notably, an increase in the electric field during 3D printing results in a decrease in generated charge [63]. The advantages of employing the 3D-printing technique include enhanced piezoelectricity, design flexibility, and the potential for large-scale manufacturing of active biomaterials for actuators and sensors. Electrospinning stands out as a leading and effective method for crafting piezoelectric scaffolds endowed with distinctive properties. Particularly crucial for scaffolds is their resemblance to the extracellular matrix of tissues, necessitating a fibrous structure. Electrospinning excels in this regard, producing an immediate piezoelectricity in the resulting fibrous structure [64]. In electrospinning, a macroscale material undergoes reduction to uninterrupted continuous nanoscale fibers. This method offers several advantages, including the ability to create substrate-free structures, a straightforward process, and high inputs. Although the NFs typically possess circular cross-sections, electrospinning

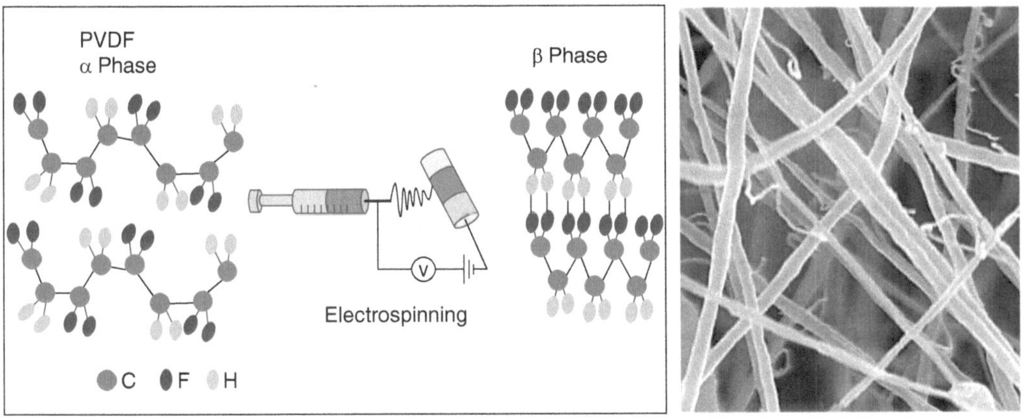

FIGURE 2.20 The β-phase generation of PVDF in electrospinning and optical FESEM illustration of PVDF nanofiber.

allows to produce diverse geometries, such as rectangular and coaxial shapes. Notably, it proves to be a suitable technique for generating PVDF NFs with a high β-phase fraction and crystallinity by manipulating molecular dipoles through the application of voltage direction. The formation of β-phase in PVDF and FESEM image is depicted in Figure 2.20.

2.2.5.5 Electroactive Phase Identification and β-Phase Nucleation

Various established techniques, including FTIR spectroscopy, XRD, and differential scanning calorimetry (DSC), are utilized for the identification and quantification of existing polymorphs in PVDF. Martins et al. extensively analyzed the α, β, and γ phases using these techniques, summarizing their findings [63]. The distinguishing characteristics of crystalline phases are outlined in Table 2.2 for easy identification. To briefly elaborate, α phases can be effectively identified through FT-IR, while β- and γ-phase vibration bands manifest in a similar manner. Conversely, α and γ phases present coincident diffraction peaks, while the β phase is distinctly visible in XRD. Therefore, the combination of FTIR and XRD proves instrumental in accurately identifying the crystalline phases. Addressing the challenge associated with PVDF, known for its stability with the nonpolar ae phase, efforts have been made to induce electro-active β phase and γ phase. Achieving the β phase in PVDF is essential for unlocking multifunctional applications such as NGs, sensors, actuators, photodetectors, magneto-electric devices, energy storage, and tissue engineering applications. In the subsequent discussion, we will delve into the various approaches employed to induce the β phase in PVDF. The electroactive phases (β+γ) present in the composite films were examining by using the Lambert–Beer Equation (2.8) [61].

$$F_{EA}\% = \frac{X_{EA}}{X_{\alpha} + X_{EA}} = \frac{A_{EA}}{\left(\frac{K_{840}}{K_{764}}\right) A_{\alpha} + A_{EA}} \times 100 \tag{2.8}$$

Here X_{α} and X_{EA} represent the crystalline mass fractions of α and electroactive (β+γ) phase. The absorbance intensities of A_{α} and A_{EA} are at 764 cm^{-1} and 840 cm^{-1}, while K_{α} and K_{EA} are the absorbance coefficients associated with the respective wavenumbers. Figure 2.21 [61] represents the XRD and FTIR spectroscopy of PVDF polymer. The characteristic peaks of α, β, and γ phases in PVDF were summarized in Table 2.2.

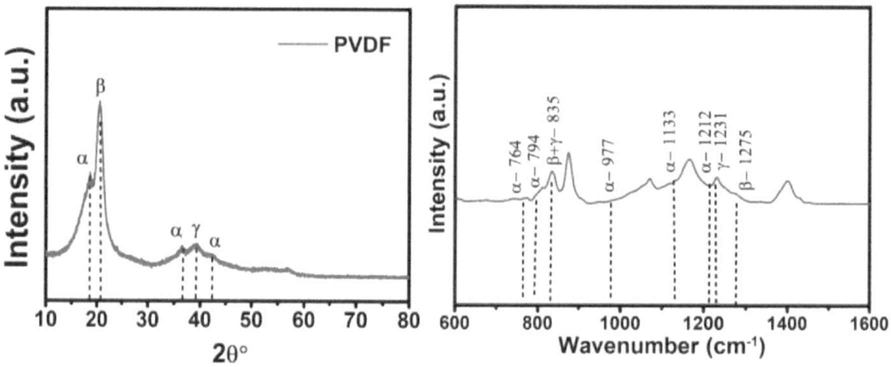

FIGURE 2.21 The X-ray diffraction and FTIR spectroscopy of PVDF polymer. (Reprinted with permission from Ref. [61]. Copyright (2024) American Chemical Society.)

TABLE 2.2
Characteristic Peaks of α-Phases, β-Phase and γ-Phase in PVDF

Polymer Characteristics	α-Phases	β-Phase	γ-Phase
X-ray diffraction (2θ° and corresponding plane)	17.6° (110), 18.3° (020), 19.9° (110), 26.5° (021)	20.6° (110/200)	18.5° (020), 19.2° (002), 20.1° (110)
FTIR absorbance bands (Wave number cm^{-1})	489 cm^{-1}, 614 cm^{-1}, 766 cm^{-1}, 795 cm^{-1}, 855 cm^{-1}, 976 cm^{-1}	510 cm^{-1}, 840 cm^{-1}, 1275 cm^{-1}	431 cm^{-1}, 512 cm^{-1}, 776 cm^{-1}, 813 cm^{-1}, 833 cm^{-1}, 840 cm^{-1}, 1234 cm^{-1}
DSC melting temperature (°C)	167–173	167–172	189–190

2.2.5.6 Piezoelectric Applications

In the preceding sections, notable progress has been achieved in the processing of piezoelectric functional materials. As a result, piezoelectric polymers and composite materials find extensive applications across various multifunctional sectors. These applications include but are not limited to energy harvesting and storage technology, sensors, actuators, photodetectors, optical sensors, and biomedical sectors. The remarkable properties of these materials contribute to their widespread adoption in these fields, and we will provide a brief overview of these applications in the following sections. Polymer-based piezoelectric materials are incorporated into devices with the aim of converting mechanical energy, such as wind, friction, acoustic vibrations, ultrasonic waves, and fluid flow, into electrical energy due to their electromechanical conversion properties. Schematic representations of polymer composite materials applications in different sectors are depicted in Figure 2.22.

The resulting electricity can power portable, smart electronic devices with power requirements typically in the mW/nW range. It's worth noting that harvesting energy directly from ambient sources is a promising approach for powering nanoscale devices without external power supplies. In this context, recent advancements in NGs for mechanical energy harvesting have been significant in building self-powered systems. The first PVDF NF-based NG was developed by Chang et al. using metallic electrodes, followed by Fang et al. who utilized nonwoven PVDF NFs with aluminum (Al)-based electrodes. Subsequently, Maity k. et al. created a device with copper (Cu) electrodes [65]. The device fabrication and PVDF NF FESEM image were depicted in Figure 2.23. The films were sandwiched securely between two aluminum foils, serving as both top and bottom

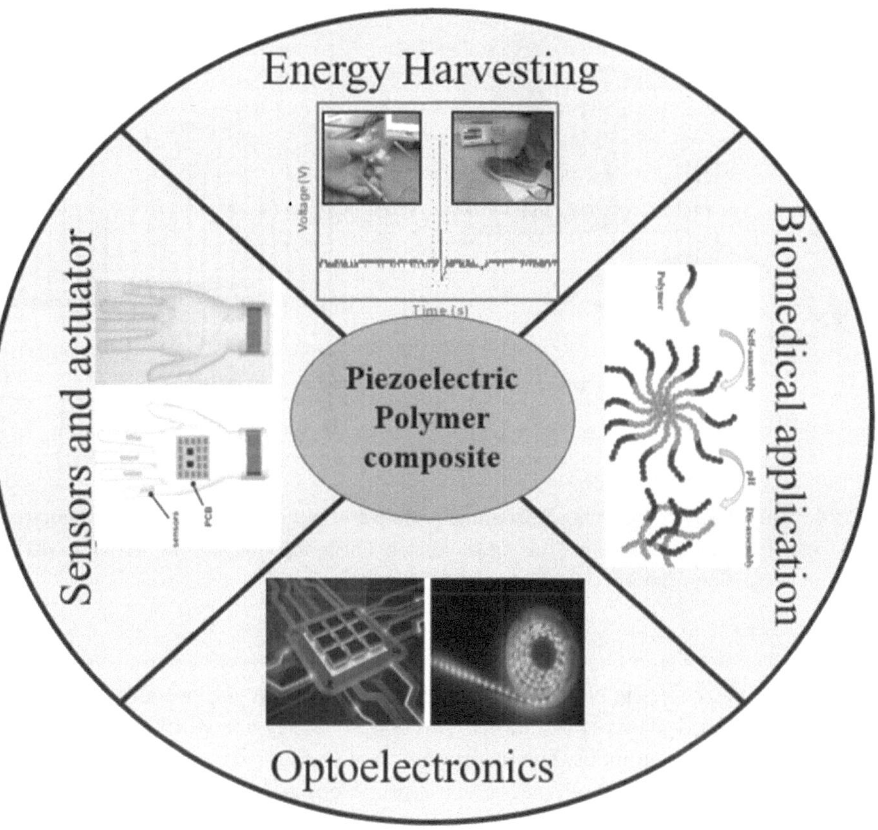

FIGURE 2.22 Schematic representation of piezoelectric polymer composite materials for multifunctional application.

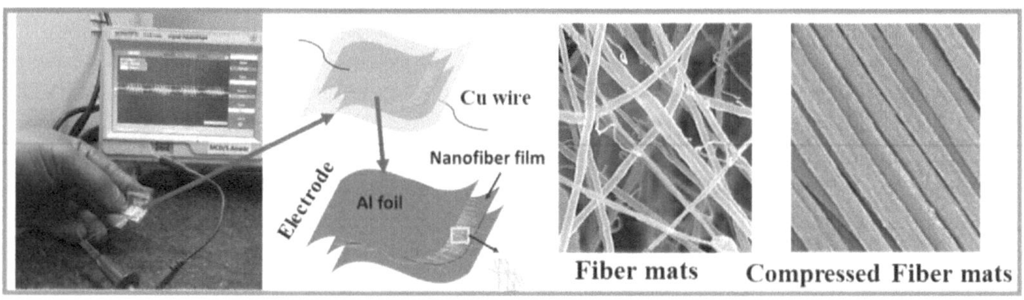

FIGURE 2.23 The schematic illustration of piezoelectric generator and device testing through Oscilloscope.

electrodes. These electrodes were intentionally made slightly smaller than the film to prevent potential short circuits. Copper wires were attached to both electrodes for easy collection of output voltage and current. The entire prototype was carefully encapsulated with Kapton tape to ensure its integrity and performance. Notably, the electric dipoles in the PVDF polymer, formed by CH_2–CF_2 bonds, have a polarization direction perpendicular to the PVDF membrane plane. This means the maximum polarization occurs along the thickness, resulting in the piezoelectric effect in the D_{33} mode.

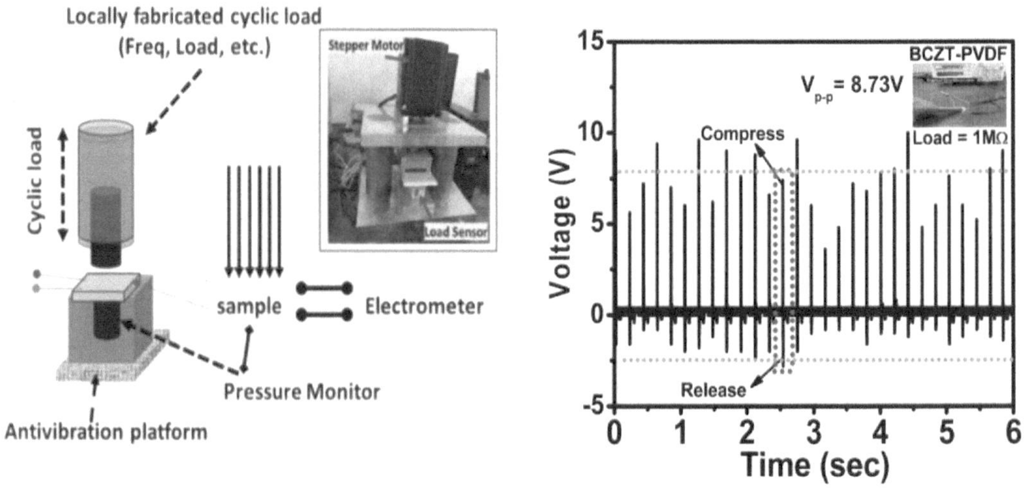

FIGURE 2.24 The schematic diagram showing the principle of piezoelectric output by prototype and PEH device and the output voltage of composite under human finger tapping motion across 1 MΩ resistance. (Reprinted with permission from Ref. [61]. Copyright (2024) American Chemical Society.)

Recent years have seen significant advancements in the field of flexible and wearable sensors and actuators. Notably, piezoelectric NGs (PNGs) with high sensitivity are proving highly suitable for the design of self-powered sensor applications. This is particularly noteworthy as traditional sensors often rely on external power supplies. The incorporation of various fillers into piezoelectric polymers is a common practice to enhance both piezoelectric properties and sensitivity. Consequently, piezoelectric polymer composites typically demonstrate distinct functional properties that are well-suited for sensor design. Garain et al. developed an acoustic sensor using Ce^{3+}-doped PVDF/graphene composite NFs with mechanical energy harvesting capabilities [65]. They demonstrated the NG's ability to generate acoustic energy under a sound pressure level (SPL) of 88 dB, illuminating several LEDs. The generated energy was stored in commercially available capacitors (1, 2.2, and 4.7 mF). The NG exhibited sensitivity to various musical vibrations, such as those from a piano, guitar, and violin. Additionally, it showed frequency sensitivity, with increased voltage observed at a specific frequency range (from 110 to 130 Hz). Bibekananda et al. were fabricated PVDF-BCZT and PVDF-MXene composite fabricated by layer-by-layer structure and the piezo-responses were examined by variation of load and resistance [61,62]. The schematic diagram in Figure 2.24 shows the principle of piezoelectric output by prototype device fabricated system, and PEH device and the output voltage of composite under human finger tapping motion across 1 MΩ resistance. Here positive and negative potentials refer as compressing and releasing pressure on the prototype device. The power density of piezoelectric energy harvester (PEH) device was calculated by using Equation (2.9).

$$\textbf{Power density} = \frac{\textbf{Current}(\textbf{I})}{\textbf{Area}(\textbf{A})} \times \frac{\textbf{Voltage}(\textbf{V})}{\textbf{Thickness}(\textbf{t})}, \tag{2.9}$$

2.3 SUMMARY

In this chapter, the basics of smart materials are explored. The discussion encompasses various avenues of smart materials like passive and active materials with a focus on materials properties, characterization, and sustainable processing methods. Smart polymer and polymer composite materials for thermoelectric and piezoelectric effect were also discussed.

The chapter begins with mainly on discussion of the active smart materials' properties and applications. The active smart materials selected for discussion were as follows: SMA, chromogenic materials, MR materials, thermoelectric materials, piezoelectric materials. In conclusion, piezoelectric materials are extensively applied as smart materials, shape memory materials address design contradictions, and giant MR materials offer high-precision actuation. Smart materials aim for enhanced safety, longer life cycles, robustness, and controllability, progressing from single-linear to composite-nonlinear control strategies. Smart structures seek higher resolution, wider coverage bandwidth, and integration of sensing and actuation and energy harvesting application. In aerospace and other sectors, SMA, chromogenic materials, MR materials, thermoelectric materials, and piezoelectric materials play vital roles in structural health monitoring, energy harvesting, vibration control, high-precision displacement output, and shape control system. Smart materials play a crucial role in modern technology, offering unique properties that adapt to external stimuli. SMAs exhibit the ability to return to their original shape when exposed to specific conditions, finding applications in robotics, aerospace, and medical devices. Chromogenic materials change color or transparency in response to stimuli, utilized in smart windows and eyewear. MR materials alter their mechanical properties under a magnetic field, employed in dampers and shock absorbers. Thermoelectric materials convert heat into electricity or vice versa, finding use in power generation and temperature control. Piezoelectric materials generate electric charge in response to mechanical stress, widely employed in sensors, actuators, and energy harvesting devices.

While considering the sustainability of these smart materials, parameters such as sourcing of material, manufacturing processes, and its life span need to be understood and analyzed. SMAs possess the unique ability to revert to a predetermined shape after deformation, offering innovative possibilities but also posing environmental challenges. To enhance sustainability, it is crucial to scrutinize the environmental impact of mining practices for SMAs' elements like nickel, titanium, and copper. Ethical sourcing practices and adopting sustainable manufacturing techniques, such as additive manufacturing or 3D printing, can contribute to an eco-friendlier production cycle. Addressing end-of-life considerations for SMAs, including recycling and repurposing alloys, is essential for minimizing environmental impact and aligning with sustainability principles in material science. Chromogenic materials, renowned for their capacity to alter color in response to external stimuli, require a thorough examination of their sustainability aspects. Regarding material acquisition, it is crucial to carefully assess the environmental and social consequences associated with sourcing components like indium, tin, or tungsten. Enhancing the sustainability of chromogenic material manufacturing involves embracing eco-friendly fabrication methods, such as energy-efficient deposition techniques, and utilizing environmentally friendly solvents and chemicals. MR materials, due to their complex composition and exploring options for repurposing or reusing these materials in various applications, contribute to a sustainable material's lifecycle. The sustainable production of thermoelectric materials requires the incorporation of energy-efficient and environmentally friendly synthesis techniques, including solution-based processes. Additionally, the consideration of non-toxic or recyclable solvents and chemicals is crucial in these manufacturing processes. In piezoelectric materials, when it comes to material sourcing, it is vital to evaluate the environmental consequences of acquiring elements like lead, zinc, or other commonly used components. It is imperative to explore sustainable alternatives or the creation of lead-free piezoelectric materials. In sustainable piezoelectric material manufacturing, the focus should be on adopting energy-efficient methods, such as hydrothermal synthesis, and exploring green synthesis approaches that involve environmentally friendly processing agents. The interdisciplinary study of these smart materials is essential for advancing applications in aerospace, biomedicine, robotics, and electronics, contributing to higher safety, longer life cycles, and enhanced control in various fields for future advancements.

REFERENCES

[1] Basheer, A. A. (2020). Advances in the smart materials applications in the aerospace industries. *Aircraft Engineering and Aerospace Technology*, 92(7), 1027–1035.

[2] Hu, J. L. (2016). Introduction to active coatings for smart textiles. *Active Coatings for Smart Textiles*, 1–7. Woodhead Publishing.

[3] Kamila. S (2013). Introduction, classification and applications of smart materials: an overview. *American Journal of Applied Sciences*, 10(8), 876–880.

[4] Rogers, C. A. (1989). *U.S. Army Research Office Workshop on Smart Materials, Structures and Mathematical Issues Held in Blacksburg*, Virginia on 15–16 September 1988. Defense Technical Information Center.

[5] Singh, M., Sharma, S., & Sharma, R. (2019). A review on fractures: ayurvedic classification in modern perspective. *International Journal of Research in Ayurveda and Pharmacy*, 10(1), 11–17.

[6] Damodharan, J., Sreedharan, A., & Ramalingam, T. (2018). A review on smart materials, types and applications. *International Journal of Engineering Technology Science and Research*,, 5, 649–653.

[7] Bahl, S., Nagar, H., Singh, I., & Sehgal, S. (2020). Smart materials types, properties and applications: a review. *Materials Today: Proceedings*, 28, 1302–1306.

[8] Mandal, S. K., Maity, A., & Narayan Ray, D. (2013). Smart materials—scopes and prospects. *Elixir Mechanical Engineering*, 65, 20154–20163.

[9] Sobczyk, M., Wiesenhütter, S., Noennig, J. R., & Wallmersperger, T. (2021). Smart materials in architecture for actuator and sensor applications: a review. *Journal of Intelligent Material Systems and Structures*, 33(3), 379–399.

[10] Mehta, K., & Gupta, K. (2018). Processing of shape memory alloys. *SpringerBriefs in Applied Sciences and Technology*, 61–81.

[11] Hosoda, H., & Inamura, T. (2009). Mechanical properties of shape memory alloys. *Shape Memory Alloys for Biomedical Applications*, 20–36. New York: CRC Press.

[12] Mohd Jani, J., Leary, M., Subic, A., & Gibson, M. A. (2014). A review of shape memory alloy research, applications and opportunities. *Materials & Design (1980–2015)*, 56, 1078–1113.

[13] Müllner, P. (1999). Shape memory alloys. Edited by K. Otsuka and C.M. Wayman, Cambridge University Press, ISBN 0-521-44487-X (hardcover), 284 pp. +xiv. *Materials Science and Engineering: A*, 268(1–2), 246–247.

[14] Quade, D. J., Jana, S. C., Morscher, G. N., & Kanaan, M. (2019). The effect of thin film adhesives on mode I interlaminar fracture toughness in carbon fiber composites with shape memory alloy inserts. *Engineering Fracture Mechanics*, 206, 131–146.

[15] Chu, C. L., Chung, C. Y., Lin, P. H., & Wang, S. D. (2004). Fabrication of porous NiTi shape memory alloy for hard tissue implants by combustion synthesis. *Materials Science and Engineering: A*, 366(1), 114–119.

[16] Duerig, T., Pelton, A., & Stöckel, D. (1999). An overview of nitinol medical applications. *Materials Science and Engineering: A*, 273–275, 149–160.

[17] Gangil, N., Siddiquee, A. N., & Maheshwari, S. (2020). Towards applications, processing and advancements in shape memory alloy and its composites. *Journal of Manufacturing Processes*, 59, 205–222.

[18] Rezvani, M. R., & Shokuhfar, A. (2012). Synthesis and characterization of nano structured Cu–Al–Mn shape memory alloy by mechanical alloying. *Materials Science and Engineering: A*, 532, 282–286.

[19] Stachiv, I., Alarcon, E., & Lamac, M. (2021). Shape memory alloys and polymers for MEMS/NEMS applications: review on recent findings and challenges in design, preparation, and characterization. *Metals*, 11(3), 415.

[20] Jeong, J., Yasir, I. B., Han, J., Park, C. H., Bok, S. K., & Kyung, K. U. (2019). Design of shape memory alloy-based soft wearable robot for assisting wrist motion. *Applied Sciences*, 9(19), 4025.

[21] Resnina, N., Belyaev, S., & Voronkov, A. (2018). Functional properties of porous Ti-48.0 at. % Ni shape memory alloy produced by self-propagating high-temperature synthesis. *Journal of Materials Engineering and Performance*, 27, 1257–1264.

[22] Cannavale, A. (2020). Chromogenic technologies for energy saving. *Clean Technologies*, 2(4), 462–475.

[23] Granqvist, C. G., Pehlivan, İ. B., & Niklasson, G. A. (2018). Electrochromics on a roll: web-coating and lamination for smart windows. *Surface and Coatings Technology*, 336, 133–138.

[24] Cui, Y., Ke, Y., Liu, C., Chen, Z., Wang, N., Zhang, L., & Long, Y. (2018). Thermochromic VO_2 for energy-efficient smart windows. *Joule*, 2(9), 1707–1746.

[25] Hakami, A, Indrakar, S, Krishnegowda, A, Huang, MY, Sahebkar, K, Biswas, PK, et al., (2021) Microencapsulation of thermochromic material by silicon oxide nanoparticles. Trans. Tech. Publ., *Key Engineering Materials*, 878, pp. 41–48.

[26] Zhang, J., Zou, Q., & Tian, H. (2013). Photochromic materials: more than meets the eye. *Advanced Materials*, 25(3), 378–399.

[27] Park, B. J., You, J. L., Choi, H. J., Park, S. Y., & Lee, B. Y. (2009). Synthesis and magnetorheological characterization of magnetite nanoparticle and poly(vinyl butyral) composite. *IEEE Transactions on Magnetics*, 45(6), 2460–2463.

[28] Ahamed, R., Ferdaus, M. M., & Li, Y. (2016). Advancement in energy harvesting magneto-rheological fluid damper: A review. *Korea-Australia Rheology Journal*, 28(4), 355–379. https://doi.org/10.1007/s13367-016-0035-2

[29] D. Chen, K. Jiang, G. Shen, (2017), New insights and perspectives into biological materials for flexible electronics. *Chemical Society Reviews*, 46, 6764.

[30] Au, B. W.-C., & Chan, K.-Y. (2022). Towards an all-solid-state electrochromic device: a review of solid-state electrolytes and the way forward. *Polymers*, 14(12) 2458, 1–16

[31] Wang, X., & Gordaninejad, F. (2008). *Magnetorheological Materials and Their Applications*. Intelligent Materials, pp. 339–385.

[32] Jinaga, R., Jagadeesha, T., Kolekar, S., & Choi, S.-B. (2019). The synthesis of organic oils blended magnetorheological fluids with the field-dependent material characterization. *International Journal of Molecular Sciences*, 20(22), 5766. https://doi.org/10.3390/ijms20225766

[33] Unuh, M. H., & Muhamad, P. (2020). A brief review of preparation method and challenges of magnetorheological fluids. *Journal of Advanced Research in Materials Science*, 74(1), 1–18. https://doi.org/10.37934/arms.74.1.118

[34] Kciuk, S., & Martynowicz, P. (2011). Special application magnetorheological valve numerical and experimental analysis. *Solid State Phenomena*, 177, 102–115.

[35] Park, B. J., Fang, F. F., & Choi, H. J. (2010). Magnetorheology: materials and application. *Soft Matter*, 6(21), 5246–5253.

[36] Ahamed, R., Choi, S. B., & Ferdaus, M. M. (2018). A state of art on magneto rheological materials and their potential applications. *Journal of Intelligent Material Systems and Structures*, 29(10), 2051–2095.

[37] Kumar, J. S., Paul, P. S., Raghunathan, G., & Alex, D. G. (2019). A review of challenges and solutions in the preparation and use of magnetorheological fluids. *International Journal of Mechanical and Materials Engineering*, 14, 1–18.

[38] Avraam, M. (2009). MR-fluid brake design and its application to a portable muscular rehabilitation device. Doktora Tezi, Faculte´ De Sciences Applique´es, Universite´ Libre De Bruxelles, Bruxelles.

[39] Afshari, F. (2020). Experimental study for comparing heating and cooling performance of thermoelectric Peltier. *Politeknik Dergisi*, 23(3), 889–894. https://doi.org/10.2339/politeknik.713600

[40] Tzounis, L. (2019). Organic thermoelectrics and thermoelectric generators (TEGs). In *Advanced thermoelectric Materials for Energy Harvesting Applications*, p. 7., IntechOpen.

[41] Singh, V., Joung, D., Zhai, L., Das, S., Khondaker, S. I., & Seal, S. (2011). Graphene based materials: past, present and future. *Progress in Materials Science*, 56, 1178–1271.

[42] Moliton, A., & Hiorns, R. C. (2004). Review of electronic and optical properties of semiconducting π-conjugated polymers: applications in optoelectronics. *Polymer International*, 53(10), 1397–1412.

[43] Bhuvana, K. P., Joseph Bensingh, R., Abdul Kader, M., & Nayak, S. K. (2018). Polymer light emitting diodes: materials, technology and device. *Polymer-Plastics Technology and Engineering*, 57, 1784–1800.

[44] Rahman, M.A., Kumar, P., Park, D.-S., & Shim, Y.B. (2008). Electrochemical sensors based on organic conjugated polymers. *Sensors*, 8, 118–141.

[45] Rösch A.G., Gall A., Aslan S., Hecht M., Franke L., Mallick M.M., Penth L., Bahro D., Friderich D., & Lemmer U. (2021). Fully printed origami thermoelectric generators for energy-harvesting. *NPJ Flexible. Electronics*, 5(1).

[46] Popli, S. A., & Patel, U. D. (2015). Electrochemical decolorization of Reactive Black 5 in an undivided cell using Ti and graphite anodes: effect of polypyrrol coating on anodes. *Journal of Electrochemical Science and Engineering*, 5(2).

[47] Hulea M., Mois G., Folea S., Miclea L., & Biscu V. (2013). Wi-Sensors: A Low Power Wi-Fi Solution for Temperature and Humidity Measurement, Proceedings of the IECON 2013—39th Annual Conference of the IEEE Industrial Electronics Society, Vienna, Austria, 10–13.

[48] Bian, M., Xu, Z., Meng, C., Zhao, H., & Tang, X. (2022). Novel geometric design of thermoelectric leg based on 3D printing for radioisotope thermoelectric generator. *Applied Thermal Engineering*, 212, 118514.

[49] Lay-Ekuakille, A., Vendramin, G., Trotta, A., & Mazzotta, G. (2009). Thermoelectric generator design based on power from body heat for biomedical autonomous devices. In *2009 IEEE International Workshop on Medical Measurements and Applications*, (pp 10–4). IEEE.

[50] Wan, C., & Bowen, C.R. (2017). Multiscale-structuring of polyvinylidene fluoride for energy harvesting: the impact of molecular-, micro- and macro-structure. *Journal of Materials Chemistry A*, 5, 3091–3128.

[51] Elorika, P., Jayasri, S., Sharmistha, A., & Anwar, S. (2022). Synthesis conditions induced disorder and its role in affecting structural, dielectric, piezoelectric, optical behavior and enhancing energy storage efficiency in $(Ba_{1-x}Ca_x)$ TiO_3 ceramics. *Ceramics International*, 48(13), 19324–19335.

[52] Ikeda, T. (1996), *Fundamentals of Piezoelectricity*. Oxford University Press, p. 1.

[53] Wang, F., Tanaka, M., & Chonan, S. (2003). Development of a PVDF piezopolymer sensor for unconstrained in-sleep cardiorespiratory monitoring. *Journal of Intelligent Material Systems and Structures*, 14, 185–190.

[54] Jayasri, S., Elorika, P., & Anwar, S. (2023). Effect of sintering time on electrical, ferroelectric, and piezoelectric performances of microwave synthesized sodium bismuth titanate ceramics. *Materials Science and Engineering: B*, 298, 116–823.

[55] Nalwa, H.S. (1995). *Ferroelectric Polymers: Chemistry: Physics, and Applications*. CRC Press.

[56] Martins, P., Lopes, A.C., & Lanceros-Mendez, S. (2014). Electroactive phases of poly (vinylidene fluoride): determination, processing and applications. *Progress in Polymer Science*, 39, 683–706.

[57] Sakamoto, W.K., Shibatta-Kagesawa, S., Kanda, D. H. F., Fernandes, S. H., Longo, E., & Chierice, G. O. (1999). Piezoelectric effect in composite polyurethane ferroelectric ceramics. *Physica Status Solidi A.*, 172, 265.

[58] David, C., Capsal, J. F., Laffont, L., Dantras, E., & Lacabanne, C. (2012). Piezoelectric properties of polyamide 11/$NaNbO_3$ nanowire composites. *Journal of Physics D: Applied Physics*, 45, 415305.

[59] Khanbareh, H., Van Der Zwaag, S., & Groen, W. A. (2015). Piezoelectric and pyroelectric properties of conductive polyethylene oxide-lead titanate composites. *Smart Materials and Structures*, 24, 045020.

[60] Sultana, A., Alam, M. M., Biswas, A., Middya, T. R., & Mandal, D. (2016). Fabrication of wearable semiconducting piezoelectric nanogenerator made with electrospun-derived zinc sulfide nanor. *Translational Materials Research*, 3, 4.

[61] Nayak, B., Anwar, S., Kumar, A., & Anwar, S. (2024). Modulating energy harvesting behaviour of PVDF piezo-polymer by incorporation of BCZT ceramic filler. *ACS Applied Electronic Materials*, 6(2), 1189–1203.

[62] Nayak, B., Anwar, S., Roy, A., & Anwar, S. (2023). Investigation of 2D-$Ti_3C_2T_X$-loaded PVDF composites for a high-performance flexible piezoelectric energy harvester and wearable sensor. *ACS Applied Electronic Materials*, 5(12), 7045–7060.

[63] Shepelin, N. A. (2019). 3D printing of poly(vinylidene fluoride-trifluoroethylene): a poling-free technique to manufacture flexible and transparent piezoelectric generators. *MRS Communications*, 9(1), 159–164.

[64] Xin, Y., Zhu, J., Sun, H., Xu, Y., Liu, T., & Qian, C. (2018). A brief review on piezoelectric PVDF nanofibers prepared by electrospinning. *Ferroelectrics*, 526(1), 140–151.

[65] Maity, K., & Mandal, D. (2018). All-organic high-performance piezoelectric nanogenerator with multilayer assembled electrospun nanofiber mats for self-powered multifunctional sensors. *ACS Applied Materials & Interfaces*, 10, 18257–18269.

3 Polymer Materials Properties and Characterization Study with Focus on Sustainability

Pramod Kumar Parida, Md. Irquam Alam, Anu Tomy, Debarupam Gogoi, Arun Kumar Pradhan, and Mihir Kumar Pandit

3.1 INTRODUCTION

Over the span of 4 billion years since the Earth's formation, elemental combinations such as carbon, hydrogen, oxygen, and nitrogen have orchestrated the intricate dance of life through the formation of complex molecules. The material basis for life, a polymer called protein, synthesized in nature from basic chemical compounds like methane and ammonia, kicked off the evolutionary journey, eventually leading to the human form. In this historical exploration, the primal human needs for covering, shelter, and defence from external forces are unveiled. Natural polymers, manifested in forms like animal skins, wood, and cotton, became early solutions to these fundamental requirements. Through civilizations, these needs evolved, with polymers adapting to meet the ever-changing demands.

Key milestones in the polymer saga include Christopher Columbus observing a bouncing game with rubber in South America, accidental discoveries like Schonbein's creation of combustible 'gun cotton,' and John Wesley Hyatt's development of 'celluloid' as a substitute for scarce ivory. Scientific breakthroughs by researchers like Leo Baekeland further propelled the polymer revolution, giving birth to Bakelite and subsequent synthetic polymers. The impact of polymers on human life has been profound, permeating every aspect of society. From textiles to building materials, polymers offer a spectrum of choices. The versatility of polymers is evident in their ability to meet diverse requirements, from high strength and lightweight to resistance to chemicals. The transformative journey of polymer, originating from natural polymers to the vast realm of synthetic counterparts, is a testament to human ingenuity and scientific progress. As we delve deeper into polymer science, the future promises even more innovative combinations and applications, making the polymer chemist the architect of a new era in materials. Polymers, in essence, are intricate and colossal molecules, distinguishing them from low molecular weight compounds like common salt. While common salt has a molecular weight of 58.5 g/mol, polymers can reach several hundred thousand. These macromolecules are formed by smaller molecules, either of one or more chemical compounds.

The term 'polymer' is derived from Greek, with 'poly' meaning 'many' and 'mer' meaning 'part.' For instance, when gaseous butadiene, with a molecular weight of 54 g/mol, combines around 4,000 times, it forms polybutadiene, a synthetic rubber with a molecular weight of nearly 200,000. The unique behaviour of polymers, compared to low molecular weight compounds, arises from their giant molecular size. Unlike simple compounds like benzene, polymers like polyethylene (PE) don't sharply melt at a specific temperature but gradually soften into a viscous molten mass with further heating. Another notable difference lies in their solubility patterns. While low molecular weight compounds, like salt, have a saturation point in water, polymers, such as polyvinyl alcohol (PVA),

DOI: 10.1201/9781003664611-4

exhibit a prolonged dissolution process. The globules absorb water, swell, and eventually dissolve, with no saturation point reached even with a large quantity added. The dissolution characteristics, namely the extended dissolution time, absence of saturation, and increased viscosity, are attributed to the considerable molecular size of polymers. In summary, the behaviour of polymers, dictated by their immense molecular size, differs significantly from that of low molecular weight compounds, influencing their melting and dissolution properties.

This chapter deals with the exploration of polymer fundamentals, classification, advantages, and constraints of various polymer types, such as thermoplastics, thermosets, elastomers, and biopolymers. The characterization techniques for polymers, like thermal analysis, X-ray diffraction (XRD), electron microscopy, spectroscopy, and mechanical testing, have also been explored in detail for understanding the structure, behaviour, and performance of polymer materials. The sustainability aspect, particularly within composite structures and smart materials, is thoroughly investigated, shedding light on renewable sources, recyclability, and biodegradability. Further research possibilities in polymers for electric vehicles (EVs) and biomedical applications are also discussed.

3.2 CLASSIFICATION OF POLYMERS

Polymers can be categorized in various ways based on sources, structures, polymerization, monomers, thermal responses, applications and physical properties, and special properties, as shown in Figure 3.1. Considering composite application and sustainability, polymers belonging to thermoplastics, thermosets, elastomers, and biopolymers are discussed in detail.

3.3 BASIC DIFFERENCES OF THERMOPLASTICS, THERMOSETS, ELASTOMERS AND BIOPOLYMERS

The basic differences in thermoplastics, thermosets, and elastomers are presented in Table 3.1. Figure 3.2 shows the difference in the polymer structures in which thermoplastics have loose polymer chains that can break apart, thermosets possess rigid, cross-linked polymer structures that, once set with heat, cannot be reshaped, and elastomers have distinct polymer chains but later develop a covalent cross-link network. The advantages and limitations associated with these polymers are presented in Table 3.2.

3.4 THERMOPLASTIC POLYMERS

Researchers have turned their focus to thermoplastic materials as matrix materials due to their distinctive characteristics like malleability, lightweight nature, recyclability, and affordability. Table 3.2 outlines the advantages and limitations associated with thermoplastic polymers. In their uncured state at standard room temperature and pressure, the viscosity of thermoplastics is 500 times higher than that of thermosetting materials. Thermoplastics, including PE, polypropylene (PP), polyvinyl chloride (PVC), and polystyrene (PS), are among the most commonly utilized polymers for commercial applications. Table 3.3 presents the mechanical properties of these thermoplastic polymers that are commonly used in composite and smart material applications, followed by a detailed discussion of the same.

3.4.1 Polyethylene

PE represents the most basic hydrocarbon structure, being colourless and combustible, and commonly employed in packaging. It is usually made up of a mix of similar ethylene polymers with different n values. This means that its properties can be changed even more by cross-linking or copolymerizing them. Notably, all forms of PE are chemically resilient and nontoxic, contributing to its extensive

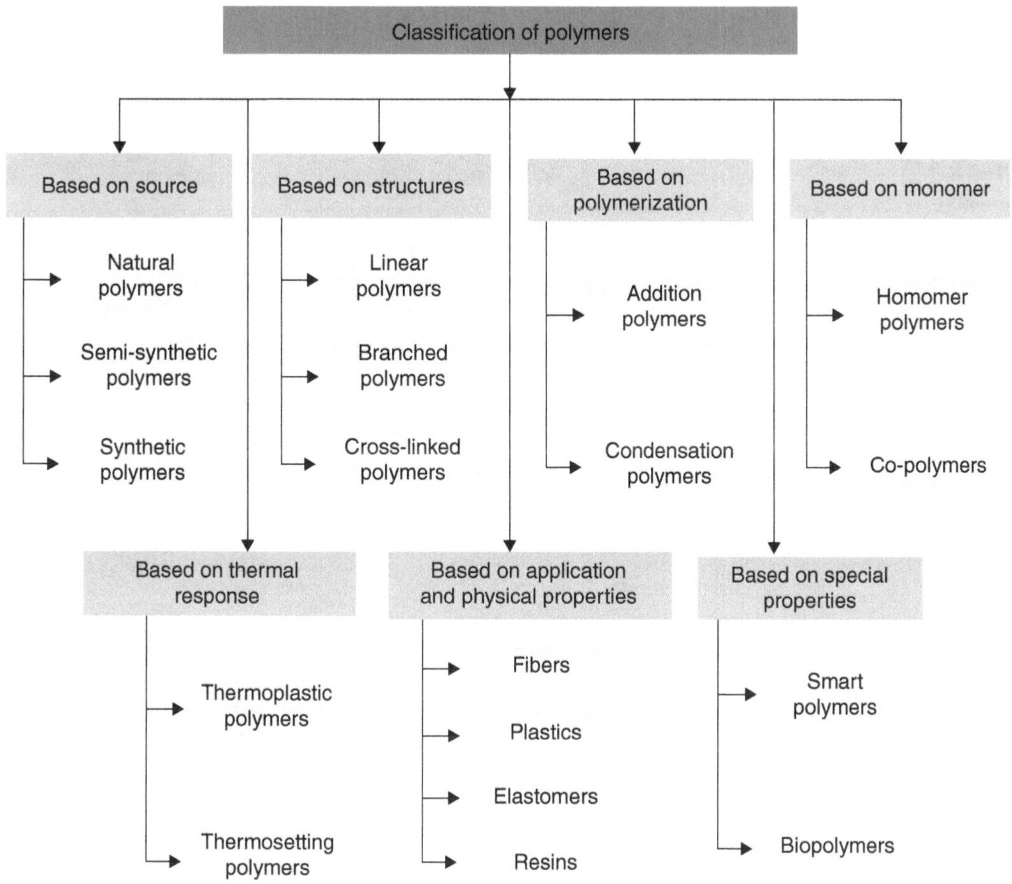

FIGURE 3.1 Classification of polymers.

TABLE 3.1
Basic Difference among Thermoplastic, Thermosets, and Elastomers

Characteristic	Thermoplastic	Thermoset	Elastomer
Polymer Structure	Loose polymer chains that can be re-melted and reshaped	Rigid, cross-linked polymer structure that cannot be reshaped	Flexible polymer chains with the ability to deform and return to their original shape
Heating Behavior	Soften when heated and harden when cooled	Hardened and cannot be reshaped once set with heat	Deform and return to the original shape when heated
Cross-Linking	No permanent cross-linking, chains can move freely	Cross-linked structure formed with heat, permanent cross-links	Cross-link network formed after polymerization, providing elasticity
Vulcanization	Not vulcanized	Vulcanized during the polymerization process	Vulcanized after the polymerization process
Glass Transition Temperature	Generally below ambient temperature	Generally above ambient temperature	Below ambient temperature
Mechanical Properties	Moderate strength and toughness	High strength and hardness	Moderate strength and flexibility
Applications	Packaging, consumer products, automotive parts	Aircraft components, electrical insulation	Tyres, seals, gaskets, and shoe soles

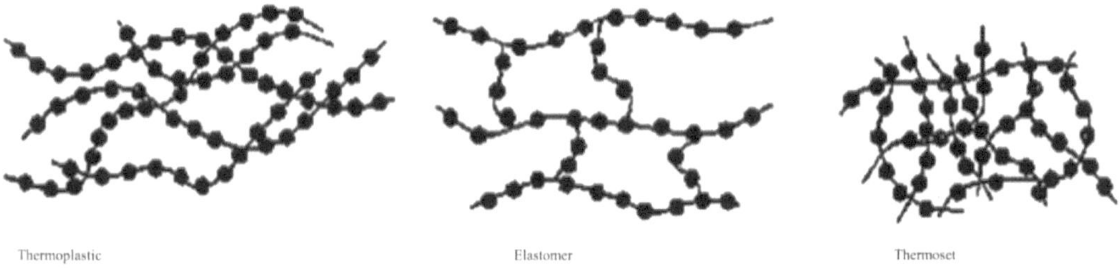

FIGURE 3.2 Differences in polymer structures.

TABLE 3.2
Advantages and Limitations of Thermoplastics, Thermosets, Elastomers, and Biopolymers

Type of Matrix	Advantages	Limitations
	Thermoplastics Polymers	
Polypropylene (PP)	High thermal and dielectric resistance, low cost, and lightweight	Comparatively high cost, difficult processing, and limited availability
Polystyrene (PS)	Moisture, fatigue, and chemical resistance	Brittle, flammable, and low impact resistance
Natural rubber	High resilience and low cost, tear and water resistance	Chances for swelling, poor resistance to hydrocarbons, complete dissolution
Polyvinyl chloride (PVC)	Low cost, fire retardant, excellent dimensional stability, advance in chemical resistance, and good tensile strength	Poor resistance to UV and temperature
Polyethylene (PE)	Good fatigue and impact strength Low moisture absorption, versatility, and high ductility	High thermal expansion, flammable, and poor weather resistance
	Thermosets Polymers	
Epoxy resin	High thermal and mechanical properties, high water resistance, durability, and low curing time	More expensive than vinyl ester, difficult to process, corrosive amine hardener
Polyester	Easy to use, lowest cost	High curing shrinkage, moderate mechanical properties, and high gas emissions
Vinyl ester	High mechanical properties than polyester, high chemical resistance	More expensive, needs post-curing, high styrene content, and curing shrinkage
Phenolic	High fire retardancy	Hard to develop
	Elastomers	
Natural rubber	Highly resilient, resistances to tearing, and good abrasion resistance	Easy to age in the air and becomes sticky when heated. Degrade in UV, ozone, sunlight, heat, petroleum oil, and fuels
Acrylonitrile butadiene rubber	Excellent abrasion resistance, good tear resistance, non-polar solvent resistance, water resistance, and oil resistance. Cheaper than fluoroelastomers	Poor ozone, sunlight, and weather resistance. Limited high temperature resistance, and poor flame resistance
Silicone	High-performance applications, versatility, low flammability, high dielectric strength, sterilizability, durability, and good overall mechanical properties	High initial cost. Untreated silicone elastomers can feel slightly sticky to the touch. Exposure to latex, sulphur, and certain other materials will inhibit curing
Neoprene	Resistance to ozone, sunlight, oils, solvents, fire and tear resistant, easily adheres to metals and fabrics	High cost, reactive to oxidizing acids and other chemicals. It may also absorb water

TABLE 3.2 (Continued)
Advantages and Limitations of Thermoplastics, Thermosets, Elastomers, and Biopolymers

Type of Matrix	Advantages	Limitations
Polyurethane	Abrasion, oil and chemical resistance. Affordable manufacturing process	Short life, hydroscopic in nature, emits odours and fumes. Durability will be adversely affected when it is exposed to direct sunlight or organic solvents
	Biopolymers	
Cellulose	Easily available, ease to adjust, inexpensive, intermediate heat and impact resistance, and eco-friendly	High moisture absorption and relative low decomposition
Starch	Biodegradable and low cost	Water sensitive, brittle, difficult to process
Poly lactic acid	High strength, nano toxic nature, high modulus, and low cost	Poor impact strength, low thermal degradation, and brittle in nature
Polyhydroxyalkanoates (PHA)	High molecular weight and biodegradable	Fragile, low decomposition, low stability, and costly.

TABLE 3.3
Mechanical Properties of Thermoplastic Polymers

Thermoplastics	Density (g/cm³)	Tensile Strength (MPa)	Young's Modulus (GPa)	Elongation (%)	Glass Transition Temperature (°C)
Polyethylene (PE)	0.9	18.0	0.5	350.0	–78.0
Low-density polyethylene (LDPE)	0.9	10.0–11.6	0.2–0.3	400.0	–110.0
High-density polyethylene (HDPE)	0.9–1.0	32.0–38.2	1.3	150.0	–110.0
Polypropylene (PP)	0.9–1.3	35.8	1.6	80.0	–20.0
Polystyrene (PS)	1.04	34.0	3.0	1.6	100.0
High impact polystyrene (HIPS)	1.0	42.0	2.1	2.5	88.0–92.0
Polyacrylonitrile (PAN)	1,184	–	–	25–50	80
Polymethyl methacrylate (PMMA)	1.1–1.2	72.4	3.0	2.5	125.0
Polyesters	1,38	57–60	1.93–3.00	50–300	68.33
Polycarbonate (PC)	1.2	69.0	2.3	200.0	147.0
Polysulphone (PSU)	1.24	80	2.6	50	185
Polyamide 6 (PA 6)	1.1	81.4	2.8	60.0	47.0
Polyethylene glycols (PEG)	1.124	–	0.00208–0.0026	2	–70
Polypropylene glycols (PPG)	1.01	–	–	–	–25
Polyacrylic acid (PAA)	1.41	0.05	–	–	106
Polymethacrylic acid (PMAA)	1.251	–	–	–	228
Polyvinyl acetate (PVAc)	1.22	–	–	–	38
Polyvinyl alcohol (PVA)	1.19–1.31	4.48	0.707	25–50	80
Polyvinyl chloride (PVC)	1.3–1.5	52.0–90.0	3.0–4.0	50.0–80.0	82.0
Polyvinylidene chloride (PVDC)	1.63–1.72	24.1–110	0.345–0.552	–	–20.15 to –10.15
Polyvinyl fluoride (PVF)	1.4	40	–	150	41
Polyvinylidene fluoride (PVDF)	1.78	40–60	2.5–3.5	10.9	–40
Polytetrafluoroethylene (PTFE)	2.3	20–196	0.390–0.750	70–600	129.4
Polyvinyl carbazole (PVK)	1.19	140	2.5–4.5	1	261
Polyvinyl pyrrolidone (PVP)	1.2	–	–	–	180

Source: Refs. [1–14].

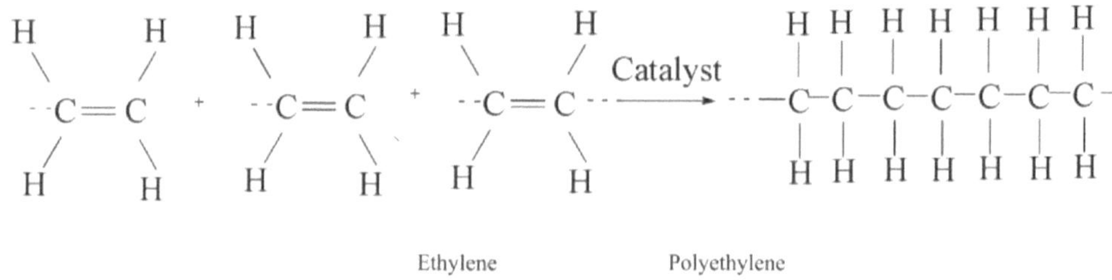

FIGURE 3.3 Polyethylene formation.

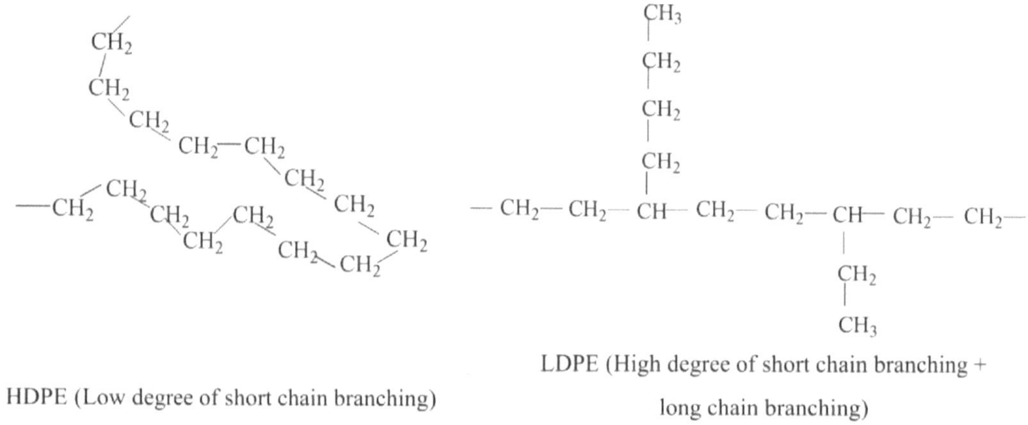

FIGURE 3.4 Structure of HDPE and LDPE.

utilization. However, its widespread application presents challenges due to its poor biodegradability, similar to other synthetic plastics, resulting in its accumulation in landfills. Nevertheless, certain species of bacteria (such as *Pseudomonas fluorescens*, *Brevibacillus borstelensis*, and *Acinetobacter*) and animals (including Indian meal moth larvae and *Galleria mellonella*) have demonstrated the ability to degrade PE. Biodegradation increased with time exposed to ultraviolet radiation. The formation structure of PE is depicted in Figure 3.3.

PE is categorized based on its density and branching, with its mechanical properties heavily influenced by factors like extent and branching type, crystal structure, and molecular weight. The two main types of PE are low density (LDPE) and high density (HDPE), illustrated in Figure 3.4. LDPE, featuring branched molecules, on the other hand HDPE structure is essentially linear. LDPE finds applications in both rigid containers and plastic film products like bags and wrap, though it is less commonly recycled despite bearing recycling symbols. It has the resin identification code '4.' HDPE, on the other hand, is utilized in the manufacturing of plastic bottles, corrosion-resistant piping, geomembranes, and plastic lumber. It has a resin identification code '2' and is commonly recycled.

PE composites, where PE serves as the matrix, are extensively utilized across various industries due to their superior mechanical and physical properties. These composites find applications in packaging, electrical, thermal energy storage, automotive, biomedical, and aerospace sectors [15–21].

3.4.2 Polypropylene

PP stands out as one of the lightest industrial polymers. It is manufactured through chain-growth polymerization of propylene monomers and falls within the polyolefin group. The structure of PP is

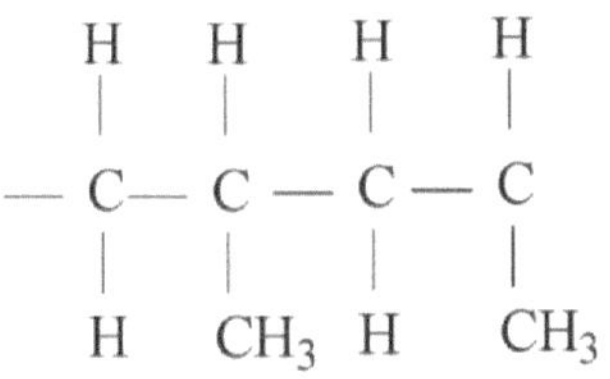

FIGURE 3.5 Polypropylene.

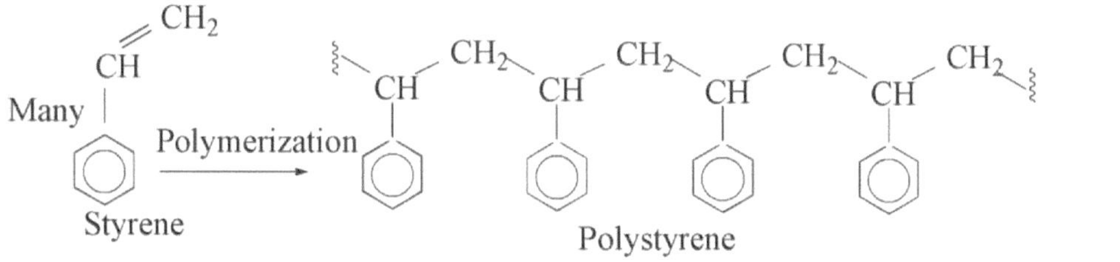

FIGURE 3.6 Polystyrene.

depicted in Figure 3.5. PP exhibits partial crystallinity and is non-polar, a white material with high chemical resistance and impressive mechanical and dielectric properties. With a typical flash point of 260°C, auto-ignition temperature of 388°C, and melting point exceeding 100°C, PP is combustible yet resilient, showcasing resistance to most solvents and adhesives, although few are specifically formulated for PP bonding, such as epoxy glue or hot-glue guns. Insoluble at room temperature in many aromatic and chlorinated hydrocarbons, PP shares similarities with PE but is slightly harder and more heat-resistant, albeit with less resistance to oxidation, heat, and light compared to PE. Due to its crystallinity, PP demonstrates high stiffness, hardness, and tensile strength, making it invaluable across a wide range of industrial applications, including living hinges, piping systems, medical devices, BOPP packaging, textiles, ropes, electrical cable insulation, roofing membranes, plastic mouldings, stationery, construction, design, EPP foam, loudspeaker units, clothing, and specialized uses such as capacitors and polymer banknotes. PP is recyclable and identified by the resin identification code '5.' Various fibres are commonly used to reinforce a PP matrix to create composites. Shubhra et al. [22] provided a comprehensive review of the mechanical properties of fibre-reinforced PP composites (FRPCs).

3.4.3 Polystyrene

PS, also referred to as polyvinyl benzene, comprises mainly linear molecules and exhibits chemical inertness. It dissolves readily in many organic solvents and possesses a low heat distortion temperature of 85°C, making it unsuitable for steam sterilization. At room temperature, PS is solid and glassy but transitions to a flowable state above 100°C, reverting to rigidity upon cooling. This characteristic facilitates processes like extrusion, moulding, and vacuum forming, enabling intricate mould detailing. Its temperature response can be managed via photo cross-linking. PS can undergo thermal depolymerization, resulting in the production of the styrene monomer and several low molecular weight compounds. The structure of PP is depicted in Figure 3.6. Although PS is widely used in various applications, it suffers from poor weather resistance and brittleness, with outdoor exposure leading to yellowing and crazing. Despite these drawbacks, it finds extensive use in manufacturing moulded containers, lids, toys, and more, indicated by its resin identification code '6.' However, PS poses environmental challenges due to its non-biodegradable nature, limited recycling options, and potential pollution from incineration. However, incineration

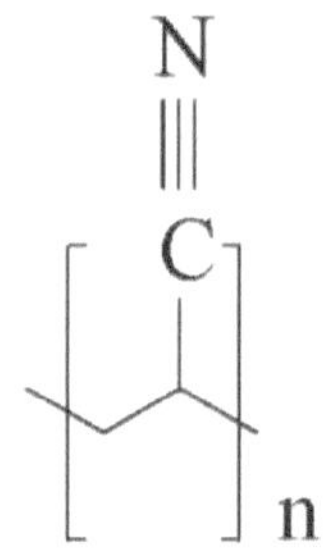

FIGURE 3.7 Polyacrylonitrile.

risks incomplete burning, generating pollutants. Incineration, if properly done, minimizes environmental impact by converting PS into water, carbon dioxide, and heat, with potential energy recovery. Recycling efforts, including compaction and upcycling methods, offer promise for mitigating its impact, yet effective waste management remains a persistent challenge. Recycled PS is valuable for producing recycled plastic pellets and is repurposed into various products, from insulation sheets to park benches. Kaseem et al. [23] presented a detailed review of PS/carbon nanotube (PS/CNT) composites. The examination of various properties in this research led to the conclusion that incorporating CNT would enhance the material performance of PS composites, rendering them advantageous for industrial applications.

3.4.4 Polyacrylonitrile (PAN)

Polyacrylonitrile (PAN) is a synthetic, semi-crystalline organic polymer resin with a density of 1.184 g/cm^3, insoluble in water but soluble in solvents like dimethyl formamide and dimethyl sulphoxide. It boasts remarkable heat resistance, enduring temperatures up to 200°C while maintaining good mechanical properties. Figure 3.7 shows the structure of PAN. In appearance, PAN is a white solid that doesn't melt under typical conditions, instead degrading before reaching a melting point; however, at heating rates exceeding 50°/min, it melts above 300°C. PAN finds diverse applications, from ultrafiltration membranes to textiles, owing to its versatility. Notably, PAN fibres serve as the precursor to high-quality carbon fibre, formed through thermal oxidation and carbonization processes. These carbon fibres are integral components in numerous high-tech and everyday products, including aircraft structures, missiles, sporting equipment, and more. The popularity of PAN arises from its capacity for high carbon yield (up to 56%), its adaptability in customizing final carbon nanofibre products, and its ease of stabilization facilitated by ladder structure formation during nitrile polymerization [24,25]. Studies have explored PAN-based nanofibres and composites, demonstrating potential applications in batteries [26,27], electromagnetic wave absorption [28], and flame-retardant materials [29,30]. An excellent review of PAN-based nanofibres was reported by Nataraj et al. [31]. Pirlot, Christophe et al. [32] performed mechanical and thermal characterization of CNT/PAN composites.

3.4.5 Polymethyl Methacrylate (PMMA)

Polymethyl methacrylate (PMMA) is a transparent synthetic polymer produced from methyl methacrylate via radical polymerization. It was initially uncovered in the 1930s by British chemists Rowland Hill and John Crawford and later utilized by German chemist Otto Rohm in 1934 [33]. The structure of PMMA is depicted in Figure 3.8. It serves as a versatile alternative to inorganic glass due to its high impact strength, lightweight nature, shatter resistance, and favourable processing characteristics [34]. Commonly referred to as acrylic or acrylic glass, PMMA is also popular with various trade names like Plexiglas, Crylux, Lucite, Hesalite, Acrylite, and Perspex. With a density

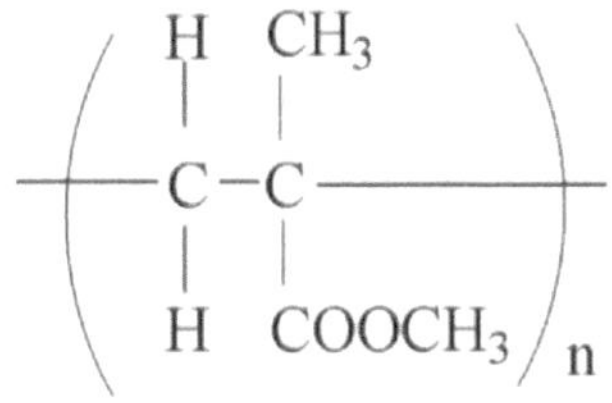

FIGURE 3.8 Polymethyl methacrylate.

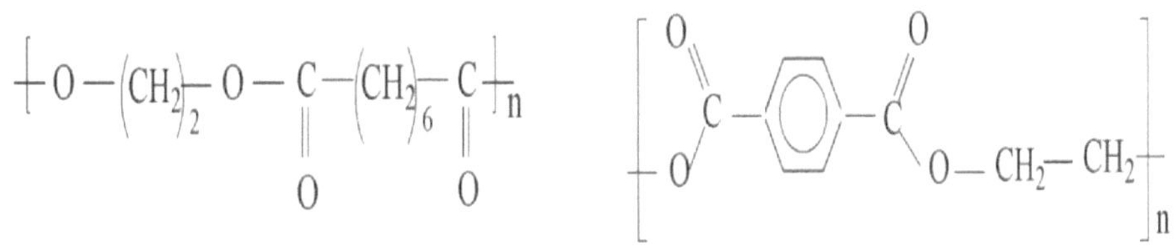

Melting point = 65°C

Aliphatic Polyester

Melting point = 265°C

Polyethylene terephthalate (PETP)

FIGURE 3.9 Polyesters.

ranging from 1.17 to 1.20 g/cm^3, PMMA undergoes ignition at 460°C and combustion produces carbon monoxide, carbon dioxide, water, and low-molecular-weight compounds, including formaldehyde. It displays a maximum water absorption ratio of 0.3–0.4% by weight and exhibits a relatively high coefficient of thermal expansion (CTE) of (5–10) $\times 10^{-5}$°C^{-1}. While resistant to many chemicals, PMMA dissolves in organic solvents such as chlorinated hydrocarbons, ketones, and esters and can be thermally depolymerized to recover the monomer. Although PMMA serves as an excellent glass substitute in many applications, it generally has poorer scratch resistance than glass. PMMA shows promise in various fields, including optics, pneumatic actuation, sensors, analytical separations, and conductive devices [35–37]. It can also be used in biomedical fields, polymer electrolytes, viscosity modification, and drug delivery systems utilizing electro-diffusion or electro-osmotic flow [38–40]. Numerous studies have explored the characterization and applications of PMMA-based composites and nanocomposites [41–49].

3.4.6 Polyesters

Polyester belongs to the category of polymers where the ester functional group is present in each repeating unit of the main chain. They are synthesized through a polycondensation reaction involving a dicarboxylic acid and diols. When both rings are aliphatic, the resulting polyester is fully aliphatic and historically held low industrial significance due to its low melting points (50–80°C). This limitation was addressed by incorporating aromatic rings into the polyester chain, thereby reinforcing the polymer backbone. Consequently, polyethylene terephthalate (PETP), a notable polyester, boasts a high melting point and ranks among the most widely used polymers, marketed as Terylene or Terene. PETP, melting around 265°C, exhibits resistance to heat and moisture, along with chemical durability. It finds application in textile fibres and wrinkle-free garments, offering robust mechanical strength up to 175°C. PETP is also utilized in films for magnetic recording tapes and aluminized sheets for specific purposes. Polyesters yield either thermoplastic or thermosetting products, distinguished by their chemical structures. Figure 3.9 shows the structure of polyesters.

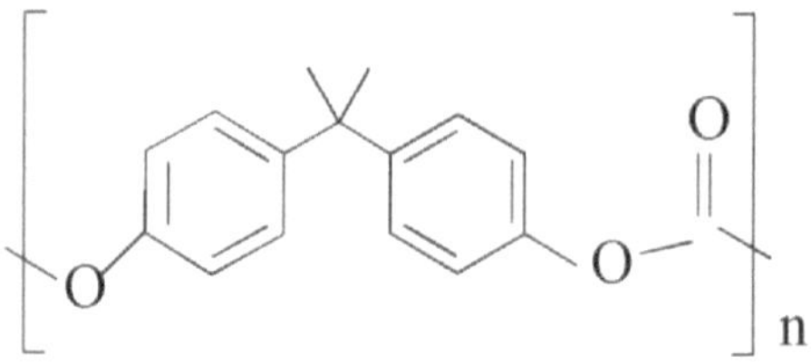

FIGURE 3.10 Polycarbonates.

Thermoplastic polyester, exemplified by PETP, serves as a synthetic fibres and plastic component. Conversely, unsaturated polyesters (aliphatic unsaturated) undergo cross-linking reactions between diacids and multifunctional alcohols, such as glycerin, forming a three-dimensional network that remains solid upon curing and heating. Typically used as a continuous matrix phase, these polyesters are reinforced with fibres and particulate materials, with reports extensively documenting their combination with cellulosic fibres like jute [50,51], sisal [52], coir [53,54], banana-cotton [55], straw [56], pineapple leaf [57], cotton-kapok [58], flax [59], and Kenaf-glass hybrids [60]. Notably, Zaghloul et al. [61] have provided a comprehensive review of advancements in polyester composite material.

3.4.7 Polycarbonates

Polycarbonate (PC) represents polyesters derived from phenols and carbonic acid, first discovered in 1898. They are synthesized by condensing bisphenol-A (BPA) with diphenyl carbonate. Figure 3.10 shows the structure of PCs. PC possesses a distinctive blend of characteristics: transparency, exceptional resistance to acids, alkalis, and mechanical stress, temperature stability, heat resistance, effective electrical insulation, and biological inertness. With a melting point of around 265°C and remarkable impact strength, it also demonstrates resistance to water and many organic compounds, albeit vulnerable to hydrolysis by alkalies. Various beneficial products such as telephone components, safety goggles, shields, and machinery housings utilize this versatile plastic, which has found favour in 3D printing, particularly in professional settings, despite inherent challenges. However, the disposal of PC raises environmental concerns. Studies indicate that at temperatures surpassing 70°C and high humidity, PC undergoes hydrolysis, releasing BPA. Leaching of BPA can occur even at ambient temperatures and normal pH levels, posing environmental risks and contributing to aquatic pollution, especially as PC components age. Exposure to UV light accelerates photo-oxidation, resulting in discolouration and the formation of compounds like ketones and phenols. Moreover, thermal degradation at elevated temperatures yields pollutants, including liquid products containing phenol derivatives categorized as volatile organic compounds (VOCs), which contribute to ground-level ozone formation and smog. Promisingly, fungi such as *Geotrichum candidum* in Belize exhibit the ability to consume PC, suggesting potential bioremediation applications. These degradation pathways underscore the importance of adopting responsible disposal practices for PC products. The reinforcement of PC with various fibres has been extensively explored [62–69] for diverse applications, with a detailed review of PC-based composite materials presented by Kharaev et al. [70].

3.4.8 Polysulphone

Polysulfone (PSU) is an amorphous polymer derived from the condensation of BPA and dichlorodiphenyl sulphone. Technically, PSUs contain an aryl-SO_2-aryl subunit. Renowned for its mechanical strength and high thermal stability across a broad temperature range (–180 to 140°C), PSU is esteemed as a high-performance plastic, prized for its rigidity, strength, and transparency,

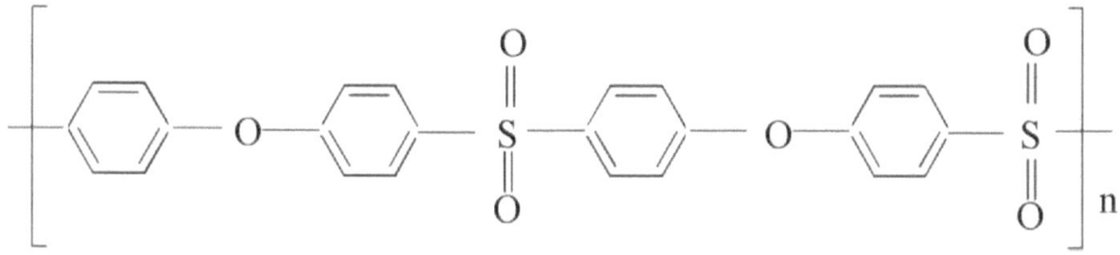

FIGURE 3.11 Polysulfones.

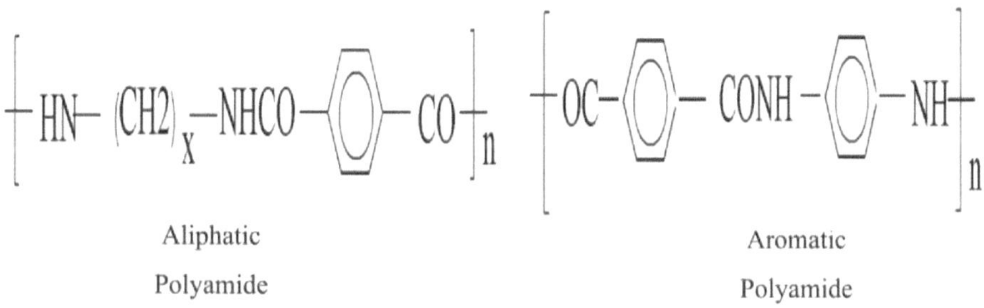

FIGURE 3.12 Polyamides.

with a glass transition temperature ranging from 190 to 230°C [71]. Chemically resilient to mineral acids, alkalis, and oxidizing agents, they exhibit susceptibility to certain organic solvents. PSUs are mechanically robust, offering high compaction resistance, making them ideal for high-pressure applications. PSU finds extensive use in membranes due to its thermal stability and strength [72], commonly applied in bioreactor membranes, fuel cell membranes, food processing equipment, as well as aviation and electronic components [73]. Additionally, PSU-based composites and nanocomposites serve various purposes, including medical applications [74–76], biomaterials [77], ultrafiltration membranes [78], and as denture bases [79]. Figure 3.11 shows the structure of PSUs.

3.4.9 Polyamides

Polyamide (PA) is a polymer characterized by repeating units linked via amide bonds. PAs are found both in nature and through synthetic means. Examples of naturally occurring PAs include proteins like wool and silk. Synthetic PAs, such as nylons, aramids, and sodium polyaspartate, can be produced through step-growth polymerization or solid-phase synthesis. PAs are commonly used in fibres, films, and sheets due to their excellent mechanical and thermal properties [80]. They exist in both aliphatic and aromatic forms [81], with aliphatic PAs commonly known as nylon and aromatic PAs referred to as aramid. Aliphatic PAs are typically prepared via melt polycondensation between dicarboxylic acids and diamines, while aramids are synthesized through self-condensation of aromatic amino acids or polycondensation between aromatic diacid chlorides and aromatic diamines. The crystal structure of PAs comprises α, β, and γ crystalline phases. The structure of PA is illustrated in Figure 3.12. A wide range of aliphatic PAs are available, including PA 6, PA 6,6, PA 6,10, PA 11, PA 12, and PA 10,10. Aliphatic PAs typically have melting points in the range of 250–300°C. PA 6, 6 exhibits good tensile strength, abrasion resistance, and toughness up to 150°C but is susceptible to dissolution by certain solvents such as formic acid, cresols, and phenols. PA 6, 10 finds applications in products like brushes and bearings, while PA 11 is commonly used as a textile fibre and is less effected by water compared to other PAs. Aromatic PAs have high melting points (>500°C), and fibres made from them boast extremely high moduli, making them suitable for applications in FRP. For instance, fibres spun from poly(paraphenylene terephthalamide), commercially known as 'Kevlar' fibres, belong to this category. PAs exhibit good transparency, hardness, strength, modulus, rigidity, thermal resistance, chemical resistance, and moisture absorption properties [82]. The physical properties of PAs can be enhanced by incorporating organic and inorganic nano fillers [83]. Hydrogen bonding interactions occur in the PA backbone between the amide linkage and the oxygen of carbonyl groups [84]. PAs and their composites find numerous technical applications in industries such as automotive, aerospace, construction, textiles, electronics, and more [85–87].

3.4.10 Polyethylene Glycols and Polypropylene Glycols

PE glycols (PEG) are waxy materials available in different molecular weight grades and are commercially known as carbowax. Ethylene oxide or propylene oxide can be polymerized with sodium methoxide or sodium hydroxide to give polymers called PEG and PP glycols (PPG). They are soluble in water and in many organic solvents. PPG shares several properties with PEG and is a liquid at room temperature. However, its solubility in water diminishes as molar mass increases. Notably, the secondary hydroxyl groups in PPG exhibit lower reactivity compared to the primary hydroxyl groups in PEG. The structures of PEG and PPG are depicted in Figure 3.13.

Additionally, PPG is less toxic than PEG, leading to its predominant use in biotechnological application. They find application in pharmaceutical preparations and cosmetics. PEG plays diverse roles across medical, chemical, biological, commercial, and even entertainment domains. PEG has garnered significant attention owing to its favourable characteristics, including nontoxicity, high latent heat capacity, appropriate phase change temperature, congruent melting, low vapour pressure, absence of supercooling, minimal or no volume change during solid–liquid phase transition, and high thermal and chemical stability over extended periods of use [88]. PEG-based composites and nanocomposites find applications across various fields such as industrial heat management, temperature-controlled environments like greenhouses and textiles, heat regulation in electronics, telecommunications, and microprocessor equipment, biomedical applications, as well as active and passive heating or cooling systems in buildings [89–94]. PPG are liquid resins and are used as diols in the preparation of urethane polymers. PPG finds utility as a surfactant, wetting agent, defoamer in drilling processes, corrosion inhibitor, and dispersant in leather finishing.

3.4.11 Polyacrylic Acid

Polyacrylic acid (PAA), commercially known as Carbomer, is a polymer represented by the formula $(CH_2\text{-}CHCO_2H)_n$, derived from acrylic acid. In a neutral pH water solution, PAA becomes an anionic polymer, with many side chains losing protons and gaining a negative charge. It is endowed with magnificent attributes such as nontoxicity, high adhesive strength, and recyclability [95]. Figure 3.14 shows the structure of PAA. Primarily, PAA is widely used in dispersants, with its molecular weight influencing rheological properties. Moreover, PAA serves to stabilize paints and cosmetics, prevent emulsion separation, and maintain consistency. Other applications of PAA are PAA-derived hydrogels for wound healing, as a thickening agent for adhesives, as a deflocculant in

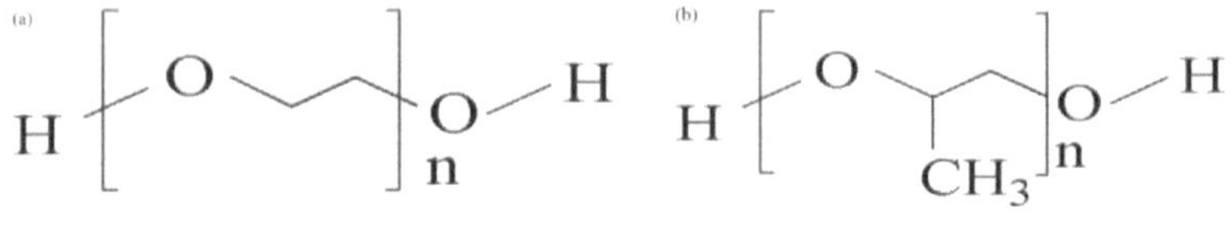

FIGURE 3.13 (a) Polyethylene glycols and (b) polypropylene glycols.

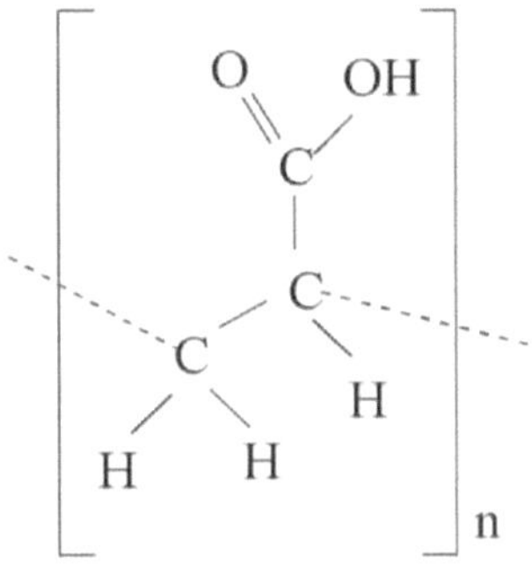

FIGURE 3.14 Polyacrylic acid.

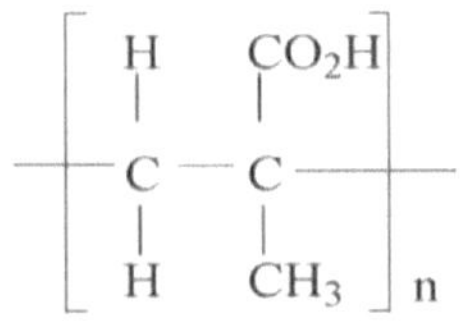

FIGURE 3.15 Polymethacrylic acid.

the oil drilling industry, and for metal quenching in metalworking processes. PPA-based composites and nanocomposites can be used in many applications, such as drug delivery, tissue engineering (TE), biosensors, soil ameliorant, thickener in textile printing, superabsorbent, and wastewater treatment in terms of heavy metal capture and removal of dyes [96–101].

3.4.12 Polymethacrylic Acid

Polymethacrylic acid (PMAA) is insoluble in its monomer but readily soluble in water. It has a limited scope as plastic. PMAA behaves as a polyelectrolyte in its solution state. PMAA-based nanocomposites exhibit enhanced thermal, optical, mechanical, and electrochemical properties compared to the conventional composites [102]. PMAA has been widely used for biomedical applications [103–107]. The structure of PMAA is shown in Figure 3.15.

3.4.13 Polyvinyl Acetate

Polyvinyl acetate (PVAc) is an aliphatic, rubbery synthetic polymer belonging to the polyvinyl ester family. Its ester groups are susceptible to base hydrolysis, slowly transforming PVAc into PVA and acetic acid. Figure 3.16 shows the structure of PVAc. PVAc is widely available and recognized for its effectiveness in bonding various materials. The glass transition temperature of PVAc falls between 30 and 45°C, depending on the molecular weight. PVAc is a colourless, usually nontoxic and biodegradable polymer [108]. It can be broken down by various microorganisms, with filamentous fungi being the most common culprits, alongside yeasts, algae, bacteria, and lichens. PVAc, typically used as an emulsion in water, serves as carpenter's glue, PVA glue, school glue, white glue, or Elmer's glue. It is a commonly used adhesive for porous materials such as paper, wood, and cloth. PVAc-based composites and nanocomposite find applications in various fields such as biomaterials, membranes for gas separation, separators for batteries, and structural applications [109–112]. It also plays a significant role in industrial coatings, as a binder in nonwovens, and in the production of polymers like polyvinyl acetate phthalate (PVAP) and PVA.

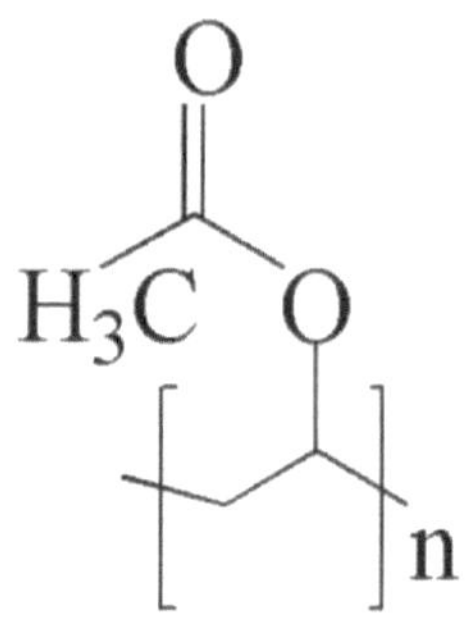

FIGURE 3.16 Polyvinyl acetate.

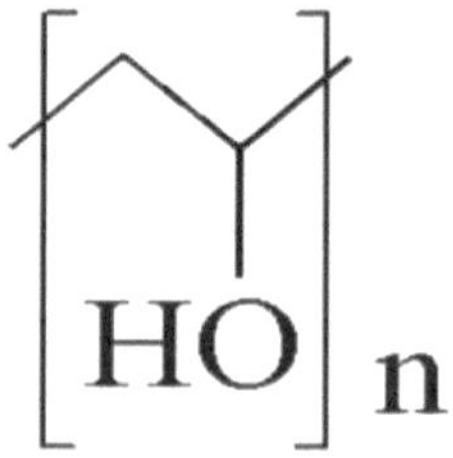

FIGURE 3.17 Polyvinyl alcohol.

3.4.14 Polyvinyl Alcohol

PVA was the inaugural synthetic colloid produced by Herrmann and Haehnel in 1924. PVA is produced through the polymerization of vinyl acetate monomer to form PVAc, which is then hydrolyzed to create PVA [113]. Figure 3.17 illustrates the structure of PVA. PVA's exceptional chemical resistance and physical qualities have led to its wide array of applications in various industries, including textile sizing, adhesives, protective colloids for emulsion polymerization, the manufacturing of PVA fibres and polyvinyl butyral, and paper sizing [114]. It is applicable in the medical and food sectors as well [115]. PVA-based composites, nanocomposites, and films find applications in various fields such as biomaterials, membranes, medical, and structural applications. Some of the work presenting a detailed discussion on these aspects can be found in open literature [5,116–121]. PVA is integral to photographic film production and is extensively employed in additive manufacturing, particularly in 3D-printed oral dosage forms in the pharmaceutical industry. Medically, PVA-based microparticles are FDA-approved for embolization in treating peripheral hypervascular tumours and uterine fibroid embolectomy (UFE).

3.4.15 Polyvinyl Chloride

PVC is the world's third most widely produced polymer produced through vinyl chloride polymerization, with an annual production of approximately 40 million tonnes, owing to its low cost, high mechanical strength, good chemical resistance, and good fire retardancy [122]. Organizations like VinylPlus focus on recycling PVC, contributing to a circular economy. The structure of PVC acid is depicted in Figure 3.18. The properties of PVC vary between rigid and flexible forms. Rigid PVC has a density of 1.3–1.45 g/cm^3, thermal conductivity of 0.14–0.28 W/(m·K), yield strength of 31–60 MPa, Young's modulus of 3.4 GPa, flexural strength of 72 MPa, compression strength of 66 MPa, and a linear CTE of 5×10^{-5} mm/(mm °C). In contrast, flexible PVC has a density of 1.1–1.35 g/cm^3, thermal conductivity of 0.14–0.17 W/(m·K), and a yield strength of 10.0–24.8 MPa. The

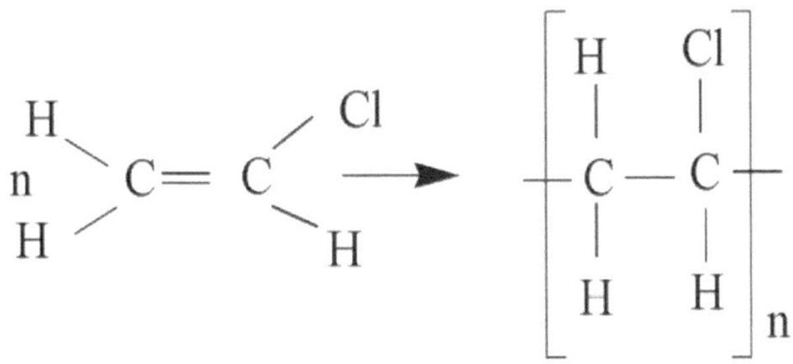

FIGURE 3.18 Polyvinyl chloride.

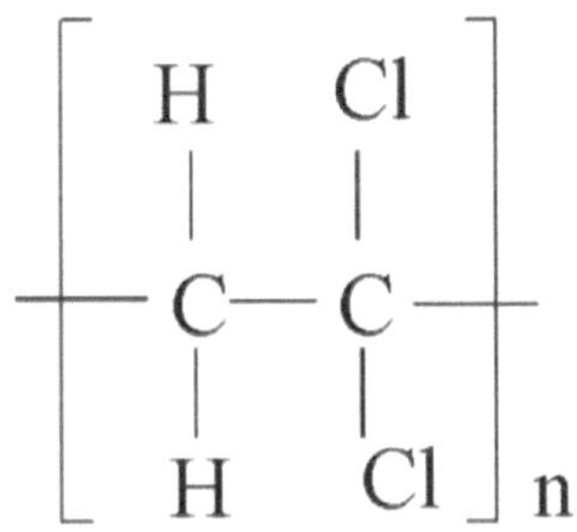

FIGURE 3.19 Polyvinylidene chloride.

resistivity and surface resistivity for rigid PVC are 10^{16} Ω m and 10^{13}–10^{14} Ω, respectively, while for flexible PVC, resistivity ranges from 10^{12} to 10^{15} Ω m, and surface resistivity is 10^{11}–10^{12} Ω. Its main applications include electric wires, pipes, window profiles, and siding. PVC-based composites, nanocomposites, and films find applications in various fields such as biomaterials, membranes, and structural applications. Some of the work presenting detailed discussion on these aspects can be found in the open literature [123–131]. In construction, it's a preferred material for pipes, vinyl siding, and low-maintenance building components. In the electric cable industry, PVC provides a fire-retardant coating. In healthcare, PVC is crucial for single-use medical items like tubing and containers. It's also employed in clothing for its water-resistant properties. Additionally, PVC is utilized in creating records, musical instruments, and automotive underbody coating.

3.4.16 Polyvinylidene Chloride

Polyvinylidene chloride (PVDC) is a homopolymer of vinylidene chloride. PVDC is prepared by radical polymerization using the emulsion technique. It is insoluble in many solvents. Its main use is in making fibres. PVDC boasts exceptional resistance to oxygen, water, and aromas, coupled with superior chemical resistance to acids and alkalis. Figure 3.19 shows the structure of polyvinylidene chloride. It remains insoluble in oil and organic solvents, possesses low moisture absorption, and is impermeable to mould, bacteria, and insects. However, it dissolves in polar solvents. When subjected to temperatures exceeding 125°C, it decomposes, yielding hydrochloric acid (HCl). Despite being highly useful in food packaging, its downside lies in thermally induced dehydrochlorination near processing temperatures. This degradation causes colour changes, leading to product loss and increased production costs. PVDC finds extensive application in packaging, serving as a water-based coating for films like polyethylene terephthalate (PET) and biaxially oriented PP (BOPP). This coating enhances the film's barrier properties, minimizing oxygen and flavour permeability, thereby prolonging the shelf life of packaged food. It also provides a glossy finish, adding aesthetic appeal and scuff resistance to printed surfaces.

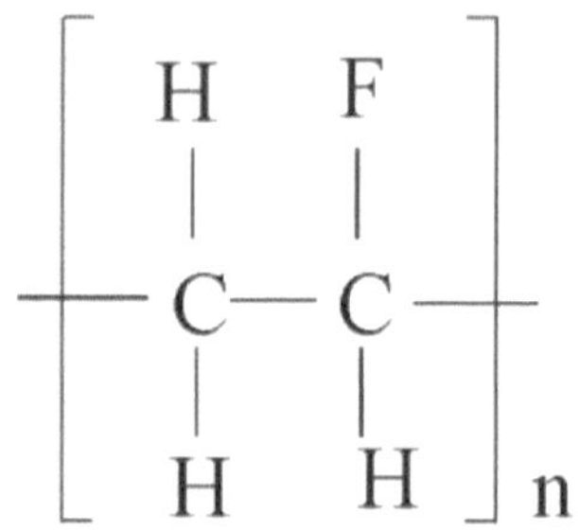

FIGURE 3.20 Polyvinyl fluoride.

3.4.17 Polyvinyl Fluoride

Polyvinyl fluoride (PVF), structurally akin to PVC, possesses repeating vinyl fluoride units. PVF has the following structure, as shown in Figure 3.20, and vinyl fluoride is obtained from acetylene. PVF is a semi-crystalline, transparent-to-opaque thermoplastic fluoropolymer primarily employed in flame-retardant coatings for aircraft interiors and as back sheets for photovoltaic modules with a melting point of around 200°C. The crystallinity, ranging from 20 to 60%, is influenced by polymerization conditions. Films made from PVF exhibit several basic characteristics: they offer resistance to UV radiation, weathering, a broad range of chemicals, solvents, and staining agents; they possess favourable mechanical properties, strength, and durability; they demonstrate good release and dielectric properties; they display piezoelectric and pyroelectric behaviour; and they serve as effective moisture barriers [132]. The monomer is a gas at ambient conditions (B.P = –27°C), and hence, free-radical polymerization using benzoyl peroxide is carried out under pressure at 85°C. PVF has excellent outdoor life, is highly resistant to chemicals, maintains good structural integrity even above 150°C, and has good thermal stability. Gas permeability is also sufficiently low. For durability, even in extremely small thicknesses, it is far better than conventionally used materials such as paints and other surface coatings. Transition phases occur below the melting point, notably at lower T_g (–15 to –20°C) and upper T_g (40–50°C). PVF is insoluble in common solvents below 100°C, becoming soluble in polar solvents at elevated temperatures. At room temperature, PVF films resist acids, bases, aliphatic, aromatic, and alcohol liquids. PVF films find application as surface-protecting laminates in the aircraft and architectural industries, enhancing chemical and UV resistance while facilitating easy cleaning. Additionally, they serve as release films in the production of fibre-reinforced composites and for transfer printing purposes.

3.4.18 Polyvinylidene Fluoride

Polyvinylidene fluoride (PVDF) is a non-linear, semi-crystalline polymer that primarily consists of the vinylidene fluoride homopolymer or copolymers of vinylidene fluoride blended with other fluorine-containing vinyl monomers [133]. Figure 3.21 shows the structure of PVDF. PVDF is a specialized, non-reactive fluoropolymer formed through vinylidene difluoride polymerization. Recognized for its high purity, nontoxicity, resistance to solvents, acids, and hydrocarbons with a low density (1.78 g/cm^3) and melting point (170°C) compared to similar fluoropolymers. It is sensitive to bases, esters, and ketones. It does not undergo creep or distortion over a wide temperature range. It has a glass transition temperature of –35°C, 50–60% crystallinity, and exists in alpha, beta, and gamma phases. PVDF is thermally stable up to 375°C. Intrinsic properties include resistance to ozone, radiation, UV damage, and microbiological elements. PVDF has gained recognition as an outstanding piezoelectric material, owing to its unique and versatile display of piezoelectric and dielectric properties. Moreover, PVDF demonstrates remarkable thermal stability and mechanical

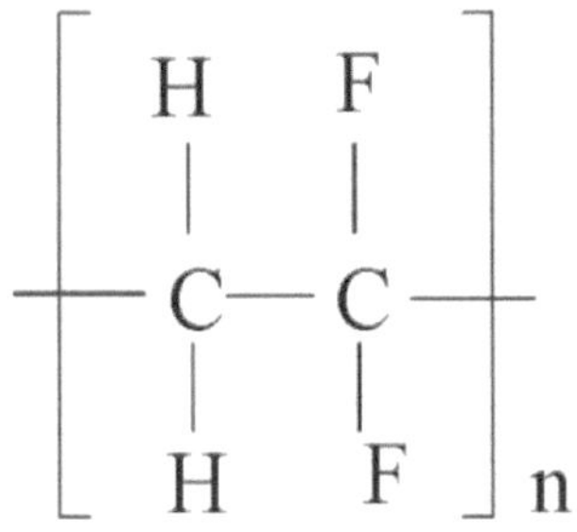

FIGURE 3.21 Polyvinylidene fluoride.

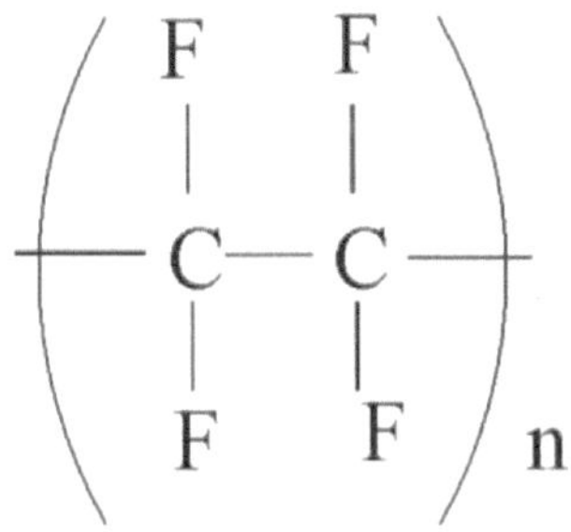

FIGURE 3.22 Polytetrafluoroethylene.

strength, along with flexible processing and low cost compared to other piezoelectric crystals and ceramics [134,135]. PVDF-based composites exhibit exceptional performance in terms of energy storage, being flexible, environmentally friendly, lightweight, and biocompatible, possessing a high dielectric constant, and high electrical strength [136,137]. They are widely used in energy harvesters [138], capacitors [139], nanogenerators [140], sensors [134], and biomedical applications [141]. These composites open up new opportunities for enhancing people's lives and bring significant overall benefits. Numerous researchers are actively investigating and exploring the mechanisms behind its energy storage and enhancement [142,143].

3.4.19 Polytetrafluoroethylene

Polytetrafluoroethylene (PTFE), commonly known as Teflon, is synthesized through the free-radical polymerization of tetrafluoroethylene (TFE). Figure 3.22 shows the structure of PTFE. This material boasts an array of exceptional properties, rendering it highly versatile across various applications. With a glass transition temperature of 114.85°C, PTFE maintains stability even at elevated temperatures. Its low thermal expansion coefficient ($112–125\times10^{-6}$ K^{-1}) and thermal diffusivity (0.124 mm^2/s) contribute to its resilience against temperature-induced deformations. Despite its flexibility, PTFE possesses a notable Young's modulus of 0.5 GPa, ensuring structural integrity. Additionally, its yield strength of 23 MPa enables it to withstand significant mechanical stress. PTFE's coefficient of friction ranges from 0.05 to 0.10, making it ideal for low-friction applications. Furthermore, it exhibits excellent dielectric properties, with a dielectric constant (ε) of 2.1. PTFE experiences decomposition at approximately 260°C, releasing fluorocarbon gases and sublimates like TFE and difluorocarbene radicals (RCF2) beyond 350°C. Widely used in the aerospace, military, medical, electronic, and pharmaceutical industries, PTFE is valued for its anti-adhesive, anticorrosive, chemically inert, thermally stable, and self-lubricating characteristics [144,145]. PTFE-based composites offer multifunctional capabilities, serving as platforms for wearable electronics, health monitoring

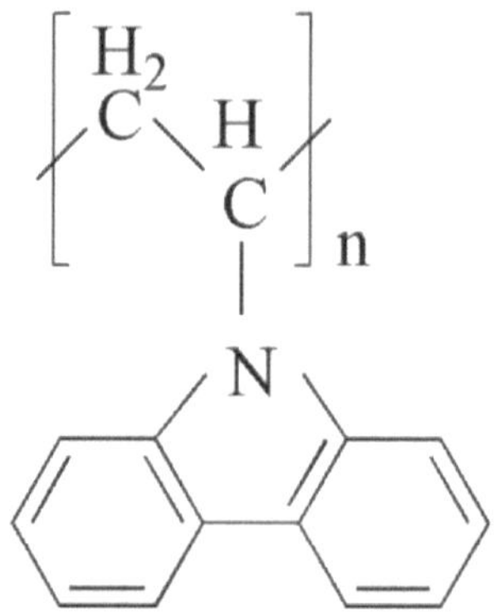

FIGURE 3.23 Polyvinyl carbazole.

structures, radio frequency identification devices [146], seals [147], smart materials [148], and high-frequency communication applications [149].

3.4.20 Polyvinyl Carbazole

Polyvinyl carbazole (PVK) is a thermoplastic polymer synthesized through the radical polymerization of N-vinylcarbazole. It serves as a blue light-emitting and hole transport organic semiconductor, boasting remarkable physicochemical properties like photoconductivity, photorefractivity, and a high refractive index. PVK exhibits robustness and temperature resistance, enduring temperatures ranging from 160 to 170°C. The structure of PVK is depicted in Figure 3.23. Its temperature resilience can be enhanced by blending it with other plastics, notably PVC. Soluble in aromatic hydrocarbons, halogenated hydrocarbons, and ketones, PVK demonstrates resistance to acids, alkalis, polar solvents, and aliphatic hydrocarbons. It finds applications in growing fields such as solar cells, optoelectronic devices, photovoltaic cells, and luminescent devices [150]. PVK-based nanocomposites are utilized in solar cells [151], electronic industries [152], and for electroluminescence purposes [153]. Additionally, PVK plays a significant role in holography, contributing to photorefractive polymer research. Another notable application involves its use in producing heat-resistant copolymers when combined with styrene.

3.4.21 Polyvinyl Pyrrolidone

Polyvinylpyrrolidone (PVP) is synthesized through the polymerization of the monomer N-vinylpyrrolidone, which is derived from acetylene chemistry. Available in a range of molecular weights and associated viscosities, PVP offers flexibility, allowing selection based on specific application requirements. Figure 3.24 shows the structure of Polyvinyl pyrrolidone. It exhibits solubility in water and polar solvents, including alcohols like methanol and ethanol. In its dry form, PVP appears as a light, flaky, hygroscopic powder capable of absorbing up to 40% of its weight in atmospheric water. In solution, PVP demonstrates excellent wetting properties, facilitating easy film formation, making it valuable as a coating or additive in coating applications. When combined with iodine, PVP forms a disinfectant complex known as povidone-iodine, which is used in various products such as solutions, ointments, and surgical scrubs. PVP is utilized in several drug delivery systems as an excipient because of its solubility and availability in various grades, providing features like film-forming, complexing, solubilizing, binding, stabilizing, suspending, and thickening [154]. Moreover, PVP is important in industries like adhesives, cosmetics, food, and textiles. It has been extensively studied in several biomedical applications, producing notable outcomes [155].

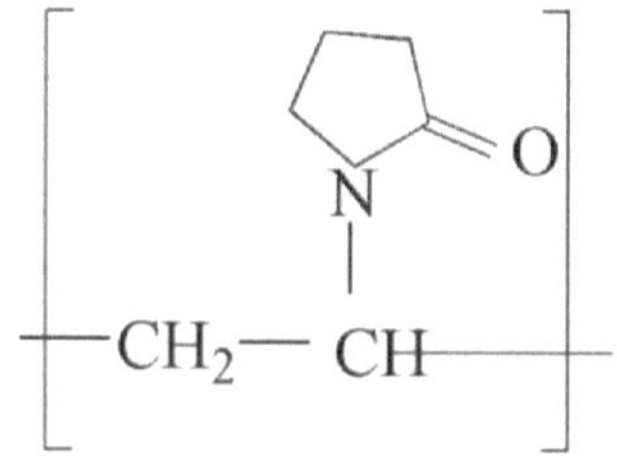

FIGURE 3.24 Polyvinyl pyrrolidone.

TABLE 3.4
Mechanical Properties of Thermoset Polymers

Thermosets	Density (g/cm³)	Tensile Strength (MPa)	Young's Modulus (GPa)	Elongation (%)	Glass Transition Temperature (°C)
Polyimides (PI)	1.37	121	3.72	6	366
Rigid thermoset polyurethane (RPU)	1.2	60.0	2.2	90.0	−150.0 to 0.0
Phenol–formaldehyde (PF)	1.2	45.0	6.5	1.2	170.8
Urea–formaldehyde (UF)	1.5–1.6	65.0	9.0	0.8	130.0
Melamine–formaldehyde (MF)	1.5–1.6	65.0	12.0	0.6	150.0
Epoxy	1.2–1.3	600.0	80.0	1.3	70.0–167
Unsaturated polyester (UPE)	1.1	60.0	3.4	2.0	110.0–180.0

PVP-based nanocomposites find applications in diverse sectors including biomaterials, medical applications, structural applications, and smart materials [156–159].

3.5 THERMOSET POLYMERS

Thermoset polymers undergo irreversible changes when subjected to heat, transforming from a fusible and soluble material into one that is infusible and insoluble. This transformation occurs through the formation of a covalently cross-linked, thermally stable network. In some thermoset polymer, network formation occurs with minimal or no heat required, as seen in the production of urethane foams and the utilization of unsaturated polyester resins. The most commonly employed thermoset resins in commercial applications include phenolic resins, epoxy resins, amino resins, urethane foams, and unsaturated polyester resins. Table 3.4 presents the mechanical properties of some thermoset polymers that are commonly used in composite and smart material applications, followed by a detailed discussion on the same.

3.5.1 Polyimides

Polyimide (PI) belongs to the class of high-performance plastics distinguished by its polymer structure containing imide groups, as shown in Figure 3.25. The synthesis of PIs involves the polycondensation of pyromellitic anhydride and p,p′-diamino diphenyl ether. Renowned for their exceptional thermal stability, remarkable chemical resistance, and robust mechanical properties, PIs are easily identifiable by their distinct orange/yellow colouration. These polymers retain their strength at elevated temperatures (up to 232°C continuously and 704°C intermittently) and display outstanding flame resistance without requiring additional flame retardants. Reinforced variants can

FIGURE 3.25 Polyimide.

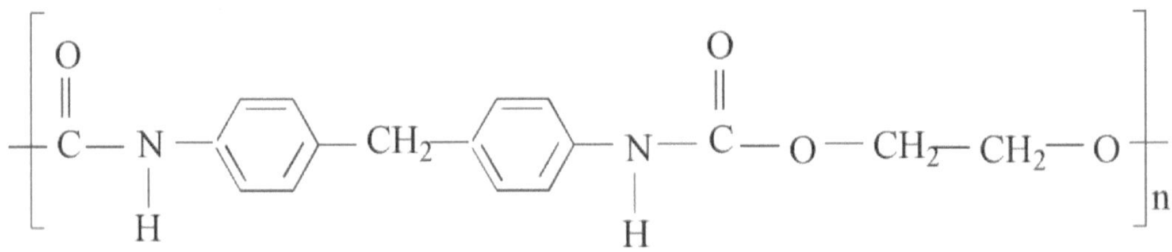

FIGURE 3.26 Polyurethanes.

achieve flexural moduli of 21 GPa and flexural strengths of up to 340 MPa. PIs also demonstrate resistance to a wide range of solvents and oils, including hydrocarbons, ethers, esters, freons, and alcohols. However, they are not suggested for use in environments containing alkalis or inorganic acids. Certain solvent-soluble PIs, such as CP1 and CORIN XLS, are renowned for their optical clarity, making them suitable for applications involving spray and low-temperature cures. PIs find extensive applications across various industries, from microelectronics to aerospace. PI-based composites offer high temperature resistance, superior insulation, flame-retardant properties, and excellent mechanical characteristics [160,161]. Consequently, they are widely employed in aerospace, electronic packaging, wind turbine blades, automobiles [162], thermal and sound insulation [163,164], and as smart materials [165].

3.5.2 Polyurethanes

Polyurethane (PU) is an organic polymer with a melting point around 150°C, formed by linking organic units through urethane molecules. Figure 3.26 shows the structure of PUs. While most PUs are thermosetting polymers, some exhibit thermoplastic properties. The preparation process involves chemical reactions between polyols and di- or tri-isocyanates. It's crucial to control the chain length and functionality of the polyol during production, as this significantly influences the properties of the final product. Longer, flexible polyol chains yield soft, elastic materials, whereas shorter polyols (high cross-linking) result in tough, rigid PUs.

The cross-linked structure forms three-dimensional networks, contributing to the high molecular weights and thermoset nature of PU. PUs, particularly those containing aromatic isocyanates, may experience discolouration upon exposure to visible light, leading to a yellowing or reddish-brown tint. UV radiation can exacerbate this effect. Hydrolysis, especially in humid conditions, can cause degradation. Microbial degradation involves enzymes, with certain fungi capable of slowly biodegrading PU. Polyester-type PUs are more susceptible to biodegradation than polyether-type PUs. PU finds application in various fields, including biomedical, automotive, building, construction,

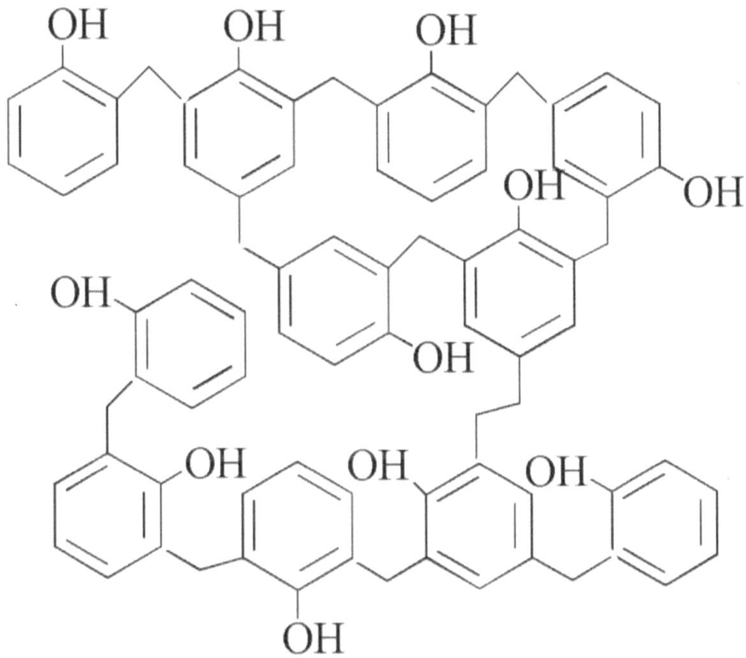

FIGURE 3.27 Phenol–formaldehyde.

textiles, and footwear industries, among others, owing to its combination of elongation, hardness, strength, and modulus [166–173].

3.5.3 Phenol–Formaldehyde

Phenol–formaldehyde (PF) resins are synthesized through the polycondensation reaction between PF, a process that can be catalysed by either acids or bases. Figure 3.27 shows the structure of PF. Acid catalysts yield Novolak or Bakelite resins, while base catalysts produce resol resins. Phenolic resins are prized for their excellent chemical and thermal resistance, dielectric strength, and dimensional stability. Products crafted from these resins exhibit inherent qualities such as low flammability, minimal moisture absorption, and robust creep resistance. Notably, PF resin plays a crucial role in ablative heat shields for atmospheric re-entry spacecraft, effectively absorbing thermal energy through pyrolysis reactions and providing insulation through ablation during re-entry. PF composites and nanocomposites have found applications across diverse sectors, including smart materials, fireproof materials, aerospace industries, and structural adhesives [174–177].

3.5.4 Urea–Formaldehyde

Urea–formaldehyde (UF), also called urea-methanal or amino resins, is a non-transparent thermosetting polymer formed from UF. The structure of UF is shown in Figure 3.28. UF resin has density (1.27–1.29 g/cm^3), viscosity (150–200 cps), and pH (7.5–8.5). UF resin boasts a high flexural modulus, tensile strength, and a notable heat-distortion temperature. It exhibits low water absorption, minimal mould shrinkage, high surface hardness, elongation at break, and impressive volume resistance. Additionally, it features a refractive index of 1.55. It finds broad applications due to its cost-effectiveness, quick reaction time, high bonding strength, moisture resistance, lack of colour,

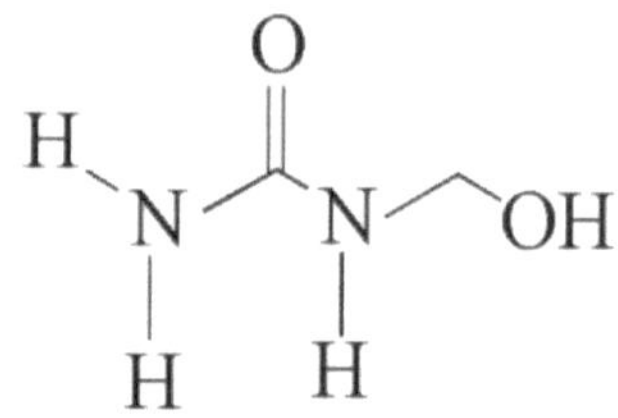

FIGURE 3.28 Urea–formaldehyde.

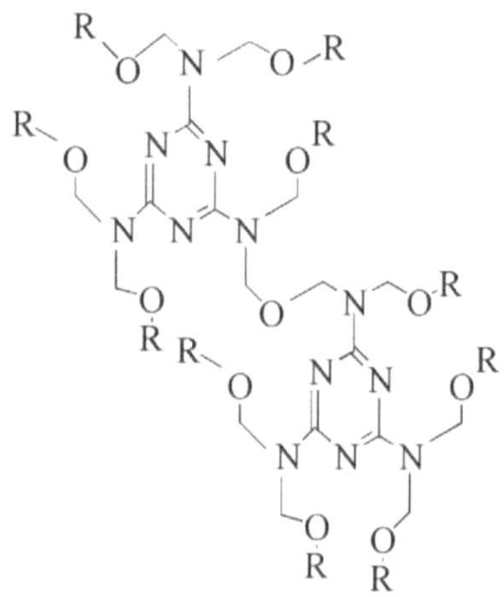

FIGURE 3.29 Melamine–formaldehyde.

and resistance to abrasion and microbes. These resins find applications in adhesives, particle boards, plywood, moulded objects, and medium-density fibreboard (MDF). In agriculture, UF compounds serve as prevalent slow-release fertilizers. Comprising 80% of the globally produced amino resins, UF and related amino resins enhance rubber bonding in automobile tyres, bolster tear strength in paper, and contribute to the moulding of various products like electrical devices and jar caps. UF resin is also used in manufacturing green composites for structural applications by reinforcing different fibres [178–180]. As a smart material, it is used for smart coating for corrosion protection, self-healing, and anticorrosion applications [181,182].

3.5.5 Melamine–Formaldehyde

Melamine–formaldehyde (MF) resin, resulting from the condensation of formaldehyde with melamine, exhibits a thermoset nature with properties influenced by various curing speeds, ranging from very slow to very fast. MF is also known as amino resin. Figure 3.29 shows the structure of MF. Notably, formaldehyde is tightly bound in MF, reducing emissions compared to UF. MF has a density of 1.22–1.24 g/cm^3 and a pH of 9.6–10.5. These resins are colourless and clearer and have higher tensile strength, hardness, heat resistance, and moisture resistance. They are water soluble and hence find use in sizing agents, textile finishing, in paper industry to improve the wet strength of paper, and in plywood industries as adhesives. In its butylated form, dissolved in n-butanol and xylene, this plastic material undergoes dehydration and cross-linking upon heating. The construction sector extensively employs MF resin as the primary constituent of high-pressure laminates like Formica and Arborite, as well as laminate flooring. Commonly found in kitchen utensils and plates, such as Melmac, it is important to note that these utensils are not microwave-safe due to the resin's high dielectric constant. MF resin plays a crucial role in cabinet and furniture making, saturating decorative paper laminated onto particle board, resulting in widely used panels known as melamine. Beyond this, it has applications in carbon capture, where it, along with formaldehyde, cyanuric acid, and DETA, binds CO_2. MF resins are also used for composite manufacturing by reinforcing various fibres for structural applications [183,184]. MF-based nanocomposites are used as coating materials [185]. MFs are also used for smart material application like forms of advanced functional design [186], radiative cooling [187], and thermal energy storage [188].

3.5.6 Epoxy Polymers

The history of epoxy resins involves their discovery in the 1930s, and the chemistry behind their production primarily relies on the reaction of compounds with acidic hydroxy groups and epichlorohydrin. The most common epoxy resins are BPA-based, resulting from the reaction of epichlorohydrin with BPA. Other types include novolaks, aliphatic, halogenated, and glycidylamine epoxy resins, each with specific properties and applications. Epoxy resins, alternatively referred to as polyepoxides, represent a category of reactive prepolymers and polymers incorporating epoxide groups (commonly known as epoxy) within their structure. These resins undergo cross-linking through catalytic homopolymerization or with various co-reactants, such as polyfunctional amines, phenols, acids, alcohols, or thiols, referred to as hardeners. Figure 3.30 shows the structure of the epoxy polymer.

The resulting thermoset polymer exhibits favourable mechanical properties and high chemical and thermal resistance. Curing is an important step that happens when epoxy reacts with different hardeners, such as amines, anhydrides, phenols, and thiols. This creates cross-linked thermoset structures that are three-dimensional. The curing process may require elevated temperatures, and accelerators or latent hardeners can be used to control cure speed. Raw materials for epoxy resin production are predominantly petroleum-derived, though there is a growing interest in plant-derived sources, such as glycerol used to make epichlorohydrin, a key component in epoxy resin manufacturing. This shift towards more sustainable and bio-based sources aligns with the broader industry trend towards environmentally friendly practices. Here's a list of some commonly used epoxy resin systems, along with their associated hardeners: LY556 & HY951, LY5052 & HY5052, Araldite 2011 & HV 2011, Araldite 2021 & HV 2021, Araldite 420 A/B & Aradur 420, EPON 828 & Versamid 125, etc. Epoxy resins are used across various fields such as metal coatings, composites, electrical components, electronics, high-tension electrical insulators, the manufacturing of paintbrushes, fibre-reinforced plastic materials, and structural adhesives. Epoxy resins are widely used in composite industries for the production of FRP composites due to their strength and thermal

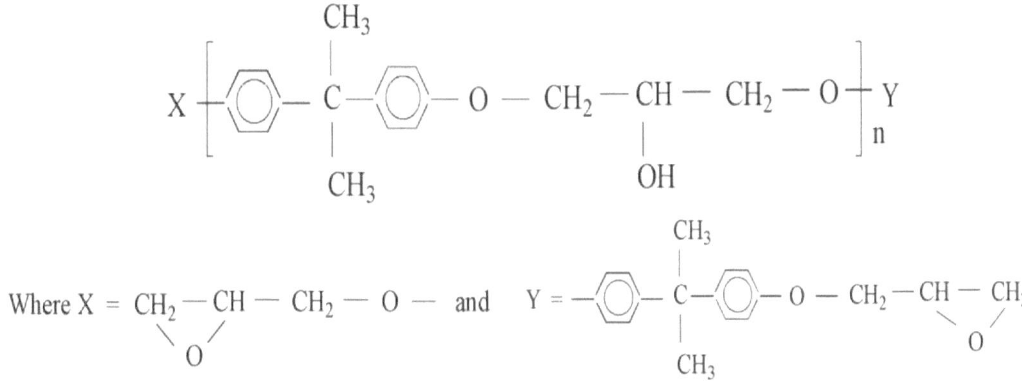

FIGURE 3.30 Epoxy.

stability. Some of common epoxy composites reinforced with both synthetic and natural fibres are jute/epoxy, flax/epoxy, hemp/epoxy, glass/epoxy, Kevlar/epoxy and carbon/epoxy composites [189–194]. Health risks associated with epoxy resins include allergic reactions and respiratory problems during exposure to vapours and dust.

3.6 ELASTOMERS

Elastomers are a class of polymers characterized by their viscosity and elasticity, collectively referred to as viscoelasticity. These materials are held together by weak intermolecular forces, resulting in a low Young's modulus and high failure strain. One defining property of elastomers is their ability to revert to their original shape and size after being significantly stretched. Elastomers exhibit various properties, including temperature sensitivity, low-temperature flexibility, hardness, ageing resistance, colour differentiation, and elongation at break. These materials find extensive applications across industries such as automotive, consumer products, construction, industrial manufacturing, wire and cable insulation, and medical devices, owing to their elasticity, flexibility, insolubility, and other favourable characteristics. Some of the elastomers commonly used for composite and smart material applications are discussed in detail.

3.6.1 Polyureas

Polyurea, initially developed in the 1980s for automotive purposes, later found widespread use in coatings for large-scale projects due to its high reactivity. In 2014, a breakthrough occurred with the introduction of self-healing polyurea elastomers capable of seamlessly melding together after being cut. This advancement holds great promise for elastic, self-repairing paints and coatings. The structure of polyuria is depicted in Figure 3.31. Polyurea is characterized by the presence of urea bonds (–NHCONH–) and is synthesized through the reaction between an isocyanate component and an amine component, with a melting point around 270°C. The resin blend usually has amine-terminated polymer resins and/or amine-terminated chain extenders in it, along with any hydroxyls that are left over from incomplete conversion. It finds diverse applications in manhole and tunnel coatings, secondary containment, truck bed liners, and tank liners, offering exceptional strength and impact resistance. Some polyureas exhibit impressive tensile strengths of up to 40 MPa and elongation exceeding 500%. Polyurea has become a preferred choice for narrowboats, supplanting traditional bitumen coatings and boasting a lifespan of up to 25–30 years. This polymer offers rapid curing, lightweight construction, excellent resistance to breakage and corrosion and is particularly well-suited for anti-shock applications. Polyurea composites, where polyurea serves as a matrix combined with other materials such as CNTs and various types of fibres (particularly glass and carbon fibres), are utilized as reinforcements and fillers due to their exceptional properties, versatility, and cost-effectiveness in the construction sector [195–199].

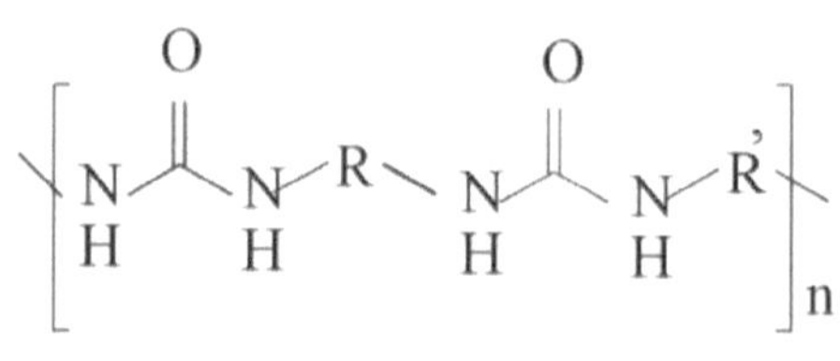

FIGURE 3.31 Polyureas.

3.6.2 Polyisoprenes

Polyisoprene, the first commercially produced elastomer worldwide, arises from the polymerization of isoprene molecules and can manifest in various forms, including cis-1, 4-polyisoprene and trans-polyisoprene. Figure 3.32 shows the structure of polyisoprenes. The trans-polyisoprene variant, historically known as gutta-percha, served as an electrical insulator and was used in producing golf balls. Natural rubber, primarily in the cis-1,4-polyisoprene form, is extracted from the latex sap of specific trees. This natural rubber, obtained from rubber trees and processed, exhibits remarkable elasticity and is applied in tyres, conveyor belts, adhesives, and various moulded rubber products. Synthetic polyisoprene, fabricated through industrial polymerization to replicate natural rubber's structure and elasticity, is similarly employed in applications ranging from adhesives to industrial rubber goods. Polyisoprene compounds, whether derived naturally (NR) or synthetically (IR), are cost-effective and possess a blend of favourable physical properties, including a low dynamic response, minimal hysteresis, and exceptional durability. They are extensively used, particularly in transportation sectors such as truck tyres, anti-vibration dampers [200], and various other diverse applications. Furthermore, its biocompatibility and resistance to various chemicals render polyisoprene suitable for medical devices like gloves, catheters, and seals. Polyisoprene-based composites and nanocomposites hold potential for a wide range of innovative applications, including electronics, marine, automotive, and smart materials [201–204].

3.6.3 Polybutadienes

Polybutadiene, classified as a polyolefin, originates from the polymerization of butadiene, a hydrocarbon monomer with the chemical formula C_4H_6 and its structure is depicted in Figure 3.33. Polybutadiene comes in different forms based on how its polymer chains are arranged. The main forms are cis-1,4-polybutadiene and trans-1,4-polybutadiene, each showcasing distinct properties. Cis-1,4-polybutadiene is known for its flexibility and elasticity, making it a crucial component in tyre manufacturing. Its high resilience enhances tread durability and overall tyre performance. In contrast, trans-1,4-polybutadiene possesses a more rigid structure, making it suitable for applications requiring increased hardness and toughness, such as certain industrial products and plastics production. Polybutadiene is a versatile polymer prized for its excellent abrasion resistance

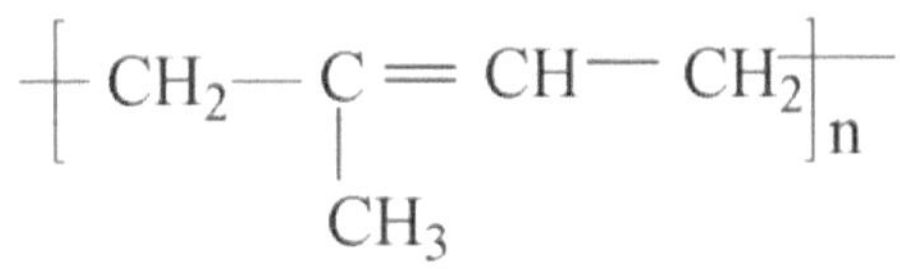

FIGURE 3.32 Polyisoprenes.

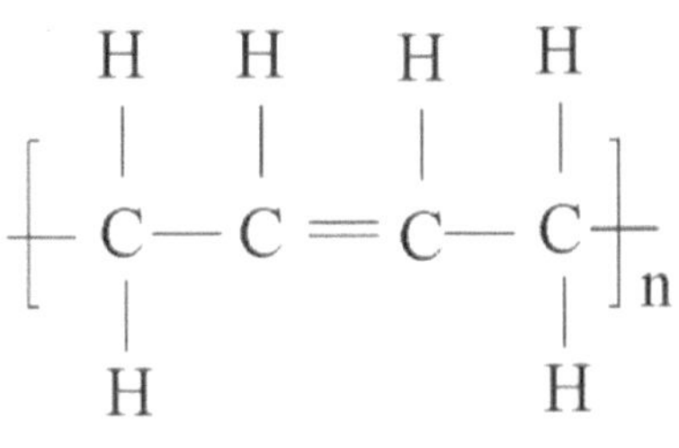

FIGURE 3.33 Polybutadiene.

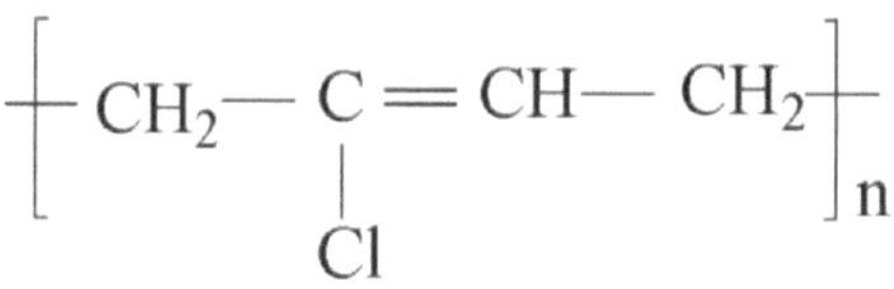

FIGURE 3.34 Polychloroprene.

and flexibility at low temperatures. Its wide-ranging applications encompass not only tyre manufacturing but also the production of hoses, belts, seals, gaskets, footwear, and various moulded and extruded rubber products. Polybutadiene-based composites, nanocomposites, and membranes hold potential for diverse innovative applications, spanning aerospace and automotive industries, textile industries, smart materials, and membrane utilization in various energy conversion systems [205–210].

3.6.4 Polychloroprene

Polychloroprene, commonly known as neoprene, is a synthetic rubber formed through the polymerization of chloroprene. The polymerization process typically employs free-radical methods, utilizing initiators like potassium persulphate. To enhance the material's properties, cross-linking agents such as bifunctional nucleophiles, thioureas, and metal oxides (e.g., zinc oxide) are often incorporated. Figure 3.34 shows the structure of polychloroprene. Neoprene is renowned for its exceptional chemical stability and flexibility across a wide range of temperatures. It is available in both closed-cell and open-cell forms, with the closed-cell variant being waterproof and less compressible. Its versatile applications span various industries, including but not limited to laptop sleeves, medical gloves, orthopaedic braces, electrical insulation, flashings, automotive fan belts, and liquid and sheet-applied elastomeric membranes. Neoprene-based composites, nanocomposites, and membranes offer innovative solutions across diverse sectors such as automotive and belt industries, textile industries, electronic and electrical industries, smart materials, medical applications, and dipped goods applications [211–216].

3.6.5 Silicone Polymer

Polymers of silicone are produced from condensation. The Si–Si linkage is thermally unstable in the case of large molecules, and hence, silicone forms polymers mainly through silicone bonds represented as –Si–O–Si–. The structure of the silicone polymer is depicted in Figure 3.35. The starting materials for producing silicone polymers are alkyl chlorosilanes, aryl chlorosilanes, or substituted esters of orthosilicic acid. In such compounds, there are groups that are easily hydrolysable to form hydroxyl groups. These compounds, when treated with water, form hydroxyl compounds called Silanols and produced by condensation reaction. These polymers are thermally very stable. These polymers are available in liquid, waxy, and rubbery states, serving various purposes. They are extensively employed in surface coatings, where attributes like water repellency and chemical resistance are paramount. Silicone polymers are also utilized in the production of laminates capable of withstanding high temperatures without degradation. Lightweight foams are produced from silicone resins using foaming agents, finding application in aircraft and missile construction. Silica, in fine powder form, is used as a filler to enhance the strength of silicone rubbers, enabling them to maintain flexibility even at extreme temperatures ranging from –90 to +150°C. Silicone polymer is used in additive manufacturing and coating for fire protection [217,218]. Silicone-based composites reinforced with sisal fibre, glass fibre, carbon fibre [219–222] have been reported in literature for structural applications.

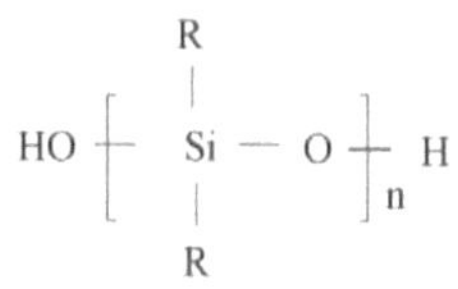

FIGURE 3.35 Silicone polymer.

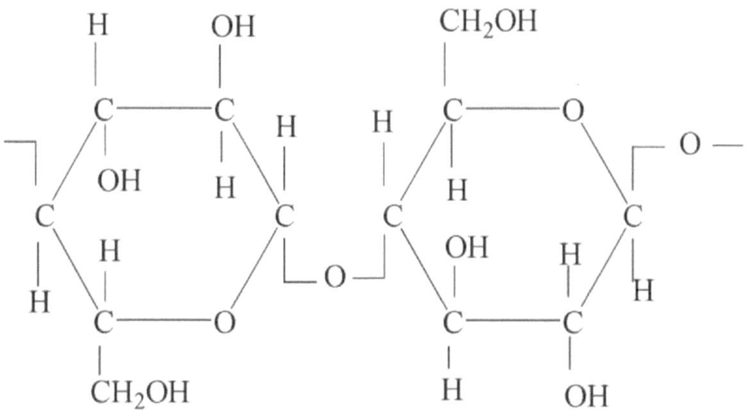

FIGURE 3.36 Cellulose.

3.7 BIOPOLYMERS

Biopolymers are polymers sourced from natural origins, either through chemical synthesis employing biological materials or entirely synthesized by living organisms. These polymers are composed of monomeric units (such as amino acids, nucleotides, sugars, and lipids) linked by covalent bonds to form larger molecular structures. In contrast to their petroleum-derived counterparts, biopolymers from renewable biological sources offer a sustainable alternative without compromising their inherent properties. This not only addresses environmental concerns associated with disposing of synthetic polymers, characterized by recurring units containing amine groups and aliphatic spacers, but also presents an economically viable solution. Examples of biopolymers include polynucleotides (DNA and RNA), polysaccharides (such as cellulose, starch, chitin, and chitosan), and polypeptides (like collagen, gelatin, gluten, and whey). These biopolymers find applications across diverse industries, from food production [223] to manufacturing [224], packaging [225], and biomedical engineering [226–229]. Ongoing research aims to incorporate nano-sized reinforcements to enhance their properties and broaden their practical applications, leading to a growing range of potential uses. The following section elucidates some of the biopolymers.

3.7.1 CELLULOSE

Cellulose, a polysaccharide, forms a linear chain comprising numerous d-glucose units with β (1→4) linkage, making it an exceptionally promising polymer due to its renewability, diverse plant-based raw materials, and inherent biodegradability. Figure 3.36 illustrates the structure of cellulose. With an annual yield surpassing 1.5×10^{12} tonnes, cellulose is regarded as an inexhaustible biomass raw material [230]. Its purity varies depending on its source, with plant-derived cellulose typically mixed with hemicellulose and lignin, while microbial cellulose tends to be purer, longer, and more hydrated. Cellulose exhibits both crystalline and amorphous regions within its molecular structure. In the crystalline region, chains are arranged in a highly ordered, repeating pattern, fostering strong intermolecular hydrogen bonding and contributing to cellulose's high strength and rigidity. Although cellulose shares structural similarities with starch, it differs in the linkage of d-glucose units through β-glucosidic bonds. Cellulose forms semi-crystalline fibres with diverse morphologies and aspect ratios across different species. Its multiple hydroxyl groups facilitate hydrogen bonding between chains, resulting in robust interchain connections and exceptional tensile strength. Cotton fabrics, with the highest cellulose content, rival synthetic fibres in strength. Industrially, cellulose is typically dissolved and reprecipitated to obtain a pure form known as regenerated cellulose. Examples of regenerated cellulose materials include rayon, cellophane, nitrocellulose (NC), cellulose acetate, cellulose propionate, cellulose acetate-propionate, carboxymethyl cellulose (CMC), ethyl cellulose, and methyl cellulose. The utilization of cellulose biopolymeric materials as matrices yields multifunctional materials with applications across various sectors, including biotechnology, biomedicine, composites, packaging, aerogels, hydrogels, textiles, coatings, membranes, and environmental applications [231–235].

3.7.2 Starch

Starch is a polysaccharide derived from cereals, legumes, and tubers represent a readily available raw material with widespread industrial applications. Starch, found in granular form, primarily comprises two types of molecules: amylose and amylopectin. Amylose, characterized by linear or sparsely branched carbohydrate chains connected by α (1–4) bonds, typically exhibits a molecular weight ranging from 10^5 to 10^6, while amylopectin, a highly branched polymer linked by both α (1–4) and α (1–6) bonds, possesses a higher molecular weight ranging between 10^7 and 10^9. The structural compositions of amylose and amylopectin are depicted in Figure 3.37.

Starch molecules possess a cyclic structure and strong hydrogen bonding, leading to a firm structure with ordered crystalline regions. This feature contributes to starch's elevated melting point and glass transition temperature, as well as its limited solubility in alcohol and cold water, solubilizing only in hot water. When exposed to water, starch granules swell and lose their semi-crystalline structure. The released amylose and amylopectin molecules gradually break down, forming a network capable of retaining water. Starch gelatinization enables the transformation of starch into a thermoplastic substance called thermoplastic starch (TPS) [236]. TPS finds application in the production of biocomposites [237] by blending it with fillers or reinforcements. These biopolymers find applications across diverse sectors, including agriculture [238,239], food packaging [240], medical [241], pharmaceutical [242], construction, automotive, and marine industries [243].

3.7.3 Chitin and Chitosan

Chitin stands out as the most widely distributed animal polysaccharide material globally, found in the exoskeletons of insects and the outer skin of fungi. Highly alkaline, chitin and chitosan are commercially significant polysaccharides due to their elevated nitrogen concentrations (5–8%) compared to synthetically substituted cellulose (1.25%). This higher nitrogen content renders chitin a valuable chelating agent. While sharing a similar linear polymer structure with cellulose, chitin's composition differs due to N-methylamide moieties at C_2 instead of hydroxyl groups. Figure 3.38 illustrates the structures of chitin and chitosan. The NH-CO bond in chitin strongly promotes intermolecular hydrogen bonding, facilitating solubility in polar solvents. However, chitin chemically degrades below its melting point, posing processing challenges. Derived from deacetylated chitin, chitosan exhibits water-swelling properties and dissolves in a water-acetic medium. This conversion transforms chitin's amide functional group into a primary amino group, yielding chitosan. Naturally occurring and synthesized by zygomycete fungi as part of their cell walls, chitosan displays pH-responsive behaviour attributed to the polycation of its amino groups. Chitosan is completely degraded by chitosanase enzymes and possesses various advantageous properties, such

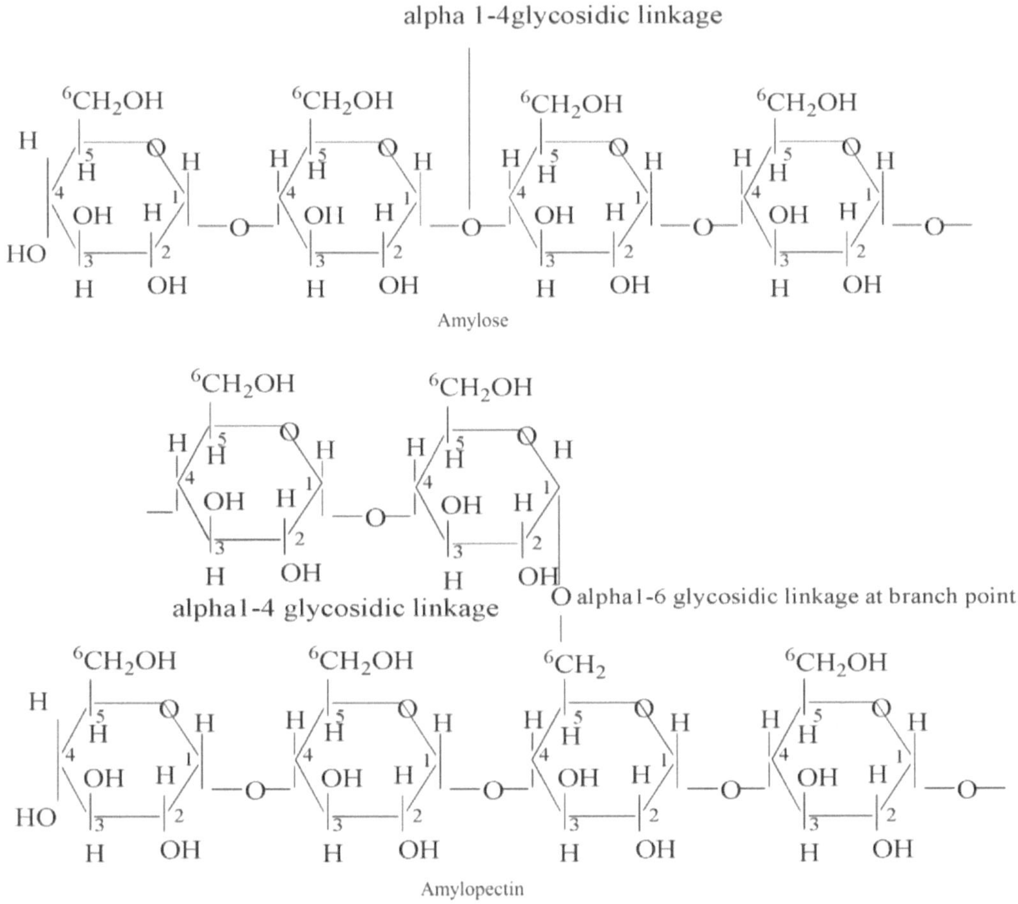

FIGURE 3.37 Amylose and amylopectin.

as hydrophilicity, biocompatibility, biodegradability, and antibacterial characteristics. Chitosan's applications in drug delivery [244], TE technology [245], wound healing [246], and other fields are currently under exploration. Some studies have also investigated chitosan as a substitute for other materials in food packaging [247] and electrical applications such as sensors, actuators, and transducers [248].

3.7.4 Poly(Lactic Acid)

Poly(lactic acid) (PLA) is a thermoplastic aliphatic polyester derived from hydroxy acids sourced from renewable materials like corn starch, roots, and sugarcane. Its biodegradability, notably in the presence of oxygen, presents challenges for recycling. Figure 3.39 illustrates the structure of PLA. PLA exists in two isomeric forms: (L-)-lactide PLA, which displays crystalline properties and a decreased melt transition temperature with a lower molecular weight, and D,L-lactide, which is amorphous and exhibits a lower melting point. The mechanical characteristics and crystallization behaviour of PLA are influenced by factors such as its molecular weight and the stereochemical configuration of its molecular backbone [249]. High molecular weight PLA is typically synthesized through ring-opening polymerization. PLA is a hydrophobic semi-crystalline polymer, with a glass transition temperature falling within the range of 40–70°C and a melting temperature ranging from 130 to 180°C. Its tensile strength typically falls between 44 and 59 MPa, and it

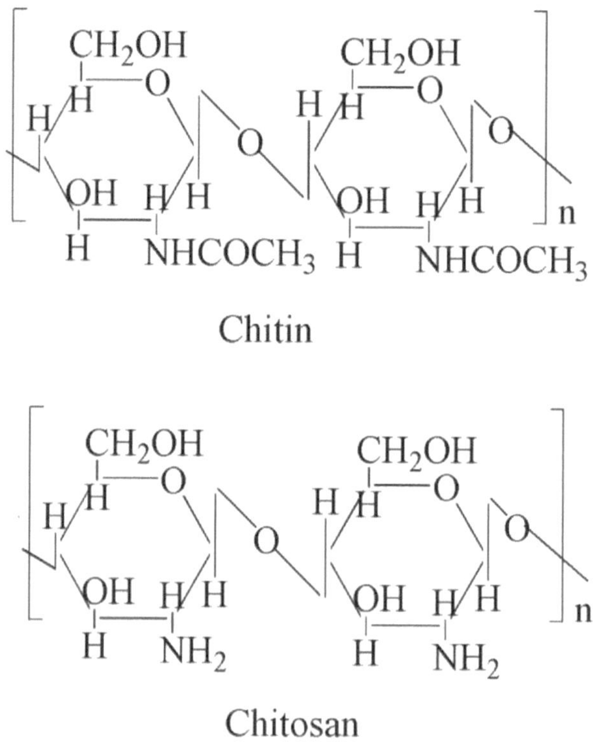

FIGURE 3.38 Chitin and chitosan.

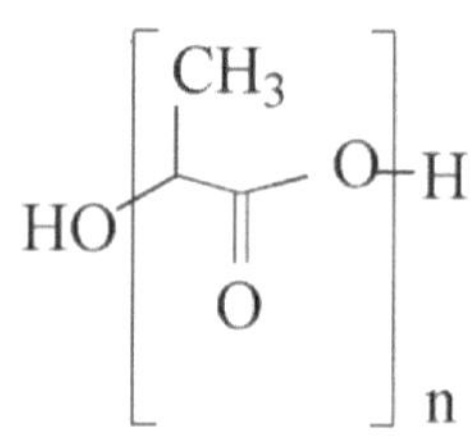

FIGURE 3.39 Poly lactic acid.

undergoes degradation over a period of 18–24 months [250]. Owing to their remarkable biocompatibility and mechanical attributes, PLA and its copolymers are extensively utilized in various biomedical applications, including implants, sutures, stent matrices, drug delivery systems, and TE [251]. Additionally, they are employed in packaging, hot melt adhesives, paints, 3D printing, waste bags, carpets, and bottles [252].

3.7.5 Poly(Glycolic Acid)

Poly(glycolic acid) (PGA) is produced through the ring-opening polymerization of glycolide and is extruded into a rigid fibre, characterized by a high melting point of 225°C and an in vitro degradation period of around 25 days. Figure 3.40 illustrates the structure of the PGA. While PGA has primarily found applications in the biomedical field, its large-scale production for non-medical industrial applications has been limited due to its comparatively higher cost compared to other biodegradable polymers.

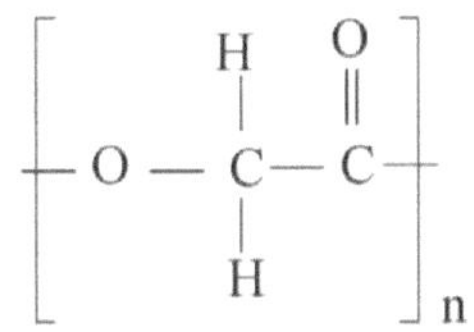

FIGURE 3.40 Poly(glycolic acid).

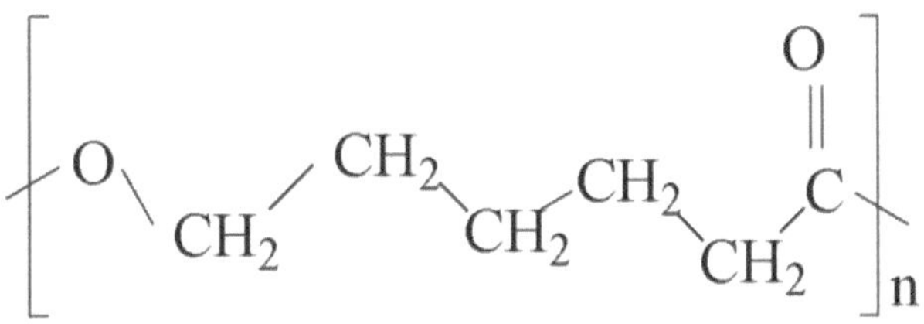

FIGURE 3.41 Poly caprolactone.

3.7.6 Poly Caprolactone (PCL)

Poly(caprolactone) (PCL), an aliphatic polyester, is a biodegradable polymer characterized by a hydrolysable aliphatic backbone. It is synthesized through the ring-opening polymerization of ε-caprolactone, as depicted in Figure 3.41. PCL has been extensively studied for its unique mechanical characteristics, compatibility with a wide variety of other polymers, and capacity to biodegrade. The physical, mechanical, and thermal characteristics of PCL are primarily influenced by its molecular weight and degree of crystallinity, which also affect its ability to degrade by hydrolysis of its ester bonds under physiological conditions. PCL is extremely hydrophobic, semi-crystalline and has great solubility at room temperature. It is easily processed because of its low melting temperature and excellent mixing compatibility. This has led researchers to explore its possible applications, particularly in the biomedical field [253].

3.8 CHARACTERIZATION TECHNIQUES OF POLYMERS

It's essential to assess the properties of polymers intended for diverse applications through sample analysis. Various characterization techniques play a crucial role in identifying functional groups, structural properties, thermal behaviours, crystallographic nature, mechanical traits, and morphological properties. The most common analysis and testing of polymers consist of thermal analysis, XRD, electron microscopy, spectroscopy, mechanical testing, and degradation studies that are elaborated below.

3.8.1 Thermal Analysis

Thermal analysis techniques such as differential scanning calorimetry (DSC) and thermogravimetric analysis (TGA) are frequently utilized to obtain data regarding various transition and degradation temperatures, respectively. Examining thermal properties is essential because these transition temperatures establish the appropriate processing temperature for a given polymer sample. Furthermore, thermal analysis techniques are employed to investigate the thermal stability of a polymer sample in diverse environments. For this purpose, various techniques like DSC, thermal gravimetric analysis (TGA), dynamic mechanical analysis (DMA), and thermo mechanical analysis (TMA) are frequently utilized. The discussion on these techniques is elaborated below.

3.8.1.1 Differential Scanning Calorimetry

DSC plays a crucial role in polymer science by analysing how polymers react to heat. DSC is instrumental in measuring and characterizing various properties of polymer materials, including heat capacity, glass transition temperature, crystallization temperature, and melting temperature. These properties are investigated based on the curve obtained from the analysis [254]. Moreover, DSC is used in the analysis of polymers to obtain the transition enthalpy, degree of crystallinity, thermal conductivity, and heat capacity of the polymer. Additionally, DSC aids in researching polymer strain release, such as mesophase transition studies, strain quality control, and purity determination. The improvement of DSC-based techniques has expanded the range of scanning rates available, from 1 μK s^{-1} to 1 MK s^{-1}, with the scanning level depending on the specific polymer being examined. Figure 3.42 illustrates the DSC curve for a four layer jute/epoxy composite, manifesting two primary events, namely endothermic and exothermic reactions. The endothermic peak is observed within the temperature range of 50–150°C for the sample, and the associated energy corresponds to the dehydration of water from the composite. Subsequently, jute composites exhibit thermal stability, as evidenced by the absence of exothermic or endothermic reactions in the temperature range of 135–220°C. A distinct exothermic peak emerges at 349°C for the composite, signifying the decomposition of a significant portion of hemicellulose and cellulose with epoxy. Notably, the composite displays enhanced thermal stability due to the release of energy (254.09 J/g) during the decomposition phase. The composite exhibits a minor exothermic peak around 427°C.

3.8.1.2 Thermo Gravimetric Analysis

TGA is a crucial technique in polymer science, allowing for the examination of weight changes in samples as they are heated or cooled under controlled conditions. By continuously monitoring these changes, TGA provides valuable insights into various processes such as decomposition, vapourization, oxidation, and reduction. The technique is versatile, accommodating solid, gel, or liquid samples with a wide weight range (1–100 mg up to 100 g). TGA data, often coupled with derivative thermograms (DTG), offer enhanced understanding of consecutive weight alterations. This

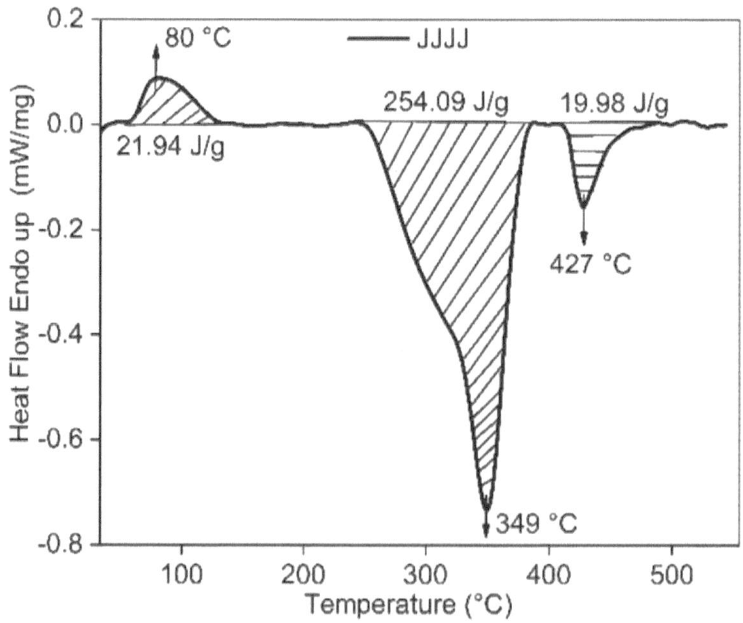

FIGURE 3.42 DSC analysis for JJJJ/epoxy composite.

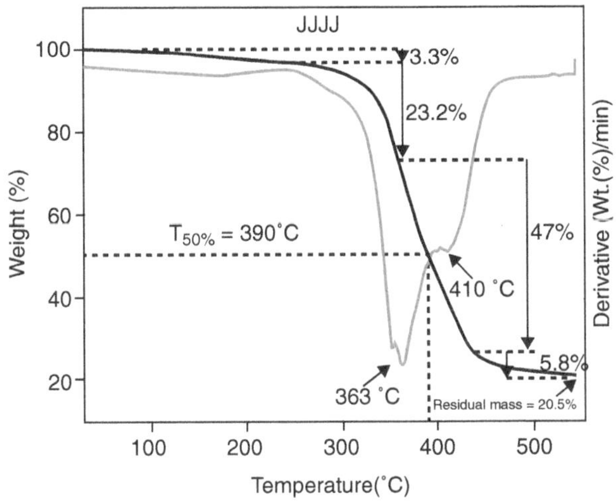

FIGURE 3.43 TGA/DTGA curves for JJJJ/epoxy composite.

analysis is particularly beneficial for studying the thermal properties and decomposition behaviour of polymeric materials. Additionally, TGA aids in assessing parameters such as thermal stability, moisture content, filler content, and compositional analysis, contributing to material selection, product performance prediction, and quality improvement [255]. Manufacturers have TGA instruments by integrating them with complementary techniques like DSC and evolved gas analysis (EGA), enabling more comprehensive characterization and shorter experiment durations. Figure 3.43 shows the pyrolysis of the jute/epoxy composite. The results reveal that composite degradation occurs in three stages. The first stage involves dehydration, associated with the physical escape of low-molecular-weight gases due to a drying effect caused by a temperature rise from room temperature. This results in rapid moisture evaporation from the composites. TGA results show that the percentage weight loss (3.3 wt.%) of moisture in the first stage occurs within temperature ranges of 30–260°C. Evidence of moisture decomposition is also observed at peaks 171°C from the DTGA plot of the corresponding composite. The second stage exhibits a pronounced decline in the weight of composites in the TGA curves, noticeable within temperature ranges of 260–460°C. The corresponding distinct peaks in the DTGA curve emerge at 363°C at the initiation of the second stage. In the last few minutes of the second stage, a small peak can be seen in the DTGA plot at 410°C. These decomposition peaks closely illustrate the pyrolysis of epoxy resin as well as fibre constituents from composites. JJJJ showed a 23.2% weight loss at the beginning and a 47% weight loss at the end. This was due to the breakdown and evaporation of hemicellulose and cellulose. This phenomenon was also noted in the DTGA plot at 363°C and 410°C, respectively. The peaks represent the maximum rate of decomposition of hemicellulose and cellulose with epoxy at 363 and 410°C, respectively. $T_{50\%}$ represents the temperature when the weight loss was 50%, indicating the onset decomposition temperature. The 50 wt.% loss of JJJJ composites occurs at 390°C. Total weight losses of 70.2% were observed for the composite during the second stage of decomposition due to the evaporation of residual solvent upon heating. During the second stage, epoxy pyrolysis took place, leading to the breakdown of ether bonds and the generation of BPA. Phenol was produced through the subsequent decomposition of BPA and/or the breakage of the fibre bond connecting the benzene ring after the breakage of the unilateral ether bond on the epoxy resin.

3.8.1.3 Thermomechanical Analyser

The International Confederation for Thermal Analysis and Calorimetry (ICTAC) recommends thermomechanical analysis (TMA) as a fundamental tool in material science, particularly for thermal analysis. It involves measuring the deformation of a sample under a constant load while subject to controlled heating or cooling conditions. TMA provides valuable insights into the structural, compositional, and application aspects of various materials, offering a comprehensive assessment of properties such as CTE, glass transition temperature, melting temperature, heat deflection, and elevated temperature creep or stress relaxation behaviour. By tracking changes in sample dimensions relative to temperature or time under a non-oscillatory load, TMA aids in characterizing both viscous and elastic properties for various materials. The technique operates within a controlled chamber, typically under inert gas atmospheres like nitrogen, helium, or argon, ensuring precise temperature control and minimizing environmental influences. TMA's applications span a wide range, including softening temperatures, melting temperatures, stress relief effects, composite delamination temperatures, and more, making it an invaluable tool in materials research, quality control, and product development across diverse industries [256]. Through various probe configurations, TMA offers flexibility to optimize test conditions for specific samples or applications, with expansion probes being commonly employed to measure dimensional changes and CTE values, particularly in materials containing fillers or cross-linked structures like composites. TMA's sensitivity surpasses that of other techniques like DSC, particularly in determining glass transition temperatures (T_g), offering enhanced precision in assessing material compatibility and thermal stress considerations crucial for product manufacturing and performance optimization. TMA remains a fundamental tool in material and polymer science, adept at detecting transitions and dimensional changes even at the nanometre level for various liquids, viscous and solid materials as a function of temperature or time under applied mechanical load (ASTM D696, ASTM D3386, ASTM E831, ISO 11359, DIN 51 005). A comparative determination of T_g from TMA, DSC, and DMA techniques is describe by Saba N. et al. [257].

Saba et al. [258] have conducted noteworthy research utilizing TMA techniques to analyse thermal expansion, yielding precise and reproducible data on the CTE, T_g, and T_m. In a recent study, they investigated the CTE of epoxy and epoxy nanocomposites by incorporating oil palm nano filler at various weights (1, 3, and 5%) using hand lay-up techniques. The resulting epoxy nanocomposites exhibited different behaviours in both rubbery and glassy regions, as depicted in Figure 3.44. The researchers observed a decrease in the CTE of epoxy composites with increasing oil palm nano filler content up to 3%, but beyond this threshold, there was a slight increase. This increase in CTE for the 5% loading is attributed to poor dispersion and agglomeration of the filler particles, which create free space for polymer segmental motion. Remarkably, the TMA graphs also indicate that all oil palm epoxy nanocomposites exhibit lower thermal expansion than pure epoxy composites in both regions. This suggests a synergistic interaction between the oil palm nano filler and epoxy polymer, leading to improved dimensional stability and CTE compared to pure epoxy composites.

3.8.1.4 Dynamic Mechanical Analyser

DMA is a thermal technique that examines material properties while subjecting them to periodic stress. DMA involves applying a varied sinusoidal stress to measure the resulting strain [259]. For purely elastic materials, the stress and strain waves exhibit a 0° phase difference, while purely viscous materials exhibit a 90° phase difference. Most materials, including polymers, are viscoelastic, showing a phase difference between these extremes. This phase difference, stress and strain wave amplitudes collectively influence key material parameters such as storage and loss modulus, tan δ, viscosity, compliance, transition temperatures, creep, and stress relaxation. DMA instruments consist of critical components, including a drive motor, drive shaft support, displacement sensor, temperature control system, and sample clamps. The DMA 2980 analyser optimizes these components, featuring a noncontact direct drive motor and an air bearing shaft support system. DMA is pivotal in evaluating materials for specific applications by providing insights into their long-term performance beyond short-term tests like deflection temperature under load (DTUL).

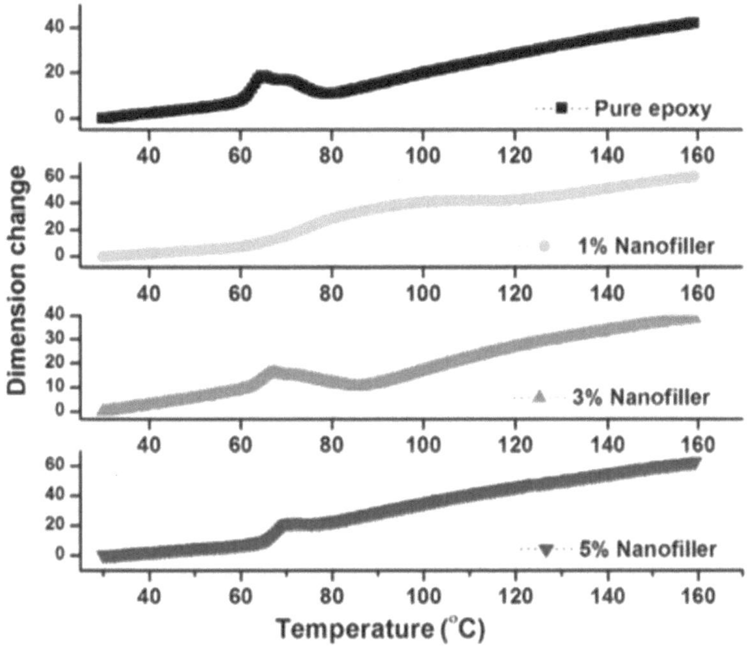

FIGURE 3.44 TMA graph of epoxy composites and epoxy nanocomposites at different filler loading. (Reprinted from Saba et al. [258] Copyright 2017, with permission from Elsevier.)

This method consistently tracks the material's modulus in relation to temperature, offering a more accurate indication of elevated temperature performance. For instance, DMA modulus curves can reveal significant differences in material behaviour beyond the DTUL, highlighting variations in long-term performance among different resins. This information is invaluable for engineers and designers tasked with selecting materials for various applications. By understanding how materials behave under dynamic stress, engineers can make informed decisions about material selection, design, and processing to ensure optimal performance and durability in their applications. Overall, DMA plays a crucial role in material characterization and performance prediction, contributing to advancements in various industries such as automotive, aerospace, electronics, and consumer goods. Figure 3.45 illustrates the DMA of an LDPE carbon-reinforced composite [260]. The storage modulus gradually increases from 30 to 80°C and then decreases abruptly at 85°C. It remains constant from 120 to 190°C. The loss modulus reaches its maximum value (351.1 MPa) at 95.31°C, while the maximum Tan δ (0.3669) is observed at 99.86°C. The storage modulus curve reflects the stiffness and rigidity of the composite structure, showing three distinct phases. In the first phase, the structure is very stiff and rigid due to tightly packed molecules and rigid polymeric chains. The second phase, above T_g, sees a decrease in storage modulus due to polymer chain movement, affecting stiffness and fibre matrix adhesion. In the third phase, there is no significant enhancement in storage modulus due to accelerated polymer chain mobility at higher temperatures. Overall, there is a noticeable shift in the temperature range, affecting the stiffness and rigidity of the composite structure.

3.8.2 X-Ray Diffraction Analysis

XRD methods like wide-angle X-ray scattering (WAXS) and small-angle X-ray scattering (SAXS) are used to analyse polymer structure and morphology. These methods are crucial for understanding crystal structure, disorder, orientation, and mesoscale structures in polymers and

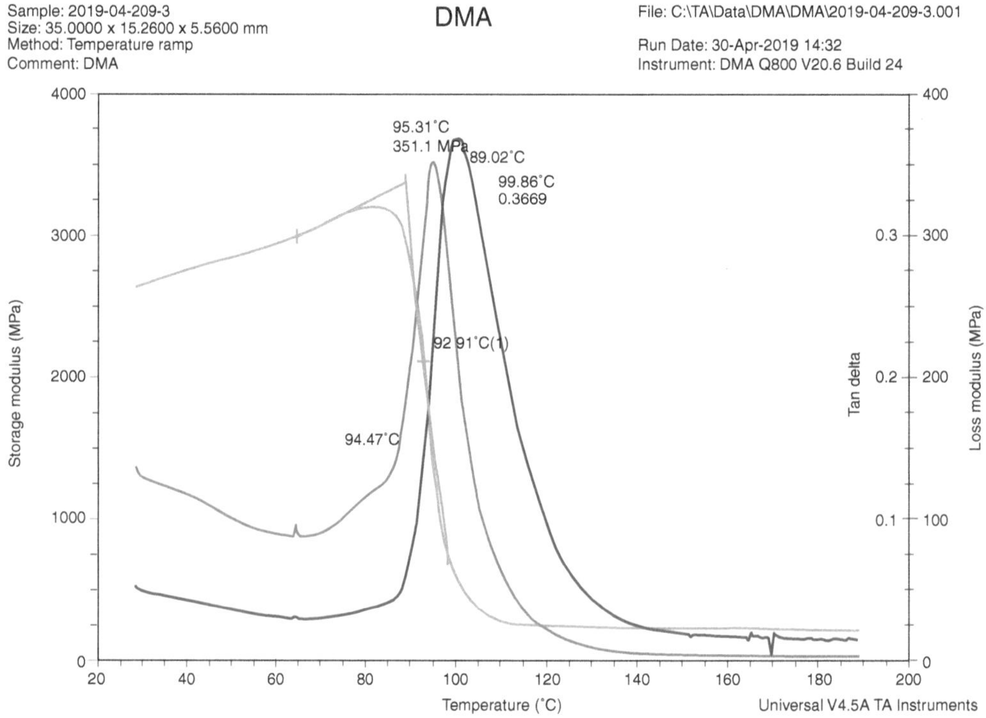

FIGURE 3.45 DMA of low density polyethylene composites reinforced with carbon fibre reinforcement. (Reprinted from Ravichandran et al. [260] Copyright 2020, with permission from Elsevier.)

polymer composites. Coherent and incoherent scattering of X-rays is utilized for structural analysis, with careful consideration given to subtracting diffuse incoherent scattering from the observed diffraction patterns, particularly in highly disordered polymers. Additionally, morphology significantly influences dimensional stability, mechanical properties, and barrier properties, which are crucial in engineering and biomaterial applications. Combination measurements such as simultaneous XRD and elongation, microbeam techniques, and grazing incidence diffraction (GID) are also utilized to map spatial variations in structure and determine structure-depth variations. By understanding these structural characteristics, researchers can gain valuable insights into polymer properties and behaviour, aiding in the development of advanced materials for various applications. Structural parameters derivable from WAXS and SAXS data in polymers in the solid state are given in [261]. The XRD plots in Figure 3.46 illustrate the structure and morphology of both pure epoxy composites and OPEFB/epoxy nanocomposites [258]. Sharp and highly intense peaks at a 2θ value of 21° are observed in the XRD pattern of pure epoxy composites, while OPEFB/epoxy nanocomposites exhibit relatively less intense peaks at lower 2θ values. This suggests that the addition of amorphous nano OPEFB filler reduces the crystallinity of the pure epoxy matrix. Specifically, the XRD graph indicates that 1 and 5% nano OPEFB filler loading results in an intercalated structure. At 1% loading, the dispersion is homogeneous but not uniform, whereas at 5% loading, particle agglomeration leads to intercalated nanocomposites. Conversely, the 3% loading shows lower peaks, indicating a more homogenous dispersion of the nano filler in the epoxy polymer, likely resulting in exfoliation nanocomposites. Overall, the XRD results demonstrate a reduction in the diffraction peak intensity and crystallinity of pure epoxy due to the addition of nano filler, thus decreasing its brittleness. However, this reduction is more pronounced for the 3% loading compared to 1 and 5% loadings.

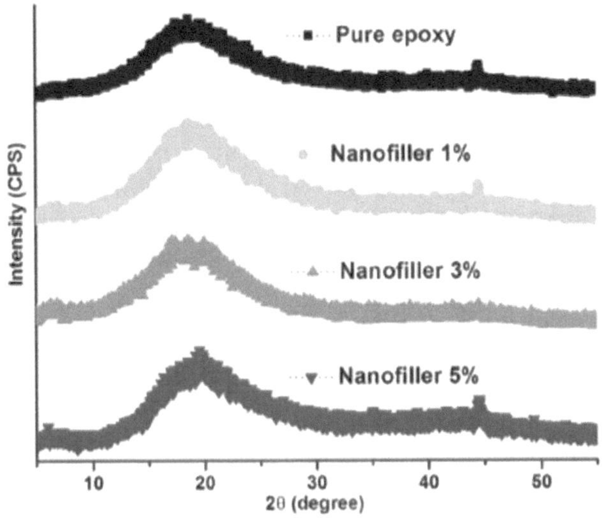

FIGURE 3.46 XRD graph of epoxy composites and epoxy nanocomposites at different filler loading. (Reprinted from Saba et al. [258] Copyright 2017, with permission from Elsevier.)

3.8.3 Electron Microscopy

Electron microscopy has been pivotal in exploring the polymer's structure and morphology, with techniques like transmission electron microscopy (TEM) and scanning electron microscopy (SEM) offering insights from macroscopic sizes down to about 0.1 nm. SEM, known for its user-friendliness and ease of specimen preparation, primarily examines surface features using a focused electron beam and can be enhanced by techniques like environmental SEM (ESEM), allowing imaging of wet samples without conductive coatings. On the other hand, TEM provides detailed structural information at atomic dimensions, although it's limited to thin samples and often requires chemical staining for contrast enhancement. Additionally, atomic force microscopy (AFM) has emerged as a valuable tool for studying polymer structures, particularly at the nanometre scale, offering advantages in surface characterization and manipulation. The properties of nanoparticles are heavily influenced by their dimensions, making the characterization of these dimensions crucial. P. Eaton et al. [262] summarize a study comparing three common methods for nanoparticle dimension characterization: AFM, SEM, and TEM. The choice of technique depends on the sample type and desired information. SEM often struggles with imaging uncoated nanoparticles, especially smaller ones, requiring gold/palladium coating for accurate measurements. AFM, TEM, and SEM produced differing images of silica samples in Figure 3.47, with AFM providing three-dimensional data. SEM images, particularly for the smaller gold nanoparticles in Figure 3.48, were noisy due to their size. Similarly, SEM struggled with small, uncoated PS nanoparticles in Figure 3.49. While all methods could characterize the samples, SEM required metal coating for accurate measurements and was less suitable for small nanoparticles compared to AFM and TEM. AFM provided high contrast and signal-to-noise ratios across all samples but was sensitive to sample cleanliness. TEM offered high throughput and was suitable for rapid characterization of numerous nanoparticles. Overall, the choice of technique depends on availability, familiarity, and cost, with AFM being the most accessible option. These methods are typically employed to analyse various aspects such as fracture propagation, the impact of filler incorporation, fibre pull-out phenomena, and more within composite structures. Additionally, it provides insights into the interfacial interactions between different components, aiding in the optimization of composite material properties and performance.

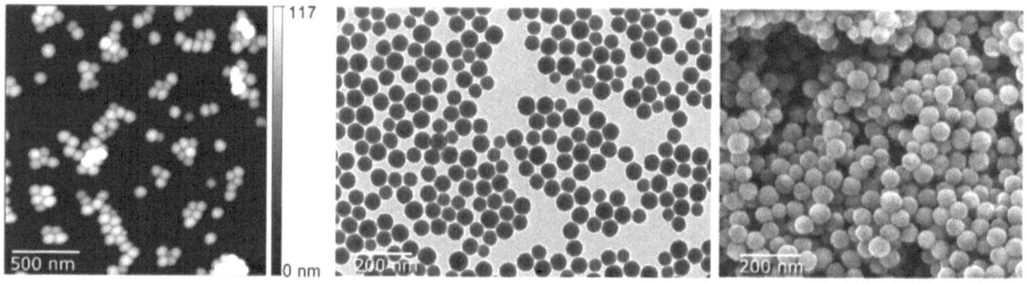

FIGURE 3.47 Examples of silica nanoparticle images from AFM (left), TEM (centre), and SEM (right). (Reprinted from P. Eaton et al. [262] Copyright 2017, with permission from Elsevier.)

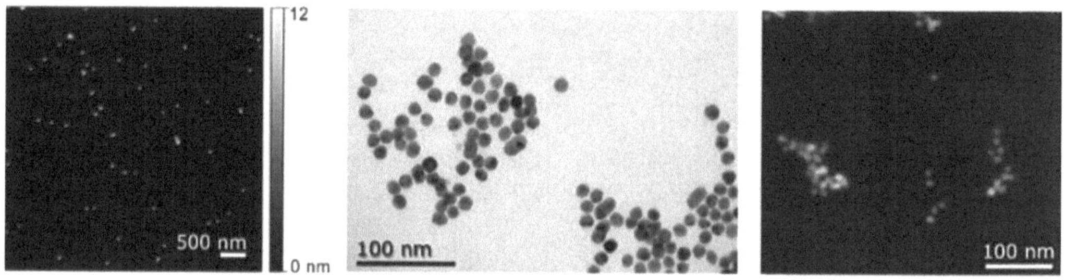

FIGURE 3.48 Examples of gold nanoparticle images from AFM (left), TEM (centre), and SEM (right). (Reprinted from P. Eaton et al. [262] Copyright 2017, with permission from Elsevier.)

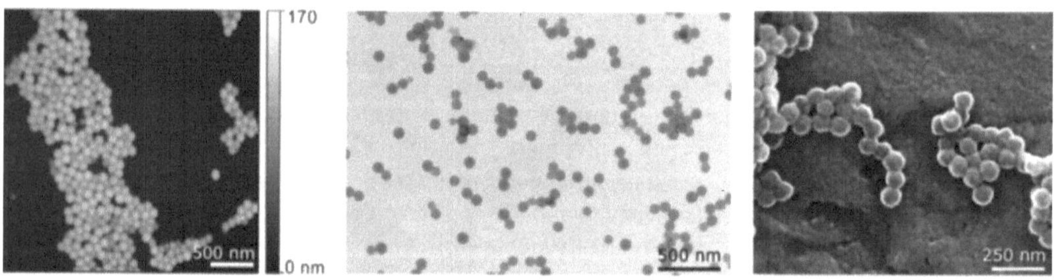

FIGURE 3.49 Examples of polystyrene nanoparticle images from AFM (left), TEM (center), and SEM (right). (Reprinted from P. Eaton et al. [262] Copyright 2017, with permission from Elsevier.)

3.8.4 Spectroscopy Analysis

3.8.4.1 Infrared Spectroscopy

The infrared (IR) spectrum of a polymer sample offers qualitative and quantitative insights into its chemical and physical structure, including the chemical nature (such as branching and end groups), steric order (cis-trans isomerism), conformational order (chain arrangement), and crystallinity. Two fundamental approaches are used for interpreting IR spectra: empirical, which focuses on the chemical nature, and theoretical, which analyses vibrational behaviour and intermolecular forces. A semi-empirical approach combines these methods, using theoretical insights to interpret the IR spectra of complex polymers based on simpler, idealized structures [263]. Rajandas et al. [264] conducted FTIR analysis on LDPE treated with oxidative nitric acid, as depicted in Figure 3.50. Except for spectrum (a), the remaining spectra (b), (c), (d), (e), and (f) exhibited the emergence of two new

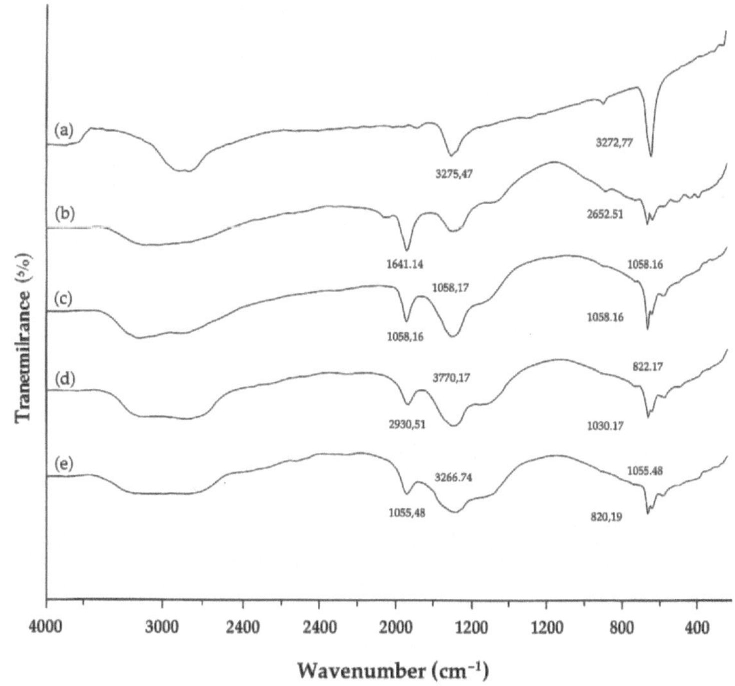

FIGURE 3.50 FTIR spectral image of nitric acid pretreated LDPE powder (a) untreated LDPE powder; (b) 2 days nitric acid pretreated LDPE powder; (c) 4 days nitric acid pretreated LDPE powder; (d) 6 days nitric acid pretreated LDPE powder; (e) 8 days nitric acid pretreated LDPE powder; (f) 10 days nitric acid pretreated LDPE powder. (Reprinted from Rajandas et al. [264] Copyright 2012, with permission from Elsevier.)

functional groups, indicated by peaks around 1630–1636 cm^{-1} and 825 cm^{-1}. These correspond to the amide (a form of carbonyl group) and nitrate ions, respectively. The presence of carbonyl compounds suggests LDPE oxidation, thereby promoting biodegradation.

3.8.4.2 Raman Spectroscopy

Raman spectroscopy (RS) is a non-destructive characterization technique that detects bond vibrations altering the polarizability of molecules, making it suitable for analysing polymers and additives. It provides identification through reference spectra called ‘fingerprints,’ offering insights into polymer identity and molecular characteristics. It reveals crucial information about polymer composition, structure, molecular interaction, orientation, stereo regularity, crystallization, and surface/interface structure [265]. The width of Raman peaks indicates stereochemical purity, with weak and wide peaks for disordered samples and strong, sharp peaks for ordered ones. Research applications include chemical structure analysis, quantitative component analysis, characterization of crystalline and amorphous phases, monitoring of crystallization processes, kinetic studies of polymer reactions, assessment of polymer orientation, examination of polymer blend compatibility and molecular interactions, monitoring of stress relaxation and strain processes in composite materials, and tracking polymerization and curing processes. Figure 3.51 shows the Raman spectrum of PS, high frequency C-H vibrations (~3000 cm^{-1} indicate aliphatic carbon chains, while low frequency C-C vibrations (~800 cm^{-1}) suggest aromatic carbon rings. Bands at ~1600 cm^{-1} denote C=C double bonds, and those at ~1000 cm^{-1} represent benzene ring vibrations. The RS of several other polymers like PC, polymethylmethacrylate, PE, PP, polyethylene terephthalate, PA are also discussed in the literature [266].

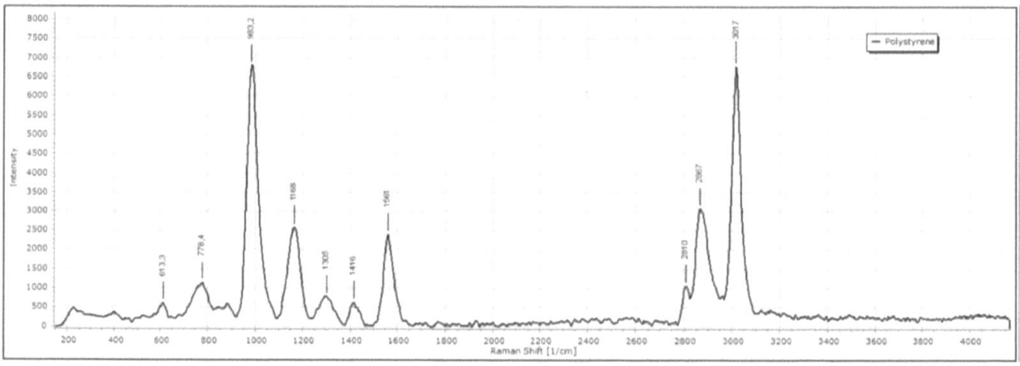

FIGURE 3.51 Polystyrene Raman spectrum [266].

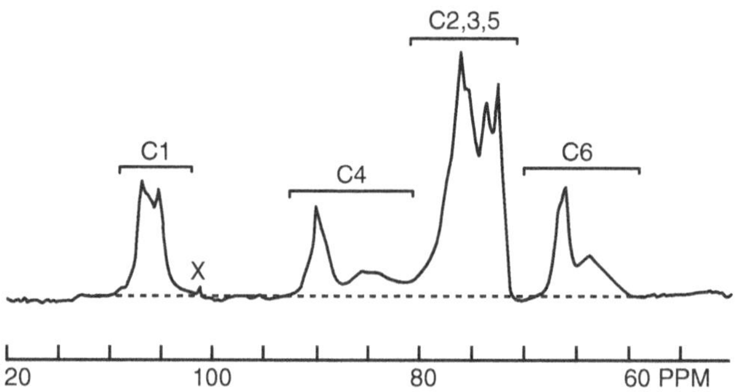

FIGURE 3.52 ^{13}C CP-MAS spectrum of cotton linters. The horizontal bars indicate the spectral ranges of the corresponding carbon sites in the anhydroglucose monomer unit of cellulose. The 'X' marks the position of the small first spinning sideband of linear polyethylene, which was added as a chemical shift reference. The polyethylene centre band (not shown) occurs at 33.63 ppm; the zero of reference for chemical shifts is liquid tetramethylsilane. Note the existence of both broad and narrow resonance features. (Reprinted with permission from Ref. [269]. Copyright (1984) American Chemical Society.)

3.8.4.3 NMR Spectroscopy

Nuclear magnetic resonance (NMR) spectroscopy stands as a fundamental and dependable method for unravelling the structural intricacies of polymers [267]. NMR spectroscopy emerges as a pivotal tool for material characterization and the exploration of polymer structure property relationships. Its significance lies in its ability to assign signals to specific atoms along the polymer backbone and side chains, offering valuable insights into polymers at an atomic scale based on local magnetic environments experienced by NMR active nuclei. It encompasses the determination of diverse polymer attributes, including tacticity, molecular weights, molecular weight distributions, copolymer compositions, monomer reactivity ratios, and monomer sequence distributions within copolymer chains [268]. Figure 3.52 depicts the NMR spectra of cotton linters, where distinct resonance regions are assigned to specific carbon atoms [269]. The spectrum is analysed from the upfield side, with the 60–70 ppm region assigned to C6, 70–81 ppm to C2, C3, and C5, 81–93 ppm to C4, and 102–108 ppm (sometimes extending down to 96 ppm) to C1. Notably, the resonance regions of C6 and C4 consist of sharper resonances overlapping broader wings. The relative intensities of the broad wings of C4 and C6 (around 85 and 64 ppm, respectively) are scaled with the crystallite surface areas, considering typical lateral crystallite dimensions in the material. Despite limited resolution, sharper

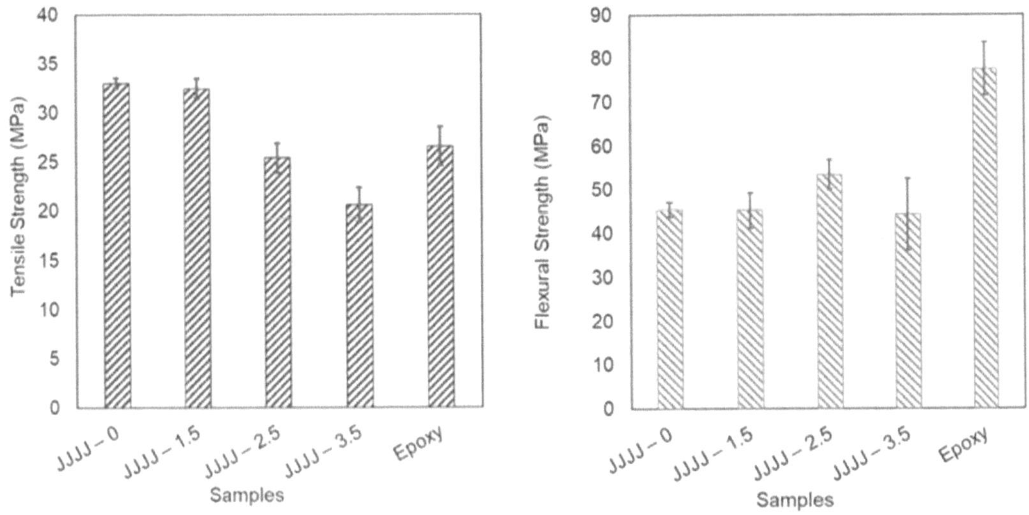

FIGURE 3.53 Tensile and flexural properties for fumed silica reinforced jute/epoxy composites.

portions of the C4 and C6 resonances (at 90 and 66 ppm) appear to consist of multiple components, possibly reflecting heterogeneity within the unit cell. The C1 resonance region (102–108 ppm) also exhibits multiplicity and sharp features, albeit with interpretation challenges due to overlapping resonances from interior crystalline and disordered surface regions. The broadness of the C4 and C6 wings primarily indicates molecular packing disorder.

3.8.5 Physico-Mechanical Properties

3.8.5.1 Tensile Strength

Mechanical properties such as tensile strength, flexural strength, bending, shear, compression, fatigue, fracture, and elongation at break are important parameters for polymer application [270–272]. Elongation at break is a valuable measure of ductility and a crucial factor for assessing the durability of packaging. The tensile and flexural strengths of the samples are determined by ASTM D638 and ASTM D790. The tensile and flexural properties of neat epoxy and all jute fibre composites with fumed silica variation are shown in Figure 3.53. The decreasing trend in tensile strength was observed in jute composites with the addition of filler (0, 1.5, 2.5, and 3.5 wt.%). The tensile strengths of JJJJ-0 and JJJJ-1.5 composites are 33.01 MPa and 32.46 MPa, respectively, showing a higher value than neat epoxy. The flexural strength of the jute composites shows an appreciable improvement in comparison to their tensile strength with the addition of filler. But the flexural strength of the jute-reinforced composites is lower than that of neat epoxy. The addition of filler in polymer-based fibre-reinforced composites affects various mechanical properties.

3.8.5.2 Tear Strength

Tear strength is a critical measure of a plastic film's resistance to tearing in packaging applications. This property is evaluated using a tear strength tester, which securely holds a sample within its jaws. A sharp blade on the machine initiates the tear from the pre-existing slit or notch. When the pendulum swings due to gravity, it imparts energy to tear the specimen along the notch. This tearing process requires energy, resulting in the pendulum having less energy than if it had fallen freely. The difference in energy is indicated by a pointer on a calibrated scale. Once released, the pendulum allows the specimen to tear completely, and the reading is recorded to determine the tear strength.

For thinner samples, tear strength is typically measured in grams. Additionally, tear strength is evaluated both in the machine direction and the transverse direction. The tear strength of high and low density PE and trans-polyisoprene was determined by Gent and Jeong [273].

3.8.5.3 Burst Strength

Plastic strength assessment involves a straightforward bursting device. The tin sample's burst strength denotes its ability to withstand progressively applied pressure perpendicular to its surface under specific conditions. This strength is identified as the pressure at the point of failure and gauges the sample's energy absorption capacity. The sample is positioned between two angular rings within this machine. The inlet valve is activated to maintain an air pressure of 110 lbs/sq inch. Subsequently, the sample undergoes controlled exposure to increasing air pressure until it reaches failure. The pressure recorded at the moment of failure, expressed in lb/sq inch, signifies the material's burst strength. The bursting strength of E-glass/epoxy composites has been studied by Hawa et al. [274].

3.9 RESEARCH AVENUES FOR POLYMERS

Polymers serve as essential elements in alternative energy applications, such as batteries and EVs [275]. The emergence of all-solid-state polymer electrolyte-based hybrid batteries signifies a significant advancement in alternative energy technology. One of the main problem faced by the EV industry is the volatile liquid lithium ion batteries, which are vulnerable to fire and chemical leakage. A promising alternative is the solid-state polymer electrolyte batteries, which provide a safer and more efficient working environment [276–278]. Hence, there is an ample opportunity to explore further in solid-state polymers for battery application. Various polymers are used in TE applications due to their tuneable biodegradation rates, mechanical properties, high porosity with various pore sizes, and high surface-to-volume ratio [279]. Poly(ε-caprolactone) (PCL), PLA, and poly(glycolic-co-lactic acid) (PLGA) are some of the biopolymers used in TE. Similarly, polymers like PEG, PVA, and poly(acrylamides) are used in the growth and proliferation of cells due to their high hydrophilic nature [280]. Before selecting a polymer for TE application, it is important to conduct safer clinical trials [281]. To facilitate the safer assessment of polymer materials for TE application, reliable and precise ex vivo studies need to be carried out.

3.10 CONCLUSION

An elaborate discussion on polymer types like thermoplastics, thermosets, elastomers, and biopolymers was carried out. Each type offers unique advantages and limitations, making them suitable for specific applications in various industries. By exploring renewable sources, recyclability, and biodegradability, researchers and engineers are directed towards a more sustainable future by incorporating biopolymers derived from renewable sources. The chapter also incorporates an in-depth analysis of various characterization techniques that are employed to elucidate the structure, behaviour, and performance of polymer materials. As we stand on the threshold of Industry 4.0, the continued advancement of polymer science promises even greater achievements in sustainability, efficiency, and performance. By harnessing the power of polymer materials and embracing innovative characterization techniques, we can shape a more sustainable and resilient future for generations to come.

REFERENCES

[1] E. Browne, Z.A. Worku, A.M. Healy, Physicochemical properties of poly-vinyl polymers and their influence on ketoprofen amorphous solid dispersion performance: A polymer selection case study, *Pharmaceutics*. 12 (2020). https://doi.org/10.3390/pharmaceutics12050433

[2] Z. Zhao, M. Zharnikov, Elastic Properties of Poly(ethylene glycol) Nanomembranes and Respective Implications, *Membranes (Basel).* 12 (2022). https://doi.org/10.3390/membranes12050509

[3] eFunda, "Polycarbonate (PC) -polymer datasheet," Available: www.efunda.com/materials/polymers/properties/polymer_datasheet.cfm?MajorID=P-TP&MinorID=1 (accessed February 26, 2024).

[4] Y. Kaplan, Role of reinforcement materials on mechanical and tribological properties of PTFE composites, *Polymer (Guildf).* 44 (2020) 436–444. https://doi.org/10.7317/pk.2020.44.4.436

[5] N. Jain, V.K. Singh, S. Chauhan, A review on mechanical and water absorption properties of polyvinyl alcohol based composites/films, *J. Mech. Behav. Mater.* 26 (2017) 213–222. https://doi.org/10.1515/jmbm-2017-0027

[6] MatWeb, "Overview of Materials for Polyvinylidinefluoride (PVDF), Molded/Extruded," www.matweb.com/search/datasheettext.aspx?matguid=a011f8ccf4b448a19246773a32085094 (accessed February 26, 2024).

[7] Y. Yan, Developments in fibers for technical nonwovens, in *Advances in Technical Nonwovens*, Elsevier Ltd, 2016: pp. 19–96. https://doi.org/10.1016/B978-0-08-100575-0.00002-4

[8] R. Dallaev, T. Pisarenko, D. Sobola, F. Orudzhev, S. Ramazanov, T. Trčka, Brief review of PVDF properties and applications potential, *Polymers (Basel).* 14 (2022) 1–29. https://doi.org/10.3390/polym14224793

[9] Petron Thermoplast, "Get in Touch – Petron Thermoplast," https://petronthermoplast.com/get-in-touch/ (accessed February 26, 2024).

[10] Poly(N-vinyl carbazole) PVK – Vinyl – Matmatch, www.ossila.com/products/poly9-vinylcarbazole (accessed February 26, 2024).

[11] AZOM, Polyvinyl Fluoride – PVF, www.azom.com/article.aspx?ArticleID=809 (accessed February 26, 2024).

[12] AZOM, Properties: Polyvinylidenechloride (PVDC) – Properties and Applications – Supplier Data by Goodfellow, www.azom.com/properties.aspx?ArticleID=2004 (accessed February 26, 2024).

[13] Polysciences, Poly(methacrylic acid) | Polysciences, Inc., www.polysciences.com/india/polymethacrylic-acid-36846 (accessed February 26, 2024).

[14] Scientific Polymer Products Inc., Poly(acrylic acid) – scipoly.com, https://scipoly.com/shop/polyacrylic-acid-11/ (accessed February 26, 2024).

[15] A Sarı, Form-stable paraffin/high density polyethylene composites as solid–liquid phase change material for thermal energy storage: Preparation and thermal properties, *Energy Convers. Manag.* 45 (2004) 2033–2042. https://doi.org/10.1016/j.enconman.2003.10.022

[16] K.M. Manu, S. Soni, V.R.K. Murthy, M.T. Sebastian, *J. Mater Sci. Mater Electron.* 24 (2013) 2098–2105. https://doi.org/10.1007/s10854-013-1064-y

[17] T. Dey, M. Tripathi, Thermal properties of silicon powder filled high-density polyethylene composites, *Thermochim. Acta.* 502 (2010) 35–42. https://doi.org/10.1016/j.tca.2010.02.002

[18] L. Fang, Y. Leng, P. Gao, Processing of hydroxyapatite reinforced ultrahigh molecular weight polyethylene for biomedical applications, *Biomaterials.* 26 (2005) 3471–3478. https://doi.org/10.1016/j.biomaterials.2004.09.022

[19] M. Wang, L. Hench, W. Bonfield, Bioglass®/high density polyethylene composite for soft tissue applications: Preparation and evaluation, *J. Biomed. Mater. Res.* 42 (1998) 577–586. https://doi.org/10.1002/(SICI)1097-4636(19981215)42:4<577::AID-JBM14>3.0.CO;2-2

[20] E. Psomiadou, I. Arvanitoyannis, C. Biliaderis, H. Ogawa, N. Kawasaki, Biodegradable films made from low density polyethylene (LDPE), wheat starch and soluble starch for food packaging applications. Part 2, *Carbohydr. Polym.* 33 (1997) 227–242. https://doi.org/10.1016/S0144-8617(97)00032-5

[21] Q. Zhang, S. Rastogi, D. Chen, D. Lippits, P. Lemstra, Low percolation threshold in single-walled carbon nanotube/high density polyethylene composites prepared by melt processing technique, *Carbon N. Y.* 44 (2006) 778–785. https://doi.org/10.1016/j.carbon.2005.09.039

[22] Q.T.H. Shubhra, A.K.M.M. Alam, M.A. Quaiyyum, Mechanical properties of polypropylene composites: A review, *J. Thermoplast. Compos. Mater.* 26 (2013) 362–391. https://doi.org/10.1177/0892705711428659

[23] M. Kaseem, K. Hamad, Y. Ko, Fabrication and materials properties of polystyrene/carbon nanotube (PS/CNT) composites: a review, *Eur. Polym. J.* 79 (2016) 36–62. https://doi.org/10.1016/j.eurpolymj.2016.04.011

[24] Y. Liang, L. Ji, B. Guo, Z. Lin, Y. Yao, Y. Li, Preparation and electrochemical characterization of ionic-conducting lithium lanthanum titanate oxide/polyacrylonitrile submicron composite fiber-based, *J. Power Sources.* 196 (2011) 436–441. https://doi.org/10.1016/j.jpowsour.2010.06.088

[25] C.E. Schildknecht, *Vinyl and Related Polymers: Their Preparations, Properties, and Applications in Rubbers, Plastics, Fibres, and in Medical and Industrial Arts*, Chapman & Hall, 1952.

[26] J. Pan, J. Nie, B. Zeng, X. Chen, Z. Chen, D. Kang, Optimized preparation of polyacrylonitrile/sulfur composite as cathode for lithium sulfur batteries, *Int. J. Electrochem. Sci.* 17 (2022) 22021. https://doi.org/10.20964/2022.02.19

[27] D. Cao, J. Deng, L. Jiang, X. Li, G. Zhang, Designing polyimide/polyacrylonitrile/polyimide sandwich composite separator for rechargeable lithium-ion batteries, *J. Energy Storage.* 55 (2022) 105496. https://doi.org/10.1016/j.est.2022.105496

[28] Y. Huang, X. Liu, M. Yu, X. Zhao, Z. Zhou, X. Sun, Polyacrylonitrile/ZIF-67 derived N-doped carbon fibers-Co_3O_4 composites as efficient electromagnetic wave absorbers, *Mater. Lett.* 352 (2023) 135151. https://doi.org/10.1016/j.matlet.2023.135151

[29] X. Yu, X. Su, Y. Liu, D. Yu, Y. Ren, X. Liu, Biomass intumescent flame retardant polyacrylonitrile composite: Flame retardancy, smoke suppression and recycling, *Compos. Part A Appl. Sci. Manuf.* (2023) 107647. https://doi.org/10.1016/j.compositesa.2023.107647

[30] Y. Liu, X. Yu, Y. Guo, Y. Ren, X. Liu, Preparation of flame retardant, smoke suppression and reinforced polyacrylonitrile composite fiber by using fully biomass intumescent flame retardant system, *Compos. Part A Appl. Sci. Manuf.* 173 (2023) 107705. https://doi.org/10.1016/j.compositesa.2023.107705

[31] S. Nataraj, K. Yang, T. Aminabhavi, Polyacrylonitrile-based nanofibers—A state-of-the-art review, *Prog. Polym. Sci.* 37 (2012) 487–513. https://doi.org/10.1016/j.progpolymsci.2011.07.001

[32] C. Pirlot, I. Willems, A. Fonseca, J.B. Nagy, J. Delhalle, Preparation and Characterization of Carbon Nanotube/Polyacrylonitrile Composites, *Adv. Eng. Mater.* 4 (2002) 109–114. https://doi.org/10.1002/1527-2648(200203)4:3<109::AID-ADEM109>3.0.CO;2-5

[33] L. Henri, Thermohygroelastic properties of polymethylmethacrylate , (2007) 11–13.

[34] M. Demir, M. Memesa, P. Castignolles, G. Wegner, PMMA/zinc oxide nanocomposites prepared by in-situ bulk polymerization, *Macromol. Rapid Commun.* 27 (2006) 763–770. https://doi.org/10.1002/marc.200500870

[35] H. Hashim, N. Adam, N.H.M. Zaki, Z.S. Mahmud, C.M.S. Said, M.Z.A. Yahya, A.M.M. Ali, Natural rubber-grafted with 30% poly (methylmethacrylate) characterization for application in lithium polymer battery, in: *Int. Conf. Sci. Soc. Res.*, 2010: pp. 485–488. https://doi.org/10.1109/CSSR.2010.5773825

[36] L. Lee, W. Chen, High-refractive-index thin films prepared from trialkoxysilane-capped poly (methyl methacrylate) – Titania materials, *Chem. Mater.* 13 (2001) 1137–1142. https://doi.org/10.1021/cm000937z

[37] B. Adhikari, S. Majumdar, Polymers in sensor applications, *Prog. Polym. Sci.* 29 (2004) 699–766. https://doi.org/10.1016/j.progpolymsci.2004.03.002

[38] J. Kost, R. Langer, Responsive polymeric delivery systems, *Adv. Drug Deliv. Rev.* 64 (2012) 327–341. https://doi.org/10.1016/J.ADDR.2012.09.014

[39] D. Beruto, R. Botter, M. Fini, The effect of water in inorganic microsponges of calcium phosphates on the porosity and permeability of composites made with polymethylmethacrylate, *Biomaterials.* 23 (2002) 2509–2517. https://doi.org/10.1016/S0142-9612(01)00385-4

[40] M. Shi, J. Kretlow, P. Spicer, Y. Tabata, Antibiotic-releasing porous polymethylmethacrylate/gelatin/antibiotic constructs for craniofacial tissue engineering, *J. Control. Release.* 152 (2011) 196–205. https://doi.org/10.1016/j.jconrel.2011.01.029

[41] M. Perween, D. Parmar, G. Bhadu, D. Srivastava, Polymer–graphite composite: a versatile use and throw plastic chip electrode, *Analyst.* 139 (2014) 5919–5926. https://doi.org/10.1039/C4AN01405G

[42] J. Gilbert, D. Ney, E.P. Lautenschlager, Self-reinforced composite poly (methyl methacrylate): Static and fatigue properties, *Biomaterials.* 16 (1995) 1043–1055. https://doi.org/10.1016/0142-9612(95)98900-Y

[43] V. Miettinen, P. Vallittu, Release of residual methyl methacrylate into water from glass fibre-poly (methyl methacrylate) composite used in dentures, *Biomaterials.* 18 (1997) 181–185. https://doi.org/10.1016/S0142-9612(96)00123-8

[44] J. Zeng, B. Saltysiak, W. Johnson, D.A. Schiraldi, Processing and properties of poly (methyl methacrylate)/carbon nanofiber composites, *Compos. Part B Eng.* 35 (2004) 245–249. https://doi.org/10.1016/j.compositesb.2003.08.009

[45] P. Vallittu, Some aspects of the tensile strength of unidirectional glass fibre–polymethyl methacrylate composite used in dentures, *J. Oral Rehabil.* 25 (1998) 100–105. https://doi.org/10.1046/j.1365-2842.1998.00235.x

[46] Y. Pan, B. Yang, N. Jia, Y. Yu, X. Xu, Y. Wang, B. Wu, Enhanced thermally conductive and thermomechanical properties of polymethyl methacrylate (PMMA)/graphene nanoplatelets (GNPs) nanocomposites for radiator of electronic components, *Polym. Test.* 101 (2021) 107237. https://doi.org/10.1016/j.polymertesting.2021.107237

[47] M. Alhasani, F. Alkhatib, A. Al-bonayan, R. Shah, Preparation of glass nanofibers to reinforce polymethyl methacrylate toward persistent photoluminescent bricks, *J. Photochem. Photobiol. A Chem.* (2023) 114988. https://doi.org/10.1016/j.jphotochem.2023.114988

[48] L. Sun, Z. Liu, H. Huang, Y. Xia, Y. Gan, X. He, Flexible Al_2O_3/polymethyl methacrylate composite nanofibers for high-performance sun shading and radiative cooling, *Mater. Today Commun.* 37 (2023) 106903. https://doi.org/10.1016/j.mtcomm.2023.106903

[49] S. Bhogal, J. D'Auria, S. Gehrmann, Polymethyl methacrylate for elbow arthroplasty: Is there another way?, *J. Hand Surg. Glob.* Online. (2024). https://doi.org/10.1016/j.jhsg.2023.11.006

[50] P.J. Roe, M.P. Ansell, Jute-reinforced polyester composites, *J. Mater. Sci.* 20 (1985) 4015–4020. https://doi.org/10.1007/BF00552393

[51] A.C. de Albuquerque, K. Joseph, L. Hecker de Carvalho, J.R.M. D'Almeida, Effect of wettability and ageing conditions on the physical and mechanical properties of uniaxially oriented jute-roving-reinforced polyester composites, *Compos. Sci. Technol.* 60 (2000) 833–844. https://doi.org/10.1016/S0266-3538(99)00188-8

[52] S.K. Pal, D. Mukhopadhyay, S.K. Sanyal, R.N. Mukherjea, Studies on process variables for natural fiber composites—Effect of polyesteramide polyol as interfacial agent, *J. Appl. Polym. Sci.* 35 (1988) 973–985. https://doi.org/10.1002/app.1988.070350412

[53] O. Owolabi, T. Czvikovszky, I. Kovács, Coconut-fiber-reinforced thermosetting plastics, *J. Appl. Polym. Sci.* 30 (1985) 1827–1836. https://doi.org/10.1002/app.1985.070300504

[54] J. Rout, M. Misra, S.S. Tripathy, S.K. Nayak, A.K. Mohanty, The influence of fibre treatment of the performance of coir-polyester composites, *Compos. Sci. Technol.* 61 (2001) 1303–1310. https://doi.org/10.1016/S0266-3538(01)00021-5

[55] K.G. Satyanarayana, K. Sukumaran, A.G. Kulkarni, S.G.K. Pillai, P.K. Rohatgi, Performance of banana fabric-polyester resin composites, in *Composite Structures 2*, Springer, 1983: pp. 535–548. https://doi.org/10.1007/978-94-009-6640-6_38

[56] N.M. White, M.P. Ansell, Straw-reinforced polyester composites, *J. Mater. Sci.* 18 (1983) 1549–1556. https://doi.org/10.1007/BF01111977

[57] L.U. Devi, S.S. Bhagawan, S. Thomas, Mechanical properties of pineapple leaf fiber-reinforced polyester composites, *J Appl Polym Sci.* 64 (1997) 1739–1748. https://doi.org/10.1002/(SICI)1097-4628(19970531)64:9<1739::AID-APP10>3.0.CO;2-T

[58] L.Y. Mwaikambo, E.T.N. Bisanda, The performance of cotton–kapok fabric–polyester composites, *Polym. Test.* 18 (1999) 181–198. https://doi.org/10.1016/S0142-9418(98)00017-8

[59] M. Baiardo, E. Zini, M. Scandola, Flax fibre–polyester composites, *Compos. Part A Appl. Sci. Manuf.* 35 (2004) 703–710. https://doi.org/10.1016/j.compositesa.2004.02.004

[60] A. Atiqah, M. Maleque, M. Jawaid, M. Iqbal, Development of kenaf-glass reinforced unsaturated polyester hybrid composite for structural applications, *Compos. Part B Eng.* 56 (2014) 68–73. https://doi.org/10.1016/j.compositesb.2013.08.019

[61] M. Zaghloul, M. Zaghloul, M. Zaghloul, Developments in polyester composite materials–An in-depth review on natural fibres and nano fillers, *Compos. Struct.* 278 (2021) 114698. https://doi.org/10.1016/j.compstruct.2021.114698

[62] C. Ozkan, N. Karsli, A. Aytac, V. Deniz, Short carbon fiber reinforced polycarbonate composites: Effects of different sizing materials, *Compos. Part B Eng.* 62 (2014) 230–235. https://doi.org/10.1016/j.compositesb.2014.03.002

[63] B. Higgins, W. Brittain, Polycarbonate carbon nanofiber composites, *Eur. Polym. J.* 41 (2005) 889–893. https://doi.org/10.1016/j.eurpolymj.2004.11.040

[64] F. Han, Y. Zhang, C. Wang, Z. Wang, H. Yue, Analysis of ballistic performance and penetration damage mechanisms of aramid woven fabric reinforced polycarbonate composites with different matrix content, *Chem. Eng. J.* 453 (2023) 139470. https://doi.org/10.1016/j.cej.2022.139470

[65] C. Chen, Z. Hou, S. Chen, J. Guo, Z. Chen, J. Hu, Photothermally responsive smart elastomer composites based on aliphatic polycarbonate backbone for biomedical applications, *Compos. Part B Eng.* 240 (2022) 109985. https://doi.org/10.1016/j.compositesb.2022.109985

[66] J. Zhang, A. Koubaa, D. Xing, H. Wang, F. Wang, X.-M. Wang, Q. Wang, Flammability, thermal stability, and mechanical properties of wood flour/polycarbonate/polyethylene bio-based composites, *Ind. Crops Prod.* 169 (2021) 113638. https://doi.org/10.1016/j.indcrop.2021.113638

[67] F. Hacioglu, U. Tayfun, T. Ozdemir, T Tincer, Characterization of carbon fiber and glass fiber reinforced polycarbonate composites and their behavior under gamma irradiation, *Prog. Nucl. Energy.* 134 (2021) 103665. https://doi.org/10.1016/j.pnucene.2021.103665

[68] Y. Baek, P. Shin, J. Kim, H. Park, K. DeVries, Thermal transfer, interfacial, and mechanical properties of carbon fiber/polycarbonate-CNT composites using infrared thermography, *Polym. Test.* 81 (2020) 106247. https://doi.org/10.1016/j.polymertesting.2019.106247

[69] X. Mu, Z. Jin, F. Chu, W. Cai, Y. Zhu, B. Yu, L. Song, High-performance flame-retardant polycarbonate composites: Mechanisms investigation and fire-safety evaluation systems establishment, *Compos. Part B Eng.* 238 (2022) 109873. https://doi.org/10.1016/j.compositesb.2022.109873

[70] A.M. Kharaev, R.C. Bazheva, A.A. Chaika, Composite materials based on polycarbonate (review), *Int. Polym. Sci. Technol.* 34 (2007) 27–33. https://doi.org/10.1177/0307174X0703401006

[71] A. Davis, Thermal stability of polysulphone, die makromol. *Chemie.* 128 (1969) 242–251. https://doi.org/10.1002/MACP.1969.021280123

[72] S. Bose, T. Kuila, T. Nguyen, N. Kim, K Lau, Polymer membranes for high temperature proton exchange membrane fuel cell: Recent advances and challenges, *Prog. Polym. Sci.* 36 (2011) 813–843. https://doi.org/10.1016/j.progpolymsci.2011.01.003

[73] G. Sur, H. Sun, S. Lyu, J. Mark, Synthesis, structure, mechanical properties, and thermal stability of some polysulfone/organoclay nanocomposites, *Polymer (Guildf).* 42 (2001) 9783–9789. https://doi.org/10.1016/S0032-3861(01)00527-4

[74] M. Hunt, Development of carbon fibre/polysulphone orthopaedic implants, *Mater. Des.* 8 (1987) 113–119. https://doi.org/10.1016/0261-3069(87)90117-8

[75] J. Chlopek, P. Rosol, A. Morawska-Chochol, Durability of polymer-ceramics composite implants determined in creep tests, *Compos. Sci. Technol.* 66 (2006) 1615–1622. https://doi.org/10.1016/j.compscitech.2005.11.016

[76] S. Saravanabhavan, S. Dharmalingam, Fabrication of polysulphone/hydroxyapatite nanofiber composite implant and evaluation of their in vitro bioactivity and biocompatibility towards the post-surgical therapy of gastric cancer, *Chem. Eng. J.* 234 (2013) 380–388. https://doi.org/10.1016/j.cej.2013.08.076

[77] N. Spanos, V. Deimede, P.G. Koutsoukos, Functionalization of synthetic polymers for potential use as biomaterials: Selective growth of hydroxyapatite on sulphonated polysulphone, *Biomaterials.* 23 (2002) 947–953. https://doi.org/10.1016/S0142-9612(01)00207-1

[78] H. Xu, M. Ding, S. Liu, Y. Li, Z. Shen, K. Wang, Preparation and characterization of novel polysulphone hybrid ultrafiltration membranes blended with N-doped GO/TiO_2 nanocomposites, *Polymer (Guildf).* 117 (2017) 198–207. https://doi.org/10.1016/j.polymer.2017.04.022

[79] P. Kemp, F. de Wet, S.J. Botha, Flexural properties of denture-base polymer reinforced with glass-fibre polysulphone composite., *SADJ J. South African Dent. Assoc.* 59 (2004) 190–192.

[80] M. Kutz, *Applied Plastics Engineering Handbook: Processing and Materials*, William Andrew, 2011.

[81] I. Page, *Polyamides as Engineering Thermoplastic Materials*, iSmithers Rapra Publishing, 2000.

[82] Y. Amintowlieh, A. Sardashti, L.C. Simon, Polyamide 6–wheat straw composites: Effects of additives on physical and mechanical properties of the composite, *Polym. Compos.* 33 (2012) 976–984. https://doi.org/10.1002/pc.22228

[83] N. Saba, P.M. Tahir, M. Jawaid, A review on potentiality of nano filler/natural fiber filled polymer hybrid composites, *Polymers (Basel).* 6 (2014) 2247–2273. https://doi.org/10.3390/polym6082247

[84] Y. Kwon, J. Leckie, Hypochlorite degradation of crosslinked polyamide membranes: II. Changes in hydrogen bonding behavior and performance, *J. Memb. Sci.* 282 (2006) 456–464. https://doi.org/10.1016/j.memsci.2006.06.004

[85] W. Lau, S. Gray, T. Matsuura, D. Emadzadeh, J. Chen, A review on polyamide thin film nanocomposite (TFN) membranes: History, applications, challenges and approaches, *Water Res.* 80 (2015) 306–324. https://doi.org/10.1016/j.watres.2015.04.037

[86] I.V. Slugin, G.B. Sklyarova, A.I. Kashirin, L.V. Tkacheva, Rusar para-aramid fibres for composite materials for construction applications, *Fibre Chem.* 38 (2006) 25–26. https://doi.org/10.1007/S10692-006-0033-0

[87] A. Kausar, Advances in carbon fiber reinforced polyamide-based composite materials, *Adv. Mater. Sci.* 19 (2019) 67–82. https://doi.org/10.2478/ADMS-2019-0023

[88] K. Pielichowski, K. Flejtuch, Differential scanning calorimetry studies on poly(ethylene glycol) with different molecular weights for thermal energy storage materials, *Polym. Adv. Technol.* 13 (2002) 690–696. https://doi.org/10.1002/pat.276

[89] I. Dincer, M.A. Rosen, *Thermal Energy Storage: Systems and Applications*, John Wiley & Sons, 2021.

[90] Q. Meng, J. Hu, A poly(ethylene glycol)-based smart phase change material, *Sol. Energy Mater. Sol. Cells.* 92 (2008) 1260–1268. https://doi.org/10.1016/j.solmat.2008.04.026

[91] A. Kürklü, Energy storage applications in greenhouses by means of phase change materials (PCMs): A review, *Renew. Energy.* 13 (1998) 89–103. https://doi.org/10.1016/S0960-1481(97)83337-X

[92] S. Mondal, Phase change materials for smart textiles – An overview, *Appl. Therm. Eng.* 28 (2008) 1536–1550. https://doi.org/10.1016/j.applthermaleng.2007.08.009

[93] Y. Xie, M. Zou, S. Xiao, W. Chen, Y. Liu, X. Hu, X. Jiang, Y. He, Q. Li, MXene-modified bio-based pitaya peel foam/polyethylene glycol composite phase change material with excellent photo-thermal conversion efficiency, thermal energy storage capacity and thermal conductivity, *J. Energy Storage.* 78 (2024) 110089. https://doi.org/10.1016/j.est.2023.110089

[94] D. Bouazzi, I. Chérif, A. Mehri, H. Touati, M. Teresa Caccamo, S. Magazù, S. Ayachi, J.M. Clacens, B. Badraoui, A joint experimental and theoretical study on structural, vibrational and morphological properties of newly synthesized nanocomposites involving hydroxyapatite-alt-polyethylene glycol (HAP/PEG), *J. Mol. Liq.* 390 (2023) 123192. https://doi.org/10.1016/j.molliq.2023.123192

[95] E. Ranjbari, S. Bazgir, M.M.A. Shirazi, Needleless electrospinning of poly(acrylic acid) superabsorbent: Fabrication, characterization and swelling behavior, *Polym. Test.* 84 (2020) 106403. https://doi.org/10.1016/j.polymertesting.2020.106403

[96] A.E. Swilem, A.H.M. Elshazly, A.A. Hamed, E.S.A. Hegazy, H.A. Abd El-Rehim, Nanoscale poly(acrylic acid)-based hydrogels prepared via a green single-step approach for application as low-viscosity biomimetic fluid tears, *Mater. Sci. Eng. C.* 110 (2020) 110726. https://doi.org/10.1016/j.msec.2020.110726

[97] M. Salami, Z. Talebpour, R. Alizadeh, Fabrication of a new SPME fiber based on Polyacrylic acid/MIL-88(Fe)-NH2 composite as a self-healing coating for the analysis of breast cancer biomarkers in the urine sample, *J. Pharm. Biomed. Anal.* 219 (2022) 114902. https://doi.org/10.1016/j.jpba.2022.114902

[98] E.S. Abdel-Halim, H.E. Emam, M.H. El-Rafie, Utilization of hydroxypropyl cellulose and poly(acrylic acid)-hydroxypropyl cellulose composite as thickeners for textile printing, *Carbohydr. Polym.* 74 (2008) 938–941. https://doi.org/10.1016/j.carbpol.2008.05.013

[99] L. Zhang, S. Tang, F. He, Y. Liu, W. Mao, Y. Guan, Highly efficient and selective capture of heavy metals by poly(acrylic acid) grafted chitosan and biochar composite for wastewater treatment, *Chem. Eng. J.* 378 (2019) 122215. https://doi.org/10.1016/j.cej.2019.122215

[100] L. Wang, J. Zhang, A. Wang, Removal of methylene blue from aqueous solution using chitosan-g-poly(acrylic acid)/montmorillonite superadsorbent nanocomposite, *Colloids Surf. A Physicochem. Eng. Asp.* 322 (2008) 47–53. https://doi.org/10.1016/j.colsurfa.2008.02.019

[101] J. Isailović, K. Vidović, S.B. Hočevar, Simple electrochemical sensors for highly sensitive detection of gaseous hydrogen peroxide using polyacrylic-acid-based sensing membrane, *Sens. Actuators B Chem.* 352 (2022) 131053. https://doi.org/10.1016/j.snb.2021.131053

[102] Y. Bao, J.Z. Ma, Polymethacrylic acid/Na-montmorillonite/SiO_2 nanoparticle composites structures and thermal properties, *Polym. Bull.* 66 (2011) 541–549. https://doi.org/10.1007/s00289-010-0334-6

[103] N.I. Georgiev, R. Bryaskova, R. Tzoneva, I. Ugrinova, C. Detrembleur, S. Miloshev, A.M. Asiri, A.H. Qusti, V.B. Bojinov, A novel pH sensitive water soluble fluorescent nanomicellar sensor for potential biomedical applications, *Bioorganic Med. Chem.* 21 (2013) 6292–6302. https://doi.org/10.1016/j.bmc.2013.08.064

[104] D.C. Martin, J.L. Semple, M. V. Sefton, Poly(methacrylic acid-co-methyl methacrylate) beads promote vascularization and wound repair in diabetic mice, *J. Biomed. Mater. Res. – Part A*. 93 (2009) 484–492. https://doi.org/10.1002/jbm.a.32528

[105] M.R. Saboktakin, R.M. Tabatabaie, A. Maharramov, M.A. Ramazanov, Synthesis and rheological properties of poly(methyl methacrylate)/polymethacrylic acid nanocomposites as denture resins, *Compos. Part B Eng.* 42 (2011) 851–855. https://doi.org/10.1016/j.compositesb.2011.01.009

[106] M. Hussain, A. Hafeez, S.P. Kushwaha, Nanoformulation mediated transdermal delivery of anti-diabetic drugs: an updated review, *Intell. Pharm.* 1 (2023) 192–200. https://doi.org/10.1016/j.ipha.2023.08.009

[107] S. Bhattacharjee, S. Goswami, S. Das, S. Bhattacharjee, S. Bhaladhare, pH-responsive, stable, and biocompatible functional nanogels based on chitosan (CS)/poly methacrylic acid (PMAA) polymers: Synthesis and characterization, *Mater. Today Commun.* 36 (2023) 106541. https://doi.org/10.1016/j.mtcomm.2023.106541

[108] A.G. Trejo, Fungal degradation of polyvinyl acetate, *Ecotoxicol. Environ. Saf.* 16 (1988) 25–35. https://doi.org/10.1016/0147-6513(88)90013-9

[109] A. Kaboorani, B. Riedl, Mechanical performance of polyvinyl acetate (PVA)-based biocomposites, in *Biocomposites: Design and Mechanical Performance*, Fourteenth, Elsevier Ltd., 2015: pp. 347–364. https://doi.org/10.1016/B978-1-78242-373-7.00009-3

[110] J. Ahmad, M.B. Hågg, Polyvinyl acetate/titanium dioxide nanocomposite membranes for gas separation, *J. Memb. Sci.* 445 (2013) 200–210. https://doi.org/10.1016/j.memsci.2013.04.052

[111] I. Hammami, K. Benhamou, H. Hammami, S. SoretoTeixeira, M. Arous, H. Kaddami, M.P.F. Graça, L.C. Costa, Electrical, morphology and structural properties of biodegradable nanocomposite polyvinyl-acetate/ cellulose nanocrystals, *Mater. Chem. Phys.* 240 (2020) 122182. https://doi.org/10.1016/j.matchemphys.2019.122182

[112] Y. Zeleke, G.K. Rotich, Design and development of false ceiling board using polyvinyl acetate (PVAc) composite reinforced with false banana fibres and filled with sawdust, *Int. J. Polym. Sci.* 2021 (2021). https://doi.org/10.1155/2021/5542329

[113] C.A. Finch, *Polyvinyl Alcohol: Properties and Applications* Wiley , 1973.

[114] F.L. Marten, Vinyl alcohol polymers, *Polym. Sci. Technol.* 8 (2002) 53–88. https://doi.org/10.1002/0471440264.pst384

[115] C.C. DeMerlis, D.R. Schoneker, Review of the oral toxicity of polyvinyl alcohol (PVA), *Food Chem. Toxicol.* 41 (2003) 319–326. https://doi.org/10.1016/S0278-6915(02)00258-2

[116] A. Timofejeva, M. D'Este, D. Loca, Calcium phosphate/polyvinyl alcohol composite hydrogels: A review on the freeze-thawing synthesis approach and applications in regenerative medicine, *Eur. Polym. J.* 95 (2017) 547–565. https://doi.org/10.1016/j.eurpolymj.2017.08.048

[117] S. Nasibi, H. Nargesi khoramabadi, M. Arefian, M. Hojjati, I. Tajzad, A. Mokhtarzade, M. Mazhar, A. Jamavari, A review of polyvinyl alcohol/carboxiy methyl cellulose (PVA/CMC) composites for various applications, *J. Compos. Compd.* 2 (2020) 69–76. https://doi.org/10.29252/jcc.2.2.2

[118] A.A. Oun, G.H. Shin, J.W. Rhim, J.T. Kim, Recent advances in polyvinyl alcohol-based composite films and their applications in food packaging, *Food Packag. Shelf Life.* 34 (2022) 100991. https://doi.org/10.1016/j.fpsl.2022.100991

[119] A.M. Al-Fa'ouri, O.A. Lafi, H.H. Abu-Safe, M. Abu-Kharma, Investigation of optical and electrical properties of copper oxide – Polyvinyl alcohol nanocomposites for solar cell applications, *Arab. J. Chem.* 16 (2023) 104535. https://doi.org/10.1016/j.arabjc.2022.104535

[120] V. Vatanpour, O.O. Teber, M. Mehrabi, I. Koyuncu, Polyvinyl alcohol-based separation membranes: a comprehensive review on fabrication techniques, applications and future prospective, *Mater. Today Chem.* 28 (2023) 101381. https://doi.org/10.1016/j.mtchem.2023.101381

[121] K. Zhang, Y. Liu, X. Shi, R. Zhang, Y. He, H. Zhang, W. Wang, Application of polyvinyl alcohol/chitosan copolymer hydrogels in biomedicine: A review, *Int. J. Biol. Macromol.* (2023) 125192. https://doi.org/10.1016/j.ijbiomac.2023.125192

[122] A.L.G. Saad, W.M. Saved, M.G.M. Ahmed, A.M. Hassan, Preparation and properties of some filled poly(vinyl chloride) compositions, *J. Appl. Polym. Sci.* 73 (1999) 2657–2670. https://doi.org/10.1002/(SICI)1097-4628(19990923)73:13<2657::AID-APP14>3.0.CO;2-7

[123] I.K. Bishay, S.L. Abd-El-Messieh, S.H. Mansour, Electrical, mechanical and thermal properties of polyvinyl chloride composites filled with aluminum powder, *Mater. Des.* 32 (2011) 62–68. https://doi.org/10.1016/j.matdes.2010.06.035

[124] H.P.S.A. Khalil, M.A. Tehrani, Y. Davoudpour, A.H. Bhat, M. Jawaid, A. Hassan, Natural fiber reinforced poly(vinyl chloride) composites: A review, *J. Reinf. Plast. Compos.* 32 (2013) 330–356. https://doi.org/10.1177/0731684412458553

[125] S. Mishra, G.B. Perumal, J.B. Naik, Studies on mechanical properties of polyvinyl chloride composites, *Polym. – Plast. Technol. Eng.* 36 (1997) 489–500. https://doi.org/10.1080/03602559708000638

[126] N. Ahmad, A. Kausar, B. Muhammad, Perspectives on polyvinyl chloride and carbon nanofiller composite: A Review, *Polym. – Plast. Technol. Eng.* 55 (2016) 1076–1098. https://doi.org/10.1080/03602559.2016.1163587

[127] S. Awad, T. Hamouda, M. Midani, Y. Zhou, E. Katsou, M. Fan, Date palm fibre geometry and its effect on the physical and mechanical properties of recycled polyvinyl chloride composite, *Ind. Crops Prod.* 174 (2021) 114172. https://doi.org/10.1016/j.indcrop.2021.114172

[128] N. Dutta, B. Bhadra, G. Gogoi, T. Kumar Maji, Development of polyvinyl chloride/waste rice husk ash/modified montmorillonite nanocomposite using epoxidized soybean oil as green additive substituting synthetic plasticizer and compatibiliser, *Clean. Mater.* 2 (2021) 100033. https://doi.org/10.1016/j.clema.2021.100033

[129] R.A. Khan, E. Mahdi, Effect of trigger mechanisms on the quasi-static axial crushing behavior of glass epoxy/polyvinyl chloride hybrid composite tubes, *Thin-Walled Struct.* 182 (2023) 110306. https://doi.org/10.1016/j.tws.2022.110306

[130] N.S. Nor Arman, R.S. Chen, S. Ahmad, Review of state-of-the-art studies on the water absorption capacity of agricultural fiber-reinforced polymer composites for sustainable construction, *Constr. Build. Mater.* 302 (2021) 124174. https://doi.org/10.1016/j.conbuildmat.2021.124174

[131] M. Safarpour, A. Safikhani, V. Vatanpour, Polyvinyl chloride-based membranes: A review on fabrication techniques, applications and future perspectives, *Sep. Purif. Technol.* 279 (2021) 119678. https://doi.org/10.1016/j.seppur.2021.119678

[132] S. Ebnesajjad, *Introduction to Fluoropolymers: Materials, Technology, and Applications*, 2nd ed., Elsevier Science, 2020.

[133] J.W. Zha, M.S. Zheng, B.H. Fan, Z.M. Dang, Polymer-based dielectrics with high permittivity for electric energy storage: A review, *Nano Energy.* 89 (2021) 106438. https://doi.org/10.1016/j.nanoen.2021.106438

[134] K. Maity, S. Garain, K. Henkel, D. Schmeißer, D. Mandal, Self-powered human-health monitoring through aligned PVDF nanofibers interfaced skin-interactive piezoelectric sensor, *ACS Appl. Polym. Mater.* 2 (2020) 862–878. https://doi.org/10.1021/acsapm.9b00846

[135] P. Saxena, P. Shukla, A comprehensive review on fundamental properties and applications of poly(vinylidene fluoride) (PVDF), *Adv. Compos. Hybrid Mater.* 4 (2021) 8–26. https://doi.org/10.1007/s42114-021-00217-0

[136] L. Wu, M. Jing, Y. Liu, H. Ning, X. Liu, S. Liu, L. Lin, N. Hu, L. Liu, Power generation by PVDF-TrFE/graphene nanocomposite films, *Compos. Part B Eng.* 164 (2019) 703–709. https://doi.org/10.1016/j.compositesb.2019.01.055

[137] L. Wu, Z. Jin, Y. Liu, H. Ning, X. Liu, Alamusi, N. Hu, Recent advances in the preparation of PVDF-based piezoelectric materials, *Nanotechnol. Rev.* 11 (2022) 1386–1407. https://doi.org/10.1515/ntrev-2022-0082

[138] M. Venkatesan, W.C. Chen, C.J. Cho, L. Veeramuthu, L.G. Chen, K.Y. Li, M.L. Tsai, Y.C. Lai, W.Y. Lee, W.C. Chen, C.C. Kuo, Enhanced piezoelectric and photocatalytic performance of flexible energy harvester based on CsZn0.75Pb0.25I3/CNC–PVDF composite nanofibers, *Chem. Eng. J.* 433 (2022) 133620. https://doi.org/10.1016/j.cej.2021.133620

[139] Z. Wang, M. Fang, H. Li, Y. Wen, C. Wang, Y. Pu, Enhanced dielectric properties in poly(vinylidene fluoride) composites by nanosized $Ba(Fe_{0.5}Nb_{0.5})O_3$ powders, *Compos. Sci. Technol.* 117 (2015) 410–416. https://doi.org/10.1016/j.compscitech.2015.07.018

[140] Z. Zhou, X. Du, J. Luo, L. Yao, Z. Zhang, H. Yang, Q. Zhang, Coupling of interface effects and porous microstructures in translucent piezoelectric composites for enhanced energy harvesting and sensing, *Nano Energy.* 84 (2021) 105895. https://doi.org/10.1016/j.nanoen.2021.105895

[141] C.M. Costa, V.F. Cardoso, P. Martins, D.M. Correia, R. Gonçalves, P. Costa, V. Correia, C. Ribeiro, M.M. Fernandes, P.M. Martins, S. Lanceros-Méndez, Smart and multifunctional materials based on electroactive poly(vinylidene fluoride): Recent advances and opportunities in sensors, Actuators,

Energy, Environmental, and Biomedical Applications, *Chem. Rev.* 123 (2023) 11392–11487. https://doi.org/10.1021/acs.chemrev.3c00196
[142] N. Meng, X. Ren, G. Santagiuliana, L. Ventura, H. Zhang, J. Wu, H. Yan, M.J. Reece, E. Bilotti, Ultrahigh β-phase content poly(vinylidene fluoride) with relaxor-like ferroelectricity for high energy density capacitors, *Nat. Commun.* 10 (2019) 1–9. https://doi.org/10.1038/s41467-019-12391-3
[143] J. Fu, Y. Hou, M. Zheng, M. Zhu, Flexible piezoelectric energy harvester with extremely high power generation capability by sandwich structure design strategy, *ACS Appl. Mater. Interfaces.* 12 (2020) 9766–9774. https://doi.org/10.1021/acsami.9b21201
[144] S. Jin, X. Qiu, B. Huang, L. Wang, Q. Zhang, Z. Fu, Dielectric properties of modified BNT/PTFE composites for microwave RF antenna applications, *J. Mater. Sci. Mater. Electron.* 27 (2016) 8378–8383. https://doi.org/10.1007/s10854-016-4849-y
[145] Y. Lv, W. Wang, G. Xie, J. Luo, Self-lubricating PTFE-based composites with black phosphorus nanosheets, *Tribol. Lett.* 66 (2018) 1–11. https://doi.org/10.1007/s11249-018-1010-5
[146] K. Zhang, D. Zhao, W. Chen, L. Zheng, L. Yao, Y. Qiu, F. Xu, Three-dimensional woven structural glass fiber/polytetrafluoroethylene (PTFE) composite antenna with superb integrity and electromagnetic performance, *Compos. Struct.* 281 (2022) 115096. https://doi.org/10.1016/j.compstruct.2021.115096
[147] F. Zhang, J. Zhang, Y. Zhu, X. Wang, Y. Jin, Microstructure and properties of polytetrafluoroethylene composites modified by carbon materials and aramid fibers, *Coatings.* 10 (2020) 1–20. https://doi.org/10.3390/coatings10111103
[148] X.L. Wu, W.M. Huang, Z.G. Seow, W.S. Chin, W.G. Yang, K.Y. Sun, Two-step shape recovery in heating-responsive shape memory polytetrafluoroethylene and its thermally assisted self-healing, *Smart Mater. Struct.* 22 (2013) 125023. https://doi.org/10.1088/0964-1726/22/12/125023
[149] H. Wei, W. He, Q. Li, Y. Yu, R. Xu, J. Zhou, J. Shen, W. Chen, Glass fiber/polytetrafluoroethylene composite with low dielectric constant and thermal stability for high-frequency application, *Ceram. Int.* 49 (2023) 28449–28456. https://doi.org/10.1016/j.ceramint.2023.06.099
[150] R.C. Sanfelice, L.A. Mercante, A. Pavinatto, N.B. Tomazio, C.R. Mendonça, S.J.L. Ribeiro, L.H.C. Mattoso, D.S. Correa, Hybrid composite material based on polythiophene derivative nanofibers modified with gold nanoparticles for optoelectronics applications, *J. Mater. Sci.* 52 (2017) 1919–1929. https://doi.org/10.1007/s10853-016-0481-8
[151] K. Mohammed Rehan, B. Anbarasu, P. Mohamed Ashfaque, S. Mohammed Safiullah, K. Anver Basha, Synthesis and characterization of poly(vinyl carbazole-co-ethoxy ethyl methacrylate) and its nanocomposites, *Mater. Today Proc.*, Elsevier Ltd. (2022) 325–330. https://doi.org/10.1016/j.matpr.2021.07.441
[152] Y. Zhou, X. Zhao, Y. Zhang, F. Dou, S. Wang, C. Wang, Ternary flash memory device based on polycarbazole with ZrO_2 composite materials, *Org. Electron.* 99 (2021) 106354. https://doi.org/10.1016/j.orgel.2021.106354
[153] S. Kumari, K.K. Kushwah, S. Dubey, M. Ramrakhiani, Studies on CdSe/PVK nanocomposites films for electroluminescent display applications, *Opt. Mater. (Amst).* 97 (2019) 109319. https://doi.org/10.1016/j.optmat.2019.109319
[154] V. Bühler, *Polyvinylpyrrolidone Excipients for Pharmaceuticals: Povidone, Crospovidone and Copovidone*, Springer Berlin Heidelberg, 2005.
[155] M. Teodorescu, M. Bercea, Poly(vinylpyrrolidone) – A versatile polymer for biomedical and beyond medical applications, *Polym. – Plast. Technol. Eng.* 54 (2015) 923–943. https://doi.org/10.1080/03602559.2014.979506
[156] M. Kurakula, G.S.N. Koteswara Rao, Moving polyvinyl pyrrolidone electrospun nanofibers and bioprinted scaffolds toward multidisciplinary biomedical applications, *Eur. Polym. J.* 136 (2020) 109919. https://doi.org/10.1016/j.eurpolymj.2020.109919
[157] G.R. Del Sorbo, G. Truda, A. Bifulco, J. Passaro, G. Petrone, B. Vitolo, G. Ausanio, A. Vergara, F. Marulo, F. Branda, Non monotonous effects of noncovalently functionalized graphene addition on the structure and sound absorption properties of polyvinylpyrrolidone (1300 kDa) electrospun mats, *Materials (Basel).* 12 (2019) 108. https://doi.org/10.3390/ma12010108
[158] X. Zhang, B. Wu, G. Chen, Y. Xu, T. Shi, Z. Huang, Y. Liu, M. Fang, X. Wu, X. Min, Preparation and characterization of flexible smart glycol/polyvinylpyrrolidone/nano-Al_2O_3 phase change fibers, *Energ. Fuel.* 35 (2021) 877–882. https://doi.org/10.1021/acs.energyfuels.0c03370

[159] S. Pokhrel, P. Karki, Peanut shells cellulose based biodegradable nanocomposites with polyvinyl pyrrolidone (PVP) and polyvinyl alcohol (PVA), *Macromol. Symp.* 409 (2023) 1–8. https://doi.org/10.1002/masy.202200183

[160] H. Zhao, C. Yang, N. Li, J. Yin, Y. Feng, Y. Liu, J. Li, Electrical and mechanical properties of polyimide composite films reinforced by ultralong titanate nanotubes, *Surf. Coatings Technol.* 360 (2019) 13–19. https://doi.org/10.1016/j.surfcoat.2019.01.013

[161] C. Yi, W. Li, S. Shi, K. He, P. Ma, M. Chen, C. Yang, High-temperature-resistant and colorless polyimide: Preparations, properties, and applications, *Sol. Energy.* 195 (2020) 340–354. https://doi.org/10.1016/j.solener.2019.11.048

[162] P. Ma, C. Dai, H. Wang, Z. Li, H. Liu, W. Li, A review on high temperature resistant polyimide films: Heterocyclic structures and nanocomposites, *Compos. Commun.* 16 (2019) 84–93. https://doi.org/10.1016/j.coco.2019.08.011

[163] S. Malakooti, S. Vivod, M. Pereira, C.R. Ruggeri, Fabric reinforced polyimide aerogel matrix composites with low thermal conductivity, high flexural strength, and high sound absorption coefficient, *Compos. Part B Eng.* 260 (2023) 110751. https://doi.org/10.1016/j.compositesb.2023.110751

[164] H. Shao, Z. Fei, X. Li, Z. Zhang, S. Zhao, K. Li, Z. Yang, Polyimide/SiO_2 composite aerogels with excellent thermal and sound insulation properties prepared by confined filling method, *Mater. Lett.* 354 (2024) 135402. https://doi.org/10.1016/j.matlet.2023.135402

[165] M. Guo, Y. Zhang, C. Huang, X. Zhao, X.P. Yan, Shape memory polyimide aerogel composites with high programming temperatures and exceptional shape recovery capability, *Compos. Part A Appl. Sci. Manuf.* 174 (2023) 107717. https://doi.org/10.1016/j.compositesa.2023.107717

[166] M. Ionescu, *Chemistry and Technology of Polyols for Polyurethanes*, iSmithers Rapra Publishing, 2005.

[167] G.T. Howard, Biodegradation of polyurethane: A review, *Int. Biodeterior. Biodegrad.* 49 (2002) 245–252. https://doi.org/10.1016/S0964-8305(02)00051-3

[168] T. Romaskevic, S. Budriene, K. Pielichowski, J. Pielichowski, Application of polyurethane-based materials for immobilization of enzymes and cells: A review, *Chemija.* 17 (2006) 74–89.

[169] A. Atiqah, M.T. Mastura, B.A Ahmed Ali, M. Jawaid, S.M Sapuan, A review on polyurethane and its polymer composites, *Curr. Org. Synth.* 14 (2017) 233–248.

[170] A. Das, P. Mahanwar, A brief discussion on advances in polyurethane applications, *Adv. Ind. Eng. Polym. Res.* 3 (2020) 93–101. https://doi.org/10.1016/j.aiepr.2020.07.002

[171] C. Kuranchie, A. Yaya, Y.D. Bensah, The effect of natural fibre reinforcement on polyurethane composite foams – A review, *Sci. African.* 11 (2021) e00722. https://doi.org/10.1016/j.sciaf.2021.e00722

[172] C. Xu, Y. Hong, Rational design of biodegradable thermoplastic polyurethanes for tissue repair, *Bioact. Mater.* 15 (2022) 250–271. https://doi.org/10.1016/j.bioactmat.2021.11.029

[173] M. Cui, Z. Chai, Y. Lu, J. Zhu, J. Chen, Developments of polyurethane in biomedical applications: A review, *Resour. Chem. Mater.* 2 (2023) 262–276. https://doi.org/10.1016/j.recm.2023.07.004

[174] W. Xu, C. Wei, J. Lv, H. Liu, X. Huang, T. Liu, Preparation, characterization, and properties of in situ formed graphene oxide/phenol formaldehyde nanocomposites, *J. Nanomater.* 2013 (2013) 1–6. https://doi.org/10.1155/2013/319840

[175] C. Liu, Z. Cheng, S. Wang, Y. Zhang, O. Hosseinaei, Effects of nanocrystalline cellulose on the micro-creep properties of phenol formaldehyde resin, *Mater. Res. Express.* 5 (2018) 125019. https://doi.org/10.1088/2053-1591/aae16f

[176] J. Tan, B. Pan, L. Hua, J. Chen, J. Liu, Y. Zhang, Q. Niu, Influence of the addition of nano-porous graphite/parafffin additive on the wear properties of phenol-formaldehyde resin composites, *J. Macromol. Sci. Part B Phys.* 58 (2019) 88–98. https://doi.org/10.1080/00222348.2018.1543826

[177] L. Hu, F. Zhang, L. Luo, L. Wang, Y. Liu, J. Leng, Design and preparation of shape memory phenol–formaldehyde foam composites with excellent thermal stability and mechanical properties, *Compos. Part A Appl. Sci. Manuf.* 174 (2023) 107738. https://doi.org/10.1016/j.compositesa.2023.107738

[178] A.S. Singha, V.K. Thakur, Effect of fibre loading on urea-formaldehyde matrix based green composites, *Iran. Polym. J.* 17 (2008) 861–873.

[179] J.B. Zhong, J. Lv, C. Wei, Mechanical properties of sisal fibre reinforced urea-formaldehyde resin composites, *Express Polym. Lett.* 1 (2007) 681–687. https://doi.org/10.3144/expresspolymlett.2007.93

[180] E. Ibe Kevin, Mechanical properties of urea formaldehyde particle board composite, *Am. J. Chem. Biochem. Eng.* 2 (2018) 10. https://doi.org/10.11648/j.ajcbe.20180201.12

[181] H.C. Yu, Y.T. Zhang, M.J. Wang, C.C. Li, Dispersion of poly(urea-formaldehyde)-based microcapsules for self-healing and anticorrosion applications, *Langmuir.* 35 (2019) 7871–7878. https://doi.org/10.1021/acs.langmuir.9b00526

[182] M. Fadil, D.S. Chauhan, M.A. Quraishi, Smart coating based on urea-formaldehyde microcapsules loaded with benzotriazole for corrosion protection of mild steel in 3.5 % NaCl, *Russ. J. Appl. Chem.* 91 (2018) 1721–1728. https://doi.org/10.1134/S107042721810021X

[183] P.-O. Hagstrand, K. Oksman, Mechanical properties and morphology of flax fiber reinforced melamine-formaldehyde composites, *Polym. Compos.* 22 (2001) 568–578. https://doi.org/10.1002/pc.10560

[184] C. Yu, W. Xu, X. Zhao, J. Xu, M. Jiang, Effects of the reaction degree of melamine-formaldehyde resin on the structures and properties of melamine-formaldehyde/polyvinyl alcohol composite fiber, *Fibers Polym.* 15 (2014) 1828–1834. https://doi.org/10.1007/s12221-014-1828-x

[185] A. Dorieh, M. Farajollah Pour, S. Ghafari Movahed, A. Pizzi, P. Pouresmaeel Selakjani, M. Valizadeh Kiamahalleh, H. Hatefnia, M.H. Shahavi, R. Aghaei, A review of recent progress in melamine-formaldehyde resin based nanocomposites as coating materials, *Prog. Org. Coatings.* 165 (2022) 106768. https://doi.org/10.1016/j.porgcoat.2022.106768

[186] B. Song, X. Zhu, W. Wang, L. Wang, X. Pei, X. Qian, L. Liu, Z. Xu, Toughening of melamine–formaldehyde foams and advanced applications based on functional design, *J. Ind. Eng. Chem.* 119 (2023) 130–152. https://doi.org/10.1016/j.jiec.2022.11.069

[187] Y. Tian, X. Liu, J. Li, A. Caratenuto, S. Zhou, Y. Deng, G. Xiao, M.L. Minus, Y. Zheng, Scalable, fire-retardant, and spectrally robust melamine-formaldehyde photonic bulk for efficient daytime radiative cooling, *Appl. Mater. Today.* 24 (2021) 101103. https://doi.org/10.1016/j.apmt.2021.101103

[188] X. Du, Y. Fang, X. Cheng, Z. Du, M. Zhou, H. Wang, Fabrication and characterization of flame-retardant nanoencapsulated n-octadecane with melamine-formaldehyde shell for thermal energy storage, *ACS Sustain. Chem. Eng.* 6 (2018) 15541–15549. https://doi.org/10.1021/acssuschemeng.8b03980

[189] N. Saba, M. Jawaid, O.Y. Alothman, M.T. Paridah, A. Hassan, Recent advances in epoxy resin, natural fiber-reinforced epoxy composites and their applications, *J. Reinf. Plast. Compos.* 35 (2016) 447–470. https://doi.org/10.1177/0731684415618459

[190] K. Malik, F. Ahmad, E. Gunister, T. Nakato, E. Mouri, M.B. Muhammad, S. Ali, A review of flax fiber reinforced thermoset polymer composites: Structure and mechanical performance, *J. Nat. Fibers.* 19 (2022) 9656–9680. https://doi.org/10.1080/15440478.2021.1990182

[191] G.S. Deshmukh, Advancement in hemp fibre polymer composites: A comprehensive review, *J. Polym. Eng.* 42 (2022) 575–598. https://doi.org/10.1515/polyeng-2022-0033

[192] P.G. Scholar, M. Engineering, M. Engineering, Mechanical behaviour of glass fiber/epoxy modified with nano composites: A review, *Int. J. Innov. Res. Sci. Eng. Technol.* 5 (2016) 76–81.

[193] T.J. Singh, S. Samanta, Characterization of Kevlar fiber and its composites: A review, in *Mater. Today Proc.*, Elsevier Ltd., 2015: pp. 1381–1387. https://doi.org/10.1016/j.matpr.2015.07.057

[194] S. Rana, R. Alagirusamy, M. Joshi, A review on carbon epoxy nanocomposites, *J. Reinf. Plast. Compos.* 28 (2009) 461–487. https://doi.org/10.1177/0731684407085417

[195] L. Cai, A. Al-Ostaz, X. Li, C. Fowler, A.H.D. Cheng, H. Alkhateb, Protection of steel railcar tank containing liquid chlorine from high speed impact by using polyhedral oligomeric silsesquioxane-enhanced polyurea, *Int. J. Impact Eng.* 75 (2015) 1–10. https://doi.org/10.1016/j.ijimpeng.2014.06.015

[196] M.A. Vulcan, C. Damian, P.O. Stanescu, E. Vasile, R. Petre, G. Hubca, Polymeric composites based on polyurea matrix reinforced with carbon nanotubes, *Mater. Plast.* 54 (2017) 41–44. https://doi.org/10.37358/mp.17.1.4781

[197] M. Barczewski, K. Biedrzycka, O. Mysiukiewicz, D. Matykiewicz, J. Andrzejewski, A. Kloziński, M. Szostak, Milled basalt fibers as reinforcement for polyurea composite spray coatings with improved thermomechanical stability and mechanical performance, *Polimery/Polymers.* 65 (2020) 184–195. https://doi.org/10.14314/POLIMERY.2020.3.3

[198] B. Shojaei, M. Najafi, A. Yazdanbakhsh, M. Abtahi, C. Zhang, A review on the applications of polyurea in the construction industry, *Polym. Adv. Technol.* 32 (2021) 2797–2812. https://doi.org/10.1002/pat.5277

[199] W. Ao, Q. Zhou, Y. Xia, Impact resistance and damage mechanism of polyurea coated CFRP laminates under quasi-static indentation and low velocity impact, *Thin-Walled Struct.* 193 (2023) 111258. https://doi.org/10.1016/j.tws.2023.111258

[200] A.D. Roberts, *Natural Rubber Science and Technology*, Oxford University Press, 1988. https://lccn.loc.gov/87030867

[201] J. Zavickis, M. Knite, G. Podins, A. Linarts, R. Orlovs, Polyisoprene-nanostructured carbon composite – A soft alternative for pressure sensor application, *Sensors Actuators A Phys.* 171 (2011) 38–42. https://doi.org/10.1016/j.sna.2011.05.035

[202] D. Ponnamma, K.K. Sadasivuni, K.T. Varughese, S. Thomas, M.A.-A. AlMa'adeed, Natural polyisoprene composites and their electronic applications, in *Flexible and Stretchable Electronic Composites*, Springer, 2016: pp. 1–35. https://doi.org/10.1007/978-3-319-23663-6_1

[203] Z. Niu, R. Wu, Y. Yang, L. Huang, W. Fan, Q. Dai, L. Cui, J. He, C. Bai, Recyclable, robust and shape memory vitrified polyisoprene composite prepared through a green methodology, *Polymer (Guildf).* 228 (2021) 123864. https://doi.org/10.1016/j.polymer.2021.123864

[204] Z. Wu, X. Lin, J. Teng, M. Li, J. Song, C. Huang, R. Wang, H. Ying, L. Zhang, C. Zhu, Recent advances of lignin functionalization for high-performance and advanced functional rubber composites, *Biomacromolecules.* 24 (2023) 4553–4567. https://doi.org/10.1021/acs.biomac.3c00606

[205] T.L. Sun, X.L. Gong, W.Q. Jiang, J.F. Li, Z.B. Xu, W.H. Li, Study on the damping properties of magnetorheological elastomers based on cis-polybutadiene rubber, *Polym. Test.* 27 (2008) 520–526. https://doi.org/10.1016/j.polymertesting.2008.02.008

[206] L.P. Jakkamputi, S. Gnanasekaran, M. Thangamuthu, J. Rakkiyannan, Y.J. Bhalerao, Dynamic behavior modeling of natural-rubber/polybutadiene-rubber-based hybrid magnetorheological elastomer sandwich composite structures, *Polymers (Basel).* 15 (2023) 4583. https://doi.org/10.3390/polym15234583

[207] V.G. Kamble, J. Mersch, M. Tahir, K.W. Stöckelhuber, A. Das, S. Wießner, Development of liquid diene rubber based highly deformable interactive fiber-elastomer composites, *Materials (Basel).* 15 (2022) 1–14. https://doi.org/10.3390/ma15010390

[208] B. Xia, Z. Li, T. Lin, M. Gao, C. Zhao, X. Wu, C. Lin, J. Wang, Self-healing, recyclable syndiotactic 1, 2-polybutadiene-based thermadapt shape memory polymers with cold-programmed shape memory effect, *ACS Appl. Polym. Mater.* 6 (2023) 102–114. https://doi.org/10.1021/acsapm.3c01717

[209] R. Fiel, I.D. Barcelos, E.R. Leite, Organo-functionalized MoS_2 as a nanofiller to enhance and control the swelling behavior of polybutadiene rubber nanocomposites, *Compos. Commun.* 30 (2022) 101053. https://doi.org/10.1016/j.coco.2021.101053

[210] C. Lin, J. Hao, J. Zhao, Y. Hou, S. Ma, X. Sui, A facile strategy for the preparation of carbon nanotubes/polybutadiene crosslinked composite membrane and its application in osmotic energy harvesting, *J. Colloid Interface Sci.* 654 (2024) 840–847. https://doi.org/10.1016/j.jcis.2023.10.084

[211] M. Ashida, Composites of polychloroprene rubber with short fibres of poly (ethylene terephthalate) and nylon, in *Short Fibre–Polymer Composites*, Woodhead Publishing, 1996: pp. 116–143.

[212] P.B. Jana, A.K. Mallick, S.K. De, Effects of sample thickness and fiber aspect ratio on EMI shielding effectiveness of carbon fiber filled polychloroprene composites in the X-band frequency range, *IEEE Trans. Electromagn. Compat.* 34 (1992) 478–481. https://doi.org/10.1109/15.179281

[213] D. Vijayan, A. Mathiazhagan, R. Joseph, Aluminium trihydroxide: Novel reinforcing filler in polychloroprene rubber, *Polymer (Guildf).* 132 (2017) 143–156. https://doi.org/10.1016/j.polymer.2017.10.058

[214] Kapgate, B.P., Das, C. (2016). Electronic applications of chloroprene rubber and its composites, in Ponnamma, D., Sadasivuni, K., Wan, C., Thomas, S., Al-Ali AlMa'adeed, M. (eds) Flexible and Stretchable Electronic Composites. Springer Series on Polymer and Composite Materials. Springer, Cham. https://doi.org/10.1007/978-3-319-23663-6_10

[215] H. Eslami, C. Tzoganakis, T.H. Mekonnen, Constructing pristine and modified cellulose nanocrystals based cured polychloroprene nanocomposite films for dipped goods application, *Compos. Part C Open Access.* 1 (2020) 100009. https://doi.org/10.1016/j.jcomc.2020.100009

[216] M.J. Lovato, L.J. del Valle, J. Puiggalí, L. Franco, Performance-enhancing materials in medical gloves, *J. Funct. Biomater.* 14 (2023) 349. https://doi.org/10.3390/jfb14070349

[217] P. Koushki, T.H. Kwok, L. Hof, R. Wuthrich, Reinforcing silicone with hemp fiber for additive manufacturing, *Compos. Sci. Technol.* 194 (2020) 108139. https://doi.org/10.1016/j.compscitech.2020.108139

[218] S. Bourbigot, B. Gardelle, S. Duquesne, Intumescent silicone-based coatings for the fire protection of carbon fiber reinforced composites, *Fire Saf. Sci.*, (2014) 781–793. https://doi.org/10.3801/IAFSS.FSS.11-781

[219] R. Karthikeyan, J. Tjong, S.K. Nayak, M.M. Sain, Mechanical properties and cross-linking density of short sisal fiber reinforced silicone composites, *BioResources.* 12 (2017) 211–227.

[220] J. Beter, B. Schrittesser, B. Lechner, M.R. Mansouri, C. Marano, P.F. Fuchs, G. Pinter, Viscoelastic behavior of glass-fiber_reinforced silicon composites exposed to cyclic loading, *Polymers (Basel).* 12 (2020) 73–98. https://doi.org/10.3390/polym12091862

[221] W. Liu, L.M. Peeke, M. Periyasamy, R.R. Campbell, M.A. Hickner, Additive manufacturing of silicone composite structures with continuous carbon fiber reinforcement, *Polym. Eng. Sci.* 63 (2023) 1716–1724. https://doi.org/10.1002/PEN.26318

[222] J.H. Koo, M.J. Miller, J. Weispfenning, C. Blackmon, Silicone polymer composites for thermal protection system: Fiber reinforcements, microstructures, and performance, in *45th AIAA/ASME/SAE/ASEE Jt. Propuls. Conf. Exhib.* (2009). https://doi.org/10.2514/6.2009-5274.

[223] S. Rahman, J. Gogoi, S. Dubey, D. Chowdhury, Animal derived biopolymers for food packaging applications: A review, *Int. J. Biol. Macromol.* (2023) 128197. https://doi.org/10.1016/j.ijbiomac.2023.128197

[224] A.C.L. de Morais, V.S.D. Voet, R. Folkersma, K. Loos, Biopolymers in additive manufacturing, in *Additive Manufacturing of Biopolymers*, Elsevier, 2023: pp. 39–63. https://doi.org/10.1016/B978-0-323-95151-7.00001-6

[225] K. Babaremu, O.P. Oladijo, E. Akinlabi, Biopolymers: A suitable replacement for plastics in product packaging, *Adv. Ind. Eng. Polym. Res.* 6 (2023) 333–340. https://doi.org/10.1016/j.aiepr.2023.01.001

[226] X. Lu, H. Jiao, Y. Shi, Y. Li, H. Zhang, Y. Fu, J. Guo, Q. Wang, X. Liu, M. Zhou, M.W. Ullah, J. Sun, J. Liu, Fabrication of bio-inspired anisotropic structures from biopolymers for biomedical applications: A review, *Carbohydr. Polym.* 308 (2023) 120669. https://doi.org/10.1016/j.carbpol.2023.120669

[227] B. Bakhshandeh, F. Mohammadipanah, M.S. Mirtaleb, M.M. Gazani, S.S. Nateghi, P. Abbasi, Bacterial biopolymers in tissue engineering and regenerative medicine; current applications and future perspective, *Biocatal. Agric. Biotechnol.* 54 (2023) 102928. https://doi.org/10.1016/j.bcab.2023.102928

[228] M.H. Syed, M.A.K.M. Zahari, M.M.R. Khan, M.D.H. Beg, N. Abdullah, An overview on recent biomedical applications of biopolymers: Their role in drug delivery systems and comparison of major systems, *J. Drug Deliv. Sci. Technol.* 80 (2023) 104121. https://doi.org/10.1016/j.jddst.2022.104121

[229] S. Bin Park, E. Lih, K.S. Park, Y.K. Joung, D.K. Han, Biopolymer-based functional composites for medical applications, *Prog. Polym. Sci.* 68 (2017) 77–105. https://doi.org/10.1016/j.progpolymsci.2016.12.003

[230] L. Szabó, R. Milotskyi, G. Sharma, K. Takahashi, Cellulose processing in ionic liquids from a materials science perspective: Turning a versatile biopolymer into the cornerstone of our sustainable future, *Green Chem.* 25 (2023) 5338–5389. https://doi.org/10.1039/d2gc04730f

[231] X. Li, G. Jiang, G. Wang, J. Zhou, Y. Zhang, D. Zhao, Promising cellulose–based functional gels for advanced biomedical applications: A review, *Int. J. Biol. Macromol.* 260 (2024) 129600. https://doi.org/10.1016/j.ijbiomac.2024.129600

[232] H. Shaghaleh, X. Xu, S. Wang, Current progress in production of biopolymeric materials based on cellulose, cellulose nanofibers, and cellulose derivatives, *RSC Adv.* 8 (2018) 825–842. https://doi.org/10.1039/c7ra11157f

[233] Y. Liu, S. Ahmed, D.E. Sameen, Y. Wang, R. Lu, J. Dai, S. Li, W. Qin, A review of cellulose and its derivatives in biopolymer-based for food packaging application, *Trends Food Sci. Technol.* 112 (2021) 532–546. https://doi.org/10.1016/j.tifs.2021.04.016

[234] S. Tanpichai, A. Boonmahitthisud, N. Soykeabkaew, L. Ongthip, Review of the recent developments in all-cellulose nanocomposites: Properties and applications, *Carbohydr. Polym.* 286 (2022) 119192. https://doi.org/10.1016/j.carbpol.2022.119192

[235] P. Kanmani, J. Aravind, M. Kamaraj, P. Sureshbabu, S. Karthikeyan, Environmental applications of chitosan and cellulosic biopolymers: A comprehensive outlook, *Bioresour. Technol.* 242 (2017) 295–303. https://doi.org/10.1016/j.biortech.2017.03.119

[236] R. Balart, D. Garcia-Garcia, V. Fombuena, L. Quiles-Carrillo, M.P. Arrieta, Biopolymers from natural resources, *Polymers (Basel).* 13 (2021) 1–9. https://doi.org/10.3390/polym13152532

[237] F. Xie, E. Pollet, P.J. Halley, L. Avérous, Starch-based nano-biocomposites, *Prog. Polym. Sci.* 38 (2013) 1590–1628. https://doi.org/10.1016/j.progpolymsci.2013.05.002

[238] D.R. Lu, C.M. Xiao, S.J. Xu, Starch-based completely biodegradable polymer materials, *Express Polym. Lett.* 3 (2009) 366–375. https://doi.org/10.3144/expresspolymlett.2009.46

[239] M. Salimi, B. eddine Channab, A. El Idrissi, M. Zahouily, E. Motamedi, A comprehensive review on starch: Structure, modification, and applications in slow/controlled-release fertilizers in agriculture, *Carbohydr. Polym.* 322 (2023) 121326. https://doi.org/10.1016/j.carbpol.2023.121326

[240] J.R.V. Matheus, P.M. de Farias, J.M. Satoriva, C.J. de Andrade, A.E.C. Fai, Cassava starch films for food packaging: Trends over the last decade and future research, *Int. J. Biol. Macromol.* 225 (2023) 658–672. https://doi.org/10.1016/j.ijbiomac.2022.11.129

[241] X. Hou, H. Wang, Y. Shi, Z. Yue, Recent advances of antibacterial starch-based materials, *Carbohydr. Polym.* 302 (2023) 120392. https://doi.org/10.1016/j.carbpol.2022.120392

[242] B. Zhao, L. Li, X. Lv, J. Du, Z. Gu, Z. Li, L. Cheng, C. Li, Y. Hong, Progress and prospects of modified starch-based carriers in anticancer drug delivery, *J. Control. Release.* 349 (2022) 662–678. https://doi.org/10.1016/j.jconrel.2022.07.024

[243] R. Phiri, S. Mavinkere Rangappa, S. Siengchin, O.P. Oladijo, H.N. Dhakal, Development of sustainable biopolymer-based composites for lightweight applications from agricultural waste biomass: A review, *Adv. Ind. Eng. Polym. Res.* 6 (2023) 436–450. https://doi.org/10.1016/j.aiepr.2023.04.004

[244] S. Wei, Y.C. Ching, C.H. Chuah, Synthesis of chitosan aerogels as promising carriers for drug delivery: A review, *Carbohydr. Polym.* 231 (2020) 115744. https://doi.org/10.1016/j.carbpol.2019.115744

[245] S. Ahmed, A. Ali, J. Sheikh, A review on chitosan centred scaffolds and their applications in tissue engineering, *Int. J. Biol. Macromol.* 116 (2018) 849–862. https://doi.org/10.1016/j.ijbiomac.2018.04.176

[246] K.S. Venkataprasanna, J. Prakash, S. Vignesh, G. Bharath, M. Venkatesan, F. Banat, S. Sahabudeen, S. Ramachandran, G.D. Venkatasubbu, Fabrication of Chitosan/PVA/GO/CuO patch for potential wound healing application, *Int. J. Biol. Macromol.* 143 (2020) 744–762. https://doi.org/10.1016/j.ijbiomac.2019.10.029

[247] S. Kumar, A. Mukherjee, J. Dutta, Chitosan based nanocomposite films and coatings: Emerging antimicrobial food packaging alternatives, *Trends Food Sci. Technol.* 97 (2020) 196–209. https://doi.org/10.1016/j.tifs.2020.01.002

[248] A.N. Raja, Recent development in chitosan-based electrochemical sensors and its sensing application, *Int. J. Biol. Macromol.* 164 (2020) 4231–4244. https://doi.org/10.1016/j.ijbiomac.2020.09.012

[249] D. Garlotta, A literature review of poly (lactic acid) a literature review of poly (lactic acid), *J. Polym. Environ.* 9 (2019) 63–84. https://doi.org/10.1023/A:1020200822435

[250] R.A. Auras, L.-T. Lim, S.E.M. Selke, H. Tsuji, *Poly (Lactic acid): Synthesis, Structures, Properties, Processing, and Applications*, John Wiley & Sons 2011.

[251] A.J.R. Lasprilla, G.A.R. Martinez, B.H. Lunelli, A.L. Jardini, R.M. Filho, Poly-lactic acid synthesis for application in biomedical devices – A review, *Biotechnol. Adv.* 30 (2012) 321–328. https://doi.org/10.1016/j.biotechadv.2011.06.019

[252] S. Mallakpour, Applications of layered double hydroxide biopolymer nanocomposites, in *Layered Double Hydroxide Polymer Nanocomposites*, Elsevier, 2020: pp. 599–676. https://doi.org/10.1016/B978-0-08-101903-0.00015-5

[253] V. Guarino, G. Gentile, L. Sorrentino, L. Ambrosio, Polycaprolactone: Synthesis, properties, and applications, *Encycl. Polym. Sci. Technol.* (2017) 1–36. https://doi.org/10.1002/0471440264.pst658

[254] A. J. Muller, R.M. Michell, Differential scanning calorimetry of polymers, in *Polymer Morphology: Principles, Characterization, and Processing*, 2016: pp. 72–99. https://doi.org/10.1002/9781118892756.ch5

[255] H.M. Ng, N.M. Saidi, F.S. Omar, K. Ramesh, S. Ramesh, S. Bashir, Thermogravimetric analysis of polymers, in *Encyclopedia of Polymer Science and Technology*, 2018: pp. 1–29. https://doi.org/10.1002/0471440264.pst667

[256] J. James, Thermomechanical analysis and its applications, in *Thermal and Rheological Measurement Techniques for Nanomaterials Characterization*, Elsevier Inc., 2017: pp. 159–171. https://doi.org/10.1016/B978-0-323-46139-9.00007-4

[257] N. Saba, M. Jawaid, A review on thermomechanical properties of polymers and fibers reinforced polymer composites, *J. Ind. Eng. Chem.* 67 (2018) 1–11. https://doi.org/10.1016/j.jiec.2018.06.018

[258] N. Saba, M. Jawaid, M.T. Paridah, O. Alothman, Physical, structural and thermomechanical properties of nano oil palm empty fruit bunch filler based epoxy nanocomposites, *Ind. Crops Prod.* 108 (2017) 840–843. https://doi.org/10.1016/j.indcrop.2017.07.048

[259] K.P. Menard, N.R. Menard, Dynamic mechanical analysis in the analysis of polymers and rubbers, in *Encyclopedia of Polymer Science and Technology*, 2002: pp. 1–33. https://doi.org/10.1002/0471440264.pst102.pub2

[260] S. Ravichandran, E. Vengatesan, A. Ramakrishnan, Synthesis and dynamic mechanical analysis of fiber reinforced low-density polyethylene hybrid polymer composites, *Mater. Today Proc.* 27 (2020) 177–180. https://doi.org/10.1016/j.matpr.2019.09.216

[261] N.S. Murthy, X-ray diffraction from polymers, in *Polymer Morphology: Principles, Characterization, and Processing*, 2016: pp. 14–36. https://doi.org/10.1002/9781118892756.ch2

[262] P. Eaton, P. Quaresma, C. Soares, C. Neves, M.P. de Almeida, E. Pereira, P. West, A direct comparison of experimental methods to measure dimensions of synthetic nanoparticles, *Ultramicroscopy.* 182 (2017) 179–190. https://doi.org/10.1016/j.ultramic.2017.07.001

[263] H.W. Siesler, K. Holland-moritz, Infrared Spectroscopy of Polymers, *Appl. Spectrosc. Rev.* 11 (1976) 1–55. https://doi.org/10.1080/05704927608081704

[264] H. Rajandas, S. Parimannan, K. Sathasivam, M. Ravichandran, L. Su Yin, A novel FTIR-ATR spectroscopy based technique for the estimation of low-density polyethylene biodegradation, *Polym. Test.* 31 (2012) 1094–1099. https://doi.org/10.1016/j.polymertesting.2012.07.015

[265] Optosky, Application of Raman Spectroscopy in Polymer Materials, www.optosky.net/application-of-raman-spectroscopy-in-polymer-materials.html (accessed February 23, 2024).

[266] PhysicsOpenLab, Polymer Analysis Using Raman Spectroscopy – PhysicsOpenLab, https://physicsopenlab.org/2022/05/08/polymer-analysis-using-raman-spectroscopy/ (accessed February 23, 2024).

[267] K. Saalwächter, Applications of NMR in polymer characterization – An introduction, in: *NMR Methods for Characterization of Synthetic and Natural Polymers*, The Royal Society of Chemistry, 2019: pp. 5–6. https://doi.org/10.1039/9781788016483-00001

[268] D. S. Bag, NMR spectroscopy for the characterization of polymers, in: *Appl. NMR Spectrosc.*, Bentham Science Publishers, 2016: pp. 108–158. https://doi.org/10.2174/9781681082875116050005

[269] D.L. VanderHart, R.H. Atalla, Studies of microstructure in native celluloses using solid-state 13C NMR, *Macromolecules.* 17 (1984) 1465–1472. https://doi.org/10.1021/ma00138a009

[270] M. Nagasawa, H. Kinuhata, H. Koizuka, K. Miyamoto, T. Tanaka, H. Kishimoto, T. Koike, Mechanical fatigue of epoxy resin, *J. Mater. Sci.* 30 (1995) 1266–1272. https://doi.org/10.1007/BF00356129

[271] B. Yang, V. Kozey, S. Adanur, S. Kumar, Bending, compression, and shear behavior of woven glass fiber-epoxy composites, *Compos. Part B Eng.* 31 (2000) 715–721. https://doi.org/10.1016/S1359-8368(99)00052-9

[272] V.K. Srivastava, P.S. Shembekar, Tensile and fracture properties of epoxy resin filled with flyash particles, *J. Mater. Sci.* 25 (1990) 3513–3516. https://doi.org/10.1007/BF00575379

[273] A.N. Gent, J. Jeong, Tear strength of oriented crystalline polymers, *J. Mater. Sci.* 21 (1986) 355–363. https://doi.org/10.1007/BF01144744

[274] A. Hawa, M.S. Abdul Majid, M. Afendi, H.F.A. Marzuki, N.A.M. Amin, F. Mat, A.G. Gibson, Burst strength and impact behaviour of hydrothermally aged glass fibre/epoxy composite pipes, *Mater. Des.* 89 (2016) 455–464. https://doi.org/10.1016/j.matdes.2015.09.082

[275] P. Gupta, B. Toksha, B. Patel, Y. Rushiya, P. Das, M. Rahaman, Recent developments and research avenues for polymers in electric vehicles, *Chem. Rec.* 22 (2022). https://doi.org/10.1002/tcr.202200186

[276] S. Ai, S. Mazumdar, H. Li, Y. Cao, T. Li, Nano-silica doped composite polymer chitosan/poly(ethylene oxide)-based electrolyte with high electrochemical stability suitable for quasi solid-state lithium metal batteries, *J. Electroanal. Chem.* 895 (2021) 115464. https://doi.org/10.1016/j.jelechem.2021.115464

[277] Z. Xie, Z. Wu, X. An, X. Yue, P. Xiaokaiti, A. Yoshida, A. Abudula, G. Guan, A sandwich-type composite polymer electrolyte for all-solid-state lithium metal batteries with high areal capacity and cycling stability, *J. Memb. Sci.* 596 (2020) 117739. https://doi.org/10.1016/j.memsci.2019.117739

[278] M.N. Pham, R. Subramani, Y.H. Lin, Y.L. Lee, J.S. Jan, C.C. Chiu, H. Teng, Acylamino-functionalized crosslinker to synthesize all-solid-state polymer electrolytes for high-stability lithium batteries, *Chem. Eng. J.* 430 (2022) 132948. https://doi.org/10.1016/j.cej.2021.132948

[279] E.J. Bolívar-Monsalve, M.M. Alvarez, S. Hosseini, M.A. Espinosa-Hernandez, C.F. Ceballos-González, M. Sanchez-Dominguez, S.R. Shin, B. Cecen, S. Hassan, E. Di Maio, G. Trujillo-De Santiago, Engineering bioactive synthetic polymers for biomedical applications: A review with emphasis on tissue engineering and controlled release, *Mater. Adv.* 2 (2021) 4447–4478. https://doi.org/10.1039/d1ma00092f

[280] A.Z. Unal, J.L. West, Synthetic ECM: Bioactive synthetic hydrogels for 3D tissue engineering, *Bioconjug. Chem.* 31 (2020) 2253–2271. https://doi.org/10.1021/acs.bioconjchem.0c00270

[281] Z. Terzopoulou, A. Zamboulis, I. Koumentakou, G. Michailidou, M.J. Noordam, D.N. Bikiaris, Biocompatible synthetic polymers for tissue engineering purposes, *Biomacromolecules.* 23 (2022) 1841–1863. https://doi.org/10.1021/acs.biomac.2c00047

4 Tensile and Piezoelectric Performance of Polyvinylidene Fluoride (PVDF), Boron Nitride Nanosheets (BNNSs), and Graphene Nanoplatelets (GNPs) Hybrid Composite

Dimple Sagar, Chandraveer Singh, Lokesh Nair, Ravikant Ravi, Hemant Kumar, and Vijyesh Kumar

4.1 INTRODUCTION

In the present world, considerable research is being carried out to come up with a solution for implantable (IP) devices. Rapid growth in the field of developing these devices is getting more and more attention in biomedical (BM) applications. Piezoelectric materials have excellent capabilities to produce IP devices for BM applications, as these can be considered self-powered in nature [1]. Scientists from different parts of the world have developed self-powered wearable devices which take the energy from the human body's activities [2]. Piezoelectric materials are now becoming pioneer materials in the field of wearable devices because of their high flexibility, lightweight, biocompatibility, and high-voltage electrical properties [3]. With the help of the piezoelectric effect, piezoelectric materials can convert mechanical movement into electric signals. These electric signals can be used for further analysis of force or stress calculations [4].

Polyvinylidene fluoride (PVDF): Polyvinylidene fluoride, often abbreviated as PVDF, is a synthetic polymer known for its remarkable piezoelectric properties. Piezoelectricity is the ability of certain materials to generate an electric charge in response to mechanical stress or deformation, and conversely, to deform when an electric field is applied. PVDF is a prominent example of a piezoelectric material, and its unique properties have led to its application in various fields ranging from electronics and sensors to energy harvesting and medical devices. A single PVDF nanofiber-built nanogenerator can generate 5 mV–30 mV of voltage and 0.5 nA–3 nA of current under stretching and releasing motions [5]. Therefore, PVDF attracts the scientific community for building flexible piezoelectric sensors as it is recorded as one of the renowned advantages of PVDF but there are certain limitations with PVDF, and one of them is limited stretchability (strain to failure) [6]. The researchers from different parts of the world collectively worked and found significant improvement in the stretchability of PVDF by reinforcing it with other materials [7].

Boron Nitride Nanosheet (BNNS): BNNSs, a two-dimensional nanomaterial, possess remarkable thermal conductivity, electrical insulation, and excellent thermal stability [1]. These

DOI: 10.1201/9781003664611-5

properties make them enticing for applications in thermal management, electronic packaging, and flexible electronics. Their compatibility with polymers enhances the potential for synergy in composite materials [5,8]. When PVDF and BNNSs are combined, their properties interact synergistically, resulting in a composite that is greater than the sum of its parts [6,9]. The enhanced thermal conductivity of BNNSs complements PVDF's mechanical strength, making the composite an excellent candidate for thermally conductive and mechanically robust materials. Furthermore, the dielectric properties of BNNSs synergize with PVDF's piezoelectric behavior, opening avenues for novel energy conversion and sensing applications [10–12].

Graphene Nanoplatelets (GNPs): The piezoelectric effect in PVDF is attributed to the alignment of its molecular dipoles under mechanical stress. The addition of GNPs can facilitate more efficient stress transfer within the composite due to their excellent mechanical properties [13,14]. Furthermore, the high electrical conductivity of GNPs enhances charge accumulation and transport, augmenting the overall piezoelectric response [15,16]. This synergy between PVDF's inherent piezoelectricity and GNPs' conductive properties presents a pathway for improving the overall performance of the composite.

4.2 MATERIALS AND METHODS

The materials which were used in this project are as follows:

a. **Dimethylformamide (DMF):** It is a solvent purchased from SRL Pvt. Ltd.
b. **PVDF:** The material was purchased from Alfa Aesar. Scanning Electron Microscope (SEM) imaging was conducted on the purchased PVDF sample as a part of the investigation process. Figure 4.1 shows the images of the PVDF powder. After analyzing the SEM images, it can be seen clearly that the PVDF particles exhibit a uniform and consistent morphology. The particle demonstrates a regular shape with minimum variation in size. From the images it can be concluded that there is a high degree of homogeneity in the PVDF powder, indicating potential advantages in terms of material processing and performance.

FIGURE 4.1 Analysis of the PVDF powder.

FIGURE 4.2 SEM images of BNNs.

FIGURE 4.3 SEM images of GNPs.

c. **BNNSs:** BNNS, obtained from Supervac, was subjected to SEM analysis to examine its structural characteristics. The SEM images, as illustrated in Figure 4.2, showcase the morphology of the BNNS powder, revealing the distinct presence of hexagonal boron nitride nanosheets. These images provide clear evidence of the regularity and near-homogeneity of the BNNS sheets, highlighting their consistent and uniform nature. Through this analysis, it becomes apparent that the BNNS powder comprises well-defined nanosheets, which are essential for understanding its potential applications and optimizing its usage in various fields.

d. **GNPs:** GNPs were purchased from **Xg Sciences** and it was subjected to SEM analysis to check its potential characteristics.

TABLE 4.1
Details of the Samples Fabricated

Code of the Sample	Details
S1	250 mg PVDF + 20 ml DMF
S2	250 mg PVDF + 20 mg mixture (GNP + BN) + 20 ml DMF

FIGURE 4.4 (a) PVDF 250 mg. (b) 20 ml DMF, (c) solution of PVDF + DMF, and (d) casting of sample S1 (PVDF) at 80°C in a Petri Dish.

The SEM images presented in Figure 4.3 unveil the irregular shapes of the GNPs, highlighting a lack of homogeneity in both particle shape and size. This observation holds significant importance in comprehending the role of GNPs in the fabrication process of the final composite material. Understanding the inherent variability in GNP morphology is crucial as it directly impacts the material's properties and performance. Factors such as dispersion, interfacial bonding, and load transfer between the GNPs and the polymer matrix heavily rely on the morphology of GNPs. Therefore, characterizing the non-uniformity in shape and size through SEM analysis aids in optimizing processing techniques and designing composite materials with enhanced mechanical, electrical, and thermal properties.

By using the materials listed in (a)–(d), the following samples were prepared. The details are shown in Table 4.,1.

Synthesis S1: A quantity of 250 mg of PVDF powder was carefully weighed using a precision scale. This measured powder was then combined with 20 ml of a solvent, specifically DMF. The resulting mixture underwent ultrasonication for a duration of 6 hours at a frequency of 25 kHz. After this ultrasonication process, the mixture was poured into a Petri Dish and allowed to settle. Subsequently, the casted solution was placed into an oven and exposed to a temperature of 80°C for a period of 8 hours. This controlled heating process facilitated the solidification of the solution, resulting in the formation of a solid material (Figure 4.4).

Synthesis of PVDF + BN + GNP (S2): The synthesis approach for sample S2 exhibits slight deviations compared to sample S1's synthesis. For this procedure, equal amounts of 20 mg were precisely measured for both BNNs and GNPs. The amalgamation of these components took place within a planetary ball mill to ensure thorough mixing, utilizing zirconium balls within the mixing chamber. The blending process of BNNs and GNPs occurred over a span of 24 hours, divided into batches of 6 hours each. This division was necessitated by the heat generated within the mixing chamber, as excessive heat could lead to the mixture melting.

Following the mixing phase, a 20 mg composite of BNNSs and GNPs was transferred to a 20 ml DMF solution. This solution was subsequently subjected to ultrasonication for 6 hours, operating at

FIGURE 4.5 (a) GNP 20 mg, (b) 20 mg BN, (c) planetary ball apparatus, (d) mixture (BN + GNP), (e) 20 mg mixture + 20 ml DMF, (f) ultrasonicate the mixture of GNP + BN and DMF at 25 kHz for 6 hours, (g) ultrasonicated mixture {20 mg (BN+GNP) + 20 ml DMF}, (h) 250 mg PVDF, (i) magnetic stirring of (g) and (h) for 1 hour, (j) casting of the sample S2 (PVDF +GNP + BN) in a Petri Dish, (k) S2 put into the oven at 80°C for 8 hours, and (l) dried S2 casted at 80°C.

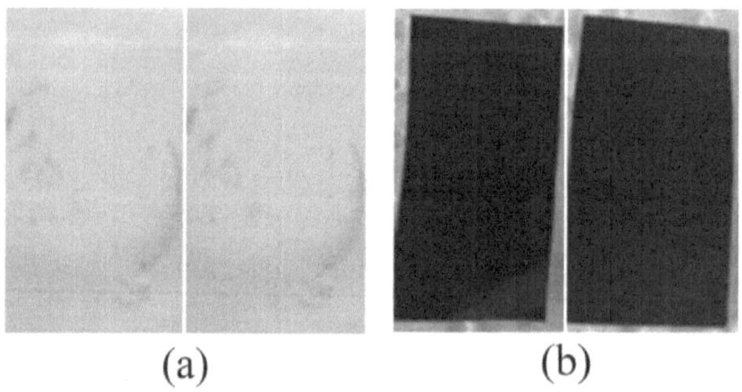

FIGURE 4.6 (a) 3 cm × 1 cm film of S1, (b) 3 cm × 1 cm film of S2.

a frequency of 25 kHz. After the 6-hour period, a precisely measured quantity of 250 mg of PVDF was introduced into the solution containing BNNSs, GNPs, and DMF. Subsequently, the remaining steps of the process mirrored those employed in the production of the preceding samples. From the samples S1 and S2, a film of 3 cm length and 1 cm width was cut off from the casted sample, as shown in Figure 4.6.

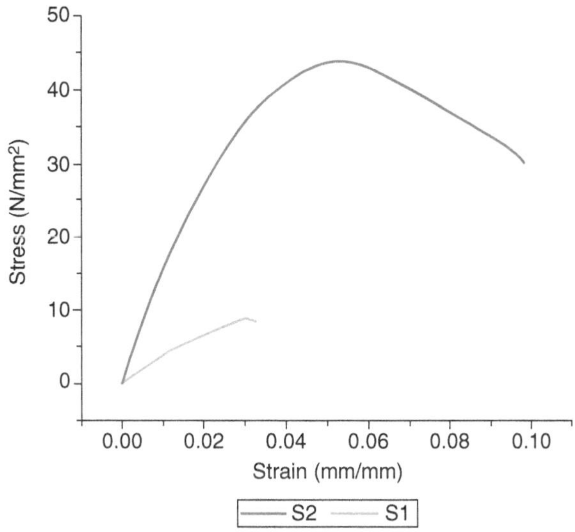

FIGURE 4.7 Stress–strain curve of the different samples (i.e. S1 and S2).

4.3 RESULTS

Figure 4.5 shows the effect of BNNSs+GNPs on the tensile properties of the PVDF composite membranes. The above reinforcement significantly enhanced the tensile strength and ductility, as shown in Figure 4.7. In particular, both the tensile strength and ductility increase with the content of BNNSs+GNPs. The strength increased from approximately 6 MPa in tensile strength to a maximum value of approximately 40 MPa. The strain of 100% was achieved when the PVDF was reinforced with the mixture of BNNSs+GNPs. In the sample S2, the Young's modulus increased from 2 MPa to 8 MPa. The tensile tests were performed on only two samples and the strain rate was taken as 2 mm/min.

In conclusion it is clear from the results that the sample S2 exhibited higher strength and ductility than the S1. Superior mechanical properties of PVDF membranes reinforced by BNNSs and GNPs are particularly important for applications that demand mechanical integrity, such as water treatment, oil–water separation, air filtration, and gas separation. The remarkable reinforcement from BNNSs and GNPs is due to their uniform distribution within the PVDF that facilitated effective load transfer between polymer matrix and BNNSs+GNPs interface. Furthermore, the addition of the hybrid reinforcement of BNNSs and GNPs delayed the failure of the composite in a significant amount. This may be due to the improved bonding between the PVDF and the reinforcement.

4.3.1 Structure of PVDF/BNNS/GNP Polymer Composite

After conducting the tensile test on the prepared specimens S1 and S2, the fractured samples were examined using field emission scanning electron microscopy (FESEM). The outcomes are illustrated in Figure 4.8, with images Figure 4.8a–4.8c corresponding to S2, while images Figure 4.8d and 4.8e pertain to S1. The incorporation of BNNSs and GNPs has led to an enlarged size of up to 200 nm in sample S2. This addition of BNNSs and GNPs has significantly enhanced the tensile properties of sample S2, a benefit not observed in S1, as indicated in Figure 4.8d and 4.8e.

The tensile behavior of sample S1 was reported as poor as compared to sample S2, which can surely be confirmed from the SEM images shown in Figure 4.8(d) and 4.8(e). The image shows poor bonding of the PVDF particle during synthesis. It is also clear from the image that the brittle nature of PVDF alone confirms its impact on its mechanical properties.

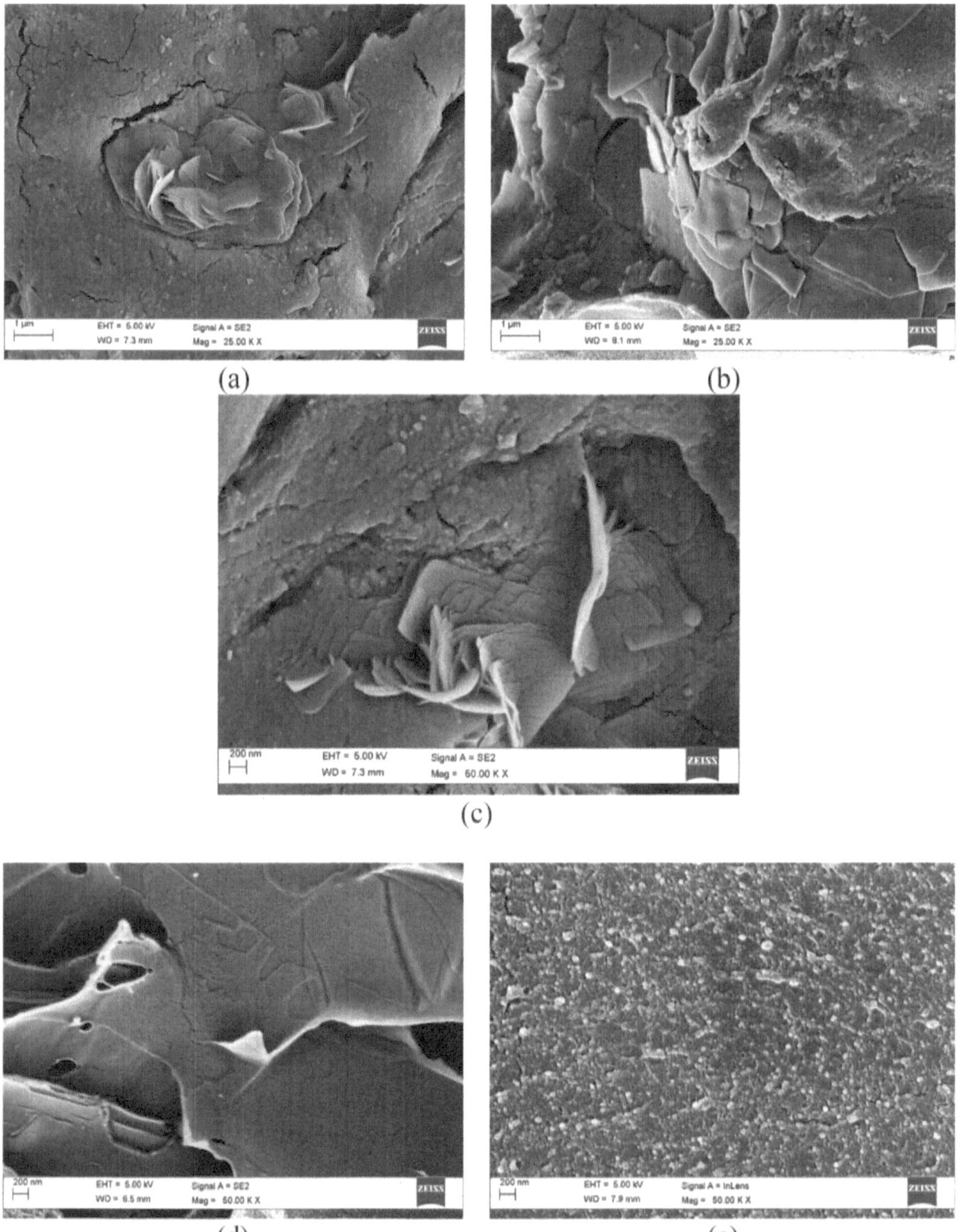

FIGURE 4.8 (a), (b), and (c) FESEM of fracture sample of S2 in tension test. Parts (d) and (e) are the images of S1.

4.3.2 Piezoelectric Performance of PVDF/BNNs/GNPs Composite

Figure 4.9 presents the electrical outputs of nanofiber membranes subjected to loads 1 N, 5 N, 10 N, and 15 N that were studied. The inclusion of BNNSs and GNPs showed slight improvement in output current and output voltage.

Figure 4.9 illustrates the electrical responses of nanofiber membranes when subjected to various loads, specifically 1 N, 5 N, 10 N, and 15 N. The impact of incorporating BNNSs and GNPs was observed, resulting in a marginal enhancement in both output current and output voltage. The composite S2 membranes that were manufactured exhibited a voltage span ranging from –40 mV to + 40 mV at a 1 N load, –80 mV to +80 mV at a 5 N load, –100 mV to +100 mV at a 10 N load, and –120 mV to +120 mV at a 15 N load. These findings indicate a noteworthy elevation in voltage levels when the S2 sample was subjected to the aforementioned loads. The comparison of voltage generation outcomes under distinct loads is depicted in Figure 4.10. The above-mentioned values were documented.

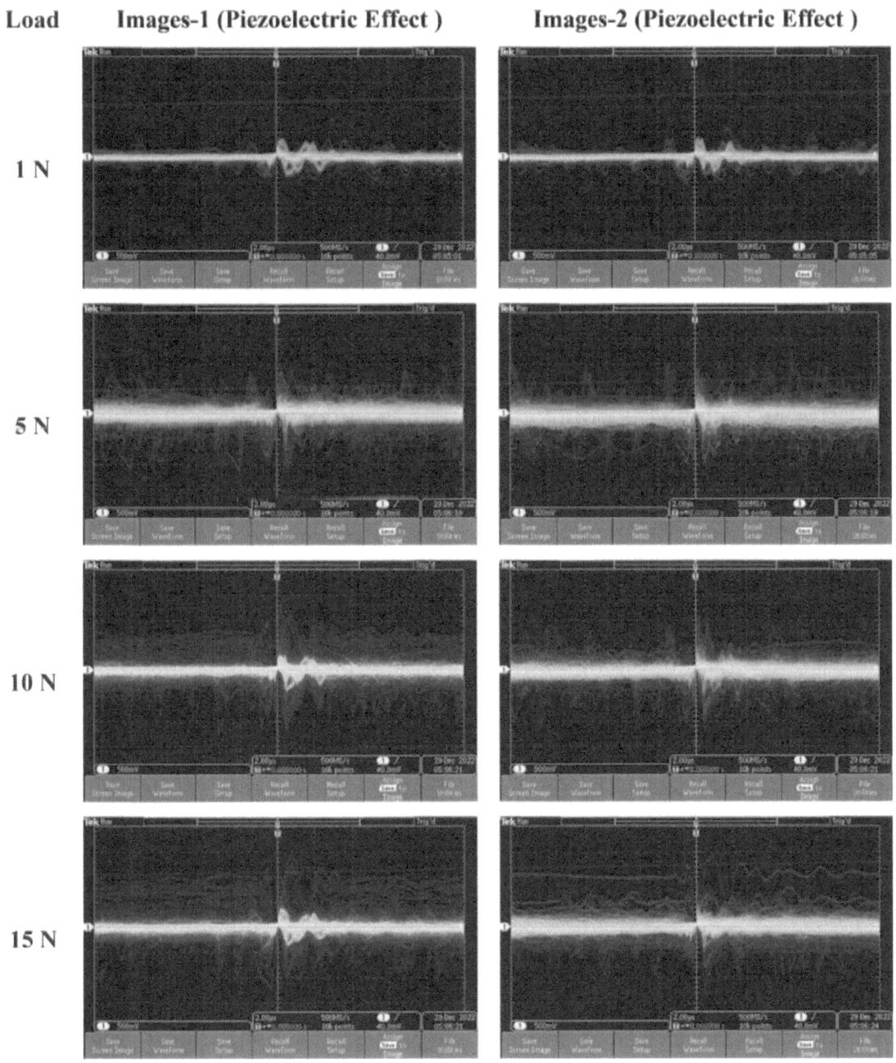

FIGURE 4.9 Piezoelectric results of samples S1 and S2.

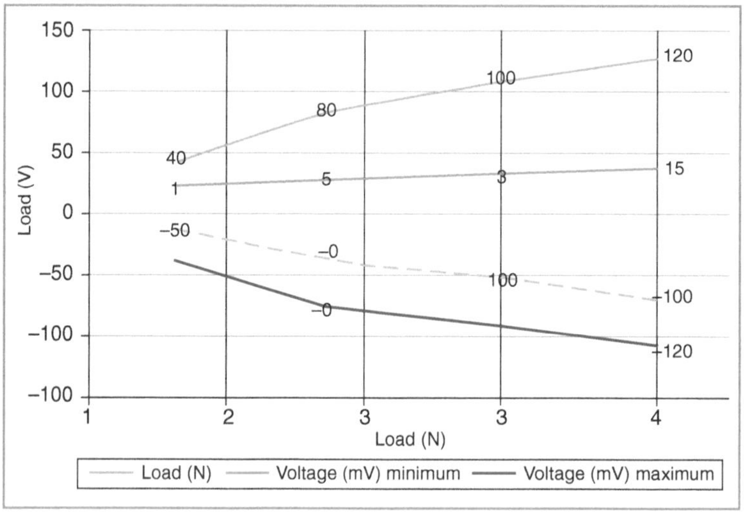

FIGURE 4.10 Effect of load on voltage generation of S2.

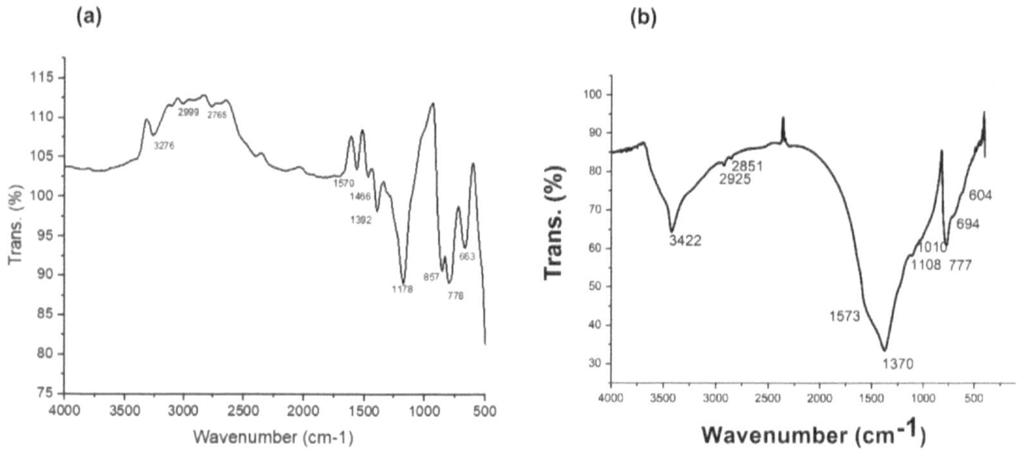

FIGURE 4.11 FTIR performance of (a) S1 sample and (b) S2.

4.3.3 Fourier Transform Infrared (FTIR) Study

Fourier transform infrared (FTIR) spectroscopy is a common technique used to analyze the chemical composition of materials. It provides the valuable information about the fabricated material. The wavenumbers 778 cm^{-1} (β phase) and 857 cm^{-1} (α phase) can be confirmed from the FTIR results, indicating the presence of both crystalline phases in the material structure. Furthermore, some relevant results for FTIR analysis of PVDF, BNNSs, and GNP composite are shown in Figure 4.11(b). To explore the bonding characteristics within nanotubes, FTIR spectroscopy was employed as well (refer to Figure 4.11). The prominent peak identified at 3422 cm^{-1} aligns effectively with the stretching bond of B–N, while the slightly less pronounced peak at 1010 cm^{-1} corresponds to the B–N–G–N bond.

SEM images of sample S2 were captured with the aim of analyzing and studying the correlation between its mechanical properties and its piezoelectric behavior. These images offer visual evidence

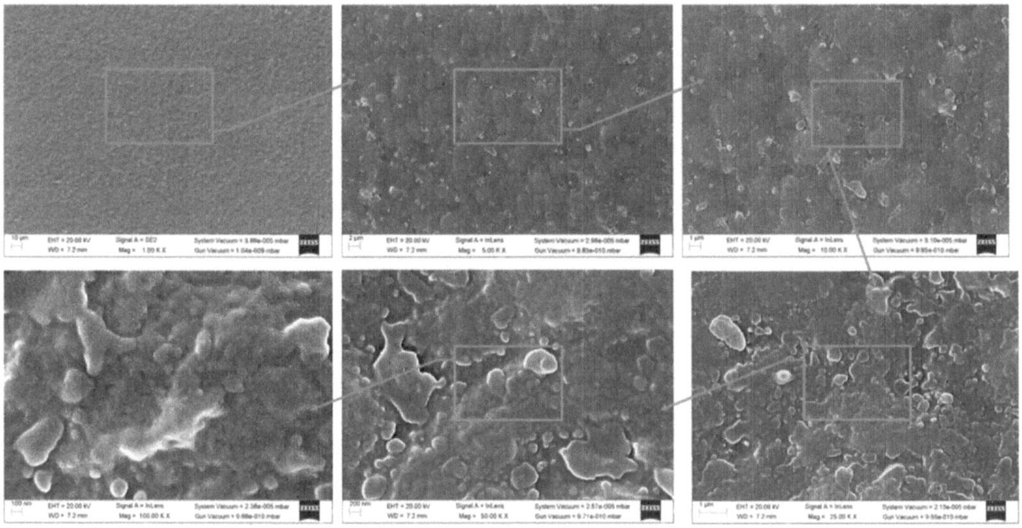

FIGURE 4.12 SEM images of the S2 sample.

of the presence of BNNSs and GNPs within the structure of the PVDF. Particularly, sample S2 exhibits negligible porosity, indicating strong bonding between the reinforcement materials and the PVDF matrix. This observation suggests that the improved mechanical properties of sample S2, particularly its higher tensile strength compared to sample S1, can be attributed to the excellent bonding between PVDF, BNNS, and GNPs, as evidenced by the SEM images (Figure 4.12).

4.4 CONCLUSIONS

The above research leads to the following conclusions:

1. **Enhanced Tensile Properties:** The incorporation of BNNSs and GNPs into the PVDF matrix led to a significant improvement in the tensile properties of the hybrid composite. The synergistic effect of these nanofillers contributed to increased strength and toughness, which is promising for applications requiring high mechanical performance. Incorporating BNNSs and GNPs into PVDF leads to a notable enhancement in its tensile strength. This enhancement is attributed to the surface chemistry of BNNS and GNP, which fosters robust interfacial connections with the polymer matrices. The BNNSs provided effective stress transfer pathways, while the GNPs acted as reinforcing agents, creating a well-dispersed and interconnected network within the composite [6].
2. **Piezoelectric Enhancement:** The introduction of BNNSs and GNPs also positively influenced the piezoelectric properties of the PVDF composite. The combined effect of BNNSs' inherent piezoelectricity and the GNPs' conductivity resulted in enhanced piezoelectric coefficients and charge generation under mechanical deformation. Both BNNS and GNP exhibit remarkable mechanical strength and exceptional electrical conductivity [6]. When these thin layers of nanoparticles are distributed throughout the PVDF matrix, they contribute to enhancing the mechanical characteristics of the material. This enhanced mechanical performance can result in amplified piezoelectric responses when the composite undergoes mechanical deformation, as the nanosheets facilitate a more efficient transmission of stress across the material.

REFERENCES

[1] R. Sun *et al.*, "Kirigami stretchable strain sensors with enhanced piezoelectricity induced by topological electrodes," *Appl. Phys. Lett.*, vol. 112, no. 25, pp. 1–6, 2018, doi: 10.1063/1.5025025

[2] L. Dong *et al.*, "Flexible electrodes and supercapacitors for wearable energy storage: A review by category," *J. Mater. Chem. A*, vol. 4, no. 13, pp. 4659–4685, 2016, doi: 10.1039/c5ta10582j

[3] Z. Mirzazadeh, Z. Sherafat, and E. Bagherzadeh, "Physical and mechanical properties of PVDF/KNN composite produced via hot compression molding," *Ceram. Int.*, vol. 47, no. 5, pp. 6211–6219, 2021, doi: 10.1016/j.ceramint.2020.10.199

[4] Y. Wu, Y. Ma, H. Zheng, and S. Ramakrishna, "Piezoelectric materials for flexible and wearable electronics: A review," *Mater. Des.*, vol. 211, p. 110164, 2021, doi: 10.1016/j.matdes.2021.110164

[5] K. B. Kim *et al.*, "Transparent and flexible piezoelectric sensor for detecting human movement with a boron nitride nanosheet (BNNS)," *Nano Energy*, pp. 91–98, 2018, doi: 10.1016/j.nanoen.2018.09.056

[6] J. Zhang *et al.*, "Mechanically stretchable piezoelectric polyvinylidene fluoride (PVDF)/Boron nitride nanosheets (BNNSs) polymer nanocomposites," *Compos. Part B Eng.*, vol. 175, no. July, p. 107157, 2019, doi: 10.1016/j.compositesb.2019.107157

[7] Q. Zheng, B. Shi, Z. Li, and Z. L. Wang, "Recent progress on piezoelectric and triboelectric energy harvesters in biomedical systems," *Adv. Sci.*, vol. 4, no. 7, pp. 1–23, 2017, doi: 10.1002/advs.201700029

[8] J. Zhang *et al.*, "Enhancing the triboelectricity of stretchable electrospun piezoelectric polyvinylidene fluoride/boron nitride nanosheets composite nanofibers," *Compos. Commun.*, vol. 22, p. 100535, 2020, doi: 10.1016/j.coco.2020.100535

[9] H. Kuang *et al.*, "Piezoelectric boron nitride nanosheets for high performance energy harvesting devices," *Nano Energy*, vol. 80, no. October 2020, p. 105561, 2021, doi: 10.1016/j.nanoen.2020.105561

[10] J. Chen, C. Ayranci, and T. Tang, "Piezoelectric performance of electrospun PVDF and PVDF composite fibers: A review and machine learning-based analysis," *Mater. Today Chem.*, vol. 30, p. 101571, 2023, doi: 10.1016/j.mtchem.2023.101571

[11] S. Panda *et al.*, "Piezoelectric energy harvesting systems for biomedical applications," *Nano Energy*, vol. 100, no. June, p. 107514, 2022, doi: 10.1016/j.nanoen.2022.107514

[12] A. Bouhamed *et al.*, "A hybrid piezoelectric composite flexible film based on PVDF-HFP for boosting power generation," *Compos. Sci. Technol.*, vol. 208, no. January, 2021, doi: 10.1016/j.compscitech.2021.108769

[13] S. S. Panicker, S. P. Rajeev, and V. Thomas, "Impact of PVDF and its copolymer-based nanocomposites for flexible and wearable energy harvesters," *Nano-Struct. Nano-Obj.*, vol. 34, p. 100949, 2023, doi: 10.1016/j.nanoso.2023.100949

[14] W. Wu *et al.*, "Thermally conductive composites based on hexagonal boron nitride nanosheets for thermal management: Fundamentals to applications," *Compos. Part A Appl. Sci. Manuf.*, vol. 169, no. February, p. 107533, 2023, doi: 10.1016/j.compositesa.2023.107533

[15] B. Yang *et al.*, "Evaluation of rheological and thermal properties of polyvinylidene fluoride (PVDF)/graphene nanoplatelets (GNP) composites," *Polym. Test.*, vol. 67, no. February, pp. 122–135, 2018, doi: 10.1016/j.polymertesting.2018.02.011

[16] M. H. Alaaeddin, S. M. Sapuan, M. Y. M. Zuhri, E. S. Zainudin, and F. M. AL- Oqla, "Physical and mechanical properties of polyvinylidene fluoride – Short sugar palm fiber nanocomposites," *J. Clean. Prod.*, vol. 235, pp. 473–482, 2019, doi: 10.1016/j.jclepro.2019.06.341

5 Emerging Trends and Prospects of Noble Metal-Based Nanocomposites for Electrochemical Sensing and Anticancer Activity

M M Sreejaya, Lakshmi Krishnaa Suresh, Simi M, Haridev M, Pournami Hareesh, Manoranjan Bera, Santhy Antherjanam, and Moumita Gangopadhyay

5.1 INTRODUCTION

Noble metal nanocomposites have found mammoth applications in the field of bioimaging, biosensing, and disease treatment [1,2]. These nanocomposites can consist a continuous inorganic phase or be grafted on any organic/polymeric matrix [2]. The formation of noble metal nanocomposites with other materials stabilizes the noble metals, improves their solubility, and regulates their physical properties [3]. Noble metals having one free electron in their valence shell can exhibit titillating optical and physicochemical properties in their nanoscale. Noble metal nanocomposites having a strongly localized surface plasmon, prominent Raman scattering, and meta-enhanced fluorescence properties can show exceptional imaging properties and prominent light-to-heat conversion in the presence of an external electromagnetic field [4] Noble metals have shown strong interaction with the composite materials, e.g., noble metal-graphene nanocomposites displayed enhanced photothermal conversion than free noble metal nanoparticles (NPs) [4]. Such improved optical properties and other physicochemical properties enable the noble metal nanocomposites to be used for diverse bioimaging techniques like magnetic resonance imaging (MRI) and X-ray computed tomography (CT). Unique physicochemical properties like light-to-heat conversion has made the noble metal nanocomposites highly suitable for cutting-edge non-invasive cancer treatments viz. photothermal therapy (PTT) and photodynamic therapy (PDT) [5,6]. Easy surface modification of the noble metal nanocomposites has also contributed to a great extent to achieve target-specific drug delivery, which plays a key role in personalized cancer treatment.

Besides prominent applications in the biomedical field, noble metal nanocomposites also found vast applications in terms of electrochemical sensors for water or environment pollution control [6,7]. Electrochemical sensors detect the changes in conductivity, potential, and current based on redox reactions taking place on the electrode. Electrochemical sensors being easily available and cheap, are widely used to check environmental pollution, gas sensing, and food quality control. Noble metal nanocomposites having high reduction potential, chemical inertness, resistance to high temperature, and size-regulated electrical properties pose wide applications in electrochemical sensing [8]. These nanocomposites have also shown prominent catalytic properties owing to their

DOI: 10.1201/9781003664611-6

strong surface adsorption properties, large number of redox-active sites on the surface, and zero-valent surface [9].

Among varieties of noble metal nanocomposites, the carbon-based ones are of particular importance because of their biocompatibility, improved solubility, and high porosity. Considering the relevance of carbon-based noble metal nanocomposites in terms of biomedical and environmental applications, this chapter will highlight the general synthetic procedures, properties, mechanisms, and application of some of the important noble metal-based nanocomposites in terms of anticancer treatment and electrochemical sensing.

5.2 SYNTHETIC METHODS OF NOBLE METAL NANOCOMPOSITES

Materials made of noble metals-elements including gold, silver, platinum, and palladium integrated with various substances at the nanoscale are referred to as noble metal nanocomposites. Noble metals are added to nanocomposites to provide them with special qualities and capabilities. Nanocomposites are materials made up of two or more components with different properties. These composites' nanoscale dimension is an important component since it offers an excellent surface area-to-volume ratio that improves their performance and characteristics. Noble metal nanocomposites have attracted a lot of interest because of their remarkable catalytic, chemical, and physical characteristics [10,11]. Among numerous examples of noble metal nanocomposites, carbon-based composites are of major importance owing to their improved biocompatibility. In the following section, some of the important synthetic methods of noble metal nanocomposites have been discussed.

5.2.1 Synthesis of Gold Nanocomposites

5.2.1.1 Reduction Method

To create materials with improved qualities or particular applications, gold NPs are commonly combined with different compounds or materials. Gold nanocomposites can be synthesized using a variety of techniques, each designed to yield composites with the required properties. One popular technique is chemical reduction, which turns gold ions into gold NPs by using a reducing agent. It is possible to regulate this process to generate desired-sized NPs. Using stabilizing agents or surfactants is an additional strategy to guarantee homogeneous dispersion and avoid nanoparticle aggregation.

The desired qualities of the gold nanocomposites determine the production technique and the other materials to be used. For example, the synthesis of graphene-oxide-gold nanocomposites involves the reduction of graphene oxide to reduced Graphene Oxide (rGO) in water, coupled with the addition of $HAuCl_4$ as a gold ion precursor [12]. The subsequent introduction of hydrazine serves a dual purpose of functioning as a reducing agent to convert Au^{3+} ions to Au(0) and acting as a stabilizer, preventing the agglomeration of gold NPs. Stirring this solution at 60°C for 2 hours ensures the controlled growth of gold NPs. Centrifugation is then employed to separate the formed nanocomposites, followed by washing and drying at 60°C, contributing to the isolation and purification of the final product. In another method, a gold-graphene oxide-based nanocomposite was synthesized by employing chloroauric acid as the gold precursor, dimethylformamide (DMF) as both a reducing agent and solvent, and polyvinylpyrrolidone (PVP) as a capping agent. Here, the final nanocomposites are formed by means of an attractive force of interaction between graphene oxide (GO) and gold NPs in the presence of an external electric field. The refluxing process at 145°C for 45 minutes ensures efficient reduction and separation of gold NPs, while subsequent agitation and extended stirring for 48 hours, followed by washing and drying at 60°C, formed stable and well-dispersed nanocomposites. The third method involves the creation of nanocomposite-chitosan film containing GO, hydroxyapatite, and gold [13]. Here, sodium citrate acts as the reducing agent, converting Au^{3+} ions to Au(0) and preventing agglomeration. The process begins with the preparation of GO through the Hummers method, followed by the addition of hydroxyapatite/gold (HAp/Au) nanocomposites and subsequent stirring. The washing and drying steps at 60°C contribute to the removal of residual solvents and the formation of a chitosan film with incorporated nanocomposites, demonstrating potential applications in biomedical or sensing devices.

Another report described the functionalization of GO with 1-Amino pyrene (1-AP) in ethanol enhancing the surface stability and providing active areas for the nucleation of gold NPs [14]. This method involved the dissolution of the 1-AP-functionalized highly reduced GO and $HAuCl_4$ separately in ethanol, followed by sonication and centrifugation at 9000 rpm. The choice of ethanol as a solvent and the application of sonication contributed to the efficient mixing and separation of the components, resulting in a product that can be dissolved in water. Overall, each procedure is meticulously crafted to control factors such as temperature, time, and reagent concentrations to achieve the desired properties of the GO-gold nanocomposites.

5.2.1.2 Green Methods

For the past few years, green synthesis has fetched much attention attributed to its sustainable usage of resources and reduced harmful side products. Thus, there have been several attempts to prepare gold-based nanocomposites following green methods. In the case of curcumin-gold NPs (CUR-AuNPs) synthesis, the process is conducted under room temperature and neutral pH, using curcumin extracted from *Curcuma pseudomontana* [15]. The use of an ethanol/water solvent system and the prevention of harsh reaction conditions ascribe to the environmentally benign nature of this method. The reduction of $HAuCl_4$ to CUR-AuNPs is evident through the characteristic color change from yellow to ruby red, indicating the successful formation of the nanocomposite. Another green synthetic process involving the use of leaf extracts from *Lawsonia inermis* also aligns with green synthesis principles [16]. The leaf extract serves as a reducing and stabilizing agent, eliminating the need for external reducing agents or toxic chemicals. The reduction of $HAuCl_4$ is monitored through color changes. The fabrication of rGO/AuNPs nanocomposite electrodes further demonstrates green synthesis [17]. Here, the graphene oxide (GO) is suspended in distilled water, and the reduction of $HAuCl_4$ ions is achieved through *Eucalyptus tereticornis* solution, which acts as the green reducing agent. The process involves refluxing under mild conditions, and the nanocomposite suspension is obtained through ultrasonication, avoiding harsh chemicals or high temperatures. The drop-casting onto a screen-printed electrode completes the green synthesis process, resulting in a homogeneous nanocomposite.

5.2.2 Synthesis of Silver-Nanocomposites

Silver-based nanocomposites are synthesized through various methods to harness the unique properties of silver NPs, such as exceptional conductivity, catalytic activity, and antibacterial effects. The synthesis processes can be broadly categorized into reduction, green synthesis, and electrochemical approaches, each offering distinct advantages for specific applications in fields like antibacterial agents, sensors, catalysis, and medical devices.

5.2.2.1 Green Method

In the synthesis of silver-GO nanocomposites, $AgNO_3$ serves as the precursor, and PVP acts as the reducing agent, preventing agglomeration [18]. The method involves mixing NPs with GO, followed by heat treatments to form fine structures. The choice of PVP as a reducing agent aligns with green synthesis principles, owing to its biodegradable nature. Another example of green synthesis of GO-silver nanocomposites utilized $AgNO_3$ as the precursor and ascorbic acid as the reducing agent [19]. This method involves the simultaneous reduction of GO and the metal precursor, creating GO-AgNP nanohybrids. The utilization of ascorbic acid as a green reductant further classifies this method as environmentally friendly, emphasizing sustainability. Another approach involves the synthesis of silver NPs decorated reduced GO nanocomposites [20]. Here, $AgNO_3$ acts as the precursor, and p-acidus fruit extract is utilized as a reducing agent. The fruit extract containing phenolic compounds and ascorbic acid acts as the green reagent. This method showcases the use of natural extracts in reducing silver ions, in line with sustainable and eco-friendly practices.

5.2.2.2 Reduction Methods

Similar to the reduction methods discussed for gold nanocomposites, silver nanocomposites also get synthesized from the mixture of silver salt and a reducing agent. For example, Ag/Cts/PEG nanocomposites are synthesized through a process involving chitosan (Cts), polyethylene glycol (PEG), and silver nitrate [21]. Chitosan in acetic acid (HAC) is mixed with a solution of PEG-2000 in double-distilled water. Furthermore, a solution of silver nitrate is added under a nitrogen atmosphere to prevent oxidation reactions. The solution is stirred at 60°C for varying times to generate silver NPs (Ag NPs), monitored by a UV-visible spectrophotometer. The resulting Ag/Cts/PEG nanocomposites exhibit a color change from pale brown to dark brown, indicating successful nanoparticle formation. Thin films are then prepared for further characterization. In another method, silver NPs are synthesized using orange peel extract, demonstrating a green synthesis approach [22]. Orange peel is crushed, and its extract is prepared in toluene through vigorous stirring and centrifugation. Silver nitrate is then dissolved in toluene, and the orange peel extract is added, leading to a color change to brown, indicating the reduction of Ag ions and the formation of silver NPs. Another example portrays the synthesis of zero-valent NPs of iron, silver, and zinc is synthesized using okra seed extract [23]. The extract is added to salt solutions of $FeCl_3$, $AgNO_3$, and $Zn(OAc)_2$ $2H_2O$, and sonication is applied. Visual color changes indicate the successful formation of iron, silver, and zinc NPs. The reaction mixture is washed several times to remove unreacted okra seed extract, and the wet sediment is dried to obtain fine powder. Nanocomposites are then prepared by dissolving polystyrene in acetone, sonication with various metal and metal oxide NPs, and evaporating acetone to form composite sheets.

5.2.3 Synthesis of Platinum-Nanocomposites

5.2.3.1 Reduction Method

In the realm of chemical reduction synthesis, various methodologies have been employed to fabricate platinum-based nanocomposites with distinct properties and applications. One such method is the synthesis of Platinum-Silver Graphene Nanocomposites [24]. In this process, $AgNO_3$ and potassium tetra chloroplatinate serve as essential precursors. The reducing agent, hydrazine hydrate, plays a dual role by effectively reducing Ag^+ and Pt^{2+} ions, ensuring the formation of platinum-silver NPs while simultaneously preventing their undesirable agglomeration. Moving forward, the chemical reduction synthesis extends its application to the realm of GO-platinum nanocomposites [25]. Here, the reduction of $PtCl_6^{2-}$ ions is orchestrated using vanillin, a crucial component acting as a reducing agent. GO, synthesized separately, is then meticulously mixed with the platinum solution. This synthesis, characterized by the reduction of metal ions facilitated by vanillin, aligns itself within the chemical reduction category. Another notable example of this category is the synthesis of platinum nanocomposite with single-walled carbon nanotubes (Pt-SWCNT) [26]. In this case, platinum-acetyl acetonate acts as the precursor, while 1,2-hexadecanediol serves as the reducing agent. The process involves the reduction of Pt ions and the controlled integration of platinum NPs (PtNPs) onto the carbon nanotube substrate. Hence, it is classified under chemical reduction synthesis, showcasing the versatility of this approach with different nanomaterial substrates. The chemical reduction synthesis category further encompasses the synthesis of Pt-Doped multi-walled carbon nanotubes (MWCNTs) (NiO-MWCNT) Nanocomposites [27]. In this intricate process, the reduction of hexa-chloroplatinate ions is executed to generate Pt NPs. Starch, acting as a stabilizing agent, contributes to preventing agglomeration and ensuring the uniform development of NPs. The synthesis involves combining Pt-doped NiO NPs and MWCNTs in a medium of DMF.

5.2.3.2 Green Methods

The green synthesis of platinum NPs (PtNPs) using an aqueous extract of *Padina* gymnosperm falls under this category [28]. The powdered form of *Padina gymnospora* is extracted in sterilized Milli-Q water and the resulting filtrate is stored for further analysis. For PtNP synthesis, the aqueous extract is treated with a platinum chloride solution. The reduction of platinum ions occurs in the presence of bioactive compounds from *P. gymnospora*, leading to the formation of PtNPs. The use of plant extract in the synthesis process aligns with green chemistry principles, emphasizing sustainability and the reduction of harmful chemical agents. The second method involves an automated system for the preparation of platinum NPs (Pt-NPs) utilizing eucalyptus leaves [29]. The leaves are washed, dried, and ground to a fine powder. The powder is mixed with deionized water, and the automated system operates through a series of steps involving temperature control, vacuum filtration, and reaction with a platinum chloride solution. The Pt-NPs are obtained through this automated system based on green chemistry principles. The resulting Pt-NPs are then utilized for the preparation of a nanocomposite with MWCNTs. Chitosan, a natural polymer, enhances the dispersion of MWCNTs and facilitates the formation of the MWCNT-Pt composite.

5.2.4 Synthesis of Palladium-Nanocomposites

5.2.4.1 Reduction Method

Palladium NPs were synthesized by reducing $PdCl_2$ with $NaBH_4$ in the presence of PVP as a stabilizer, followed by centrifugation to remove excess reactants [30]. These NPs were then combined with GO, sonicated, refluxed, and subjected to swift heavy ion irradiation to form a PD-GO nanocomposite. GO was sonicated in ethylene glycol, mixed with palladium and cerium precursors, and ultrasonically dispersed [31]. After adding ascorbic acid and adjusting pH to 12, the mixture was refluxed, then washed and dried to obtain the Pd–CeO_2/rGO nanocomposite. Palladium(II) acetate was dissolved in dichloromethane, reacted with a ligand, stirred, and then filtered to yield a palladium complex with 73% efficiency [32]. For nanocomposite fabrication, MWCNTs and reduced graphene oxide (rGO) were mixed in water, dried, and then combined with palladium or silver complexes and doxorubicin (DOX) in a water-ethanol solution, under dark conditions to protect DOX's fluorescence. The mixture was then stirred with PEG-200, centrifuged to remove excess DOX and complexes, and dried at room temperature. GO was synthesized from graphite powder using the Hummers and Offeman method, involving treatment with H_2SO_4, $KMnO_4$, and $NaNO_3$, followed by washing and freeze-drying [33]. Pd NPs on graphene (Pd/G) and PdO NPs on graphene oxide (PdO/GO) were prepared by mixing GO with $Pd(OAc)_2$, sonication, and stirring, then subjected to thermal treatments under H_2 and O_2 gases for reduction and oxidation, respectively. The metal content on the graphene support was quantified using inductively coupled plasma mass spectrometry (ICP-MS). GO was synthesized from graphite using the modified Hummers method, involving sonication and stirring with chemicals, then repeatedly washed and dialyzed for purity [34]. Chemically derived graphene and palladium NPs were prepared through reduction and chemical reaction respectively, with the latter involving ascorbate and palladium nitrate. The nanocomposite was deposited on surfaces using a layer-by-layer technique, alternating between graphene and palladium nanoparticle suspensions. In another synthesis, MWCNTs were first combined with a pyrrole monomer to create a composite, which involved stirring the mixture with cetyltrimethylammonium bromide and then adding ammonium persulfate for polymerization [35]. This composite served as a base for adding palladium chloride ($PdCl_2$), which was subsequently reduced to palladium NPs using sodium borohydride, anchoring them onto the MWCNT composite. The final product was then filtered, washed, and dried to obtain the MWCNT/polymer/palladium nanocomposite.

5.2.4.2 Green Methods

Graphene oxide (GO) was mixed with melamine in a 1:1 ratio, stirred in deionized water, and dried to prepare for solar exfoliation into graphene (SG) and nitrogen-doped graphene (N-SG) using a convex lens [36]. $PdCl_2$ was then added to the GO or GO-melamine mixture, stirred, and dried to achieve $PdCl_2$ composites, which were also solar exfoliated to create palladium nanoparticle-decorated graphene (Pd/SG) and nitrogen-doped graphene (Pd/N-SG). Palladium acetate and *Cynomorium songaricum* polysaccharides (CSP) were mixed in deionized water, stirred, and then treated with KOH before being heated in a microwave oven [37]. The resulting Pd–CSP compound was combined with carbon powder, ultrasonically dispersed, and stirred to ensure thorough mixing. Finally, the mixture was filtered, washed, and dried in an oven to yield the Pd–CSP/C nanocomposite. GO was synthesized using a modified Hummers' method and mixed with cerium nitrate, followed by sonication and the addition of sodium hydroxide to form a CeO_x/GO nanocomposite [38]. Pd quantum dots on carbon (PdQDs@C) were then grafted onto the CeO_x/RGO composite by mixing with $PdCl_2$ and ascorbic acid, followed by microwave irradiation. This method offers a green synthesis approach to carbon-palladium nanocomposites.

5.3 PROPERTIES OF NOBLE METAL NANOCOMPOSITES

Noble metal nanocomposites are a fascinating group of compounds that display unique properties. They often show improved catalytic, optical, and electrical properties, rendering them relevant in various applications, including sensing [30], imaging [31], and energy conversion [32]. A comprehensive study of the intricate properties of these entities is crucial to fully grasp their potential in improving technology and addressing challenges across various sectors, including medicine and environmental science [33]. We aim to understand their underlying mechanisms and applications by investigating their unique properties. These nanocomposites are of great importance in shaping the future of material research [34]. Noble-metal semiconductor nanocomposites hold significant potential for energy, optics, and electronics applications. The noble metals can be precisely adjusted such that their inclusion into a nanocomposite material maximizes its efficiency for a specific application [35]. These effects involve energy absorption, electron transfer processes, and the generation of holes. Photocatalysis has emerged as a fascinating technology for breaking down organic pollutants, facilitating redox processes for fuel production, and catalyzing specific organic reactions [37]. These nanomaterials are continuously advancing to function as highly effective photocatalysts. In addition, they make a valuable contribution to sustainable energy by efficiently harnessing solar energy for utilization in photovoltaic cells [38]. Undoubtedly, these nanocomposites hold immense potential for advancing materials used in energy, optical, and electronic devices. Consequently, continued study on these materials remains both intriguing and groundbreaking [39].

Silver-based nanocomposites have attracted much interest over the years because of their distinct characteristics and wide range of uses. They are commonly known for their antimicrobial activities against various microorganisms [40]. The electrical, optical, and antibacterial properties are dependent on their size, shape, surface area, and distribution within these nanocomposites, which requires several characterization techniques such as Transmission Electron Microscope (TEM), Scanning Electron Microscope (SEM), and X-ray Diffraction (XRD) to understand and augment these characteristics ultimately. In addition, the physical attributes of these nanocomposites, such as electrical conductivity and chemical stability, can be exploited for numerous applications in electronics and catalysis. Nanocomposites with GO-AgNP can serve as photocatalysts in both industrial and environmental applications [38]. The fibers made of cellulose/Ag nanocomposites demonstrated acceptable mechanical characteristics and thermal stability. Furthermore, studies done to assess their antibacterial capabilities showed that the composite fibers containing silver NPs exhibited exceptional antimicrobial efficacy against *Staphylococcus aureus* [41]. A recent study showed substantial enhancements in the electrical conductivity and dielectric characteristics of the polyaniline/silver (PANI/Ag) nanocomposite [42]. The PANI/Ag-based sensor's response is quick and reversible, exhibiting the highest sensing capabilities and superior long-term stability.

Gold nanocomposites possess several important physical properties, including radioactivity [43], a high X-ray absorption coefficient [44], and localized surface plasmon resonance (LSPR). These properties can be exploited in detecting and treating tumors. Compared to other NPs, gold NPs are distinguished by their unique capacity to form long-lasting chemical interactions with groups that contain sulfur and nitrogen substances [45]. Gold nanocomposites are able to bind to a variety of chemical ligands or polymers, which makes them useful for modifying electrodes in diagnostic applications. They also exhibit exceptional drug delivery properties [46], targeting ability, and biocompatibility [47]. Spherical gold NPs have several favorable properties, such as optoelectronic features that depend on their size and shape, a high ratio of surface area to volume, remarkable biocompatibility, and barely any toxicity [48]. Surface plasmon resonance and the ability to inhibit fluorescence are two of the important physical features which mark their significance in the field of bio-nanotechnology [49]. In water, spherical gold NPs exhibit a variety of hues, including brown, orange, red, and purple, when their size increases from 1 to 100 nm [50].

Furthermore, Pt nanocomposites have been developed for various applications ranging from biological to electrochemical applications [51]. Pt nanocomposites have also been used for drug delivery, with various systems designed to deliver Pt anticancer drugs [33]. Daunorubicin, a well-known anthracycline anticancer chemotherapy medication, can be determined using highly sensitive DNA biosensors that were recently designed and produced by modifying a glassy carbon electrode (GCE) with Pt/SWCNTs nanocomposite.

Water-soluble Pt complexes have been synthesized for bioimaging, providing strong green luminescence and allowing penetration into cancer and normal cells. Platinum is a promising electrocatalyst in applications such as alkaline fuel cells and dye-sensitized solar cells. The quality of the Pt nanocomposite depends on the synthesis method, necessitating optimization for desired results.

Palladium (Pd) nanocomposites exhibit desirable electrical conductivity and hydrogen interaction, rendering them potentially valuable for sensing applications. The addition of Pd NPs to nanocomposites leads to variations in color, elasticity, and glass transition temperature. In addition, Pd NPs immobilized on garlic skin-derived activated carbons (GACs), exhibited exceptional catalytic reduction of Cr (VI) in the presence of formic acid. This eco-friendly, economical, and efficient catalyst system, fabricated via microwave-assisted synthesis, offers potential applications for efficient toxic metal reduction and large-scale fabrication [52]. The NPs engage with the nanocomposites, causing them to become more plasticized and so enhancing their oxidative thermal stability [53]. In general, the presence of Pd NPs enhances the thermal stability and performance of nanocomposites. Various test results indicated that the addition of tiny palladium particles to the surface of nanotubes enhances their sensitivity in detecting high-temperature hydrogen gas [54].

5.4 ADVANTAGES OF NOBLE METAL NANOCOMPOSITES

Noble metal nanocomposites are synthesized by various methodologies, which include chemical vapor deposition, template-assisted techniques, and sol-gel procedures. These synthetic processes are having a significant impact on the fundamental properties and advantages of the nanocomposites [36]. To comprehend the structural arrangement and establish an association with the resulting beneficial properties, scientists employ characterization techniques such as (XRD), (TEM), and other spectroscopic methods. These techniques provide insights into the physical structure and crystallographic attributes of carbon-based noble metal nanocomposites [39]. The combination of noble metals and carbon-based materials has brought about significant transformations in the field of nanomaterials, offering an extensive range of possibilities for the advancement of technology. The outstanding blend of noble metal NPs and various carbon allotropes has been recognized as a crucial area of research because of their exceptional properties and many possible applications. The numerous advantages of these nanocomposites including their increased catalytic activity, electronic properties, and their wide range of potential applications across different fields will be explained in this chapter ranging from energy storage to medicinal applications.

The catalytic activity possibly arises due to the synergistic interactions between carbon and noble metal constituents. By incorporating noble metals into carbon matrices, greater catalytic performance is achieved, especially in hydrogen evolution processes and redox reactions. Graphene-supported platinum nanocomposites demonstrate enhanced catalytic efficiency in fuel cells due to the combination of graphene's exceptional conductivity and huge surface area, which facilitates the dispersion of platinum NPs [33]. Graphene additionally offers robust mechanical reinforcement, hence augmenting the longevity and stability of platinum NPs during catalytic processes. In carbon nanotube-supported palladium, the distinctive structure of carbon nanotubes enhances the stability and distribution of palladium NPs. This unique structure enables these nanocomposites to function as highly efficient catalysts in a range of reactions, including hydrogenation reactions and Suzuki coupling reactions [55, 56]. Carbon nanotubes combined with noble metal NPs exhibit excellent electrical conductivity, facilitating efficient electron transfer in catalytic reactions. In addition, carbon quantum dots (CQDs) embellished with silver NPs have notable characteristics such as intense luminescence, exceptional biocompatibility, and antibacterial efficacy. These characteristics make them appropriate for bioimaging applications and prospective utilization in biological domains. Moreover, these can be utilized for several biological applications, including administering drugs, imaging, and therapeutic interventions [57].

Advanced electronic properties including enhanced conductivity, electron transfer kinetics, and electron transport processes demonstrate the potential of noble metal nanocomposites for their use in electronic devices and electrochemical sensors. The enhanced sensitivity of gold NPs on carbon nanotubes has the ability to amplify the electrochemical signals causing increased sensitivity to detect target analytes. These nanocomposites become significantly more beneficial when their characteristics are altered [58]. For example, the composition and structure of graphene-based noble metal nanocomposites can be modified, which is not only useful in fine tuning sensor properties but also for increased selectivity. Similarly, Improved selectivity of hydrogen sensors can be achieved by incorporating palladium NPs on carbon support by means of improved hydrogen absorption and reducing interferences from other gases. Carbon-based platinum nanocomposites have the ability to accelerate the rate of electrocatalytic reactions in gas sensors which in turn results in faster response with respect to environment [59]. Moreover, these can also be used in energy storage devices, such as batteries and supercapacitors, to enhance the energy storage capacity, cycling stability, and overall performance of these devices.

The considerable potential of carbon-based noble metal nanocomposites is opening prospects for pioneering ideas as we continue to delve further into the diverse world of material science.

5.5 APPLICATIONS OF NOBLE METAL NANOCOMPOSITES IN ELECTROCHEMICAL SENSING

5.5.1 Mechanism of Electrochemical Sensing

Electrochemical sensors composed of noble metals such as gold, silver, platinum, and palladium are highly appropriate for modifying the surface of electrodes [60–64]. This is because these metals possess notable qualities such as low cytotoxicity, biocompatibility, ease of surface functionalization, high reduction potential, and electrical characteristics that vary with size. Their diverse forms, exceptional electrical and photochemical properties, stability, and large surface area enable them to be well-suited for the production of electrochemical sensors [65–68].

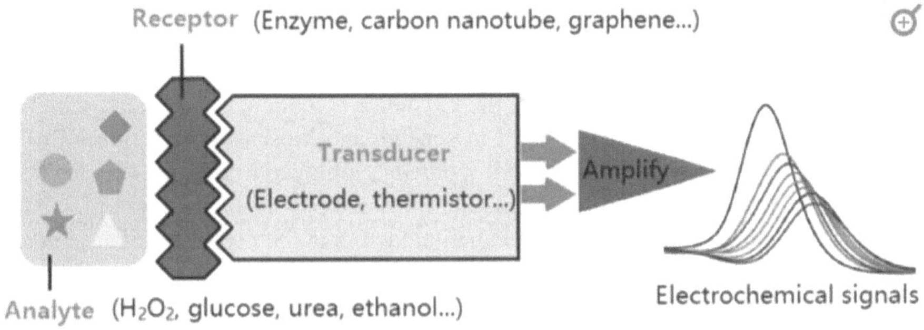

FIGURE 5.1 Schematic representation of electrochemical sensing. (Reused from Chen et al. [83] Creative Commons Attribution (CC BY) license (https://creativecommons.org/licenses/by/4.0/).)

Electrochemical sensing is one of the most popular and economical methods for the analytical detection of a wide variety of analytes including biomarkers, synthetic food colors, environmental contaminants, and pharmaceuticals [69–74]. Due to the rapid progressions in technology, electrochemical sensing gained more attention in the analytical area of research. The devices that can produce electronic signals based on the electrochemical reduction or oxidation of the analytes are electrochemical sensors. The electronic signals thus produced by the electrochemical sensors are proportional to the number of analytes present in the sample and hence can utilized for the quantification of various analyte species [75]. Those analytes which are electroactive can be successfully analyzed using the electrochemical sensing method [76]. The basic components of an electrochemical sensor are transducer and receptor [77]. The receptor is the part of the sensor, where the analyte interactions and it may be biomolecules like enzymes, DNAs, nanomaterials and nanocomposites, polymers, and supramolecules [78–81]. The interaction of the analyte with the receptor produces an electrochemical response and this is transferred to the transducer. The transducer converts the electrochemical signal into a measurable electrical signal (current or voltage) [82]. An amplifier then amplifies the electrical signal and conveys the signal to the processor for a readable output like a voltammogram (Figure 5.1).

Among all other analytical techniques such as chromatography and spectrophotometry for the analytical determination of various analytes, electrochemical sensors are more selective, sensitive, and economical [84]. According to the electrochemical method by which the analyte is being detected, the electrochemical sensors are mainly classified into voltammetric sensors potentiometric sensors, and amperometry sensors. In all these electrochemical sensors, voltammetric sensors are more adaptable for fabricating miniaturized devices and also more reliable with quick electrochemical response [85]. With the advancements in technology, numerous voltammetric sensors were reported with various modified electrodes. The chemically modified electrodes would facilitate the selective oxidation or reduction of analytes and thus the performance of the sensor would be reliable. Hence, these modifying materials are crucial elements in the electrochemical sensor, where the analyte-sensor interaction is occurring. Hence, this chapter emphasizes to deliver the significance of nanomaterials composed of noble metals such as Au, Ag Pd, and Pt in the development of voltammetric sensors with a special case of noble metal-based carbon nanocomposites.

5.5.2 Gold-Based Nanocomposites

Gold nanocomposite-based electrochemical sensors are gaining popularity because of their ease of use, simple preparation procedures, low cost, excellent selectivity, high sensitivity, and quick response [86]. The development of electrochemical methods is significantly impeded by the poor conductivity and catalytic activity of the bare electrodes [54]. A range of materials with exceptional conductivity and catalytic activity, such as metal NPs, graphene [87], carbon dots [55, 66], and carbon nanospheres have been incorporated into electrodes to improve their performance. Since gold nanomaterials offer unique qualities such as large catalytic surface areas, improved diffusion and adsorption toward target molecules, and the existence of conduction centers, which facilitate electron transfer, they have been widely used in sensors and biosensors [88]. Au NPs (Au-NPs), one of the more promising new materials, were typically employed to enhance the electrochemical properties of the electrodes [89]. The consistent chemical and physical resources, effective catalytic activities, and small size of Au-NPs were fundamental factors in their selection. These benefits have led to the demonstration that adding Au-NPs to electrodes increases electrode sensitivity, promotes charge transfer, and reduces the analyte detection limit [39]. Several electrochemical sensors were developed, considering the above-mentioned properties.

Epinephrine is a crucial neurotransmitter and biomarker that has a significant impact on our neurological system and overall body function. By adding carbon nanofiber (CNF) and gold nanoparticle (AuNP) nanocomposite to a GCE, Sipuka et al. made an electrochemical sensor that can detect epinephrine [67]. Compared to the bare electrode, the modified electrode shows enhanced sensitivity within a detection limit of 1.70 μM. Square-wave voltammetry (SWV) was used to determine epinephrine in human serum quantitatively.

In another work, a novel electrochemical sensor for detecting hydroxychloroquine in pharmaceutical and clinical samples [90]. The GCE was modified with an MWCNT and AuNP nanocomposite through green synthesis. The synergistic effect between AuNPs and functionalized MWCNTs (f-MWCNTs) increased the sensitivity with a detection limit of 0.0093 μmol/L (Figure 5.2).

Indole-3-acetic acid (IAA) is an essential plant growth hormone. A cost-effective, and portable electrochemical sensor based on AuNP and three-dimensional (3D) reduced graphene oxide (rGO) modified screen-printed carbon electrode (SPCE) was developed by Shao and coworkers to detect IAA in plants [91]. The modified sensor detects IAA with a wide linear range and has a lower detection level. The unique pore structure of 3D rGO increased the surface area and the binding site of AuNPs improved the electrical conductivity of the sensor. In order to electrochemically detect methylparaben (MP) in cosmetic samples, Natchaya and colleagues used a pencil graphite electrode

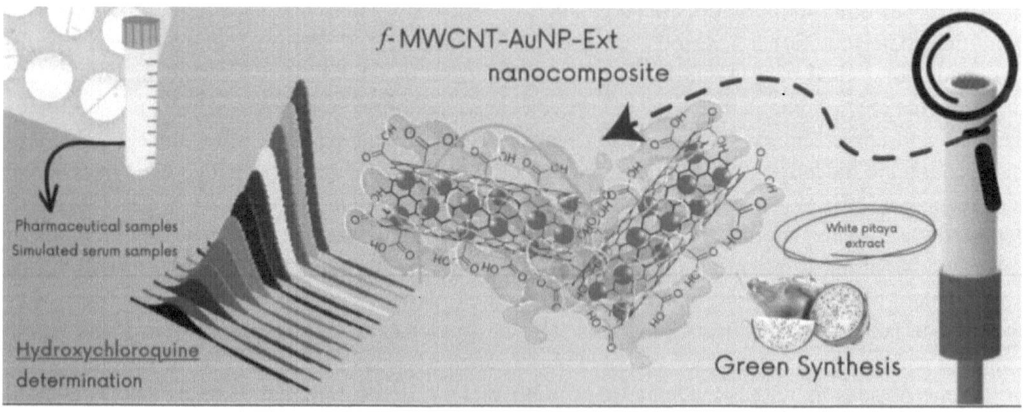

FIGURE 5.2 Graphical explanation of the electrochemical detection of hydroxychloroquine using f-MWCNTs and Au nanocomposite. (Reprinted from Journal of Pharmaceutical and Biomedical Analysis, 236, Nathalia Nardi, Luan Gabriel Baumgarten, Juliana Priscila Dreyer, Edson Roberto Santana, João Paulo Winiarski, Iolanda Cruz Vieira, Nanocomposite based on green synthesis of gold nanoparticles decorated with functionalized multi-walled carbon nanotubes for the electrochemical determination of hydroxychloroquine, 115681, 2024, with permission from Elsevier [90].)

modified with GO and AuNP nanocomposite [92]. With an improved detection limit, excellent repeatability, and reproducibility, the modified electrode outperforms the unmodified electrode in terms of response stability.

An inexpensive, and highly sensitive electrochemical sensor was created by Nasiri et al. to monitor diclofenac in water, pharmaceutical, and clinical samples [93]. The sensor was fabricated by adding AuNPs and f-MWCNTs/rGO nanocomposite layer on the gold electrode. The different aspects, including pH, electrodeposition duration, and amount of nanocomposite, were investigated to study the effect on the sensing performance of the electrode. The suggested sensor exhibited outstanding catalytic activity for the oxidation of diclofenac molecules using differential pulse voltammetry (DPV).

An electrochemical sensor with high sensitivity was developed in another work by Feng et al. for the detection of nitrite, utilizing a one-pot green manufacturing approach [94]. The study utilized Xylan-based CQDs on gold NPs (Au/CQDs) and MXene. This process resulted in the development of a sensor that exhibits great sensitivity, stability, a wide linear range, and exceptional selectivity. The sensor was effectively utilized to detect nitrite in tap water and salty vegetables, showcasing a novel approach to nitrite detection.

Recently, a novel electrochemical sensor was devised by Kavitha and her coworkers for the purpose of detecting nicotine in samples of saliva [95]. The sensor consists of a GCE that has been enhanced with graphene, iron, phthalocyanine, and gold NPs. It demonstrates a linear reaction to nicotine within the concentration range of 0.5–27 μM, with a minimum detectable concentration of 17 nm. The sensor exhibited exceptional selectivity and sensitivity, even when exposed to commonly encountered interfering chemicals.

5.5.3 Silver-Based Nanocomposites

The electrochemical, optical, magnetic, catalytic, and electronic capabilities of AgNPs make them one of the most used NPs in various areas of research [48]. To make the NPs more stable, nevertheless, extra capping agents such as polymers may be necessary for AgNPs [10]. This led to its widespread use in environmental, biomedical, and industrial uses. Shape diversity [85], size-adjustable characteristics [97], crystallinity [98], and simplicity of synthesis are the benefits of AgNPs. Several electrochemical sensors were developed by modifying bare electrodes with silver NPs, which are discussed later in this section.

Shekher et al. fabricated a simple and disposable electrochemical sensor for the antiviral drug valganciclovir (VGC) detection by directly electrodepositing AgNPs onto a conducting polymer (CP) modified with CNT on (AgNPs/CP-CNT/SPCE) [96]. A novel technique using DPV was devised to determine VGC, demonstrating a direct correlation between the peak current and the concentration.

Another recent development in this area is the fabrication of an electrochemical sensor to detect the presence of sunset yellow and tartrazine food colors [99]. The sensor utilizes CNFs that have been enhanced with silver NPs and are supported by a carbon paste electrode (CPE). The sensor has the capability to detect dyes at extremely low concentrations of nanomolar levels, covering a broad linear range. The sensor demonstrated remarkable selectivity in the presence of interferents, along with commendable reproducibility, repeatability, and stability. The accuracy of the sensor was confirmed through a computational investigation using a discrete Fourier transform.

By incorporating AgNPs, MWCNTs, and folic acid (FA) into a CPE, Fekry et al. created a highly sensitive electrochemical sensor that could detect creatinine with greater accuracy [100]. In a universal buffer solution, the sensor oxidized creatinine electrocatalytically. The electrode showed high selectivity, stability, and rapid response for creatinine at pH 7.0, with a detection level of 0.008 μM. The sensor's low detection limit, excellent sensitivity, and reproducibility make it an effective device for the electrochemical monitoring of creatinine.

Zhang et al. fabricated an electrochemical device for the detection of cyanide in food products using a composite of Ag NPs and CNTs on a GCE [65]. The electrodes exhibited great density, porosity, and aspect ratio, along with a distinct and consistent oxidation peak for cyanide detection. The amperometry results demonstrated that the electrode exhibited high sensitivity, selectivity, and stability toward cyanide. Later, the sensor was redesigned which exhibited superior linear range, sensitivity, and detection limit compared to literature reports.

The most recent innovation is an electrochemical method that can quickly and accurately measure the concentrations of three organophosphorus compounds: diazinon (DZ), malathion (MT), and chlorpyrifos (CLP). The method involves utilizing a modified pyrolytic graphite electrode (PGE). Nanocomposite consisting of functionalized CNTs and AgNPs was used to modify the PGE. This was used in conjunction with a batch injection analysis system that incorporates multiple pulse amperometric detection (BIA-MPA). The schematic representation of the analytical determination of OPs using this sensor is given in Figure 5.3. The sensor exhibits a linear range of 0.1–20 μM for DZ, 1.0–30 μM for MT, and 0.25–50 μmol L^{-1} for CLP. This approach has a notable level of sensitivity direct identification of these OPs in water and food samples [101].

In order to detect the presence of Imidacloprid (IMC) pesticide in agriculture, Rashed et al. designed an innovative electrochemical sensor that employs Ag NPs coated onto a nanocomposite derived from naturally produced mesoporous carbon/hematite ore [102]. The sensor exhibits exceptional sensitivity and detection limitations, encompassing a broad linear sweep voltammetry range and a detection limit [103] as low as 1.06 μM. The sensor electrode demonstrates distinctive selectivity characteristics as well as exceptional reproducibility, repeatability, and storage stability.

Wang and his coworkers fabricated a highly responsive electrochemical sensor in 2020 for the detection of pentachlorophenol (PCP) by altering GCE with silver-reduced GO nanocomposites [103]. The sensor exhibited exceptional electrocatalytic performance in oxidizing PCP, boasting a detection limit of 0.001 μM, surpassing the United States Environmental Protection Agency's suggested threshold of 0.0037 μM. The sensor was effectively utilized to determine the presence of PCP in vegetables and was verified by high-performance liquid chromatography, thereby establishing a novel platform for detecting PCP in actual samples.

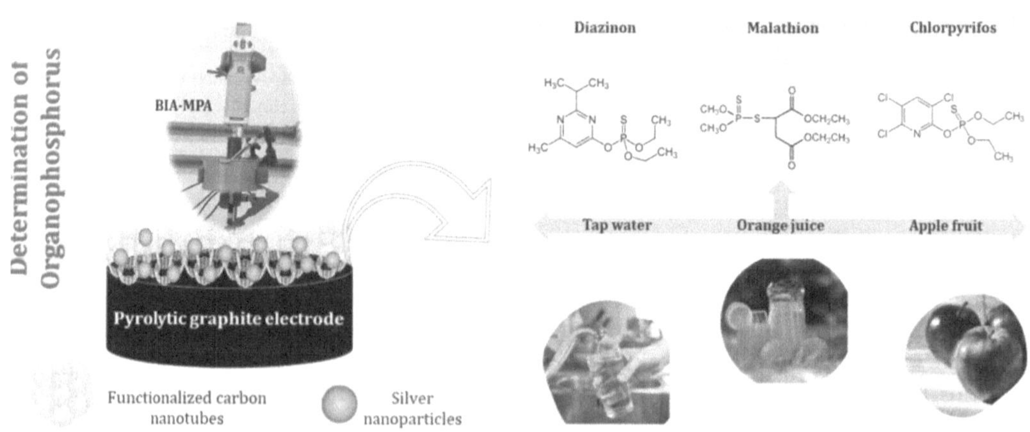

FIGURE 5.3 Schematic illustration of the electrochemical sensing of OPs using CNTs and AgNPs nanocomposite. (Reprinted from Talanta, 246, Laís Sales Porto, Lucas Franco Ferreira, Wallans Torres Pio dos Santos, Arnaldo César Pereira, Determination of organophosphorus compounds in water and food samples using a non-enzymatic electrochemical sensor based on silver nanoparticles and carbon nanotubes nanocomposite coupled with batch injection analysis., Copyright (2022), with permission from Elsevier [101].)

5.5.4 Platinum-Based Nanocomposites

Pt is yet another noble metal and the nanocomposites of Pt have been widely utilized in the development of various electrochemical sensors [104]. Jing Li and their coworkers explored a Pt-based nanocomposite for the detection of dopamine secreted from the cells by electrochemical method. Pt NPs were combined with MWCNTs functionalized with polydiallyl dimethylammonium chloride (PDDA) to form a composite and the synthesized composite was used to modify a screen-printed electrode (SPE). The schematic representation of the fabrication and the sensing mechanism of the electrode is given in Figure 5.4. The Pt-MWCNTs/SPE showed an excellent electrochemical response for the electrooxidation of dopamine with good selectivity compared to unmodified SPE. Also, the sensor demonstrated a wide linear range of 0.005–1μM with a detection limit of 2 nm. This suggests that Pt-MWCNTs/SPE could be a good analytical tool for the electrochemical assay of dopamine in real samples.

Another electrochemical sensor was fabricated using Pt-Ni nanocomposite anchored on an activated carbon scaffold (Pt-Ni/C) for the sensing of the analyte, glucose in serum samples [106]. The Pt-Ni/C was coated on a GCE to study the electrochemical behavior of glucose in 0.1 M NaOH as the supporting electrolyte. The synthesized composite was characterized using TEM, XRD, XPS, and Raman Spectra. The amperometry assay of glucose on the Pt-Ni/C/GCE exhibited a linear range of 0.025–12 mM with good anti-interference properties. Recently another glucose monitoring device was developed using a Pt- and SWCNT-based nanocomposite on a GCE [107]. The ultrasonication method was applied for the synthesis of Pt/NiO/SWCNT composite. The Pt/NiO/SWCNT displayed an excellent electrocatalytic nature for the electrochemical detection of glucose due to the synergetic effect of Pt NPs, NiO, and SWCNT in the composite matrix. The prepared composite materials were fruitfully used for the amperometry determination of glucose with a dynamic concentration range of

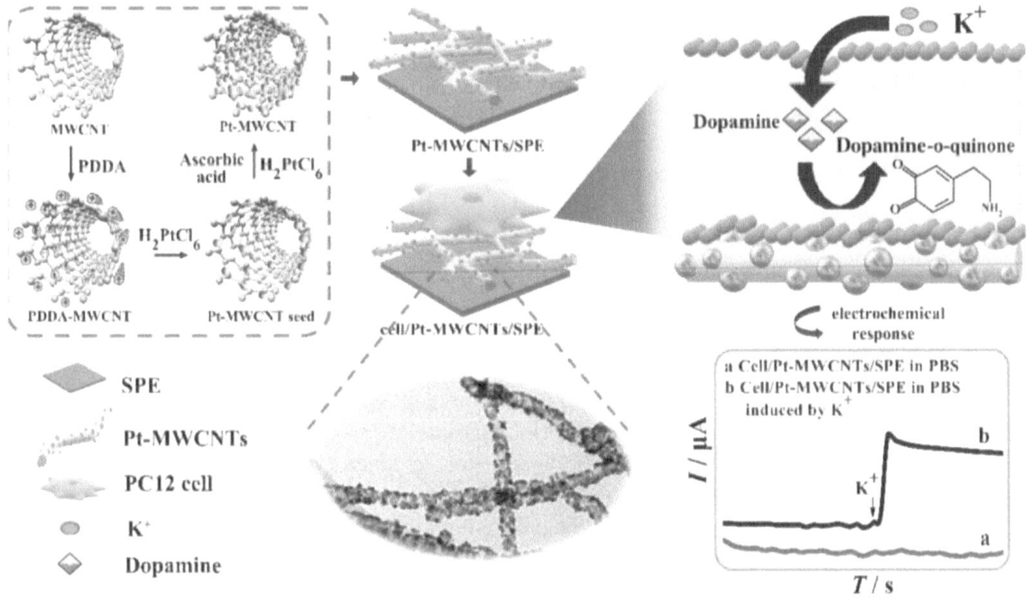

FIGURE 5.4 Representation of the synthesis of nanocomposite and the sensing mechanism of electrode (Reprinted from Sensors and Actuators B: Chemical, Vol 330, Jing Li, Xin Huang, Weishan Shi, Mengyuan Jiang, Liang Tian, Mengjie Su, Jie Wu, Quan Liu, Chunmei Yu, Haiying Gu, Pt nanoparticle decorated carbon nanotubes nanocomposite based sensing platform for the monitoring of cell-secreted dopamine, Copyright (2021), with permission from Elsevier [105].)

0.05–2.70 mM. Hence, Pt-Ni/C/GCE was proved to be a promising sensor the glucose monitoring in biological fluids in the future.

A cost-effective electrochemical sensing strategy for the detection of favipiravir (FV), an antiviral drug was proposed by Wiem Bouali et al. using the nanocomposite of Pt and reduced GO [108]. FV is an approved drug used for the treatment of coronavirus, [109] and hence, sensor for the detection of FV in biological samples and pharmaceutical samples is important in the present scenario. The Pt and graphene-oxide-based nanocomposites were proved to be good catalytic materials for electrochemical sensing according to the previous reports, [110,111]. Using SWV, the developed sensor detected FV from 3.16 μM to 100 μM with good reproducibility. Under optimized conditions, the analytical application of the sensor was done in serum samples as well as in pharmaceutical formulations with satisfactory recovery.

A CPE modified with a nanocomposite of Pt, SWCNT, and ionic liquid was employed for the electrochemical determination of bisphenol A, a toxic compound [112]. The nanocomposite was synthesized and characterized by TEM, SEM, and EDX analysis. SWV was used to determine the amount of bisphenol A and the sensor showed a detection limit of 0.2 nm. The Pt-grafted nanocomposite drastically reduced the redox overpotential compared to the unmodified CPE. Hence, the composite exhibited electrocatalytic properties toward the electrochemical determination of bisphenol A. Stability and selectivity studies were also conducted and the real-time application of the device was done in canned food samples by standard addition method.

An electroanalytical approach for the determination of nevirapine, an antiretroviral drug using another Pt/MWCNT-based nanocomposite was done by Fredrick O. Okumuy et al. in 2020 [113]. The composite was synthesized using NPs of Pt, and Ag combined with SWCNT. A GCE was modified with Pt-Ag/MWCNT hybrid material to evaluate the electrochemical response of nevirapine using DPV. The addition of the NPs of the noble metals with the MWCNT increased the porosity, conductivity, and surface area of the nanocomposites than MWCNT alone, which improved the performance of the sensor to a great extent. The sensor manifested a detection limit of 0.021 μM and was analytically applied for the detection of nevirapine in human urine and milk samples.

In another work, the simultaneous electrochemical detection of two drugs, piroxicam (PXM) and amlodipine (APE) was achieved using a nanocomposite composed of Pt, NiO, and MWCNTs [102]. NiO was doped with Pt by sol-gel method followed by combining it with MWCNT by ultrasonication. The nanocomposite was then coated on a GCE. The Pt/NiO/MWCNTs-GCE was then used for the simultaneous electroanalysis of PXM and APE. The modified nanocomposite-based GCE showed a linear detection range of 0.6–320.0 μM and 1.0–250.0 μM for PXM and APE in citrate buffer as supporting electrolytes. The PXM and APE detection in biological samples was achieved with good selectivity using the Pt/NiO/MWCNTs modified sensor.

5.5.5 Palladium-Based Nanocomposites

Pd and its various nanocomposites were highly explored in the area of electrochemical sensors for various years [115]. This section of the chapter will give a comprehensive account of the applications of Pd- and carbon-based nanocomposites in electrochemical sensing of different analytes [68]. A SPE-based sensor modified with Pd and carbon nanocomposite was introduced by Pitchaimani Veerakumar and coworkers for the determination of a drug, dimetridazole (DLE) [116]. A nanocomposite synthesized using carbon derived from teak leaves (TLC) and an alloy of Cu and Pd NPs were employed for the voltammetric determination of DLE with good analytical performance. The pictorial representation of the synthesis of the nanocomposite and the sensing of DLE is given in Figure 5.5.

In another work, a nanocomposite was synthesized using Pd and Co NPs on an activated carbon (AC) for the detection of dopamine [117]. Using DPV, a picomolar level of detection limit was achieved for the electrochemical determination of dopamine on the Pd/Co-AC modified sensor. Yi

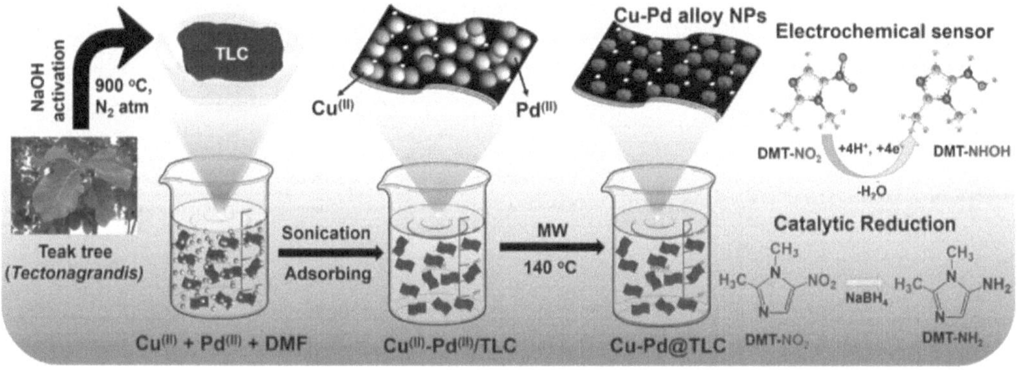

FIGURE 5.5 The illustration of the synthesis of the Cu-Pd/TLC nanocomposite and the electrochemical detection of DLE. (Reprinted from Journal of Alloys and Compounds, 931, Pitchaimani Veerakumar, Chelliah Koventhan, Shen-Ming Chen, Copper-palladium alloy nanoparticles immobilized over porous carbon for voltammetric determination of dimetridazole, Copyright (2024), with permission from Elsevier [116].)

Wu et al. reported the utility of another Pd- and MWCNT-based sensor for the analytical determination of acetaminophen in pharmaceutical samples [118]. The effect of Pd in the nanocomposite greatly influenced the sensor performance by enhancing the oxidation peak current of acetaminophen.

The significance of Pd and GO-based nanocomposite in the simultaneous determination of eight different biomolecules was reported recently by Celia Ferrag and the co-researchers [119]. The simultaneous electrochemical sensing of more than two biomolecules with good selectivity is an arduous task, as these biomolecules coexist together in the biological samples. Also, the simultaneous resolution of the oxidation peaks is very difficult as many of the biomolecules oxidize at almost the same or nearby potential. However, this difficulty was overcome by simultaneously detecting eight biomolecules such as guanine, 8-hydroxyguanine, adenine, cytosine, thymine, uric acid, ascorbic acid, and dopamine utilizing a nanocomposite composed of Pd and Fe nanoparticle and graphene oxide aerogel. This is a very interesting work, which opens up the room for more studies on bimetallic Pd and carbon-based nanocomposites in the future for electrochemical applications. Also, this is the pioneering work reported for the determination of eight biomolecules at a time using a single electrode and this would be a good analytical tool in the future for clinical applications.

A comparison of the reported electrochemical sensors based on the noble metal-based carbon nanocomposites is given in Table 5.1.

5.6 APPLICATIONS OF NOBLE METAL NANOCOMPOSITES IN ANTICANCER ACTIVITY

Cancer, a complex and pervasive group of diseases characterized by uncontrolled cell growth and spread, poses a formidable challenge in the realm of healthcare. Conventional methods of cancer treatment, such as surgery, chemotherapy, and radiation therapy, have been instrumental in managing the disease. However, they are accompanied by inherent disadvantages, prompting the exploration of novel approaches for more effective and targeted interventions. In this context, nanocomposites have emerged as a promising frontier in cancer treatment offering distinct advantages over traditional methods. Noble metal NPs exhibit unique mechanical and chemical properties distinct from their bulk counterparts, setting the stage for innovative applications in the realm of anticancer activity. Their exceptional attributes comprise high thermal stability, catalytic performance, expansive surface area, outstanding electrical conductivity, and commendable biocompatibility [124].

TABLE 5.1
Comparison Table for Some of the Electrochemical Sensors Reported with the Noble Metals and Carbon-Based Nanocomposites as the Sensing Scaffold

Type of Noble Metals	Modified Electrode	Electrochemical Method	Analyte	Reference
Gold	CNF/AuNPs/GCE	SWV	Epinephrine	[67]
	MWCNT/AuNPS/GCE	SWV	Hydrochloroquine	[90]
	rGO/AuNPs/SPCE	DPV	Indole -3- acetic acid	[91]
	GO/AuNPs/PGE	SWV	Methylparaben	[92]
	fMWCNT/GO/AuNPs/Au	DPV	Diclofenac	[93]
	CQDs/AuNPs/GCE	DPV	Nitrite	[94]
	GO/AuNPs/GCE	Amperometry	Nicotine	[120]
Silver	f-CNT/AgNPs/SPCE	SWV	Valganciclovir	[96]
	CNF/AgNPs/CPE	SWV	Sunset yellow and Tartrazine	[121]
	MWCNT/AgNPS/CPE	DPV	Creatinine	[122]
	CNT/AgNPs/PGE	Amperometry	Organophosphorus compounds	[101]
	rGO/AgNPs/GCE	DPV	Pentachlorophenol	[103]
	meso-C/AgNPs/GCE	Amperometry	Imidacloprid	[102]
Platinum	MWCNT/PtNPs/SCPE	SWV	Dopamine	[123]
	AC/PtNPs/GCE	Amperometry	Glucose	[106]
	rGO/PtNPs/GCE	SWV	Favipiravir	[108]
	SWCNT/PtNPs/CPE	SWV	Bisphenol A	[112]
	MWCNT/PtNPs/GCE	DPV	Nevirapine	[113]
	MCNT/Pt-NiO/GCE	SWV	Piroxicam and Amlodipine	[114]
Palladium	TLC/Pd-Cu/SPCE	Linear sweep voltammetry	Dimetridazole	[116]
	AC/PdCo/GCE	DPV	Dopamine	[117]
	MWCNT/Pd/GCE	DPV	Acetaminophen	[118]
	GO/Pd-Fe/GCE	SWV	Biomolecules	[119]

5.6.1 Mechanism of Anticancer Activities by the Noble Metals Nanocomposites

Although a plethora of research is being done in the field of cancer treatment, it still remains one of the most detrimental diseases of the present day. Late diagnosis of the disease is one of the primary reasons behind its fatality rates. Therefore, new treatment modalities involving prominent imaging agents can prove to be promising choices for effective remedies for cancer. One of the primary keys to successfully restricting cancer is its early diagnosis and application of target-specific treatment methodologies. Noble metals being non-reactive to air, water, or acid, have shown fascinating applications in the field of cancer therapy. The superiority of noble metal NPs compared to other metal NPs is attributed to their intriguing properties like high surface-to-volume, easy surface functionalization, fascinating optical properties, and facile synthetic methods [55,125,126]. Such extraordinary physicochemical properties of noble metal nanocomposites enable them with prominent imaging properties. Owing to characteristics like surface-enhanced Raman scattering (SERS), localized surface plasmon resonance (LSPR), and metal-enhanced fluorescence (MEF), noble metal-based nanocomposites behave as efficient imaging agents, thereby useful for early-stage tumor detection [127–129]. Besides imaging, noble metal composites have shown pivotal roles in various modern cancer treatments, e.g., target-specific drug administration, photothermal activities, PDT, radiation therapy, and chemotherapy. Some of the mechanistic details of these modern treatment methodologies by the noble metal nanocomposites have been discussed in the following section.

5.6.1.1 Imaging and Early Diagnosis

The electronic configurations of noble metals clearly indicate the presence of a free electron at the outermost s-orbital, which plays a key role in activating various optical properties in the noble metal composites. These electrons can lead to coherent oscillation when subjected to an external electromagnetic oscillating field like light. Such electronic oscillation is comparable to the excitation of a photon confined in a small nanostructure. This electronic oscillation in noble metal nanocomposites takes place only when the energy gap of the noble metal composite matches with that of the external light source. This phenomenon of electronic oscillation in the presence of a light source at a specific frequency is often termed LSPR [130–133]. The surface plasmon of the oscillating electron in the noble metal composites can de-excite in either a radiative form in terms of light scattering or via a non-radiative thermal form [134]. This trait is also influenced by the different shapes and nature of the noble metal nanocomposites, e.g., gold NPs within 10 nm size absorb at 520 nm; however, silver NPs of the same size absorb around 390 nm [131]. The high ductility of the noble metal nanocomposites has shown a wide range of shapes and sizes leading to fascinating easily tunable radiative nature. One more important contribution of noble metal composites in bioimaging techniques is to enhance the fluorescence quantum yield of the otherwise weak fluorophores. This phenomenon of increasing fluorescence properties of the weak fluorophores in close vicinity of the noble metal surface can be attributed to the MEF effect [135,136]. MEF stems from the interaction of the dipole of the fluorophore with the surface plasmon of the noble metal composites [135]. This effect not only enhances the fluorescence but also reduces fluorescence lifetime, thereby reducing the rate of photobleaching. This property of noble metal nanocomposites enables a wide range of fluorophores, which were otherwise ineffective in terms of bioimaging and drug-tracking, to be utilized in real-time monitoring of drug action. For example, recently Choi and his co-workers reported a gold nanorod-based recognition tool for miRNAs, which act as biomarkers for different diseases including cancer [137]. In their work, porous gold nanorods were functionalized with specific molecular beacons in a quenched state; however, upon interaction with the miRNAs the stem parts opened up achieving the optimum distance required for MEF with a prominent increase in fluorescence signals. In addition to these methodologies, noble metal nanocomposites are also utilized in other optical imaging instruments like MRI and X-ray CT [138].

5.6.1.2 Target-Specific Drug Delivery

Smooth internalization and accumulation specifically at the diseased site are the two most important criteria in the field of drug design and discovery. Facile target-specific internalization and accumulation help reduce undesired side effects from the drug. Targeting mechanism can be of two types; (i) active and (ii) passive. Active targeting involves functionalization with specific antibodies, peptides, FA, and other molecular beacons recognizable by the cancer cell-membrane receptors [139]. Easy surface modification methods associated with the noble metal nanocomposites make them great choices for the development of targeted drug delivery systems. The small size of noble metal nanocomposites is also another contributory factor in the development of drug delivery systems as the cancer cell vasculature ranges from hundred to several hundred nanometers [140]. In one of the recent studies, Bariana et al. developed porous gold nanoframeworks for the selective delivery of antilymphoma drug IT848 in the myeloma and lymphoma cell lines [141]. The pores of gold NPs were utilized to load drug molecules and hyaluronic acid. Hyaluronic acid acted as the targeting agent with respect to the glycoprotein CD44 overexpressed in the cell transmembrane. Passive targeting can be introduced in any drug delivery system by incorporating internal stimuli-responsive functional groups in the noble metal nanocomposites [142]. Cancer cells have different characteristic pH, temperature, polarity, and reducing nature compared to normal cells. Thus, noble metal nanocomposites are often modified with functional groups, which are sensitive to such heterogeneous microenvironments of cancer cells. For example, Yim et al. reported a gold nanocomposite loaded with chemotherapeutic drug doxorubicin for pH-responsive drug delivery

[143]. In this study, the drug was attached to the AuNP surface via a pH-sensitive covalent amide linkage. Cancer cells have high oxidative stress compared to normal cells leading to a strong reducing intracellular environment. One recent study from the Li group with cystamine-functionalized gold nanocomposites displayed systematic release of the anticancer drug doxorubicin in response to the intracellular glutathione present in excess in the breast cancer cells. It was attributed to the cleavage of the disulfide bond by the excess glutathione [144]. Gold and silver NPs have also been shown to generate reactive oxygen species (ROS), e.g., H_2O_2 and O_2^-, which in turn can damage the cancer cell DNA restricting cell proliferation [145,146]. Thus, the acidic pH and reducing microenvironment of the cancer cells as compared to normal cells are often exploited to synthesize target-specific noble metal nanocomposites.

5.6.1.3 Photothermal Therapy (PTT)

PTT is one of the most prominent cancer treatments owing to its non-invasive and localized nature. Cancer cells are poor heat resistant attributing to increased rate of cell proliferation, slow blood flow to the cells, vascular distortion, etc. [147]. Therefore, a rise in temperature inside the cancer cell can lead to successful apoptosis and annihilate the cancer cells. One of the most captivating traits of noble metal nanocomposites is the conversion of light energy to thermal energy [134]. This property stems from the LSPR effect of noble metal nanocomposites as discussed earlier. The non-radiative decay of the surface plasmon in the presence of an electromagnetic field, light, takes place via the dissipation of energy in terms of heat [134]. Such elevated temperature inside the cancer cells induces denaturation of proteins and permanent damage to the cancer cell membranes [5]. Light of longer wavelength for example of the near-infrared (NIR) region can penetrate deep inside the tissue. Hence, NIR light in the range of 650–900 nm is the most useful light source to initiate the photothermal effect in noble metal nanocomposites [148]. To increase the light-to-heat conversion, noble metals are often combined with carbon-based nanocomposites. For example, Vischio et al. reported gold NPs grown on nanographene oxide, which exhibited excellent photothermal activity [149]. Graphene oxide NPs themselves generate heat upon photoexcitation. Thus formation of nanocomposite of gold with nanographene oxide showed improved PTT compared to gold NPs alone. Light-to-heat conversion by the noble metal nanocomposites has also been found to induce controlled drug release. This resulted in a multimodal therapeutic modality combining PTT and the release of chemotherapeutic drugs [150].

Other than the treatments mentioned above, noble metal nanocomposites also bring about cytotoxic effects by directly binding with the cancer cell DNA and ceasing the replication process, thereby restricting the rapid cell division in the cancer cells, [151,152].

5.6.1.4 Photodynamic Therapy (PDT)

PDT is another non-invasive cancer treatment. This treatment modality is gaining the utmost interest because of spatio-temporal control over the drug action. PDT involves the excitation of the organic chromophore, often known as photosensitizer (PS) to its triplet excited state upon absorption of light. From, the triplet excited state, the PSs dissipate their excess energy to the nearby ground state molecular oxygen leading to the generation of singlet oxygen, which induces cellular apoptosis in cancer cells [153]. The efficiency of the PDT process depends on the triplet excited state lifetime, facile singlet to triplet transition, etc. Noble metal nanocomposites exhibit a very high triplet state lifetime owing to the heavy atom effect [154]. The heavy atom effect facilitates singlet to triplet intersystem crossing via strong spin-orbit coupling [154]. Noble metal nanocomposites also show higher molar extinction efficiency compared to regular organic PSs [155]. Furthermore, organic PS like photofrin, which is an FDA-approved PDT drug, shows absorption around 630 nm, which lies below the biologically permissible window. On the contrary, noble metal nanocomposites absorb even in the range of 650–1350 nm, which is well within the biologically permissible limit [155]. Moreover, the noble metal nanocomposites can adsorb molecular oxygen onto their porous surface, which makes the energy transfer process faster [156]. Thus, noble metal nanocomposites have shown improved results in terms of PDT-mediated cancer treatment compared to organic PSs.

In the following sections, we have tried to exemplify the use of noble metal nanocomposites based on Au, Ag, Pd, Pt, etc. in effective cancer diagnosis and treatment.

5.6.2 Applications of Gold Nanocomposites in Anticancer Activity

Gold nanocomposites formed by integrating gold NPs (AuNPs) with other nanomaterials such as carbon nanotubes represent an innovative approach to overcome the limitations of conventional methods of cancer treatment. Their ability to encapsulate drugs, imaging agents, and targeting moieties within a single nanoscale structure provides a comprehensive solution for precision medicine in cancer treatment. Gold nanocomposites can be utilized for PTT [157,158]. Localized heat generation from gold-based nanocomposites can selectively destroy cancer cells while minimizing damage to surrounding healthy tissues. They can serve as drug carriers, facilitating targeted delivery of anticancer drugs to specific cells. Functionalized gold NPs enhance drug stability, bioavailability, and controlled release, improving therapeutic efficacy [159]. Gold nanocomposites are employed as contrast agents in imaging techniques such as CT and photoacoustic imaging. Their high X-ray absorption and strong imaging contrast aid in cancer detection and monitoring [160]. Elevated levels of ROS, often associated with apoptosis, serve as a promising indicator of the gold nanocomposite's effective induction of programmed cell death, highlighting its potential in anti-cancer activity [14,161]. Gold nanocomposites can be functionalized for gene delivery, assisting in gene therapy approaches for cancer treatment. They can efficiently transport therapeutic genes to cancer cells, contributing to targeted and specific interventions [162]. They can be utilized in biosensors for the detection of cancer-specific biomarkers, allowing for early diagnosis. For example, elevated levels of Prostate Specific Antigen (PSA) are associated with prostate cancer, and gold nanocomposites can be employed for sensitive detection [163]. Some recent examples of applications of noble metal-based nanocomposites in cancer biomarkers detection are given in Table 5.2 (Figure 5.6).

5.6.3 Applications of Silver Nanocomposites in Anticancer Activity

Silver nanocomposites refer to materials that incorporate silver NPs into a matrix or support structure, resulting in a composite material with unique properties. In anticancer treatments, silver nanocomposites can function as drug carriers, PTT and PDT agents, imaging agents, biosensors for cancer biomarkers, antiangiogenic agents, etc. Silver nanocomposites can serve as carriers for anticancer drugs, enhancing drug delivery to cancer cells [142,164]. The unique properties of silver NPs such as their small size and large surface area, make them suitable candidates for enhancing drug delivery specifically to cancer cells. Additionally, the targeted delivery of anticancer drugs is facilitated by the ability of silver nanocomposites to interact with cancer cells, selectively. PTT makes use of the photothermal properties of silver nanocomposites to target and eliminate cancer cells [165]. In this approach, silver NPs, known for their strong light absorption and efficient conversion of light into heat, are incorporated into nanocomposites. This precise and controlled method allows for targeted ablation of cancer cells while minimizing damage to surrounding healthy tissues. Silver nanocomposites are applied in PDT, where they aid in the generation of ROS [166]. ROS, such as singlet oxygen, which can cause oxidative damage to cellular components, ultimately causes the destruction of cancer cells while sparing healthy tissues. Silver nanocomposites can increase the efficiency of radiotherapy [167,168]. They can enhance the impact of radiation on cancer cells. Silver nanocomposites exhibit antiangiogenic effects by inhibiting the formation of blood vessels within tumors [169,170]. These nanocomposites can interfere with angiogenesis, a crucial process for tumor growth and progression, by targeting key molecular pathways involved in blood vessel formation. By restricting the tumor's blood supply, silver nanocomposites contribute to preventing

TABLE 5.2
Recent Applications of Noble Metal-Based Nanocomposites in Cancer Biomarkers Detection

Sl No	Nanocomposite	Application	Cancer Type	Biomarker Detected	References
Gold	GO@Au-NS	miRNA Detection	Colorectal Cancer	miR-223	[184]
	Fe-Ni@rGO, GQD-Ag, GNS	miRNA Detection	Breast Cancer	miR-155	[185]
	PPI/GO	CEA Measurement	Colorectal Cancer	Carcinoembryonic Antigen (CEA)	[186]
	GQD/AuNPs	MTX Measurement	Anti-Cancer Drug (Methotrexate)	Methotrexate (MTX)	[187]
	AuNPs@rGO	Fluorescent Immunosensor	CEA Detection	Carcinoembryonic Antigen (CEA)	[188]
	AuNRs/GQDs	Photothermal Therapy	Cancer Treatment	Not specified	[189]
	g-C_3N_4/Fe_3O_4/PANI	Ovarian Cancer Detection	Ovarian Cancer	CA125	[190]
Silver	CQD@Ag NCs	Fluorescent Imaging of Superoxide Anion	General	Intracellular $O_2^{\bullet -}$	[191]
	Ag/CQD Nanocomposites	Fabrication of Silver Carbon Quantum Dots	General	Not specified	[192]
	Ag/CQD Nanocomposites	Sustainable Synthesis of Silver Nanocomposite	Breast cancer	Not specified	[193]
	Graphene Oxide (GO)	Drug Delivery Using Graphene Oxide Carrier	Hepatocellular Carcinoma	Methotrexate	[194]
Platinum	CNTs/PEG@M/CPTMS@ CisPt(II)	Drug Delivery for Platinum Anticancer Drugs	General	Not specified	[177]
	rNGO/β-Lg@OXP	Encapsulation, Loading, and Release of Oxaliplatin	Colon Tumor Cells	Oxaliplatin	[195]
	oxCNTs-HA-CPT	Drug Nanovector Conjugation for Lung Cancer Treatment	Lung Cancer	Carboplatin	[196]
	MWCNT/PBI/Pt-NPs	Targeting Cancer Stem Cells with Platinum Nanoparticles	Breast Cancer Stem Cells	Not specified	[197]
Palladium	Halloysite nanotube/carbon composite decorated with Pd nanoparticles (HNT/C@Pd NPs)	Electrochemical aptasensor for tumor marker detection	Breast Cancer	Human epidermal growth factor receptor 2 (HER2)	[178]
	Pd/carbon nanofibers/ ionic liquid	Determination of anti-cancer drug pemetrexed	General	Not specified	[179]

TABLE 5.2 (Continued)
Recent Applications of Noble Metal-Based Nanocomposites in Cancer Biomarkers Detection

Sl No	Nanocomposite	Application	Cancer Type	Biomarker Detected	References
	Chitosan/WS2/Pd composite (CS/WS2/Pd)	Nano-drug carrier for cancer treatment	Breast cancer	Not specified	[180]
	MWCNT modified with polypyrrole (PPy) matrix with the incorporation of palladium nanoparticles (NPs)	Antibiofilm, anticancer and osteoblast adhesion activities	Not specified	Not specified	[35]
	Chitosan/Pd nanocomposite	Drug delivery system	General	Not specified	[182]
	Poly(ionic crosslinked chitosan-palladium) nanocapsules	Hybrid drug delivery system	Colon cancer	Not specified	[183]

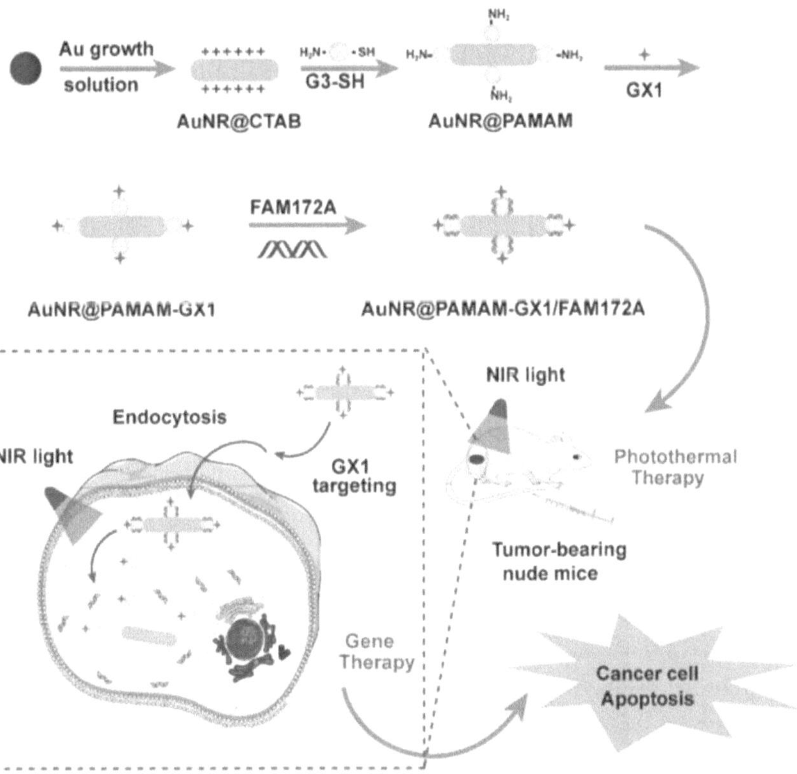

FIGURE 5.6 Schematic portraying the use of combined gene therapy and photothermal therapy with GX1 targeting nanoparticles to treat colon cancer [110].

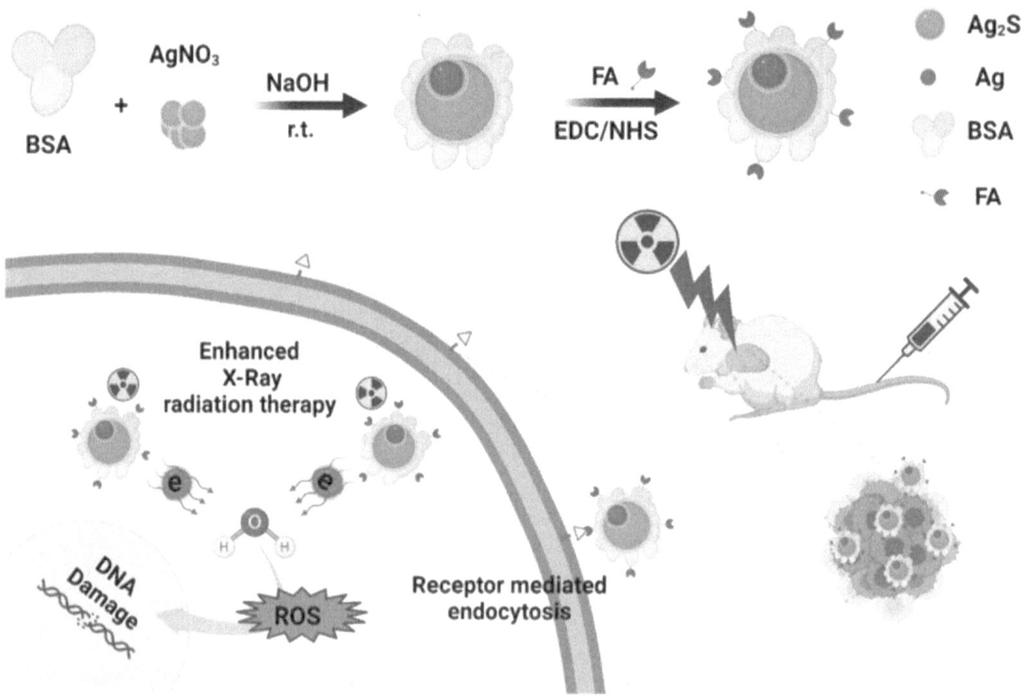

FIGURE 5.7 Diagrammatic illustration of the radio-sensitizing function and synthesis of Ag-Ag2S@BSA-FA and Ag-Ag2S@BSA-NPs [128].

the nourishment of cancer cells, which can potentially hinder tumor growth and metastasis. Silver nanocomposites can act as adjuvants in immunotherapy by enhancing the body's immune response against cancer cells [171]. These nanocomposites can be employed to stimulate the immune system, making it more effective in recognizing and targeting cancer cells. Functionalized silver NPs play a crucial role in biosensors designed for detecting specific cancer biomarkers, contributing to early cancer diagnosis. By incorporating biological molecules onto the surface of silver nanocomposites, these biosensors can selectively interact with cancer-related biomolecules present in our bodies. For example, the silver NPs coated with luminol, act as a key component to enhance the electrochemiluminescent signal, providing a specific method for carcinoembryonic antigen (CEA) detection which is a biomarker for breast cancer [172] (Figure 5.7).

5.6.4 Applications of Platinum Nanocomposites in Anticancer Activity

Platinum nanocomposites involve the integration of platinum NPs into different matrices or carrier systems, exploiting the distinct characteristics of platinum on the nanoscale for a wide range of applications. In anticancer treatment platinum nanocomposites can function as PDT agents, PTT agents, radiosensitizers, chemotherapy enhancers, imaging agents, etc. [173]. Cisplatin, a coordination complex containing a central platinum atom with two chlorides and two ammonia molecules in the cis position, primarily inhibits tumor growth by blocking DNA synthesis [174]. Platinum nanocomposites have the potential for targeted cancer treatment by disrupting cancer cell viability. For example, graphene oxide platinum NPs induced cell death of cancer cells through the elevation of ROS, malondialdehyde, nitric oxide, and carbonylated protein levels, leading to mitochondrial dysfunction and apoptosis [175]. The cancer stem cells can renew themselves and grow into tumors even after treatment. So the early detection of these cells is essential. The integration of platinum

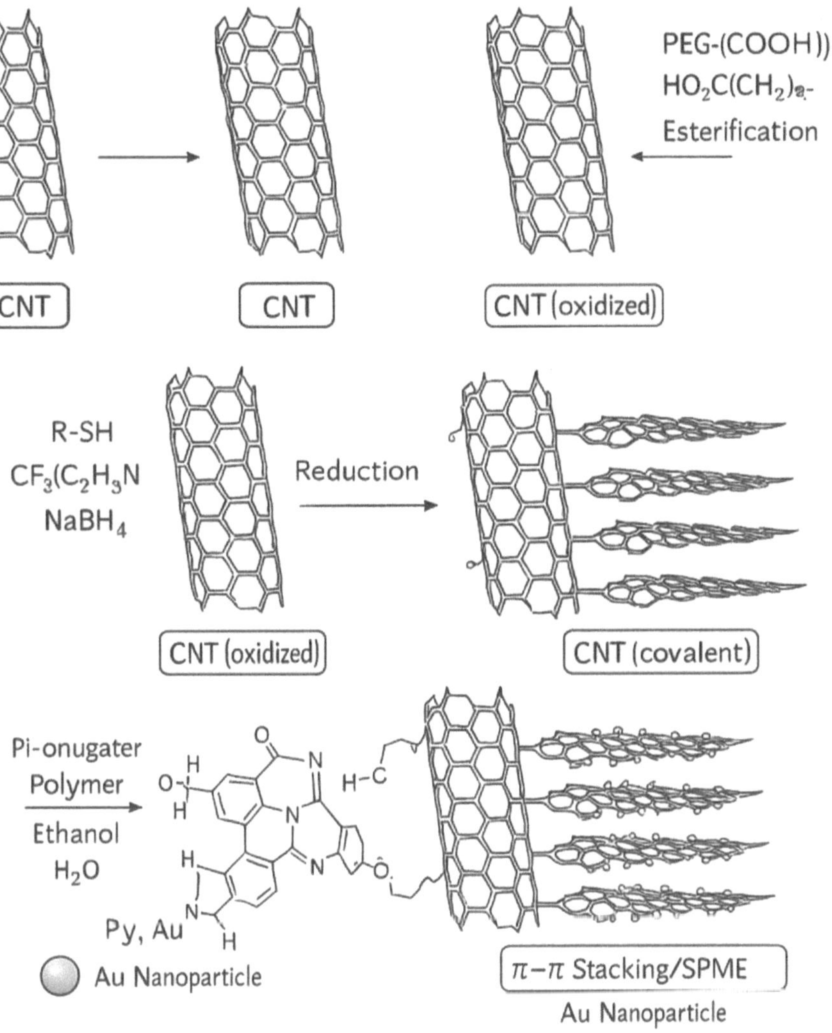

FIGURE 5.8 CNTs/PEG@M/CPTMS@CisPt(II) nanocomposites are shown schematically [177].

NPs and CD133+ in a nanobiocomposite offers an advanced method for improving the detection of cancer stem cells [176]. CD133+, also known as prominin-1, is a pentaspan transmembrane protein that has been identified as a stem cell marker (Figure 5.8).

5.6.5 Applications of Palladium Nanocomposites in Anticancer Activity

A recent study introduced a novel biosensing platform leveraging the unique properties of halloysite nanotube/carbon composite decorated with palladium NPs (HNT/C@Pd NPs) to create an electrochemical aptasensor aimed at detecting human epidermal growth factor receptor 2 (HER2), a key biomarker for breast cancer [178]. The integration of HNT/C provides a substantial specific surface area, while the conductivity is significantly enhanced by the palladium NPs, enabling the aptasensor to achieve high sensitivity and selectivity toward tumor markers. Another research outlines the development of a sensitive voltammetric technique for determining pemetrexed, utilizing a novel electrode made from palladium/CNFs and an ionic liquid-modified CPE, with Nafion as the binder [179]. The electrode's selectivity, repeatability, reproducibility, and stability were thoroughly evaluated, and it was successfully applied to quantify pemetrexed in cancerous human plasma, pharmaceutical, and health urine samples, showcasing its potential for clinical diagnostics and monitoring. Furthermore, palladium nanocomposite functionalizing WS_2 nanosheets with chitosan (CS) demonstrated significant cell inhibition, with a notable effect on MCF-7 breast cancer cells, achieving a 73.24% inhibition rate [180]. These findings highlight the potential of CS/WS_2/Pd composites in the development of new approaches for cancer therapy, showcasing the effectiveness of 2D nanomaterials in targeting cancer cells. Additionally, multifunctional material composed of MWCNTs modified with polypyrrole (PPy) matrix and incorporated with palladium NPs [181]. The composite exhibited notable anti-biofilm activity, cytotoxicity against Vero and HeLa cell lines, and promoted adhesion and proliferation of human osteosarcoma cells (HOS), indicating its biocompatibility. These findings suggest the composite's potential in biomedical applications, especially in combating bacterial infections and cancer, and enhancing bone cell growth. Another study focused on a dual drug delivery system using a chitosan/palladium (CS/Pd) nanocomposite for the controlled release of curcumin (CUR) and 5-Fluorouracil (5-FU), aiming to overcome cancer cell resistance [182]. The release kinetics followed a zero-order model, indicating a constant release rate that is beneficial for reducing dosage frequency. The cytotoxicity tests on colon cancer cells showed the composite's effectiveness in inhibiting cancer cell growth, highlighting its potential as a promising approach for cancer therapy. In another research effort, a novel hybrid drug delivery system composed of palladium NPs (PdNPs) and polyionic cross-linked chitosan (PICCS@Pd) was developed for the delivery of doxorubicin (DOX) and 5-FU, both separately and in combination [183]. The release profiles indicated a sustained and prolonged release of the drugs. In vitro cytotoxicity studies demonstrated the nanocapsules' significant inhibitory effects on tumor cell proliferation (MCF-7 and HT-29) with minimal hemolytic activity, showcasing the advantage of using a co-drug strategy over monotherapy in cancer treatment.

5.7 CONCLUSION AND FUTURE PERSPECTIVES

Noble metal-based NPs are one of the most attractive materials with a wide range of applications. The research in exploring the utility of noble-metal-based nanocomposites in various areas of scientific research is vastly growing due to their unique physical, chemical, optical properties, etc. As mentioned in this chapter, noble metals and their nanocomposites have been widely used in numerous electrochemical and biomedical applications. This chapter sheds light on the several possibilities of nanocomposites of noble metals with a focus on carbon-based nanocomposites in electrochemical sensing and anticancer activity. This chapter elaborated on the synthesis, properties, and recent research based on nanocomposites of Au, Ag, Pd, and Pt with carbon nanomaterials, which would enhance the exploration of these fascinating materials in various analytical applications in the future. Even though several research reports were documented on this topic, more studies in this area should be needed to have sustainable research with good societal impacts. Hence, this chapter will enlighten the scientific community regarding the plethora of opportunities for noble metal-based nanocomposites in electrochemical and biomedical applications to a great extent.

REFERENCES

[1] R. Zhao, J. Xiang, B. Wang, L. Chen, and S. Tan, "Recent advances in the development of noble metal NPs for cancer therapy," *Bioinorganic Chemistry and Applications*, vol. 2022, 2022, doi: 10.1155/2022/2444516

[2] M. Kutz, *Mechanical Engineers' Handbook, Volume 3 – Manufacturing and Management* (4th Edition), 2015.

[3] E. E. Elemike, D. C. Onwudiwe, L. Wei, L. Chaogang, and Z. Zhiwei, "Noble metal – Semiconductor nanocomposites for optical, energy and electronics applications," *Solar Energy Materials and Solar Cells*, vol. 201, p. 110106, Oct. 2019, doi: 10.1016/J.SOLMAT.2019.110106

[4] P. K. Jain, X. Huang, I. H. El-sayed, and M. A. El-sayed, "Noble metal – Semiconductor nanocomposites for optical, energy and electronics applications," *Accounts of Chemical Research*, vol. 41, no. 12, pp. 1578–1586, 2008.

[5] Z. Zhou, Y. Yan, K. Hu, et al., "'Autophagy inhibition enabled efficient photothermal therapy at a mild temperature,'" *Biomaterials*, vol. 141, pp. 116–124, 2017.

[6] A. Karthika, S. Nikhil, A. Suganthi, and M. Rajarajan, "A facile sonochemical approach based on graphene carbon nitride doped silver molybdate immobilized nafion for selective and sensitive electrochemical detection of chromium (VI) in real sample," *Advanced Powder Technology*, vol. 31, pp. 1879–1890, 2020.

[7] M. R. Mahmoudian, W. J. Basirun, Y. Alias, and P. MengWoi, "Investigating the effectiveness of g-C3N4 on Pt/g-C3N4/polythiophene nanocomposites performance as an electrochemical sensor for Hg^{2+} detection," *Journal of Environmental Chemical Engineering*, vol. 8, p. 104204, 2020.

[8] P. Nasr-Esfahani and A. A. Ensafi, "Noble metals and nonnoble metal oxides based electrochemical sensors, Editor(s): Chaudhery Mustansar Hussain, Jamballi G. Manjunatha, In *Functionalized Nanomaterial-Based Electrochemical Sens*, Woodhead Publishing Series in Electronic and Optical Materials, Woodhead Publishing, 2022, Pages 115–140, ISBN 9780128237885.

[9] J. Wang, and H. Gu, "Novel metal nanomaterials and their catalytic applications," *Molecules*, vol. 20, pp. 17070–17092, 2015, doi: 10.3390/molecules200917070

[10] L.-F. Hsu, K. Natarajan, C. Karuppiah, and C.-C. Yang, "Green reduction and uniform assembly of silver nanoparticles on polydopamine functionalized sulfur-doped graphitic carbon nitride nanosheets for highly sensitive H_2O_2 detection via nonenzymatic approach," 2023, doi: 10.1016/j.colsurfa.2023.132383

[11] S. P. Nayak, S. S. Ramamurthy, and J. K. Kiran Kumar, "Green synthesis of silver nanoparticles decorated reduced graphene oxide nanocomposite as an electrocatalytic platform for the simultaneous detection of dopamine and uric acid," *Materials Chemistry and Physics*, vol. 252, p. 123302, Sep. 2020, doi: 10.1016/J.MATCHEMPHYS.2020.123302

[12] A. Joy *et al.*, "Hybrid gold/graphene oxide reinforced polycaprolactone nanocomposite for biomedical applications," *Surfaces and Interfaces*, vol. 40, p. 103000, Aug. 2023, doi: 10.1016/j.surfin.2023.103000

[13] J. Prakash, D. Prema, K. S. Venkataprasanna, K. Balagangadharan, N. Selvamurugan, and G. D. Venkatasubbu, "Nanocomposite chitosan film containing graphene oxide/hydroxyapatite/gold for bone tissue engineering," *International Journal of Biological Macromolecules*, vol. 154, pp. 62–71, Jul. 2020, doi: 10.1016/j.ijbiomac.2020.03.095

[14] S. F. Adil *et al.*, "Enhanced apoptosis by functionalized highly reduced graphene oxide and gold nanocomposites in MCF-7 breast cancer cells," *ACS Omega*, vol. 6, no. 23, pp. 15147–15155, Jun. 2021, doi: 10.1021/acsomega.1c01377

[15] N. Muniyappan, M. Pandeeswaran, and A. Amalraj, "Green synthesis of gold nanoparticles using *Curcuma pseudomontana* isolated curcumin: Its characterization, antimicrobial, antioxidant and anti-inflammatory activities," *Environmental Chemistry and Ecotoxicology*, vol. 3, pp. 117–124, 2021, doi: 10.1016/j.enceco.2021.01.002

[16] P. Kumari and A. Meena, "Green synthesis of gold nanoparticles from *Lawsonia inermis* and its catalytic activities following the Langmuir-Hinshelwood mechanism," *Colloids and Surfaces A: Physicochemical and Engineering Aspects*, vol. 606, no. June, p. 125447, 2020, doi: 10.1016/j.colsurfa.2020.125447

[17] S. Nazarpour, R. Hajian, and M. H. Sabzvari, "A novel nanocomposite electrochemical sensor based on green synthesis of reduced graphene oxide/gold nanoparticles modified screen printed electrode for determination of tryptophan using response surface methodology approach," *Microchemical Journal*, vol. 154, p. 104634, May 2020, doi: 10.1016/j.microc.2020.104634

[18] S. Kumari *et al.*, "A novel synthesis of the graphene oxide-silver (GO-Ag) nanocomposite for unique physiochemical applications," *ACS Omega*, vol. 5, no. 10, pp. 5041–5047, Mar. 2020, doi: 10.1021/acsomega.9b03976

[19] M. Cobos, I. De-La-Pinta, G. Quindós, M. J. Fernández, and M. D. Fernández, "Graphene oxide – Silver nanoparticle nanohybrids: Synthesis, characterization, and antimicrobial properties," *Nanomaterials*, vol. 10, no. 2, p. 376, Feb. 2020, doi: 10.3390/nano10020376

[20] S. P. Nayak, S. S. Ramamurthy, and J. K. Kiran Kumar, "Green synthesis of silver nanoparticles decorated reduced graphene oxide nanocomposite as an electrocatalytic platform for the simultaneous detection of dopamine and uric acid," *Materials Chemistry and Physics*, vol. 252, p. 123302, Sep. 2020, doi: 10.1016/j.matchemphys.2020.123302

[21] M. Bin Ahmad, M. Y. Tay, K. Shameli, M. Z. Hussein, and J. J. Lim, "Green synthesis and characterization of silver/chitosan/polyethylene glycol nanocomposites without any reducing agent," *International Journal of Molecular Sciences*, vol. 12, no. 8, pp. 4872–4884, Aug. 2011, doi: 10.3390/ijms12084872

[22] M. A. Awad *et al.*, "Green synthesis, characterization, and antibacterial activity of silver/polystyrene nanocomposite," *Journal of Nanomaterials*, vol. 2015, pp. 1–6, 2015, doi: 10.1155/2015/943821

[23] M. W. Yaseen, M. A. Asghar, E. M. Bakhsh, K. Akhtar, S. B. Khan, and M. Iqbal, "Plant-mediated synthesis of thermally stable metal-polystyrene nanocomposites for antioxidant and antibacterial food packaging," *Industrial Crops and Products*, vol. 209, p. 117959, Mar. 2024, doi: 10.1016/j.indcrop.2023.117959

[24] N. S. Anuar, W. J. Basirun, M. Shalauddin, and S. Akhter, "A dopamine electrochemical sensor based on a platinum–silver graphene nanocomposite modified electrode," *RSC Advances*, vol. 10, no. 29, pp. 17336–17344, 2020, doi: 10.1039/C9RA11056A

[25] J. Rieffel, U. Chitgupi, and J. F. Lovell, "Recent advances in higher-order, multimodal, biomedical imaging agents," *Small*, vol. 11, no. 35, pp. 4445–4461, Sep. 2015, doi: 10.1002/SMLL.201500735

[26] H. Karimi-Maleh *et al.*, "Guanine-based DNA biosensor amplified with Pt/SWCNTs nanocomposite as analytical tool for nanomolar determination of daunorubicin as an anticancer drug: A docking/experimental investigation," *Industrial & Engineering Chemistry Research*, vol. 60, no. 2, pp. 816–823, Jan. 2021, doi: 10.1021/acs.iecr.0c04698

[27] A. Dehdashti and A. Babaei, "Designing and characterization of a novel sensing platform based on Pt doped NiO/MWCNTs nanocomposite for enhanced electrochemical determination of epinephrine and tramadol simultaneously," *Journal of Electroanalytical Chemistry*, vol. 862, p. 113949, Apr. 2020, doi: 10.1016/j.jelechem.2020.113949

[28] V. S. Ramkumar *et al.*, "Synthesis of platinum nanoparticles using seaweed Padina gymnospora and their catalytic activity as PVP/PtNPs nanocomposite towards biological applications," *Biomedicine & Pharmacotherapy*, vol. 92, pp. 479–490, Aug. 2017, doi: 10.1016/j.biopha.2017.05.076

[29] S. O. Ibrahim, K. U. Isah, A. S. Abdulkareem, U. Ahmadu, J. O. Tijani, and W. D. Roos, "Synthesis and characterization of platinum multi-walled carbon nanotubes nanocomposite film electrode," *Journal of Materials Science: Materials in Electronics*, vol. 32, no. 10, pp. 12800–12811, May 2021, doi: 10.1007/s10854-020-03550-0

[30] Manan Mehta, Deepti, A.K. Sinha, Shikha Wadhwa, Arvind Kumar, D.K. Avasthi, "Change in the gas sensing mechanism and improved sensing response on ion-irradiated palladium-graphene oxide nanocomposite," *International Journal of Hydrogen Energy*, vol. 48, no. 85, pp. 33372–33381, Oct. 2023, doi: 10.1016/j.ijhydene.2023.05.071

[31] G. Venkata Prasad *et al.*, "Fine-tuning of Pd–CeO_2/rGO nanocomposite: A facile synergetic strategy for effective electrochemical detection of dopamine in pharmaceutical and biological samples," *Journal of Electroanalytical Chemistry*, vol. 941, p. 117544, Jul. 2023, doi: 10.1016/j.jelechem.2023.117544

[32] M. Safarkhani *et al.*, "Bioengineered smart nanocarriers for breast cancer treatment: Adorned carbon-based nanocomposites with silver and palladium complexes for efficient drug delivery," *ACS Omega*, vol. 9, no. 1, pp. 1183–1195, Jan. 2024, doi: 10.1021/acsomega.3c07432

[33] M. Lee, B.-H. Kim, Y. Lee, B.-T. Kim, and J. B. Park, "Synthesis and characterization of graphene and graphene oxide based palladium nanocomposites and their catalytic applications in carbon-carbon cross-coupling reactions," *Bulletin of the Korean Chemical Society*, vol. 35, no. 7, pp. 1979–1984, Jul. 2014, doi: 10.5012/bkcs.2014.35.7.1979

[34] U. Lange, T. Hirsch, V. M. Mirsky, and O. S. Wolfbeis, "Hydrogen sensor based on a graphene – Palladium nanocomposite," *Electrochimica Acta*, vol. 56, no. 10, pp. 3707–3712, Apr. 2011, doi: 10.1016/j.electacta.2010.10.078

[35] B. Murugesan, J. Sonamuthu, S. Samayanan, S. Arumugam, and S. Mahalingam, "Highly biological active antibiofilm, anticancer and osteoblast adhesion efficacy from MWCNT/PPy/Pd nanocomposite," *Applied Surface Science*, vol. 434, pp. 400–411, Mar. 2018, doi: 10.1016/j.apsusc.2017.10.142

[36] B. P. Vinayan, R. Nagar, and S. Ramaprabhu, "Solar light assisted green synthesis of palladium nanoparticle decorated nitrogen doped graphene for hydrogen storage application," *Journal of Materials Chemistry A*, vol. 1, no. 37, p. 11192, 2013, doi: 10.1039/c3ta12016c

[37] F. Wang, X. Niu, W. Wang, W. Jing, Y. Huang, and J. Zhang, "Green synthesis of Pd nanoparticles via extracted polysaccharide applied to glucose detection," *Journal of the Taiwan Institute of Chemical Engineers*, vol. 93, pp. 87–93, Dec. 2018, doi: 10.1016/j.jtice.2018.08.022

[38] R. Kannan, H.-R. Jang, E.-S. Yoo, H.-K. Lee, and D. J. Yoo, "Facile green synthesis of palladium quantum dots@carbon on mixed valence cerium oxide/graphene hybrid nanostructured bifunctional catalyst for electrocatalysis of alcohol and water," *RSC Advances*, vol. 5, no. 45, pp. 35993–36000, 2015, doi:10.1039/C5RA04226G

[39] M. Alsaiari, J. Ahmed, M. Faisal, and F. A. Harraz, "Biomass-derived carbon decorated by gold nanoparticles as efficient methanol electrochemical sensor," *Materials Science in Semiconductor Processing*, vol. 146, no. March, p. 106693, 2022, doi: 10.1016/j.mssp.2022.106693

[40] A. M. Sabrina *et al.*, "Exploiting synergies in catalysis and gas sensing using noble metal-loaded oxide composites," pp. 864–880, 2018, doi: 10.1002/cctc.201701545

[41] Y. Bai, C. Liu, Y. Shan, T. Chen, Y. Zhao, and C. Yu, "Metal-organic frameworks nanocomposites with different dimensionalities for energy conversion and storage," vol. 2100346, pp. 1–52, 2022, doi: 10.1002/aenm.202100346

[42] G. Chang *et al.*, "Synthesis of highly dispersed Pt nanoclusters anchored graphene composites and their application for non-enzymatic glucose sensing," *Electrochimica Acta*, vol. 157, pp. 149–157, Mar. 2015, doi: 10.1016/J.ELECTACTA.2015.01.085

[43] T. Zhao, W. Li, R. Shang, Y. Lei, Y. Liu, and C. Ma, "Experimental and theoretical study on one-step synthesis of AuAg alloy nanoparticle catalytic layer as highly stable non-enzymatic glucose sensing interface," *Journal of Electroanalytical Chemistry*, vol. 950, no. September, p. 117898, 2023, doi: 10.1016/j.jelechem.2023.117898

[44] J. Tang, D. Chen, Q. Yao, J. Xie, and J. Yang, "Recent advances in noble metal-based nanocomposites for electrochemical reactions," *Materials Today Energy*, vol. 6, pp. 115–127, 2017, doi: 10.1016/j.mtener.2017.09.005

[45] T. Zhao, W. Li, R. Shang, Y. Lei, Y. Liu, and C. Ma, "Experimental and theoretical study on one-step synthesis of AuAg alloy nanoparticle catalytic layer as highly stable non-enzymatic glucose sensing interface," *Journal of Electroanalytical Chemistry*, vol. 950, p. 117898, Dec. 2023, doi: 10.1016/j.jelechem.2023.117898

[46] L. Zhou, H. Gu, C. Wang, J. Zhang, M. Lv, and R. He, "Study on the synthesis and surface enhanced Raman spectroscopy of graphene-based nanocomposites decorated with noble metal nanoparticles," *Colloids and Surfaces A. Physicochemical and Engineering Aspects*, vol. 430, pp. 103–109, 2013, doi: 10.1016/j.colsurfa.2013.04.012

[47] K. Yang, L. Feng, X. Shi, and Z. Liu, "Nano-graphene in biomedicine: Theranostic applications," *Chemical Society Reviews*, vol. 42, no. 2, pp. 530–547, Dec. 2013, doi: 10.1039/C2CS35342C

[48] Y. B. Wankhede, S. B. Kondawar, S. R. Thakare, and P. S. More, "Synthesis and characterization of silver nanoparticles embedded in polyaniline nanocomposite," *Advanced Materials Letters*, vol. 4, no. 1, pp. 89–93, 2013, doi: 10.5185/amlett.2013.icnano.108

[49] K. Shrivastava *et al.*, "Silver nanoparticle incorporated calcium crosslinked hydrogel composite for reduction of methylene blue dye and nitrite sensing," *Journal of Photochemistry and Photobiology A: Chemistry*, vol. 447, no. October 2023, p. 115256, 2024, doi: 10.1016/j.jphotochem.2023.115256

[50] A. Gao, H. Chen, A. Hou, and K. Xie, "Materials Science & Engineering C E ffi cient antimicrobial silk composites using synergistic e ff ects of violacein and silver nanoparticles," vol. 103, no. July 2018, 2019, doi: 10.1016/j.msec.2019.109821

[51] M. M. Abdelhamied and A. Atta, "Synthesis and optical properties of PVA/PANI/Ag nanocomposite films," *Journal of Materials Science: Materials in Electronics*, vol. 31, no. 24, pp. 22629–22641, 2020, doi: 10.1007/s10854-020-04774-w

[52] H. Xu, L. Zeng, S. Xing, G. Shi, Y. Xian, and L. Jin, "Electrochemistry communications microwave-radiated synthesis of gold nanoparticles/carbon nanotubes composites and its application to voltammetric detection of trace mercury (II)," *Electrochemistry Communications*, vol. 10, no. 12, pp. 1839–1843, 2008, doi: 10.1016/j.elecom.2008.09.030

[53] M. C. M. Higgins and J. V Rojas, "Applied surface science X-ray radiation enhancement of gold-TiO_2 nanocomposites," *Applied Surface Science*, vol. 480, no. February, pp. 1147–1155, 2019, doi: 10.1016/j.apsusc.2019.02.234

[54] M. Shehata, M. Zaki, and A. M. Fekry, "New Au/chitosan nanocomposite modified carbon paste sensor for voltammetric detection of nicotine," *Scientific Reports*, vol. 13, no. 1, pp. 1–11, 2023, doi: 10.1038/s41598-023-47703-7

[55] G. Lin *et al.*, "Non-viral gene therapy using multifunctional nanoparticles: Status, challenges, and opportunities," *Coordination Chemistry Reviews*, vol. 374, pp. 133–152, 2018, doi: 10.1016/j.ccr.2018.07.001

[56] H. Huang and J. F. Lovell, "Advanced functional nanomaterials for theranostics," *Advanced Functional Materials*, vol. 27, no. 2, Jan. 2017, doi: 10.1002/ADFM.201603524

[57] J. Liu, Q. Ma, Z. Huang, G. Liu, and H. Zhang, "Recent progress in graphene-based noble-metal nanocomposites for electrocatalytic applications," *Advanced Materials*, vol. 31, no. 9, pp. 1–20, 2019, doi: 10.1002/adma.201800696

[58] H. Ammari, Y. Deng, and P. Millien, "Surface plasmon resonance of nanoparticles and applications in imaging," vol. 220, pp. 109–153, 2016, doi: 10.1007/s00205-015-0928-0

[59] O. T. Vu, Q. H. Nguyen, T. T. T. Luong, K. Eersels, P. Wagner, and L. T. Ngoc Truong, "Highly sensitive molecularly imprinted polymer-based electrochemical sensors enhanced by gold nanoparticles for norfloxacin detection in aquaculture water," *ACS Omega*, vol. 8, 2022, doi: 10.1021/acsomega.2c04414

[60] Ş. Aysun, H. Ayd, K. Cellat, and Ş. Fatih, "A novel high performance non-enzymatic electrochemical glucose biosensor based on activated carbon-supported Pt-Ni nanocomposite," *Journal of Molecular Liquids*, vol. 300, 2020, doi: 10.1016/j.molliq.2019.112355

[61] C. Yu, C. Fan, and X. Meng, "A novel Ag/BiOBr nanoplate catalyst with high photocatalytic activity in the decomposition of dyes,", *Reaction Kinetics, Mechanisms, and Catalysis*, pp. 141–151, 2011, doi: 10.1007/s11144-011-0291-6

[62] N. S. Youssef and K. H. Hegab, "Synthesis and characterization of some transition metal complexes of thiosemicarbazones derived from 2-acetylpyrrole and 2-acetylfuran," *Synthesis and Reactivity in Inorganic, Metal-Organic and Nano-Metal Chemistry*, vol. 35, no. 5, pp. 391–399, 2005, doi: 10.1081/SIM-200059215

[63] S. Z. Haeri, B. Ramezanzadeh, and M. Asghari, "Journal of colloid and interface science a novel fabrication of a high performance SiO_2 -graphene oxide (GO) nanohybrids: Characterization of thermal properties of epoxy nanocomposites filled with $_{SiO2}$ -GO nanohybrids," *Journal of Colloid and Interface Science*, vol. 493, pp. 111–122, 2017, doi: 10.1016/j.jcis.2017.01.016

[64] M. Bagherzadeh *et al.*, "Magnetic carbon – Based nanocomposite decorated with palladium complex for co-delivery of DOX/pCRISPR," *Journal of Drug Delivery Science and Technology*, vol. 78, no. October, p. 103917, 2022, doi: 10.1016/j.jddst.2022.103917

[65] H. Zhang, D. Sun, and T. Cao, "Electrochemical sensor based on silver nanoparticles/multi-walled carbon nanotubes modified glassy carbon electrode to detect cyanide in food products," *International Journal of Electrochemical Science*, vol. 15, no. 4, pp. 3434–3444, 2020, doi: 10.20964/2020.04.32

[66] W. Zhang *et al.*, "An ultrastable virus-like particle with a carbon dot core and expanded sequence plasticity," *Small*, vol. 17, no. 35, 2021, doi: 10.1002/smll.202101717

[67] D. S. Sipuka *et al.*, "Electrochemical sensing of epinephrine on a carbon nanofibers and gold nanoparticle-modified electrode," *Electrocatalysis*, vol. 14, no. 1, pp. 9–17, 2023, doi: 10.1007/s12678-022-00769-9

[68] L. Vanyorek, G. Halasi, P. Pekker, F. Kristály, and Z. Kónya, "Characterization and catalytic activity of different carbon supported Pd nanocomposites," *Catalysis Letters*, vol. 146, pp. 2268–2277, 2016.

[69] S. Antherjanam, B. Saraswathyamma, and S. Murugesan Senthil Kumar, "Simultaneous electrochemical determination of the tumour biomarkers homovanillic acid and vanillylmandelic acid using a modified pencil graphite electrode," *Microchemical Journal*, vol. 190, p. 108659, Jul. 2023, doi: 10.1016/j.microc.2023.108659

[70] B. Sajeevan, Sreelekshmi, M. G. Gopika, A. Santhy, M. Govindasamy, and B. Saraswathyamma, "Polyterephthalic acid modified pencil graphite electrode for the electrochemical determination of vitamin B6," *Ionics (Kiel)*, vol. 29, no. 10, pp. 4389–4396, Oct. 2023, doi: 10.1007/s11581-023-05156-y

[71] S. Antherjanam and B. Saraswathyamma, "A disposable voltammetric sensor for the determination of diphenylamine using modified pencil graphite electrode," *Analytical Sciences*, vol. 40, no. 1, pp. 163–174, Jan. 2024, doi: 10.1007/s44211-023-00440-3

[72] S. Antherjanam and B. Saraswathyamma, "Simultaneous electrochemical determination of hydrazine and hydroxylamine on a thiadiazole derivative modified pencil graphite electrode," *Materials Chemistry and Physics*, vol. 275, p. 125223, Jan. 2022, doi: 10.1016/j.matchemphys.2021.125223

[73] A. Santhy, S. Beena, G. Krishnan Rajasree, and S. Greeshma, "A commercially viable electrochemical sensor for the immunosuppressant drug mycophenolate mofetil utilizing pencil graphite electrode," *IOP Conference Series: Materials Science and Engineering*, vol. 872, no. 1, p. 012127, Jun. 2020, doi: 10.1088/1757-899X/872/1/012127

[74] J. Zhao, Y. Zhang, K. Wu, J. Chen, and Y. Zhou, "Electrochemical sensor for hazardous food colourant quinoline yellow based on carbon nanotube-modified electrode," *Food Chemistry*, vol. 128, no. 2, pp. 569–572, Sep. 2011, doi: 10.1016/j.foodchem.2011.03.067

[75] H. A. Saputra, "Electrochemical sensors: Basic principles, engineering, and state of the art," *Monatshefte für Chemie – Chemical Monthly*, vol. 154, no. 10, pp. 1083–1100, Oct. 2023, doi: 10.1007/s00706-023-03113-z

[76] M. M. Shanbhag, G. Manasa, R. J. Mascarenhas, K. Mondal, and N. P. Shetti, "Fundamentals of bioelectrochemical sensing," *Chemical Engineering Journal Advances*, vol. 16, p. 100516, Nov. 2023, doi: 10.1016/j.ceja.2023.100516

[77] H. Karimi-Maleh, F. Karimi, M. Alizadeh, and A. L. Sanati, "Electrochemical sensors, a bright future in the fabrication of portable kits in analytical systems," *The Chemical Record*, vol. 20, no. 7, pp. 682–692, Jul. 2020, doi: 10.1002/tcr.201900092

[78] J. Baranwal, B. Barse, G. Gatto, G. Broncova, and A. Kumar, "Electrochemical sensors and their applications: A review," *Chemosensors*, vol. 10, no. 9, p. 363, Sep. 2022, doi: 10.3390/chemosensors10090363

[79] O. S. Kwon, H. S. Song, T. H. Park, and J. Jang, "Conducting nanomaterial sensor using natural receptors," *Chemical Reviews*, vol. 119, no. 1, pp. 36–93, Jan. 2019, doi: 10.1021/acs.chemrev.8b00159

[80] X. Sun and T. D. James, "Glucose sensing in supramolecular chemistry," *Chemical Reviews*, vol. 115, no. 15, pp. 8001–8037, Aug. 2015, doi: 10.1021/cr500562m

[81] H. M. N. Ahmad *et al.*, "Stimuli-responsive templated polymer as a target receptor for a conformation-based electrochemical sensing platform," *ACS Applied Polymer Materials*, vol. 3, no. 1, pp. 329–341, Jan. 2021, doi: 10.1021/acsapm.0c01120

[82] L. Fu, A. Yu, and G. Lai, "Conductive hydrogel-based electrochemical sensor: A soft platform for capturing analyte," *Chemosensors*, vol. 9, no. 10, p. 282, Oct. 2021, doi: 10.3390/chemosensors9100282

[83] K. Chen, W. Chou, L. Liu, Y. Cui, P. Xue, and M. Jia, "Electrochemical sensors fabricated by electrospinning technology: An overview," *Sensors*, vol. 19, no. 17, p. 3676, Aug. 2019, doi: 10.3390/s19173676

[84] R. G. Krishnan, R. Rejithamol, and B. Saraswathyamma, "Non-enzymatic electrochemical sensor for the simultaneous determination of adenosine, adenine and uric acid in whole blood and urine," *Microchemical Journal*, vol. 155, p. 104745, Jun. 2020, doi: 10.1016/j.microc.2020.104745

[85] A. Santhy, B. Saraswathyamma, A. Parvathy Krishnan, and L. Luscious, "Nanomaterials incorporated electrochemical sensors for the monitoring of pyridoxine: A mini review," *Materials Today: Proceedings*, vol. 46, pp. 2998–3004, 2021, doi: 10.1016/j.matpr.2020.12.703.

[86] D. Wu, Y. Yu, D. Jin, M. M. Xiao, Z. Y. Zhang, and G. J. Zhang, "Dual-aptamer modified graphene field-effect transistor nanosensor for label-free and specific detection of hepatocellular carcinoma-derived microvesicles," *Analytical Chemistry*, vol. 92, no. 5, pp. 4006–4015, 2020, doi: 10.1021/acs.analchem.9b05531.

[87] L. Zhou, H. Gu, C. Wang, J. Zhang, M. Lv, and R. He, "Study on the synthesis and surface enhanced Raman spectroscopy of graphene-based nanocomposites decorated with noble metal nanoparticles," *Colloids and Surfaces A: Physicochemical and Engineering Aspects*, vol. 430, pp. 103–109, 2013, doi: 10.1016/j.colsurfa.2013.04.012

[88] M. Alsaiari, J. Ahmed, M. Faisal, and F. A. Harraz, "Biomass-derived carbon decorated by gold nanoparticles as efficient methanol electrochemical sensor," *Materials Science in Semiconductor Processing*, vol. 146, no. March, p. 106693, 2022, doi: 10.1016/j.mssp.2022.106693

[89] L. Luo *et al.*, "An electrochemical aptasensor for detection of carbofuran using gold nanoparticles decorated hierarchical porous carbon as an effective sensing platform," *Chemosphere*, vol. 341, no. August, 2023, doi: 10.1016/j.chemosphere.2023.140033
[90] N. Nardi, L. G. Baumgarten, J. P. Dreyer, and I. C. Vieira, E. R. Santana, J. P. Winiarski, and I. C. Vieira, "Nanocomposite based on green synthesis of gold nanoparticles decorated with functionalized multi-walled carbon nanotubes for the electrochemical determination of hydroxychloroquine," *Journal of Pharmaceutical and Biomedical Analysis*, vol. 236, 2023.
[91] J. Liang, F. Yan, C. Jiang, L. Xie, Y. Wang, T. Li, L. Zheng, J. Wang, D. Ning, L. Tang and Y. Ya, "In situ one-step electrochemical preparation of mesoporous molecularly imprinted sensor for efficient determination of indole-3-acetic acid," *Journal of Electroanalytical Chemistry*, vol. 905, 2022.
[92] B. Shao, Y. Ai, L. Yan, B. Wang, Y. Huang, Q. Zou, H. Fu, X. Niu, and W. Sun, "Wireless electrochemical sensor for the detection of phytoregulator indole-3-acetic acid using gold nanoparticles and three-dimensional reduced graphene oxide modified screen printed carbon electrode," *Talanta*, vol. 253, 124030.
[93] N. Malarat, W. Oin, K. Kanjana, F. Makkliang, M. Siaj, and S. Poorahong, "Electrochemical platform based on activated carbon/graphene oxide-gold nanoparticle composites for the electrochemical sensing of methylparaben in cosmetic samples," *Microchemical Journal*, vol. 188, 2023.
[94] X. Feng, G. Han, J. Cai, and X. Wang, "Au@Carbon quantum Dots-MXene nanocomposite as an electrochemical sensor for sensitive detection of nitrite," *Journal of Colloid and Interface Science*, vol. 607, pp. 1313–1322, 2022, doi: 10.1016/j.jcis.2021.09.036
[95] Kamalasekaran, Kavitha, Vasanth Magesh, Raji Atchudan, Sandeep Arya, and Ashok K. Sundramoorthy., "Development of electrochemical sensor using iron (III) phthalocyanine/gold nanoparticle/graphene hybrid film for highly selective determination of nicotine in human salivary samples," *Biosensors,* vol.13, 2023.
[96] S. Kummari, L. Panicker, K. V. Gobi, R. Narayan, and Y. G. Kotagiri, "Electrochemical strip sensor based on a silver nanoparticle-embedded conducting polymer for sensitive in vitro detection of an antiviral drug," *ACS Applied Nano Materials*, vol. 6, no. 13, pp. 12381–12392, 2023, doi: 10.1021/acsanm.3c02080
[97] N. Jayarambabu, K. Saraswathi, A. Akshaykranth, N. Anitha, T. Venkatappa Rao, and R. Rakesh kumar, *Inorganic Chemistry Communications*, vol. 147, no. November 2022, p. 110249, 2023, doi: 10.1016/j.inoche.2022.110249
[98] S. A. Love, M. A. Maurer-Jones, J. W. Thompson, Y. S. Lin, and C. L. Haynes, "Assessing nanoparticle toxicity," *Annual Review of Analytical Chemistry*, vol. 5, pp. 181–205, Jul. 2012, doi: 10.1146/ANNUREV-ANCHEM-062011-143134
[99] Y. F. Ascencio-Flores *et al.*, "Simultaneous square wave voltammetry detection of azo dyes using silver nanoparticles assembled on carbon nanofibers," *Electrochimica Acta*, vol. 441, no. December 2022, 2023, doi: 10.1016/j.electacta.2022.141782
[100] A. M. Fekry, S. A. Abdel-Gawad, R. H. Tammam, and M. A. Zayed, "An electrochemical sensor for creatinine based on carbon nanotubes/folic acid/silver nanoparticles modified electrode," *Measurement (Lond)*, vol. 163, p. 107958, 2020, doi: 10.1016/j.measurement.2020.107958
[101] L. S. Porto, L. Franco Ferreira, W. T. Pio dos Santos, and A. César Pereira, "Determination of organophosphorus compounds in water and food samples using a non-enzymatic electrochemical sensor based on silver nanoparticles and carbon nanotubes nanocomposite coupled with batch injection analysis," *Talanta*, 2022, 123477.
[102] Md.A. Rashed, M. Faisal, S.A. Alsareii, Mabkhoot Alsaiari, Mohammed Jalalah, Farid A. Harraz, "Highly sensitive and selective electrochemical sensor for detecting imidacloprid pesticide using novel silver nanoparticles/mesoporous carbon/hematite ore ternary nanocomposite," *Journal of Environmental Chemical Engineering*, vol. 10, 2022.
[103] Ling Wang, Xiao Li, Ran Yang, Jian-Jun Li & Ling-Bo Qu, "A highly sensitive and selective electrochemical sensor for pentachlorophenol based on reduced graphite oxide-silver nanocomposites," *Food Analytical Methods*, vol. 13, pp. 2050–2058, 2020.
[104] A. John, L. Benny, A. R. Cherian, S. Y. Narahari, A. Varghese, and G. Hegde, "Electrochemical sensors using conducting polymer/noble metal nanoparticle nanocomposites for the detection of various analytes: A review," *Journal of Nanostructure in Chemistry*, vol. 11, no. 1, pp. 1–31, Mar. 2021, doi: 10.1007/s40097-020-00372-8

[105] A. John, L. Benny, A. R. Cherian, S. Y. Narahari, A. Varghese, and G. Hegde, "Electrochemical sensors using conducting polymer/noble metal nanoparticle nanocomposites for the detection of various analytes: A review," *Journal of Nanostructure in Chemistry*, vol. 11, no. 1, pp. 1–31, Mar. 2021, doi: 10.1007/s40097-020-00372-8

[106] A. Şavk, H. Aydın, K. Cellat, and F. Şen, "A novel high performance non-enzymatic electrochemical glucose biosensor based on activated carbon-supported Pt-Ni nanocomposite," *Journal of Molecular Liquids*, vol. 300, p. 112355, Feb. 2020, doi: 10.1016/j.molliq.2019.112355

[107] M. A. Rashed *et al.*, "A sensitive, selective non-enzymatic electrochemical detection and kinetic study of glucose over Pt nanoparticles/SWCNTs/NiO ternary nanocomposite," *Journal of the Taiwan Institute of Chemical Engineers*, vol. 151, p. 105113, Oct. 2023, doi: 10.1016/j.jtice.2023.105113

[108] W. Bouali, N. Erk, G. Kholafazadehastamal, M. Naser, and G. Tiris, "Low-cost voltammetric sensor based on reduced graphene oxide anchored on platinum nanoparticles for robust determination of Favipiravir in real samples," *Diamond and Related Materials*, vol. 131, p. 109609, Jan. 2023, doi: 10.1016/j.diamond.2022.109609

[109] U. Agrawal, R. Raju, and Z. F. Udwadia, "Favipiravir: A new and emerging antiviral option in COVID-19," *Medical Journal Armed Forces India*, vol. 76, no. 4, pp. 370–376, Oct. 2020, doi: 10.1016/j.mjafi.2020.08.004

[110] N. Puri, S. K. Mishra, A. Niazi, A. K. Srivastava, Rajesh, "Physicochemical characteristics of reduced graphene oxide based Pt-nanoparticles-conducting polymer nanocomposite film for immunosensor applications," *Journal of Chemical Technology & Biotechnology*, vol. 90, no. 9, pp. 1699–1706, Sep. 2015, doi: 10.1002/jctb.4480

[111] L. Ma, X. Wang, Q. Zhang, X. Tong, Y. Zhang, and Z. Li, "Pt catalyzed formation of a Ni@Pt/reduced graphene oxide nanocomposite: Preparation and electrochemical sensing application for glucose detection," *Analytical Methods*, vol. 10, no. 31, pp. 3845–3850, 2018, doi: 10.1039/C8AY01275J

[112] F. H. Moghadam, M. A. Taher, and H. Karimi-Maleh, "A sensitive and fast approach for voltammetric analysis of bisphenol a as a toxic compound in food products using a Pt-SWCNTs/ionic liquid modified sensor," *Food and Chemical Toxicology*, vol. 152, p. 112166, Jun. 2021, doi: 10.1016/j.fct.2021.112166

[113] F. O. Okumu, B. Silwana, and M. C. Matoetoe, "Application of MWCNT/Ag-Pt nanocomposite modified GCE for the detection of nevirapine in pharmaceutical formulation and biological samples," *Electroanalysis*, vol. 32, no. 12, pp. 3000–3008, Dec. 2020, doi: 10.1002/elan.202060374

[114] A. Dehdashti and A. Babaei, "Highly sensitive electrochemical sensor based on Pt doped NiO nanoparticles/MWCNTs nanocomposite modified electrode for simultaneous sensing of piroxicam and amlodipine," *Electroanalysis*, vol. 32, no. 5, pp. 1017–1024, May 2020, doi: 10.1002/elan.201900580

[115] U. Lange, T. Hirsch, V. M. Mirsky, and O. S. Wolfbeis, "Hydrogen sensor based on a graphene – Palladium nanocomposite," *Electrochimica Acta*, vol. 56, no. 10, pp. 3707–3712, Apr. 2011, doi: 10.1016/j.electacta.2010.10.078

[116] P. Veerakumar, C. Koventhan, and S.-M. Chen, "Copper-palladium alloy nanoparticles immobilized over porous carbon for voltammetric determination of dimetridazole," *Journal of Alloys and Compounds*, vol. 931, p. 167474, Jan. 2023, doi: 10.1016/j.jallcom.2022.167474

[117] F. Ameen *et al.*, "Synthesis and characterization of activated carbon supported bimetallic Pd based nanoparticles and their sensor and antibacterial investigation," *Environmental Research*, vol. 221, p. 115287, Mar. 2023, doi: 10.1016/j.envres.2023.115287

[118] Y. Wu *et al.*, "A sensitive sensing platform for acetaminophen based on palladium and multi-walled carbon nanotube composites and electrochemical detection mechanism," *Materials Chemistry and Physics*, vol. 239, p. 121977, Jan. 2020, doi: 10.1016/j.matchemphys.2019.121977

[119] C. Ferrag, M. Noroozifar, and K. Kerman, "Ultralight 3D graphene oxide aerogel decorated with Pd–Fe nanoparticles for the simultaneous detection of eight biomolecules," *ACS Applied Materials & Interfaces*, vol. 15, no. 23, pp. 27502–27514, Jun. 2023, doi: 10.1021/acsami.3c01600

[120] K. Kamalasekaran, V. Magesh, R. Atchudan, S. Arya, and A. K. Sundramoorthy, "Development of electrochemical sensor using iron (III) phthalocyanine/gold nanoparticle/graphene hybrid film for highly selective determination of nicotine in human salivary samples," *Biosensors (Basel)*, vol. 13, no. 9, p. 839, Aug. 2023, doi: 10.3390/bios13090839

[121] Y. F. Ascencio-Flores *et al.*, "Simultaneous square wave voltammetry detection of azo dyes using silver nanoparticles assembled on carbon nanofibers," *Electrochimica Acta*, vol. 441, no. October 2022, 2023, doi: 10.1016/j.electacta.2022.141782

[122] A. M. Fekry, S. A. Abdel-Gawad, R. H. Tammam, and M. A. Zayed, "An electrochemical sensor for creatinine based on carbon nanotubes/folic acid/silver nanoparticles modified electrode," *Measurement (Lond)*, vol. 163, p. 107958, 2020, doi: 10.1016/j.measurement.2020.107958

[123] J. Li *et al.*, "Pt nanoparticle decorated carbon nanotubes nanocomposite based sensing platform for the monitoring of cell-secreted dopamine," *Sensors and Actuators B: Chemical*, vol. 330, p. 129311, Mar. 2021, doi: 10.1016/j.snb.2020.129311

[124] K. Kasinathan *et al.*, "Synthesis of biocompatible chitosan functionalized Ag decorated biocomposite for effective antibacterial and anticancer activity," *International Journal of Biological Macromolecules*, vol. 178, pp. 270–282, May 2021, doi: 10.1016/j.ijbiomac.2021.02.127

[125] H. Mok and T. G. Park, "Hybrid polymeric nanomaterials for siRNA delivery and imaging," *Macromolecular Bioscience*, vol. 12, no. 1, pp. 40–48, Jan. 2012, doi: 10.1002/MABI.201100169

[126] T. K. Sau, A. L. Rogach, F. Jäckel, T. A. Klar, and J. Feldmann, "Properties and applications of colloidal nonspherical noble metal nanoparticles," *Advanced Materials*, vol. 22, no. 16, pp. 1805–1825, Apr. 2010, doi: 10.1002/adma.200902557

[127] M. P. Melancon, M. Zhou, and C. Li, "Cancer theranostics with near-infrared light-activatable multimodal nanoparticles," *Accounts of Chemical Research*, vol. 44, no. 10, pp. 947–956, Oct. 2011, doi: 10.1021/ar200022e

[128] C. Zhao, R. Shi, J. Wu, X. Luo, and X. Liu, "Point-of-care detection of salivary nitrite based on the surface plasmon-assisted catalytic coupling reaction of aromatic amines," *Biosensors (Basel)*, vol. 11, no. 7, p. 223, Jul. 2021, doi: 10.3390/bios11070223

[129] S. Pawar, A. Bhattacharya, and A. Nag, "Metal-enhanced fluorescence study in aqueous medium by coupling gold nanoparticles and fluorophores using a bilayer vesicle platform," *ACS Omega*, vol. 4, no. 3, pp. 5983–5990, Mar. 2019, doi: 10.1021/acsomega.9b00036

[130] P. K. Jain, K. S. Lee, I. H. El-Sayed, and M. A. El-Sayed, "Calculated absorption and scattering properties of gold nanoparticles of different size, shape, and composition: Applications in biological imaging and biomedicine," *The Journal of Physical Chemistry B*, vol. 110, no. 14, pp. 7238–7248, Apr. 2006, doi: 10.1021/jp057170o

[131] K. L. Kelly, E. Coronado, L. L. Zhao, and G. C. Schatz, "The optical properties of metal nanoparticles: The influence of size, shape, and dielectric environment," *Journal of Physical Chemistry B*, vol. 107, no. 3, pp. 668–677, Jan. 2003, doi: 10.1021/jp026731y

[132] U. Kreibig and M. Vollmer, *Optical Properties of Metal Clusters*, vol. 25. Springer Series in Materials Science, vol. 25. Springer Berlin Heidelberg, 1995. doi: 10.1007/978-3-662-09109-8

[133] S. Link and M. A. El-Sayed, "Optical properties and ultrafast dynamics of metallic nanocrystals," *Annual Review of Physical Chemistry*, vol. 54, no. 1, pp. 331–366, Oct. 2003, doi: 10.1146/annurev.physchem.54.011002.103759

[134] P. K. Jain, X. Huang, and I. H. El-sayed, "Review of some interesting surface plasmon resonance-enhanced properties of noble metal nanoparticles and their applications to biosystems," pp. 107–118, 2007, doi: 10.1007/s11468-007-9031-1

[135] J. W. C. J. H. Choi, T. Ha, M. Shin, and S. N. Lee, "Nanomaterial-based fluorescence resonance energy transfer (FRET) and metal-enhanced fluorescence (MEF) to detect nucleic acid in cancer diagnosis," *Biomedicines*, vol. 9, p. 928, 2021.

[136] J. H. Choi, J. Lim, M. Shin, S. H. Paek, and J. W. Choi, "CRISPR-Cas12a-based nucleic acid amplification-free DNA biosensor via au nanoparticle-assisted metal-enhance fluorescence and colorimetric analysis" *Nano Letters*, vol. 21, no. 1, pp. 693–699, 2021. DOI: 10.1021/acs.nanolett.0c04303

[137] J.-H. C. Min Yu Choi and D. H. Park, "Multiplex metal enhanced fluorescence (MEF) effect on porous Au nanorod for highly sensitive multi-microRNA (miRNA) detection," *Sensors and Actuators B: Chemical*, vol. 393, p. 134280, 2023.

[138] E. K. Lim, T. Kim, S. Paik, S. Haam, Y. M. Huh, and K. Lee, "Nanomaterials for theranostics: Recent advances and future challenges," *Chemical Reviews*, vol. 115, no. 1, pp. 327–394, Jan. 2015, doi: 10.1021/CR300213B

[139] Neha Desai, Momin, M., Khan, T., Gharat, S., Ningthoujam, R. S., & Omri, A. "Metallic nanoparticles as drug delivery system for the treatment of cancer," Expert Opinion on Drug Delivery, vol. 18, no. 9, pp. 1261–1290, 2021. https://doi.org/10.1080/17425247.2021.1912008

[140] H. Maeda, J. Wu, T. Sawa, Y. Matsumura, and K. Hori, "'Tumor vascular permeability and the EPR effect in macromolecular therapeutics: A review,'" *Journal of Controlled Release*, vol. 65, 2000.

[141] J. L. Z. Manpreet Bariana, B. Zhang, J. Sun, W. Wang, J. Wang, E. Cassella, F. Myint, S. A. Anuncio, S. Ouk, H.-C. Liou, M. Tan, and H. Wang, "Targeted lymphoma therapy using a gold nanoframework-based drug delivery system," *ACS Applied Materials & Interfaces*, vol. 15, pp. 6312–6325, 2023.

[142] D. H. Hanna, M. H. El-Mazaly, and R. R. Mohamed, "Synthesis of biodegradable antimicrobial pH-sensitive silver nanocomposites reliant on chitosan and carrageenan derivatives for 5-fluorouracil drug delivery toward HCT116 cancer cells," *International Journal of Biological Macromolecules*, vol. 231, p. 123364, Mar. 2023, doi: 10.1016/j.ijbiomac.2023.123364

[143] E. K. R. Min Su Yim, Y. S. Hwang, J. K. Bang, D.-W. Jung, J. M. Kim, G.-R. Yi, G. Lee, "Morphologically homogeneous, pH-responsive gold nanoparticles for non-invasive imaging of HeLa cancer," *Nanomedicine: Nanotechnology, Biology, and Medicine*, vol. 34, p. 102394, 2021.

[144] Y. Guo, Y. Fan, G. Li, Z. Wang, X. Shi, and M. Shen, "'Cluster bomb' based on redox-responsive carbon dot nanoclusters coated with cell membranes for enhanced tumor theranostics," *ACS Applied Materials & Interfaces*, vol. 13, no. 47, pp. 55815–55826, 2021, doi: 10.1021/acsami.1c15282

[145] W. A. Sasidharan and N. A. Monteiro-Riviere, "Biomedical applications of gold nanomaterials: opportunities and challenges," *Interdisciplinary Reviews: Nanomedicine and Nanobiotechnology*, vol. 7, no. 7, 2015.

[146] A. T. R. P. P. Siritongsuk, N. Hongsing, S. Thammawithan, S. Daduang, S. Klaynongsruang, *PLoS One*, vol. 11, e0168098, 2016.

[147] T. O. H. Okuda, M. Ohnishi, H. Takahashi, C. Takagi, and N. Takada, "'Long-term efficacy and toxicity of concurrent chemoradiotherapy with nedaplatin and S-1 for head and neck squamous cell carcinoma,'" *Auris Nasus Larynx*, vol. 46, no. 6, pp. 882–888, 2019.

[148] R. Weissleder, "'A clearer vision for in vivo imaging,'" *Nature Biotechnology*, vol. 19, no. 4, pp. 316–317, 2001.

[149] Fabio Vischio, Livianna Carrieri, Giuseppe Valerio Bianco, Francesca Petronella, Nicoletta Depalo, Elisabetta Fanizza, Maria Principia Scavo, Luciano De Sio, Antonella Calogero, Marinella Striccoli, Angela Agostiano, Gianluigi Giannelli, Maria Lucia Curri, Chiara Ingrosso, "Au nanoparticles decorated nanographene oxide-based platform: Synthesis, functionalization and assessment of photothermal activity," *Biomaterials Advances*, vol. 145, p. 213272, 2023.

[150] Haisong Tan, Nan Hou, Yanlei Liu, Bin Liu, Wen Cao, Dachao Zheng, Wenzhi Li, Yushan Liu, Bin Xu, Zhong Wang, Daxiang Cui, "CD133 antibody targeted delivery of gold nanostars loading IR820 and docetaxel for multimodal imaging and near-infrared photodynamic/photothermal/chemotherapy against castration resistant prostate cancer," *Nanomedicine: Nanotechnology, Biology, and Medicine*, vol. 27, p. 102192, 2020.

[151] D. Wang and S. J. Lippard, "'Cellular processing of platinum anticancer drugs,'" *Nature Reviews Drug Discovery*, vol. 4, no. 4, pp. 307–320, 2005.

[152] P. J. Shiny, A. Mukherjee, and N. Chandrasekaran, "'DNA damage and mitochondria-mediated apoptosis of A549 lung carcinoma cells induced by biosynthesised silver and platinum nanoparticles,'" *RSC Advances*, vol. 6, no. 33, pp. 27775–27787, 2016.

[153] Cai-Ping Tan, Yan-Mei Zhong , Liang-Nian Ji and Zong-Wan Mao, "Phosphorescent metal complexes as theranostic anticancer agents: combining imaging and therapy in a single molecule," *Chemical Science*, vol. 12, 2357–2367, 2021.

[154] Luke K. McKenzie, Helen E. Bryant, and Julia A. Weinstein, "Transition metal complexes as photosensitisers in one- and two-photon photodynamic therapy," *Coordination Chemistry Reviews*, vol. 379, 2–29, 2019.

[155] J. Krajczewski, K. Rucińsk, H. E. Townley, and A. Kudelski, "Role of various nanoparticles in photodynamic therapy and detection methods of singlet oxygen," *Photodiagnosis and Photodynamic Therapy*, vol. 26, pp. 162–178, 2019.

[156] U. V. L. Vattuone, M. Rocca, and C. Boragno, "Initial sticking coefficient of O2 on Ag (110)," *The Journal of Chemical Physics*, vol. 101, no. 1, pp. 713–725, 1994, https://doi.org/10.1063/1. 468127.

[157] Y. Zhao, X. Wang, Y. Li, Y. Liu, J. Hou, and Y. Guo, "Preparation and photothermal therapy of gold nanorods modified by *Belamcanda chinensis* (L.) DC polysaccharide," *International Journal of Biological Macromolecules*, vol. 255, p. 127854, Jan. 2024, doi: 10.1016/j.ijbiomac.2023.127854

[158] L. Ye, Y. Chen, J. Mao, X. Lei, Q. Yang, and C. Cui, "Dendrimer-modified gold nanorods as a platform for combinational gene therapy and photothermal therapy of tumors," *Journal of Experimental & Clinical Cancer Research*, vol. 40, no. 1, p. 303, Dec. 2021, doi: 10.1186/s13046-021-02105-3

[159] X. Meng *et al.*, "Gold-crowned bismuth-based nanocomposites for sonodynamic, photothermal, and chemotherapeutic cancer therapy," *ACS Applied Materials & Interfaces* , vol. 15, no. 50, pp. 58041–58053, Dec. 2023, doi: 10.1021/acsami.3c08071

[160] L. Zhang *et al.*, "A core-shell Au@Cu_{2-x}Se heterogeneous metal nanocomposite for photoacoustic and computed tomography dual-imaging-guided photothermal boosted chemodynamic therapy," *Journal of Nanobiotechnology*, vol. 19, no. 1, p. 410, Dec. 2021, doi: 10.1186/s12951-021-01159-x

[161] D. Musib *et al.*, "La(iii)–curcumin-functionalized gold nanocomposite as a red light-activatable mitochondria-targeting PDT agent," *Inorganic Chemistry Frontiers*, vol. 9, no. 4, pp. 686–701, 2022, doi: 10.1039/D1QI01045J

[162] Y. Hu, Y. Zhou, N. Zhao, F. Liu, and F.-J. Xu, "Multifunctional pDNA-conjugated polycationic Au nanorod-Coated Fe_3O_4 hierarchical nanocomposites for trimodal imaging and combined photothermal/gene therapy," *Small*, vol. 12, no. 18, pp. 2459–2468, May 2016, doi: 10.1002/smll.201600271

[163] L. Suresh, P. K. Brahman, K. R. Reddy, and J. S. Bondili, "Development of an electrochemical immunosensor based on gold nanoparticles incorporated chitosan biopolymer nanocomposite film for the detection of prostate cancer using PSA as biomarker," *Enzyme and Microbial Technology*, vol. 112, pp. 43–51, May 2018, doi: 10.1016/j.enzmictec.2017.10.009

[164] R. Fathi and R. Mohammadi, "Preparation of pH-responsive magnetic nanocomposite hydrogels based on k-carrageenan/chitosan/silver nanoparticles: Antibacterial carrier for potential targeted anticancer drug delivery," *International Journal of Biological Macromolecules*, vol. 246, p. 125546, Aug. 2023, doi: 10.1016/j.ijbiomac.2023.125546

[165] S. S. Murugan *et al.*, "The therapeutic efficacy of silver loaded rhenium disulfide nanoparticles as a photothermal agent for cancer eradication," *Journal of Photochemistry and Photobiology B: Biology*, vol. 250, p. 112831, Jan. 2024, doi: 10.1016/j.jphotobiol.2023.112831

[166] L. Ye *et al.*, "Oxygen-loaded lipid nanobubbles for biofilm eradication by combined trimodal treatment of oxygen, silver, and photodynamic therapy," *ACS Applied Nano Materials*, vol. 6, no. 13, pp. 11715–11724, Jul. 2023, doi: 10.1021/acsanm.3c01677

[167] Z. Liu *et al.*, "Enhancement of radiotherapy efficacy by silver nanoparticles in hypoxic glioma cells," *Artificial Cells, Nanomedicine, and Biotechnology*, vol. 46, no. sup3, pp. 922–930, Nov. 2018, doi: 10.1080/21691401.2018.1518912

[168] M. Ghaffarlou *et al.*, "Facile preparation of silver based radiosensitizers via biomineralization method for enhanced in vivo breast cancer radiotherapy," *Scientific Reports*, vol. 13, no. 1, p. 15131, Sep. 2023, doi: 10.1038/s41598-023-40763-9

[169] M. Buabeid, E.-S. A. Arafa, H. S. Yaseen, M. I. Umar, and G. Murtaza, "Anti-inflammatory effect of simvastatin by impeding TNF-α and interleukin-1ß pathways: Antiangiogenic activity of simvastatin and simvastatin-loaded silver nanoparticles," *Artif Cells Nanomed Biotechnol*, vol. 50, no. 1, pp. 208–217, Dec. 2022, doi: 10.1080/21691401.2022.2098306

[170] A. Pasha *et al.*, "Role of biosynthesized Ag-NPs using Aspergillus niger (MK503444.1) in antimicrobial, anti-cancer and anti-angiogenic activities," *Frontiers in Pharmacology*, vol. 12, Jan. 2022, doi: 10.3389/fphar.2021.812474

[171] X. Kuang *et al.*, "Ag nanoparticles enhance immune checkpoint blockade efficacy by promoting of immune surveillance in melanoma," *Journal of Colloid and Interface Science*, vol. 616, pp. 189–200, Jun. 2022, doi: 10.1016/j.jcis.2022.02.050

[172] S. Akbari Nakhjavani *et al.*, "Electrochemiluminescent immunosensor for detection of carcinoembryonic antigen using luminol-coated silver nanoparticles," *Microchimica Acta*, vol. 190, no. 2, p. 77, Feb. 2023, doi: 10.1007/s00604-023-05656-8

[173] M. Patel and A. Prabhu, "Smart nanocomposite assemblies for multimodal cancer theranostics," *International Journal of Pharmaceutics*, vol. 618, p. 121697, Apr. 2022, doi: 10.1016/j.ijpharm.2022.121697

[174] A. W. Prestayko, J. C. D'Aoust, B. F. Issell, and S. T. Crooke, "Cisplatin (cis-diamminedichloroplatinum II)," *Cancer Treatment Reviews*, vol. 6, no. 1, pp. 17–39, Mar. 1979, doi: 10.1016/S0305-7372(79)80057-2

[175] S. Gurunathan, M. Jeyaraj, M.-H. Kang, and J.-H. Kim, "Graphene oxide–platinum nanoparticle nanocomposites: A suitable biocompatible therapeutic agent for prostate cancer," *Polymers (Basel)*, vol. 11, no. 4, p. 733, Apr. 2019, doi: 10.3390/polym11040733

[176] S. Sadi *et al.*, "Early stage evaluation of cancer stem cells using platinum nanoparticles/CD133+ enhanced nanobiocomposite," *Cancer Nanotechnology*, vol. 14, no. 1, p. 55, Dec. 2023, doi: 10.1186/s12645-023-00208-4

[177] F. Moniriyan, S. Jamal Tabatabaei Rezaei, and S. Javad Sabounchei, "Pegylated magnetic carbon nanotubes for efficient and safe delivery of poorly water-soluble platinum anticancer drugs," *Results in Chemistry*, vol. 7, p. 101275, Jan. 2024, doi: 10.1016/j.rechem.2023.101275

[178] L. Bi, Y. Teng, M. Baghayeri, and J. Bao, "Employing Pd nanoparticles decorated on halloysite nanotube/carbon composite for electrochemical aptasensing of HER2 in breast cancer patients," *Environmental Research*, vol. 237, p. 117030, Nov. 2023, doi: 10.1016/j.envres.2023.117030

[179] M. Afzali, A. Mostafavi, R. Nekooie, and Z. Jahromi, "A novel voltammetric sensor based on palladium nanoparticles/carbon nanofibers/ionic liquid modified carbon paste electrode for sensitive determination of anti-cancer drug pemetrexed," *Journal of Molecular Liquids*, vol. 282, pp. 456–465, May 2019, doi: 10.1016/j.molliq.2019.03.041

[180] K. Kasinathan, K. Marimuthu, B. Murugesan, S. Samayanan, Y. Cai, and C. Rathinam, "Facile synthesis of highly biologically active chitosan functionalized 2D WS2 nanocomposite anchored with palladium nanoparticles for antibacterial and anticancer activity: In-vitro biomedical evaluation," *Journal of Molecular Liquids*, vol. 335, p. 116582, Aug. 2021, doi: 10.1016/j.molliq.2021.116582

[181] B. Murugesan, J. Sonamuthu, S. Samayanan, S. Arumugam, and S. Mahalingam, "Highly biological active antibiofilm, anticancer and osteoblast adhesion efficacy from MWCNT/PPy/Pd nanocomposite," *Applied Surface Science*, vol. 434, pp. 400–411, Mar. 2018, doi: 10.1016/j.apsusc.2017.10.142

[182] S. Dhanavel, E. A. K. Nivethaa, V. Narayanan, and A. Stephen, "In vitro cytotoxicity study of dual drug loaded chitosan/palladium nanocomposite towards HT-29 cancer cells," *Materials Science and Engineering: C*, vol. 75, pp. 1399–1410, Jun. 2017, doi: 10.1016/j.msec.2017.03.058

[183] Y. A. Hassan, M. Y. Alfaifi, A. A. Shati, S. E. I. Elbehairi, R. F. M. Elshaarawy, and I. Kamal, "Co-delivery of anticancer drugs via poly(ionic crosslinked chitosan-palladium) nanocapsules: Targeting more effective and sustainable cancer therapy," *Journal of Drug Delivery Science and Technology*, vol. 69, p. 103151, Mar. 2022, doi: 10.1016/j.jddst.2022.103151

[184] A. Akbari *et al.*, "Detection of plasma miR-223 by a novel label-free graphene oxide/gold nanocomposite immunosensor in colorectal cancer patients: An electrochemical biosensor approach," *Biosens Bioelectron X*, vol. 14, p. 100331, Sep. 2023, doi: 10.1016/j.biosx.2023.100331

[185] F. Khosravi, M. Rahaie, M. R. Ghaani, M. Azimzadeh, and E. Mostafavi, "Ultrasensitive electrochemical miR-155 nanocomposite biosensor based on functionalized/conjugated graphene materials and gold nanostars," *Sensors and Actuators B: Chemical*, vol. 375, p. 132877, Jan. 2023, doi: 10.1016/j.snb.2022.132877

[186] C. Tao and J. Rouhi, "A biosensor based on graphene oxide nanocomposite for determination of carcinoembryonic antigen in colorectal cancer biomarker," *Environmental Research*, vol. 238, p. 117113, Dec. 2023, doi: 10.1016/j.envres.2023.117113

[187] M. Zamani, N. Tavakkoli, and N. Soltani, "Electrochemical sensor based on graphene quantum dot/gold nanoparticles and thiol-containing organic compound for measuring methotrexate anticancer drug," *Diamond and Related Materials*, vol. 136, p. 109954, Jun. 2023, doi: 10.1016/j.diamond.2023.109954

[188] A. Kalkal, R. Pradhan, and G. Packirisamy, "Gold nanoparticles modified reduced graphene oxide nanosheets based dual-quencher for highly sensitive detection of carcinoembryonic antigen," *International Journal of Biological Macromolecules*, vol. 242, p. 125157, Jul. 2023, doi: 10.1016/j.ijbiomac.2023.125157

[189] T. C. Lebepe *et al.*, "NIR-II window absorbing graphene oxide-coated gold nanorods and graphene quantum dot-coupled gold nanorods for photothermal cancer therapy," *Nanotechnology Reviews*, vol. 12, no. 1, Jun. 2023, doi: 10.1515/ntrev-2022-0541

[190] A. Foroozandeh, M. Abdouss, H. SalarAmoli, M. Pourmadadi, and F. Yazdian, "An electrochemical aptasensor based on g-C_3N_4/Fe_3O_4/PANI nanocomposite applying cancer antigen_125 biomarkers detection," *Process Biochemistry*, vol. 127, pp. 82–91, Apr. 2023, doi: 10.1016/j.procbio.2023.02.004

[191] H. Liang, H. Liu, B. Tian, R. Ma, and Y. Wang, "Carbon quantum Dot@Silver nanocomposite – Based fluorescent imaging of intracellular superoxide anion," *Microchimica Acta*, vol. 187, no. 9, p. 484, Sep. 2020, doi: 10.1007/s00604-020-04359-8

[192] M. Saikia, T. Das, and B. K. Saikia, "A novel rapid synthesis of highly stable silver nanoparticle/carbon quantum dot nanocomposites derived from low-grade coal feedstock," *New Journal of Chemistry*, vol. 46, no. 1, pp. 309–321, 2022, doi: 10.1039/D1NJ04039A

[193] U. Suresh, S. Subhadra, and S. Sivaramakrishnan, "Green hydrothermal synthesis of carbon dot-silver nanocomposite from Chondrococcus hornemanni (marine algae): An application of mosquitocidal, anti-bacterial, and anti-cancer (MDA-MB-231 cells)," *Biomass Convers Biorefin*, Apr. 2023, doi: 10.1007/s13399-023-04214-9

[194] "Graphene oxide as a carrier for drug delivery of methotrexate," *Biointerface Research in Applied Chemistry*, vol. 11, no. 6, pp. 14726–14735, Mar. 2021, doi: 10.33263/BRIAC116.1472614735

[195] Y. Q. Almajidi *et al.*, "Nanocomposite of reduced nanographene oxide with β-lactoglobulin protein (rNGO/β-Lg) as a carrier of the anticancer drug oxaliplatin (Eloxatin)," *Diamond and Related Materials*, vol. 136, p. 110015, Jun. 2023, doi: 10.1016/j.diamond.2023.110015

[196] D. Salas-Treviño *et al.*, "Hyaluronate functionalized multi-wall carbon nanotubes filled with carboplatin as a novel drug nanocarrier against murine lung cancer cells," *Nanomaterials*, vol. 9, no. 11, p. 1572, Nov. 2019, doi: 10.3390/nano9111572

[197] M. R. Berber, H. Elkhenany, I. H. Hafez, A. El-Badawy, M. Essawy, and N. El-Badri, "Efficient tailoring of platinum nanoparticles supported on multiwalled carbon nanotubes for cancer therapy," *Nanomedicine*, vol. 15, no. 8, pp. 793–808, Apr. 2020, doi: 10.2217/nnm-2019-0445

6 Mechanical Properties and Water Absorption Behavior of Jute/Nettle Hybrid Epoxy Composite

Vikas Upadhyay, Ravi Kumar, and Chaitanya Sharma

6.1 INTRODUCTION

A rapid growth in the design, development, fabrication, and application of composite materials was the well-deserved outcome of the long-pressing need to develop materials whose properties can be tailored. Composite materials have offered remarkable properties, resulting in widespread application [1,2]. However, growing environmental concerns have forced the global community to develop environmentally friendly and easily disposable composites. Natural fibers may be used as an alternative to traditional reinforcements/fibers in composites, as different types of natural fibers are easily available around the globe depending upon the climatic condition of that particular region. Natural fibers are now widely used in various application areas, such as the automobile industry (for components like bumpers, roofs, and inside door panels), aerospace (for wings and tails), and other structural purposes. These applications often expose the composites to both moisture and mechanical loads throughout their operational lifespan. However, extensive research is still going on to fully explore the potential of natural fibers by exploring various fibers, fiber combinations, secondary reinforcements, and various polymer matrix materials [3,4].

A natural fiber jute composite refers to a material made by combining jute fibers with a matrix material, often a polymer like epoxy. Jute, a natural plant fiber, is utilized as a reinforcement in the composite, providing strength and other desirable properties. The resulting composite material combines the advantages of jute fibers, such as being lightweight and environmentally friendly, with the matrix material's characteristics, leading to a versatile material with enhanced mechanical properties [5]. Natural fiber jute composites find applications in various industries, including automotive, construction, and packaging, due to their eco-friendly nature and potential for replacing traditional synthetic composites in certain applications. Moisture absorption in jute fiber composites refers to the ability of these materials to absorb and retain moisture from their surroundings. Jute fibers, being natural plant fibers, tend to absorb water [6]. The moisture absorption in jute fiber composites can have implications for their mechanical properties and overall performance. The extent of moisture absorption depends on factors such as the composition of the composite, the matrix material used (e.g., epoxy), and the environmental conditions to which the composite is exposed. Excessive moisture absorption may lead to swelling of the fibers, affecting the structural integrity of the composite and potentially leading to a decrease in mechanical properties such as strength and stiffness [7]. Similar to other natural fiber composites, nettle fiber composites are known for their lightweight nature, biodegradability, and eco-friendly profile. They find applications in various industries, including automotive, textiles, construction, and packaging. The mechanical properties of nettle

DOI: 10.1201/9781003664611-7

fiber composites, such as tensile strength and modulus, are influenced by factors like fiber orientation, fiber content, and the choice of matrix material [8]. One of the major problems with polymeric composites is their degradation when exposed to natural environments such as sunlight, humidity, and water. Degradation increases with exposure time and adversely affects mechanical properties [9]. Several studies have already been published regarding the moisture behavior and mechanical strength of hybrid composites incorporating natural fibers.

Athijayamani et al. [10] analyzed the mechanical behavior of sisal and roselle-based hybrid composites under dry and wet conditions. It was reported that moisture exposure leads to deterioration of the fiber–matrix interface, resulting in significantly reduced mechanical properties. Venkateshwaran et al. [11] reported that adding sisal fiber by up to 50% by weight can increase mechanical properties and minimize moisture absorption in banana/epoxy composites. Dhakal et al. [12] studied the effect of carbon fiber hybridization on the water absorption behavior of thermal and mechanical properties on cross-ply (CP) and unidirectional (UD) cellulose flax epoxy composites. The water absorption behavior, tensile strength, flexural property, and thermal stability of hybrid samples were markedly improved compared to those without hybridization. The hybrid composite of bamboo bast and sisal fibers was treated with 10% sodium hydroxide solution for 24 hours and compared with an untreated hybrid composite by Venkatesh et al. [13]. They reported a 30%, 27.4%, and 36.9% increase in tensile, flexural, and impact strength, respectively. Shrivastava et al. [14] also reported that the chemical treatment of coir fiber by sodium hydroxide reduces the moisture absorption of coir/glass fiber hybrid-epoxy-composites. Chaudhary et al. [15] reported that the effect of water immersion on the hardness of hybrid bio-composites was almost negligible, whereas tensile and flexural properties were severely affected. Ray et al. (2020) [16] investigated the influence of varying glass/jute/sisal fiber loadings in polypropylene hybrid composite. The hybrid composite consisting 10% of each fiber experienced the highest water absorption. Yorseng et al. [17] examined individual kenaf and sisal, kenaf/sisal hybrid epoxy composites. They reported that sisal/kenaf/sisal hybrid epoxy composites presented the best properties and retained the thermo-mechanical properties with the lowest water absorption. Sekar et al. [18] reported that the hybridization of Calotropis gigantea fiber (CGFs) and Palmyra fiber (PFs) was effective at higher fiber volume percentage composites and water absorption increased with increase in fiber volume percentage. The moisture absorption and mechanical properties of hybrid composites with chemically treated abaca, kenaf, and carbon fiber reinforcement were experimentally investigated by Venkatasudhahar et al. [19]. They reported that the treated fiber hybrid composite provided higher mechanical strength and absorbed less moisture. Sanjeevi et al. [20] fabricated CGF/areca fine fibers (AFF)/Phenol–Formaldehyde hybrid composites by varying the fiber concentrations. They reported lower mechanical properties in wet than in dry composite conditions. Nasimudeen et al. [21] fabricated hybrid composites having jute, kenaf, and banana in different stacking orders in a vinyl ester matrix and reported improved tensile strength and lower moisture absorption in hybrid fiber composites. Ghori et al. [22] employed date palm and kenaf fiber to fabricate a hybrid composite. Treated 30% date palm and 70% kenaf (T-3DP7K) provided the highest tensile strength and modulus among all hybrid composites. In a recent study, a hybrid composite using natural fiber (enset/sisal) and polyester matrix has been prepared by Bekele et al. [23] to investigate the mechanical and swelling behavior. Fibers were chemically treated using NaOH solution at different concentrations (5% and 10%). The outcomes demonstrated that the 5% NaOH-treated fiber outperformed the 10% NaOH-treated or untreated fiber.

From the aforementioned research, it is evident that certain studies have explored the moisture and mechanical properties of natural fiber hybrid composites. To the best of the author's knowledge, no hybrid composite involved both jute and nettle fibers and examined its moisture and mechanical properties. In the current study, jute and nettle fibers were chosen due to their superior mechanical properties and lower moisture absorption rates, as well as their widespread availability. Composite materials were fabricated by blending jute and nettle fibers with an epoxy matrix at varying weight percentages (100/0, 75/25, 50/50, 25/75, and 0/100). This work focuses on reducing water intake of natural fiber hybrid composite by using jute fiber and nettle fiber in varying interply hybrid configurations and simultaneously studying the degradation in mechanical properties with increased exposure time.

6.2 MATERIALS AND EXPERIMENTAL METHODS

In this work, two natural fibers, jute and nettle, were used in varying proportions with the epoxy to fabricate hybrid composites. Nettle fibers were obtained from Uttarakhand Bamboo and Fiber Development Board, Dehradun, whereas jute fibers were procured from local commercial markets. Figure 6.1 shows the images of jute and nettle fiber used in this work for fabricating different composite materials. Epoxy resin and hardener (diethylene triamine) were purchased from Savita Scientific & Plastic Products, Jaipur. Plastic sheets, rollers, and releasing agents were used in the fabrication process. Jute fibers were cut into 20 cm lengths and arranged in the proper orientation. In contrast, nettle fibers were kept in random orientation as they cannot be given a UD orientation in a carding state, i.e., short, small pieces. Matrix and fibers were mixed in a ratio of 85:15. In contrast, epoxy and hardener were mixed in a ratio of 10:1. The weight proportion of fibers was varied in a step of 25%, keeping the total fiber weight proportion constant, i.e., 15% of the total weight of the composite.

Demerits of individual fiber were overcome by hybridizing jute fiber with nettle fiber. Five different composite materials were fabricated with jute and nettle fibers in most commonly weight (jute/nettle) ratios of 100/0, 75/25, 50/50, 25/75, and 0/100 [24–26]. In hybrid composites, jute fibers were kept between the outer and inner layers of nettle fibers. The traditional hand layup technique was used to fabricate composites, and the entire process was performed at room temperature. The releasing agent was used on the plastic sheet to easily remove composites from the plastic sheet after its curing. A roller was used to remove the trapped air from the uncured composite. A weight of 20 kg was applied to the uncured composite for 24 hours. After curing, the mold was opened, and the fabricated composites were taken out of the mold. The same process was repeated for the fabrication of all types of composites. The specimens were then cut from the composite sheets in the required number and size per various tests' requirements.

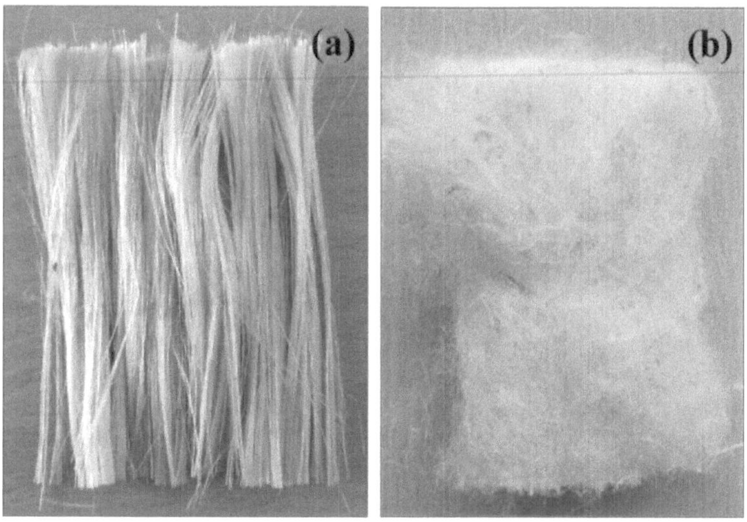

FIGURE 6.1 The appearance of natural fibers used for composite fabrication (a) Jute and (b) Nettle.

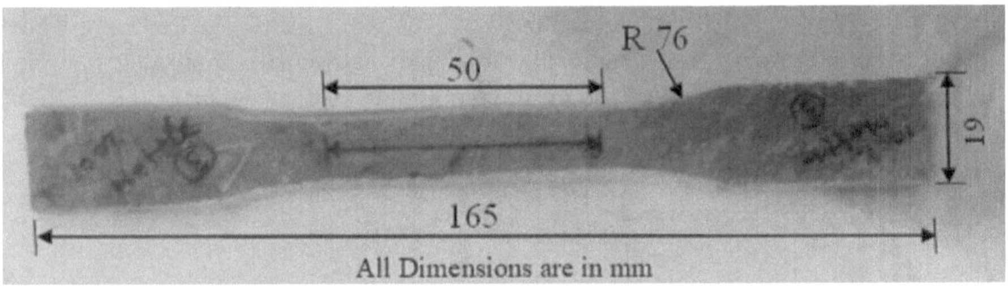

FIGURE 6.2 Fabricated tensile test specimen of natural fiber composite.

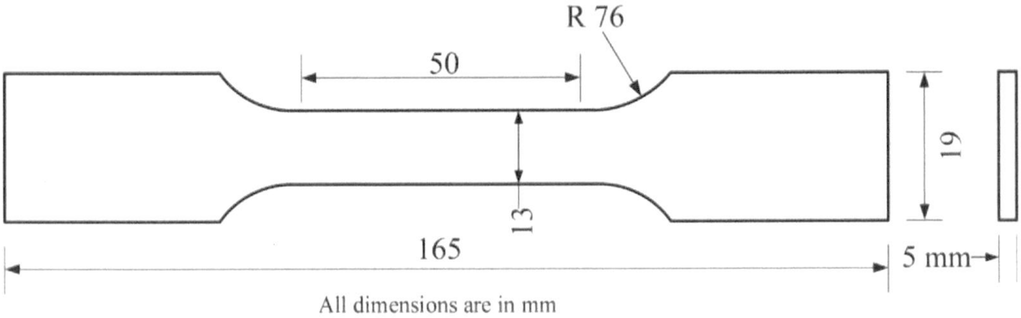

FIGURE 6.3 Schematic showing the dimensions of tensile test specimen.

After the fabrication of all composite materials with a varying weight proportion of jute and nettle, water absorption and tensile tests were performed to investigate the influence of varying weight proportions on the characteristics of natural hybrid composites. Composites were cut per ASTM D570 specifications for the moisture absorption test and dipped in distilled water. After every 12 hours, specimens were removed from the water, wiped with tissue paper, and weighed. Each specimen was measured immediately after taking out from the distilled water to prevent losses by evaporation. A moisture absorption test was performed at room temperature for short (24 hours) and long duration (240 and 350 hours). The following formula was used to calculate the moisture gain, i.e., water absorption in percentage [4].

$$\text{Moisture absorption}(\%) = \left[\left(W - W_0\right)/W_0\right] \times 100 \quad (1)$$

where
W = weight of composite after moisture gain
W_0 = weight of composite before moisture gain

The tensile test specimens were crafted as per ASTM D 638 standards, and are illustrated in Figure 6.2. Additionally, a depicting the specimen dimensions is available in Figure 6.3. Specimens were prepared by hand using a hacksaw and filing. Three specimens of each of the five types of composite material were subjected to a tensile test as per ASTM guidelines at a test speed of 5 mm/min, and the average values are used for the discussion. The fracture surfaces of the tested tensile specimens were examined by a field emission scanning electron microscope (FE-SEM) (FEI-Quanta 200®, Thermo Fisher Scientific, USA) to study the bonding and structure of composites and know the fracture mode of each type of composite material/surface.

6.3 RESULTS AND DISCUSSION

6.3.1 Water Absorption Test

The large amount of water absorption in natural (plant) fiber composites compared to synthetic fiber composites is detrimental to composite performance. It is mainly attributed to cellulose, which is hygroscopic [27]. Although epoxy resin covers the fibers and prevents direct contact with moisture/water, it cannot completely isolate the fibers due to microcracks that provide a route for water to reach the fibers through capillary action. Large-duration exposure of natural fiber composites to moisture/water leads to swelling of fibers, adversely affecting the interface properties, leading to poor interface bonding, cracking, and failure of composites. Therefore, the fabricated composites were subjected to a water absorption test to study the behavior of composite with varying weight proportions of jute and nettle fibers. Figure 6.4 reveals hybrid composites' moisture absorption behavior in distilled water for up to 24 hours.

The composite with 100% jute had the highest water absorption, whereas the composite with 100% nettle showed the least water absorption. This is due to jute fibers that are primarily composed of cellulose, hemicellulose, and lignin. Cellulose is hydrophilic, meaning it has a high affinity for water and has a more open and porous microstructure, which can facilitate water penetration and absorption [5,6]. On the other hand, nettle fiber also contains cellulose, but they may have different proportions of other components as well. Nettle fibers may have a denser or more closed microstructure compared to jute [28]. All the other hybrid composites had water absorption in between them. The addition of a nettle decreases water absorption drastically. The slope of the moisture absorption curve of hybrid composites up to 16 hours is not very sharp, indicating a nominal water absorption after initial 6 hours of rapid water absorption. The initial increase up to 6 hours may be attributed to the absorption by the surface irregularities and voids on the composite surface. Beyond 6 hours, the diffusion of water molecules has occurred inside the gaps and voids in the composites. After 16 hours, the steep slope of the curve showed an increase in water absorption because of the initiation of the capillary transport mechanism between the composites' voids/microcracks and fibers [29]. The nettle/jute (75/25) hybrid composite absorbs the least moisture as compared to rest two hybrid

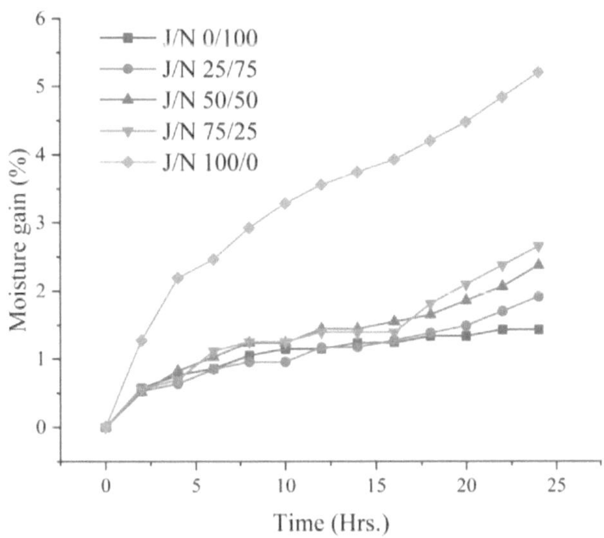

FIGURE 6.4 Effect of weight proportion of nettle and jute on short-term water absorption behavior of hybrid composites.

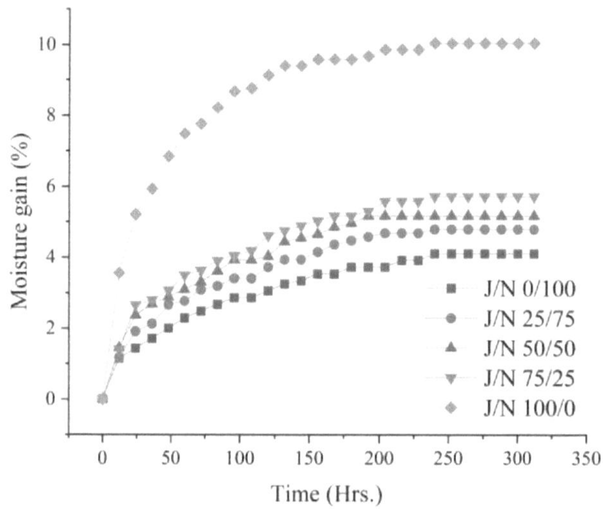

FIGURE 6.5 Effect of weight proportion of nettle and jute on long-term water absorption behavior of hybrid composites.

composites (50/50 and 27/75) as the nettle fiber absorbs less moisture as shown in the graph of jute v/s nettle composite moisture test.

However, exact reasoning cannot be provided about this initial trend based on fiber types. The duration is too short to assume any water absorption by fibers and may be contributed by surface cracks/irregularities. Hence, a long-term study was carried out up to saturation point, and the continued water intake may be attributed to the absorption of water by the fibers, leading to swelling of fibers after 240 hours. Water absorption by the specimens was reduced, and then moisture absorption became constant for up to 312 hours. Moisture absorption data is presented in Figure 6.5 by plotting the absorption percentage on the ordinate and square root of time on the abscissa.

The water absorption in composites with higher jute percentage was higher, i.e., nettle/jute hybrid composites (75/25) absorbed less moisture than nettle/jute 25/75 and 50/50 composites. Moisture absorption of composites reduces at 240 hours. After that, it was constant due to the saturation of the specimen. A better saturation point test was carried out for up to 312 hours. It is imperative to note that sandwiching jute fiber between nettle fibers has significantly reduced water absorption. Jute contains 67% cellulose, 16% hemicellulose, 9% lignin, 0.2% pectin, and 0.5% waxes, whereas nettle contains 71.4% cellulose, 16.7% hemicellulose, 4.5% lignin, 2.3% pectin, and 2.8% fat and wax [21,22]. The less water absorption by nettle may be attributed to the comparatively high percentage of fat and wax that significantly reduces water uptake.

Moreover, further addition of nettle fibers does not yield a proportional reduction in water absorption. The higher water absorption at higher jute weight % is obvious. This is due to its irregular cross-sectional geometry containing well-defined lumens separated by cell walls and rough surfaces with random micropores [30]. Lumen and micropores make the jute fiber porous, while a rough surface facilitates water retention. Both these factors result in more water absorption [23]. The composite with 25/75 weight % of jute/nettle outperformed other composites in water absorption.

6.3.2 Mechanical Properties

Tensile tests were performed to evaluate the effect of varying amounts of jute and nettle on hybrid composites' tensile strength, Young modulus, and stiffness. Adding jute to the nettle in a step of 25 weight % enhanced all the mechanical properties evaluated in this study. As evident from Figure 6.6,

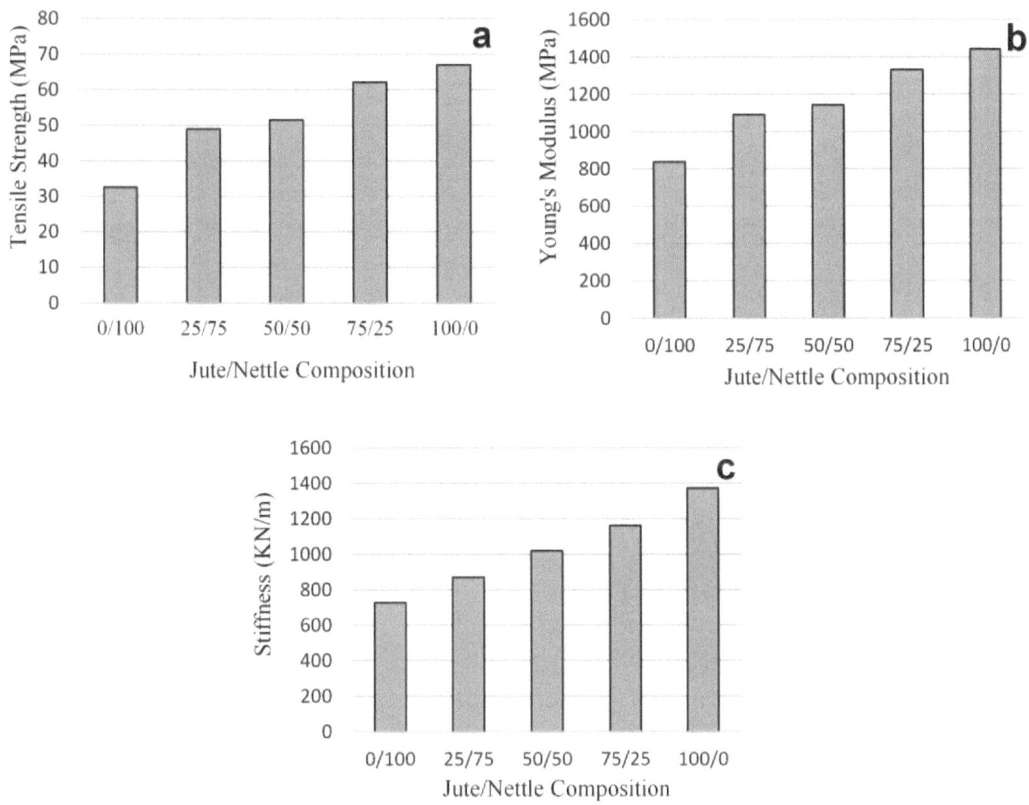

FIGURE 6.6 Variation of (a) tensile strength, (b) Young's modulus, (c) stiffness with jute and nettle composition.

tensile strength, Young modulus, and stiffness increased with the augmentation of jute weight % and vice versa with nettle.

Maximum tensile strength (65.92 MPa) was offered by 100% jute composite, whereas minimum tensile strength (32.156 MPa.) was observed for 100% nettle composite. In hybrid composite, jute/nettle (75/25) showed better tensile strength (62 MPa) as compared to 50/50 jute/nettle (51.5 MPa) and 25/75 jute/nettle (48.9 MPa), offering a 7.4%, 23.5%, and 26.9% reduction in tensile properties, respectively, in comparison to 100% jute composite with each replacement of 25% of jute with nettle. Composites with more jute fiber had better tensile strength because jute fiber had greater strength than nettle fiber and better bonding with epoxy than nettle [31]. Young's modulus has also followed a trend similar to tensile strength, accounting for a 7.7%, 20.8%, and 24.4% decrease in comparison to 100% jute composite in successive replacement of 25% of jute with nettle fiber. However, the influence of substituting jute fibers on stiffness was uniformly distributed, resulting in reductions of 15.33%, 25.73%, and 36.55% relative to the stiffness observed in a 100% jute configuration specimen. Jute is more beneficial than nettle for improving the mechanical properties of jute–nettle hybrid composites. Replacement of jute with nettle has a more serious effect on the stiffness than Young's modulus and tensile strength.

6.3.3 Fractography

A general view of the failure surface of natural hybrid composites with varying weight proportions of nettle and jute fiber is shown in Figure 6.7. Fiber pull-out, debonding, open cell, fractured fibers, and resin residues were observed on the tensile fracture surfaces. The presence of cells leads to a porous

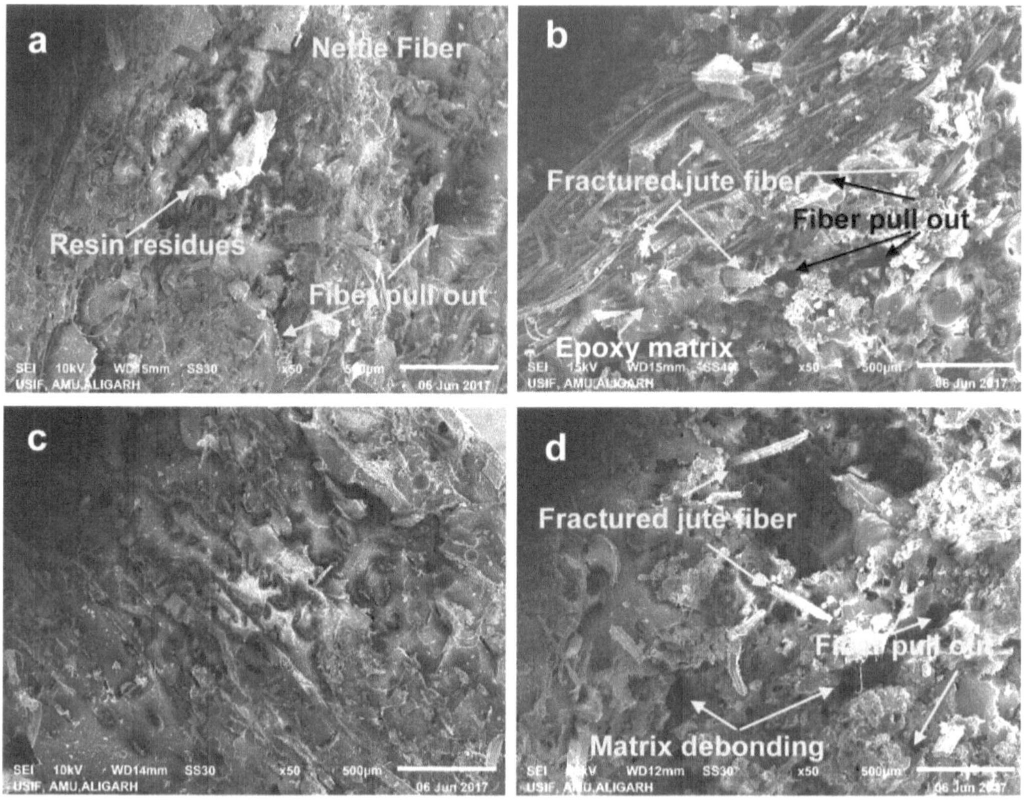

FIGURE 6.7 Microstructure variation of hybrid natural fiber composite with jute and nettle (a) 100% nettle (b) 75% nettle and 25% jute (c) 50% nettle and 50% jute (d) 25% nettle and 75% jute.

and weak epoxy matrix, from where a bundle of fibers is pulled out along with epoxy residues. In the case of composite with 100% nettle, agglomeration of nettle fiber with epoxy residue can be seen on the fracture surface. It is difficult to see fractured nettle fiber; however, fiber pull-out is seen at a few locations. Some cracks were also present formed due to brittle fracture of the epoxy matrix (Figure 6.7a). A Bunch of jute fibers, properly oriented, were observed in the case of composite with 25% jute and 75% nettle fibers. Abundant fractured and pull-out fibers can also be seen on the fractured surface. Fiber's pull-out and debonding from the epoxy matrix may be due to poor interfacial adhesion between the epoxy matrix and fibers for the first two composites (Figure 6.7a and b). Properly bonded fibers were observed in composites with 50% jute and 50% nettle (Figure 6.7c). At higher weight % of jute (75% and 100%), epoxy matric is bonded to jute fiber strongly and thoroughly, reducing fiber pull-out and fracture. A porous structure owing to lumens and rough surface with random micropores is ideal for the infiltration of epoxy into the jute fiber, therefore, a strong bonding resulting in higher mechanical properties [32,33]. At higher weight % of jute, bonding is strong, and load transfer from jute fiber to the epoxy matrix is better, increasing tensile strength, Young's modulus, and stiffness of these composites.

6.4 CONCLUSIONS

A natural hybrid composite of nettle and jute fiber was made using the hand layup method with varying weight % of each. The moisture absorption, tensile strength, Young modulus, and stiffness were determined. The following points were observed from the experimental study.

- Jute–epoxy composite absorbed more moisture than nettle–epoxy composite because jute fiber contains much more cellulose, hemicellulose, and lignin besides lumens and rough surface with micropores.
- In jute/nettle (25/75) hybrid composite, absorbed moisture was less as compared to the remaining two hybrid composites (50/50, 75/25) because the quantity of jute fiber in jute/nettle (25/75) was less.
- 100% jute fiber composite offers the highest tensile strength due to better bonding between fiber and matrix, followed by 75/25, 50/50, and 25/75 jute/nettle hybrid composite.
- In hybrid composites, jute/nettle (75/25) offers better tensile strength than jute/nettle (25/75 and 50/50).

So, it can be concluded from the above results that a 50/50 jute/nettle composite can be used for general-purpose applications as a 50/50 hybrid composite has a balanced value of moisture absorption, tensile strength, and delamination factor. However, wherever the tensile strength requirement dominates the moisture requirement, jute/nettle (75/25) is recommended. A high nettle content composite must be used when moisture absorption reduction is crucial. Owing to low water absorption, excellent strength, low density, small thickness, and insulating properties natural jute and nettle fiber hybrid composites are ideal materials for outdoor as well as indoor applications in automotive, construction, aerospace, consumer industries for specific and potential applications viz. interior panels, dashboards, bumpers, sports equipments, furniture, fan blades, roofing, insulating pipes, etc.

REFERENCES

[1] Nabi Saheb, D., and Jog, J. P., "Natural Fiber Polymer Composites: A Review," *Advances in Polymer Technology*, Vol. 18, No. 4, 1999, pp. 351–363. https://doi.org/10.1002/(SICI)1098-2329(199924)18:4<351::AID-ADV6>3.0.CO;2-X

[2] Sanjay, M. R., Madhu, P., Jawaid, M., Senthamaraikannan, P., Senthil, S., and Pradeep, S., "Characterization and Properties of Natural Fiber Polymer Composites: A Comprehensive Review," *Journal of Cleaner Production*, Vol. 172, 2018, pp. 566–581. https://doi.org/10.1016/j.jclepro.2017.10.101

[3] Wang, H., Memon, H., A. M . Hassan, E., Miah, M. S., and Ali, M. A., "Effect of Jute Fiber Modification on Mechanical Properties of Jute Fiber Composite," *Materials*, Vol. 12, No. 8, 2019, p. 1226. https://doi.org/10.3390/ma12081226

[4] Prasad, L., Kumain, A., Patel, R. V., Yadav, A., and Winczek, J., "Physical and Mechanical Behavior of Hemp and Nettle Fiber-Reinforced Polyester Resin-Based Hybrid Composites," *Journal of Natural Fibers*, Vol. 19, No. 7, 2022, pp. 2632–2647. https://doi.org/10.1080/15440478.2020.1821284

[5] Dhanasekaran, S., and Balachandran, G., "Structural Behavior of Jute Fiber Composites – A Review," SAE Technical Paper 2008-01-2653, 2008. https://doi.org/10.4271/2008-01-2653

[6] Gassan, J., and Bledzki, A. K., "Effect of Moisture Content on the Properties of Silanized Jute-Epoxy Composites," *Polymer Composites*, Vol. 18, No. 2, 1997, pp. 179–184. https://doi.org/10.1002/pc.10272

[7] Islam, S., Karim, F.E. and Islam, M.R., "Assessing the consequences of water retention on the structural integrity of jute fiber and its composites: a review," *SPE Polymers*, Vol. 5, No. 4, 2024, pp. 457–480. https://doi.org/10.1002/pls2.10142

[8] Viju, S., and Thilagavathi, G., "Characterization of Surface Modified Nettle Fibers for Composite Reinforcement," *Journal of Natural Fibers*, Vol. 19, No. 5, 2022, pp. 1819–1827. https://doi.org/10.1080/15440478.2020.1788491

[9] Fakhrul, T., and Islam, M. A., "Degradation Behavior of Natural Fiber Reinforced Polymer Matrix Composites," *Procedia Engineering*, Vol. 56, 2013, pp. 795–800. https://doi.org/10.1016/j.proeng.2013.03.198

[10] Athijayamani, A., Thiruchitrambalam, M., Natarajan, U., and Pazhanivel, B., "Effect of Moisture Absorption on the Mechanical Properties of Randomly Oriented Natural Fibers/Polyester Hybrid

Composite," *Materials Science and Engineering A*, Vol. 517, Nos. 1–2, 2009, pp. 344–353. https://doi.org/10.1016/j.msea.2009.04.027

[11] Venkateshwaran, N., ElayaPerumal, A., Alavudeen, A., and Thiruchitrambalam, M., "Mechanical and Water Absorption Behaviour of Banana/Sisal Reinforced Hybrid Composites," *Materials and Design*, Vol. 32, No. 7, 2011, pp. 4017–4021. https://doi.org/10.1016/j.matdes.2011.03.002

[12] Dhakal, H. N., Zhang, Z. Y., Guthrie, R., MacMullen, J., and Bennett, N., "Development of Flax/Carbon Fibre Hybrid Composites for Enhanced Properties," *Carbohydrate Polymers*, Vol. 96, No. 1, 2013, pp. 1–8. https://doi.org/10.1016/j.carbpol.2013.03.074

[13] Prasanna Venkatesh, R., Ramanathan, K., and Srinivasa Raman, V., "Tensile, Flexual, Impact and Water Absorption Properties of Natural Fibre Reinforced Polyester Hybrid Composites," *Fibres and Textiles in Eastern Europe*, Vol. 24, No. 3, 2016, pp. 90–94. https://doi.org/10.5604/12303666.1196617

[14] Shrivastava, R., Telang, A., Rana, R. S., and Purohit, R., "Mechanical Properties of Coir/ G Lass Fiber Epoxy Resin Hybrid Composite," *Materials Today: Proceedings*, Vol. 4, No. 2, 2017, pp. 3477–3483. https://doi.org/10.1016/j.matpr.2017.02.237

[15] Chaudhary, V., Bajpai, P. K., and Maheshwari, S., "Effect of Moisture Absorption on the Mechanical Performance of Natural Fiber Reinforced Woven Hybrid Bio-Composites," *Journal of Natural Fibers*, Vol. 17, No. 1, 2020, pp. 84–100. https://doi.org/10.1080/15440478.2018.1469451

[16] Ray, K., Patra, H., Swain, A. K., Parida, B., Mahapatra, S., Sahu, A., and Rana, S., "Glass/Jute/Sisal Fiber Reinforced Hybrid Polypropylene Polymer Composites: Fabrication and Analysis of Mechanical and Water Absorption Properties," *Materials Today: Proceedings*, Vol. 33, No. xxxx, 2020, pp. 5273–5278. https://doi.org/10.1016/j.matpr.2020.02.964

[17] Yorseng, K., Rangappa, S. M., Parameswaranpillai, J., and Siengchin, S., "Influence of Accelerated Weathering on the Mechanical, Fracture Morphology, Thermal Stability, Contact Angle, and Water Absorption Properties of Natural Fiber Fabric-Based Epoxy Hybrid Composites," *Polymers*, Vol. 12, No. 10, 2020, pp. 1–14. https://doi.org/10.3390/polym12102254

[18] Sekar, S., Suresh Kumar, S., Vigneshwaran, S., and Velmurugan, G., "Evaluation of Mechanical and Water Absorption Behavior of Natural Fiber-Reinforced Hybrid Biocomposites," *Journal of Natural Fibers*, 2020, pp. 1–11. https://doi.org/10.1080/15440478.2020.1788487

[19] Venkatasudhahar, M., Ravichandran, A. T., and Dilipraja, N., "Effect of Stacking Sequence on Mechanical and Moisture Absorption Properties of Abaca-Kenaf-Carbon Fiber Reinforced Hybrid Composites," *Journal of Natural Fibers*, Vol. 19, No. 13, 2022, pp. 1–12. https://doi.org/10.1080/15440478.2021.1944434

[20] Sanjeevi, S., Shanmugam, V., Kumar, S., Ganesan, V., Sas, G., Johnson, D. J., Shanmugam, M., Ayyanar, A., Naresh, K., Neisiany, R. E., and Das, O., "Effects of Water Absorption on the Mechanical Properties of Hybrid Natural Fibre/Phenol Formaldehyde Composites," *Scientific Reports*, Vol. 11, No. 1, 2021, pp. 1–11. https://doi.org/10.1038/s41598-021-92457-9

[21] Nasimudeen, N. A., Karounamourthy, S., Selvarathinam, J., Kumar Thiagamani, S. M., Pulikkalparambil, H., Krishnasamy, S., and Muthukumar, C., "Mechanical, Absorption and Swelling Properties of Vinyl Ester Based Natural Fibre Hybrid Composites," *Applied Science and Engineering Progress*, Vol. 14, No. 4, 2021, pp. 680–688. https://doi.org/10.14416/j.asep.2021.08.006

[22] Ghori, S. W., Rao, G. S., and Rajhi, A. A., "Investigation of Physical, Mechanical Properties of Treated Date Palm Fibre and Kenaf Fibre Reinforced Epoxy Hybrid Composites," *Journal of Natural Fibers*, Vol. 20, No. 1, 2023, pp. 1–12. https://doi.org/10.1080/15440478.2022.2145406

[23] Bekele, A. E., Lemu, H. G., and Jiru, M. G., "Study of the Effects of Alkali Treatment and Fiber Orientation on Mechanical Properties of Enset/Sisal Polymer Hybrid Composite," *Journal of Composites Science*, Vol. 7, No. 1, 2023, pp. 2–11. https://doi.org/10.3390/jcs7010037

[24] Ornaghi, H. L., da Silva, H. S. P., Zattera, A. J., and Amico, S. C., "Hybridization Effect on the Mechanical and Dynamic Mechanical Properties of Curaua Composites," *Materials Science and Engineering: A*, Vol. 528, No. 24, 2011, pp. 7285–7289. https://doi.org/10.1016/j.msea.2011.05.078

[25] Boopalan, M., Niranjanaa, M., and Umapathy, M. J., "Study on the Mechanical Properties and Thermal Properties of Jute and Banana Fiber Reinforced Epoxy Hybrid Composites," *Composites Part B: Engineering*, Vol. 51, 2013, pp. 54–57. https://doi.org/10.1016/j.compositesb.2013.02.033

[26] Gupta, M. K., Ramesh, M., and Thomas, S., "Effect of Hybridization on Properties of Natural and Synthetic Fiber-reinforced Polymer Composites (2001–2020): A Review," *Polymer Composites*, Vol. 42, No. 10, 2021, pp. 4981–5010. https://doi.org/10.1002/pc.26244
[27] Deo, C., and Acharya, S. K., "Effect of Moisture Absorption on Mechanical Properties of Chopped Natural Fiber Reinforced Epoxy Composite," *Journal of Reinforced Plastics and Composites*, Vol. 29, No. 16, 2010, pp. 2513–2521. https://doi.org/10.1177/0731684409353352
[28] Mahendrakumar, N., Thyla, P. R., Mohanram, P. V., Sabareeswaran, A., Manas, R. B., and Srivatsan, S., "Mechanical and Dynamic Properties of Nettle-Polyester Composite," *Materials Express*, Vol. 5, No. 6, 2015, pp. 505–517. https://doi.org/10.1166/mex.2015.1263
[29] Kord, B., and Roohani, M., "Water Transport Kinetics and Thickness Swelling Behavior of Natural Fiber-Reinforced HDPE/CNT Nanocomposites," *Composites Part B: Engineering*, Vol. 126, 2017, pp. 94–99. https://doi.org/10.1016/j.compositesb.2017.06.008
[30] Gholampour, A., and Ozbakkaloglu, T., "A Review of Natural Fiber Composites: Properties, Modification and Processing Techniques, Characterization, Applications," Springer US, 2020. https://doi.org/10.1007/s10853-019-03990-y
[31] Patel, R. V., Yadav, A., and Winczek, J., "Physical, Mechanical, and Thermal Properties of Natural Fiber-Reinforced Epoxy Composites for Construction and Automotive Applications," *Applied Sciences (Switzerland)*, Vol. 13, No. 8, 2023. https://doi.org/10.3390/app13085126
[32] Alves Fidelis, M. E., Pereira, T. V. C., Gomes, O. D. F. M., De Andrade Silva, F., and Toledo Filho, R. D., "The Effect of Fiber Morphology on the Tensile Strength of Natural Fibers," *Journal of Materials Research and Technology*, Vol. 2, No. 2, 2013, pp. 149–157. https://doi.org/10.1016/j.jmrt.2013.02.003
[33] Zhang, Z., Cai, S., Li, Y., Wang, Z., Long, Y., Yu, T., and Shen, Y., "High Performances of Plant Fiber Reinforced Composites—A New Insight from Hierarchical Microstructures," *Composites Science and Technology*, Vol. 194, No. November 2019, 2020, p. 108151. https://doi.org/10.1016/j.compscitech.2020.108151

7 Utilizing Artificial Intelligence for Characterization and Quantification of Damage in Fiber Reinforced Composites

Advanced Techniques and Applications

Alok Behera, Tiju Thomas, Janaki Dehury, and Kishore Kumar Mahato

7.1 INTRODUCTION

21st-century fiber reinforced polymer composites (FRPCs) have supplanted traditional materials across various industries. The definition says, "A composite material contains two or more distinct constituents with significantly different macroscopic behavior and a distinct interface between each constituent" [1]. The matrix transfers stresses between reinforcement fibers/particles and shields them from damage, while the fibers/particles enhance mechanical properties like strength and stiffness. This enhanced performance has spurred rapid adoption in the automotive, construction, aerospace, marine, sports equipment, defense, and electrical industries [2–4]. FRPCs have been utilized since the early 20th century, initially embraced by aerospace and defense industries for aircraft structures, bulletproof jackets, spacecraft, missiles, tanks, and lightweight armor. Leading aerospace manufacturers such as Boeing and Airbus employ advanced synthetic fiber composites in over 50% of their aircraft structures. These materials contribute to a remarkable 25% increase in fuel efficiency by reducing weight, streamlining maintenance efforts, and enhancing overall performance. The Airbus A350 XWB notably integrates 53% FRPCs, showcasing tangible performance and fuel efficiency improvements. Moreover, India's indigenous fighter aircraft, LCA Tejas, and Boeing 787 domestic aircraft employ FRPCs for 40% and 50% of their structure, respectively [5,6]. The volume fraction of composites in aerospace applications has been steadily increasing, with components like wings, fuselages, and empennages incorporating higher percentages [7–10].

In the late 20th and early 21st centuries, the automotive industry began adopting FRPCs to reduce vehicle weight and improve fuel efficiency. Carbon fiber composites became popular in high-performance sports cars, racing vehicles, luxury automobiles, and sports equipment due to their lightweight and high-strength characteristics, such as carbon fiber bicycle frames, tennis rackets, golf clubs, and fishing rods [11–14]. In recent years, FRPCs have been extensively used in infrastructure and construction projects like bridges, building facades, reinforcement of concrete

 DOI: 10.1201/9781003664611-8

structures, and seismic retrofitting. The renewable energy industry uses these in wind turbine blades, solar panels, consumer electronics, and gadgets.

Similarly, FRPCs are finding applications in the medical field for orthopedic implants, prosthetics, and dental materials due to their biocompatibility, strength, and lightweight properties [15]. Overall, the use of FRPCs continues to expand across various industries and applications as researchers and engineers develop new materials, manufacturing techniques, and design methodologies further to enhance their performance, durability, and cost-effectiveness. The widespread utilization of FRPCs underscores the critical need for continuous structural health and damage monitoring [16]. Artificial intelligence (AI) integration has become pivotal in identifying significant damages within composite materials, estimating their mechanical properties, and analyzing micrographs and results through AI-based post-processing techniques. The emergence of several in-situ microstructural characterization machines equipped with built-in AI tools represents a significant advancement in this field, pushing capability boundaries to new heights. This chapter encompasses the recent strides made in implementing AI and data science methodologies in damage analysis, mechanical testing, and data post-processing relevant to FRPCs.

7.2 TYPES OF DAMAGE IN FRPCS

FRPCs are susceptible to damage based on manufacturing faults, environmental exposure, and loading conditions. In FRPCs, common damage types include the following:

i. **Delamination:** Delamination is the separation of layers within a composite structure, as shown in Figure 7.1(a). It happens when there is a breakdown in the fiber layers (laminae) and matrix interface, which reduces mechanical properties and structural integrity [17–20].
ii. **Fiber Breakage:** When exposed to intense loads, the reinforcing fibers within the composite fracture, as depicted in Figure 7.1(b). High tensile or compressive stresses higher than the fiber's strength can cause this damage, resulting in localized failure and a reduction in the fiber's ability to bear loads [21–23].
iii. **Matrix Cracking:** Matrix cracking is the process of fissures appearing encircling the reinforcing fibers, as shown in Figure 7.1(c). Tensile, compressive, or shear stresses, environmental variables, and thermal cycling can all result in matrix cracking. It may result in the composite losing some of its strength, stiffness, and durability [24–26].
iv. **Fiber–Matrix Debonding:** This results from a weakening or loss of the link between the fibers and the matrix around it. Concentrated interfacial tension, heat cycling, moisture absorption, or manufacturing flaws can lead to this damage, as shown in Figure 7.1(d). Reduced fiber-to-matrix load transfer caused by fiber–matrix debonding may result in diminished mechanical qualities and delamination [27,28].
v. **Impact Damage:** Impact damage results from sudden, high-energy stresses applied to the composite material, such as collisions or impacts from foreign objects. Depending on the impact's force and length, its damage may appear as indentation, cracking, delamination, fiber breakage, or a combination of many, like through holes, as shown in Figure 7.1(e) [29,30].
vi. **Environmental Degradation:** FRPCs eventually degrade as they are exposed to environmental factors in the actual working conditions, such as moisture, UV light, temperature changes, and chemical exposure. Degradation of the environment can cause changes in the qualities of materials, such as decreased stiffness, strength, and resistance to corrosion and fatigue [31–33]. Studies indicate that prolonged exposure to high temperatures of moisture aging, known as hygrothermal aging, can result in fiber–matrix interfacial debonding and matrix removal, even without any external mechanical loading. These defects cause

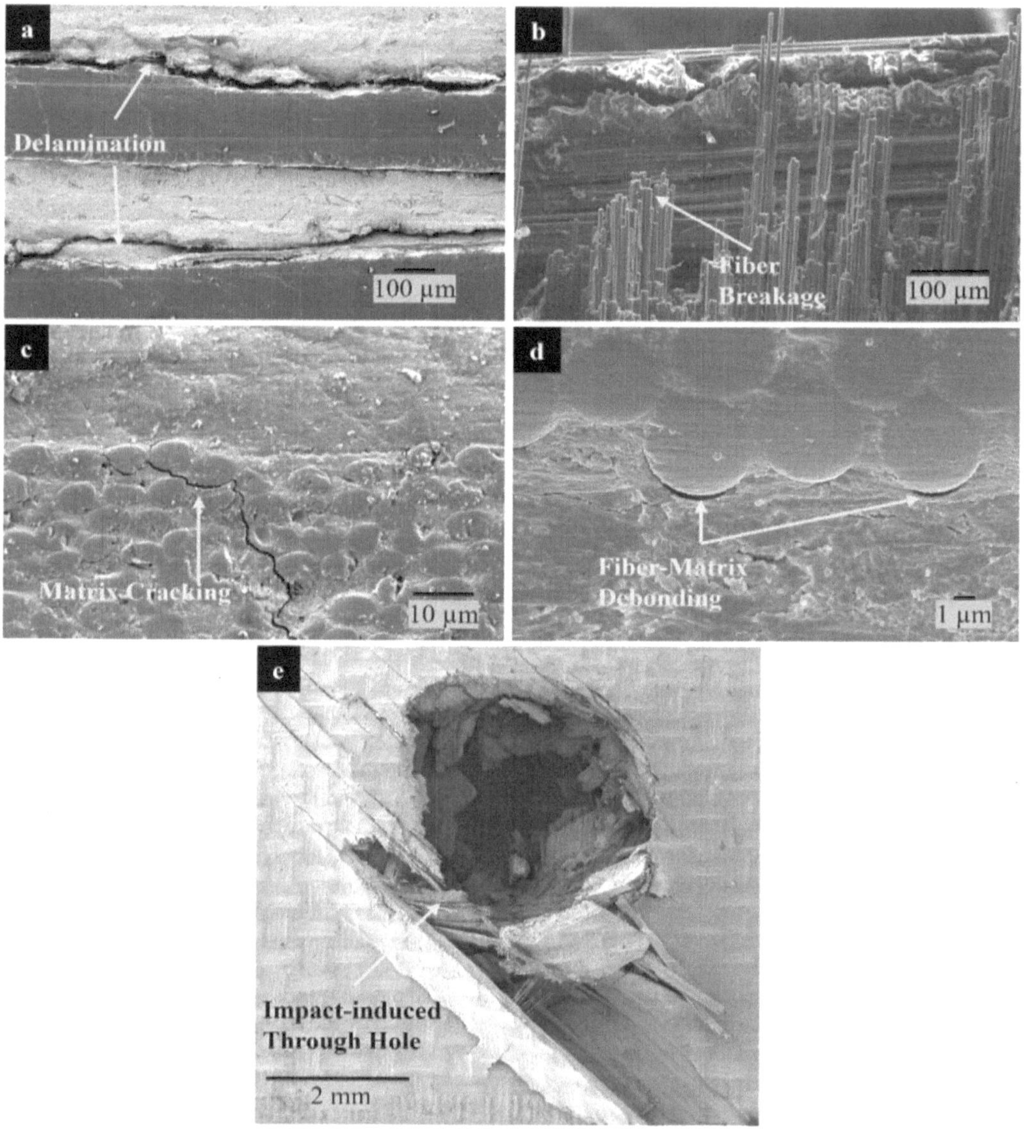

FIGURE 7.1 SEM micrographs of composite damages: (a) delamination, (b) fiber breakage, (c) matrix cracking, (d) fiber–matrix debonding, and (e) impact-induced through hole.

irreversible property degradation and may contribute to premature failure under actual loading conditions [33–35].

vii. **Fatigue Damage:** When FRPCs are repeatedly loaded cyclically, fatigue damage develops, gradually deteriorating mechanical properties. Delamination, stiffness, and strength loss, matrix cracking, and fiber breaking are examples of fatigue damage. To evaluate structural integrity, forecast performance under various operating conditions, and put appropriate maintenance and repair strategies into place, it is imperative to comprehend the damage that FRPCs can sustain [35]. The overall strength under fatigue loading is contingent upon the alignment of fibers with the loading direction and the severity of the fatigue stress ratio. Preexisting damage, such as free volume, defects at the time of molding, micro-voids, inadequate fiber–matrix bond strength, or interfacial debonding, can become critical under cyclic loading, leading to significant failure mechanisms such as delamination, ultimately resulting in complete structural failure [34].

The fusion of conventional characterization methods with cutting-edge AI and machine learning (ML) approaches presents significant potential for deepening our comprehension of damages in FRPCs and streamlining damage quantification procedures across diverse applications and industries.

7.3 EXPERIMENTAL CHARACTERIZATION TECHNIQUES FOR POLYMER COMPOSITES WITH AI/ML ML-DRIVEN ADVANCEMENTS

Experimental characterization techniques for composites amenable to ML and AI include acoustic emission (AE) measurements, infrared and laser thermography, nanoindentation tests, and physical experiments to characterize stress–strain curves. ML algorithms and AI models can be trained using the data acquired from these techniques. Even the ML algorithms can predict stress–strain curves of polymer composites based on composition, processing, and environmental conditions. For example, Kek et al. studied the application of fiber optic sensors to measure AE in biocomposites [36]. Neural networks and extreme learning machines ML methods were used to categorize the source material based on the extracted feature. This section delved into numerous techniques and models that contribute to the progress of automated design and characterization of composites.

7.3.1 Digital Image Correlation and Machine Learning

Structural health monitoring (SHM) employing digital image correlation (DIC) and ML is undergoing a significant transformation due to the advancements in drone technology. Traditionally, DIC required cumbersome set-ups with fixed cameras, limiting its applicability to easily accessible areas. However, the emergence of high-resolution cameras integrated with drones has revolutionized data acquisition for DIC. Drones offer unprecedented flexibility, enabling inspection of large, complex structures and hard-to-reach places like bridges, dams, and wind turbines with minimal human intervention [37].

Integrating drones with DIC facilitates the acquisition of high-resolution images from various angles and distances, enhancing the accuracy and comprehensiveness of strain measurements. When used with ML algorithms, these can provide a wealth of information for damage detection and assessment [38,39]. For instance, researchers have used drone-based DIC to detect and quantify cracks in concrete structures. Moreover, ML algorithms can leverage the spatiotemporal data from drone-captured DIC images to track the evolution of damage over time, enabling predictive maintenance strategies [40].

The synergy of DIC, ML, and drone technology has the potential to transform SHM into a proactive and efficient process [41]. Drones can autonomously inspect structures at regular intervals, collecting DIC data that ML algorithms can analyze for early signs of damage. This early detection allows for timely repairs, preventing catastrophic failures and extending the lifespan of critical infrastructure. While integrating drones with DIC and ML offers numerous advantages, challenges remain. Ensuring the stability and accuracy of DIC measurements from a moving drone platform requires robust image stabilization and processing techniques. Additionally, the large volumes of data generated by drone-based DIC necessitate efficient data management and workflow processing. Nevertheless, ongoing research addresses these challenges and pushes the boundaries of SHM.

7.3.2 Acoustic Emission Measurement

The non-destructive AE measurement technique enables the detection and analysis of various composite damage phenomena by capturing and analyzing AE parameters such as the number of hits, counts, amplitude, and signal energy [42]. Piezoelectric sensors can be embedded in CFRP to continuously measure AE without disrupting hydro- or aerodynamics. These embedded sensors can accurately identify AE sources and measure degradation in CFRP specimens under different loading conditions. AE signal processing techniques, like fast Fourier transform (FFT) and wavelet packet transform (WPT), can be used to analyze AE signals and identify different failure modes in composite materials [43]. To improve comprehension of composite behavior and structural integrity, AE measurements can offer important information regarding failure mechanisms and stress limits [44].

AE measurement and ML techniques are used in composite materials for defect detection and damage identification. AE signals generated from failure processes in composites can be classified using ML methods [45]. Different metrics, such as signal energy, can be utilized to detect defects in composite specimens [46]. When applied to AE datasets, ML techniques like convolutional neural networks (CNNs) and long short-term memory (LSTM) neural networks provide good results for defect detection and energy prediction [47]. Different damage modes in CFRP composites can be identified by wavelet packet analysis of AE signals and ML methods such as support vector machine (SVM) [42]. ML has employed unsupervised clustering techniques to classify AE occurrences based on temporal factors, such as k-means clustering, which has shed light on the damage mechanisms in composite materials.

In their research, Qiao et al. explored using AE signals and ML techniques to comprehend damages in CFRP composite subjected to bending loads [47]. They used wavelet packet analysis to identify four main damage modes identified by specific frequency bands in the AE signals: matrix cracking, delamination, fiber/matrix debonding, and fiber breaking. Furthermore, they developed a SVM classifier to categorize these signals. Evaluation conducted on pre-cracked and untreated specimens demonstrated the SVM classifier's promising efficacy in monitoring the health of composite structures, exhibiting considerable similarity to outcomes obtained through the k-means algorithm. Similarly, Rubio-González et al. explored the combined use of AE and self-sensing enabled by carbon nanotube networks to monitor damage progression in glass fiber epoxy composites under flexural loading, as demonstrated in Figure 7.2(a) and (b) [48]. Acoustic signals recorded during testing were analyzed using k-means and principal component analysis (PCA) to identify damage mechanisms like matrix cracking, fiber/matrix debonding, delamination, and fiber breakage. Additionally, changes in electrical resistance signal and selected AE parameters captured specific failure events during composite damage evolution.

7.3.3 Infrared and Laser Thermography

Infrared thermography is a non-contact technique that can determine temperature during curing and defects in composite by monitoring variations in emissivity and thermal diffusivity [49]. It can be used for wrinkle detection in composite GFRP structures, regardless of the type of fabric weave [50]. Internal defects in aircraft composite structures can also be detected using infrared thermal wave imaging technology [51]. Thermal non-destructive testing is also used for impact damage detection in polymer composite materials, considering the presence of a coating, the type of thermal loading source, and software proficiency. Additionally, infrared thermography is used to analyze FRPC curing mechanisms at the microscale, providing insights into the influence of fillers on the curing process [52]. Using infrared pulsed thermography to assess composite damage composite materials, thermal datasets are analyzed using multi-label classification algorithms.

Laser-based non-destructive testing of composites involves using laser thermography to detect defects in CFRP samples with precision and efficiency. This technique utilizes controlled heat excitation via a laser, enabling accurate measurements, and data processing. Automatic defect identification is achieved through regression methods and clustering algorithms [53]. Laser-line

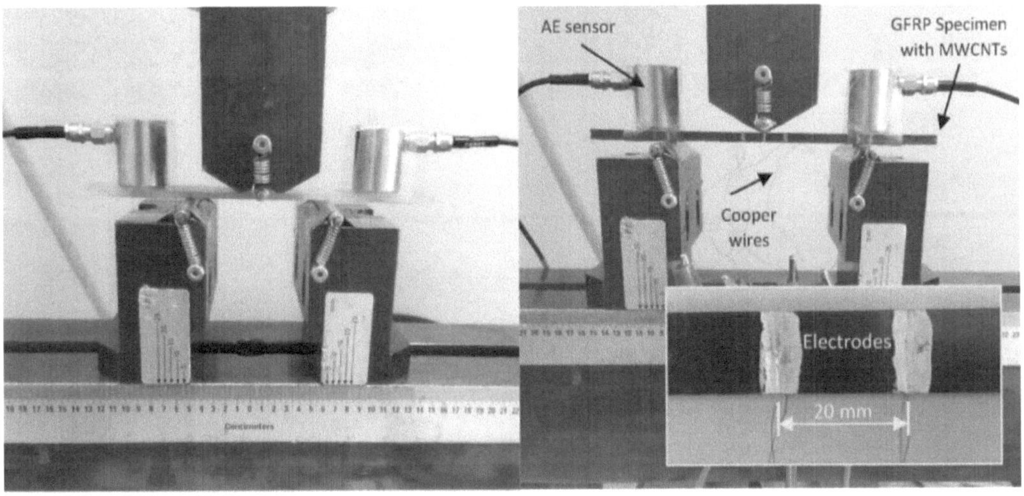

(a) (b)

FIGURE 7.2 Bending experimental set-up, (a) AE monitoring on samples without CNT, (b) AE and change of resistance monitoring on samples with CNT. (Adapted with permission form Rubio-González et al. [48].)

thermography is compared with lock-in thermography to enhance defect identification and imaging quality. Additionally, a non-contact approach employing guided waves and high-frequency bulk waves is proposed for inspecting carbon fiber reinforced structures, reducing sample preparation time, and ensuring precise damage characterization [54]. Rahman et al. used ML algorithms to analyze thermal images of graphite/graphene-based polymer composites, identifying filler size, shape, and location and facilitating quantification of sample dispersion [49]. Figure 7.3 illustrates thermal images processed manually to remove artifacts or blurriness that act as ML model input to determine true fabrication for various composite samples Figure 7.3(a–d).

Image processing methods are used to analyze thermograms and extract features for classification purposes, utilizing ML techniques like multilayer perceptron and SVM. Inverse heat transfer problems can be solved using the k-nearest neighbors (kNN) ML algorithm to predict composite defects. Multi-label classification algorithms, such as random forest, can handle thermal datasets associated with different data characteristics and detect damaged composite materials [55]. ML-integrated pulsed thermography can detect and classify composite structural damage, providing a reliable estimate of material conditions.

7.3.4 Electrical Resistance Tomography

Electrical resistance tomography (ERT) is a non-destructive evaluation technique [56] used to assess structural integrity by measuring internal conductivity variations. ERT technique has been employed to monitor damage accumulation in CFRP aircraft components. It is difficult to differentiate between various damage modes in CFRP laminates, such as delamination and matrix cracking, using ERT alone. To address this, ML algorithms such as neural networks, random forests, kNN, and SVMs have been combined with ERT to classify and characterize delamination and matrix cracking damage in CFRP laminates with high accuracy [57]. Low spatial and contrast resolution pictures can be recovered using electrical impedance tomography (EIT), a less expensive and non-invasive imaging method.

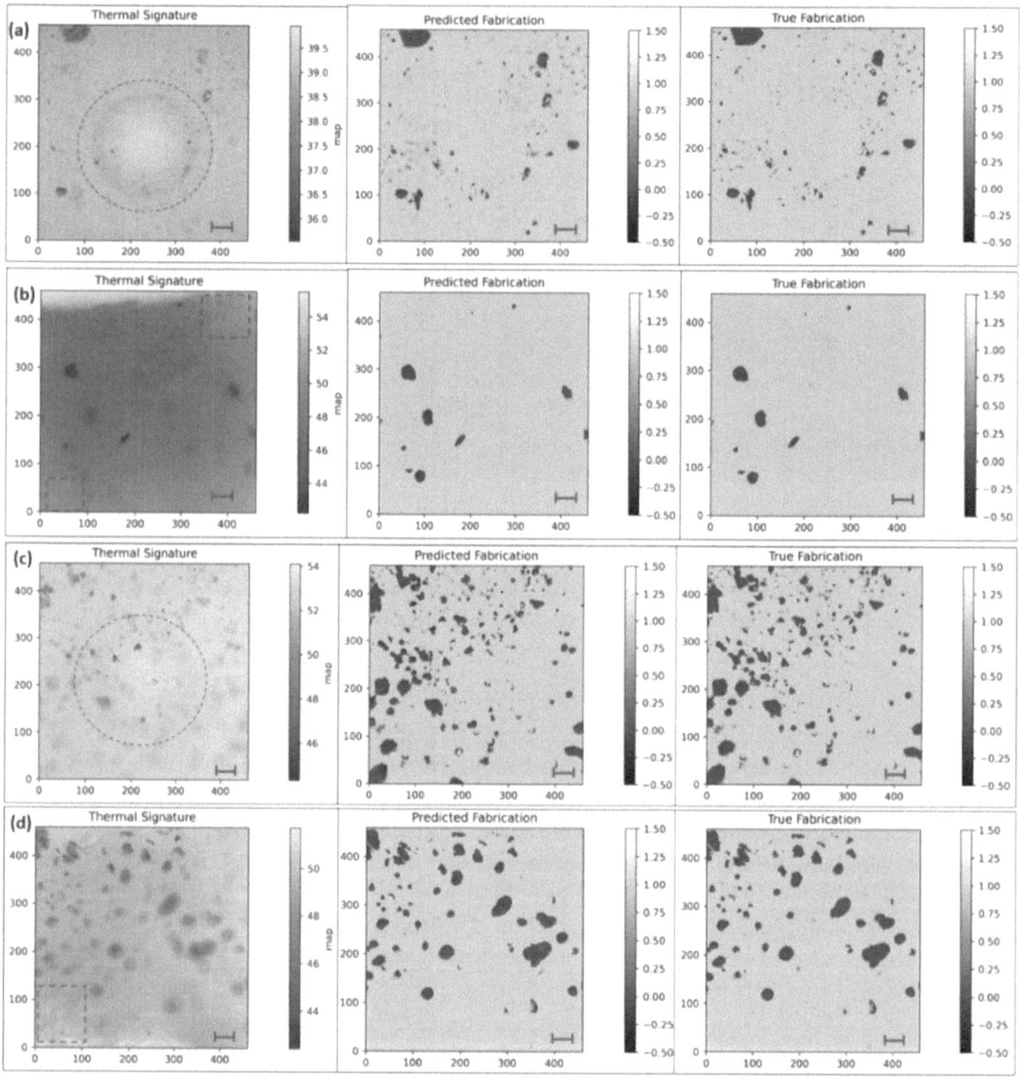

FIGURE 7.3 (a–d) Thermal images processed to eliminate artifacts, then analyzed by an ML model to identify true fabrication across different composite samples. (Adapted from Rahman et al. [49] (Open access).)

7.3.5 Nanoindentation

Nanoindentation is a valuable method for determining the hardness and elastic modulus of the composite at the nanoscale [58,59]. It allows the measurement of hardness and elastic modulus properties on a nanoscale level. This is also used for nanomechanical characterization of silica nanoparticles, carbon nanotubes, or nanofibers FRPC. Nanoindentation has been used to determine the distribution of hardness and modulus and evaluate the interfacial properties [60]. Jeong et al. used a finite element-ANN (FE-ANN) modeling technique to predict uniaxial tensile curves using actual indentation data. The procedure for obtaining the tensile curve from experimental indentation data is shown in Figure 7.4. ANN, which links input data representing monotonic loading portions of the load-depth curve, is trained using simulation results to extract material flow parameters [61].

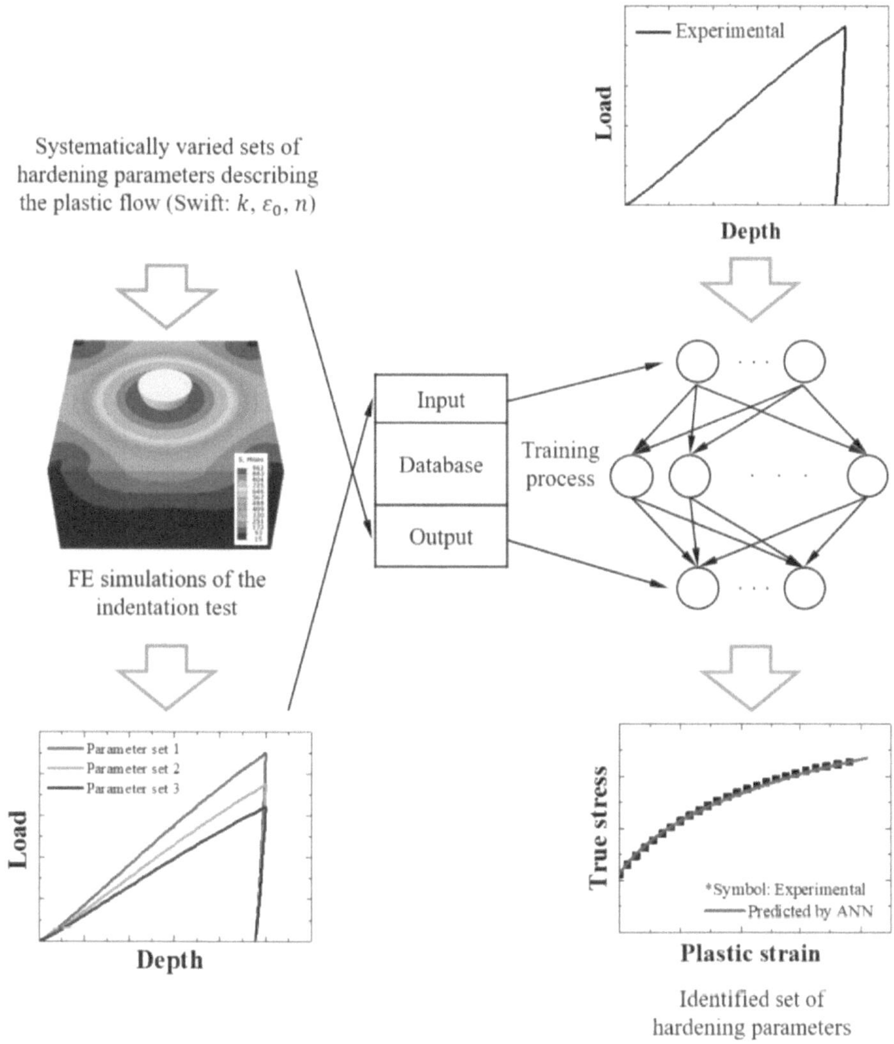

FIGURE 7.4 Procedure for deriving uniaxial tensile flow from experimental indentation data using. (Adapted from Jeong et al. [61] (Open access).)

Nanoindentation combined with ML has been used for composite characterization [62,63]. These techniques were used to investigate the effects of nanofiller hybridization on mechanical properties reduction of acid-aged specimens [64]. ML models trained on nanoindentation parameters have also been used to model the resistance to plastic deformation in CFRPs [62]. Furthermore, a combined approach of electrostatic force microscopy (EFM) and ML has been used to measure interfacial permittivity in polymer nanocomposites [65].

7.3.6 Bending, Compression, and Tension Tests

Bending, compression, and tension tests are commonly used for the mechanical characterization of polymer composites. For example, Duan et al. proposed a stress–strain constitutive model to describe polymer-bonded composite's tension and compression behavior at different temperatures [66]. Similarly, Hao et al. investigated the non-linear behavior of FRPCs under transverse loading and studied the thermomechanical response of the polymers using advanced glassy polymer models [67]. Researchers have focused on characterizing FRPCs under compressive force and examined several process parameters that impact compressive strength. Tensile and compressive tests were used by Malalli et al. to assess the elastic and strength characteristics of flax, jute, and hemp FRPCs [68]. These studies highlight the importance of bending, compression, and tension tests in characterizing the mechanical behavior of polymer composites.

In addition, ANN ML algorithms have been developed to predict stress–strain curves of polymer composites based on their composition, processing, and environmental conditions [69]. ML algorithms have also been used to predict material behavior, optimize composite properties, and conduct sensitivity analyses. Material scientists can reap significant benefits by applying ML to forecast and characterize the properties of FRPCs according to their components, production methods, and anticipated lifetime [70].

7.4 POST-PROCESSING USING SEM, TEM, AND AFM

Over the past five years, AI and ML have surged in identifying defects in FRPCs, presenting numerous prospects for composite companies. These advancements enable detailed microstructure analysis, input parameter optimization, resource utilization, and labor-intensive task automation. Exploiting vast datasets, often petabyte-scale, this section explores AI/ML-driven automation in SEM, focused ion beam SEMs (FIB-SEM), TEM, and scanning TEM (STEM).

7.4.1 AI-Driven Advancements in SEM

AI-driven SEM, utilizing advanced technologies like deep learning for score prediction, has streamlined operations by reducing operator control parameters that impact micrograph quality. Image processing automation has addressed focus control, live feed clarity during stage movement, and in-situ surface analysis. Unlike autofocus techniques with limitations, AI-SEM ensures optimal focus, brightness, and contrast across various specimen types. Integration of AI/ML has primarily impacted system calibration and process automation, while data analytics continue to undergo extensive research and development [15].

System calibration involves maintaining and enhancing electron microscope performance, including tool alignment, predictive maintenance, fleet management, and image optimization. Tool alignment utilizes computer vision and algorithms for high-quality data capture [71,72]. This calibration significantly improves the structural characterization of polymer matrix composites, polymers, and fibers. Process automation boosts productivity by optimizing metrology, data assumption, and sample preparation. Automated workflows, end-pointing, and ROI navigation increase accuracy and efficiency. Technologies like JEOL's LIVE-AI enhance live image quality and focus [73–76]. Quijada-Pioquinto et al. applied a neural network to SEM micrographs for accurate fiber counting and structural analysis [77]. The Mask R-CNN neural network specializes in computer vision tasks like object identification and segmentation [78–83]. Wu et al. explored UD FRPCs' elastic properties using the stochastic multiscale approach on SEM microstructures [82].

7.4.2 AI-Driven Advancements in TEM

Integrating AI is promising to enhance TEM across various domains, including image processing, artifact detection, automated data interpretation, real-time monitoring, data fusion, and accelerating image acquisition [83]. Trained deep learning models efficiently recognize structures and patterns within TEM micrographs of FRPCs, facilitating rapid and precise analysis. AI effectively tackles common artifacts like charging effects, specimen drift, and beam-induced damage while enabling automatic defect detection and analysis of defect densities. Moreover, AI algorithms enable

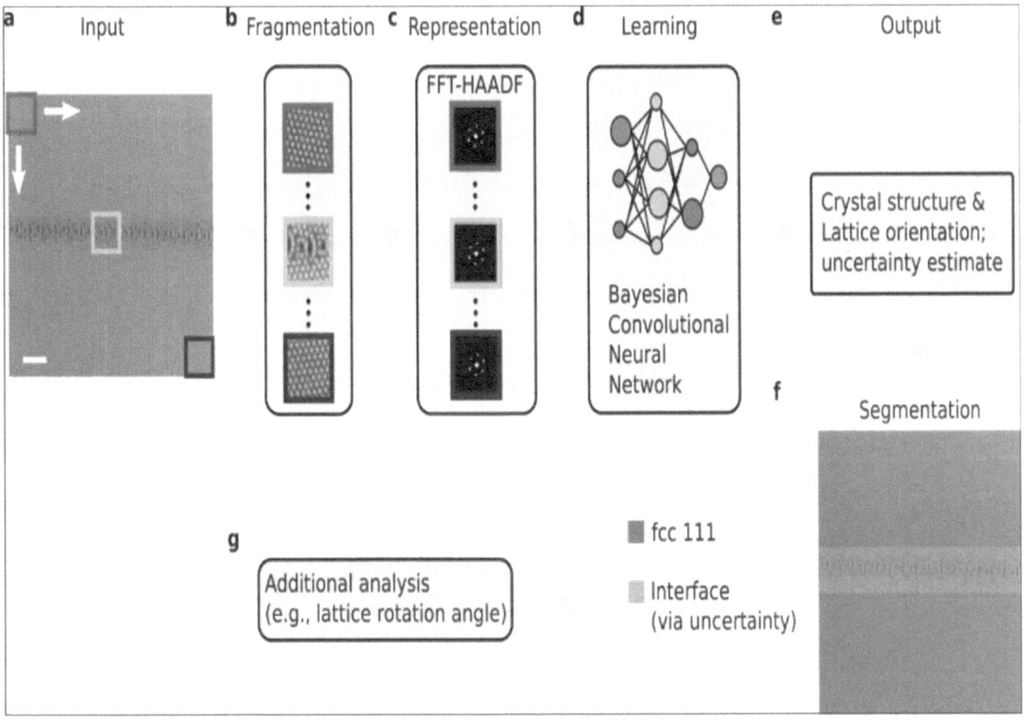

FIGURE 7.5 Schematic of the AI-STEM for analyzing experimental STEM image showing (a) Input, (b) Fragmentation, (c) Representation, (d) Learning, (e) Output, (f) Segmentation, and (g) Additional Analysis. (Adapted from Leitherer et al. [85] (Open access).)

real-time monitoring and control of TEM experiments by adjusting imaging parameters based on live data streams, preemptively addressing potential issues. Techniques such as generative adversarial networks (GANs) and compressed sensing accelerate image acquisition by reducing required measurements or reconstructing high-quality images from sparse data, streamlining TEM analyses [84]. Leitherer et al. introduced an automated AI technique, AI-STEM, which extracts meaningful information from atomic-resolution micrographs of polycrystalline materials. Using a Bayesian CNN, the model recognizes crystal structures, lattice orientation, and interface positions in synthetic and experimental micrographs. It can recognize interfaces, crystal flaws, and bulk regions, and its segmentation capabilities allow the examination of local lattice orientation, as shown in Figure 7.5 [85]. Olszta et al. proposed a closed-loop instrument control platform guided by emerging sparse data analytics [86]. Their concept envisions a centralized controller empowered by ML insights, enabling real-time experimental decision-making, automating material feature analysis, and unlocking possibilities for high-throughput research. The control system architecture, depicted in Figure 7.6, consists of three main components: (a) open-loop control, generating search grid data; (b) few-shot ML feature classification for potential use; and (c) closed-loop control, facilitating complete automation.

7.4.3 AI-Driven Advancements in Atomic Force Microscopy

Atomic force microscopy (AFM) offers exceptional resolution down to the atomic scale, enabling researchers to observe materials in three dimensions. Using a scanning probe with a sharp tip fixed to a cantilever, AFM captures nanoscale details in various materials. By detecting the movement of the cantilever at sub-nanometer resolution, AFM utilizes surface scanning techniques to measure

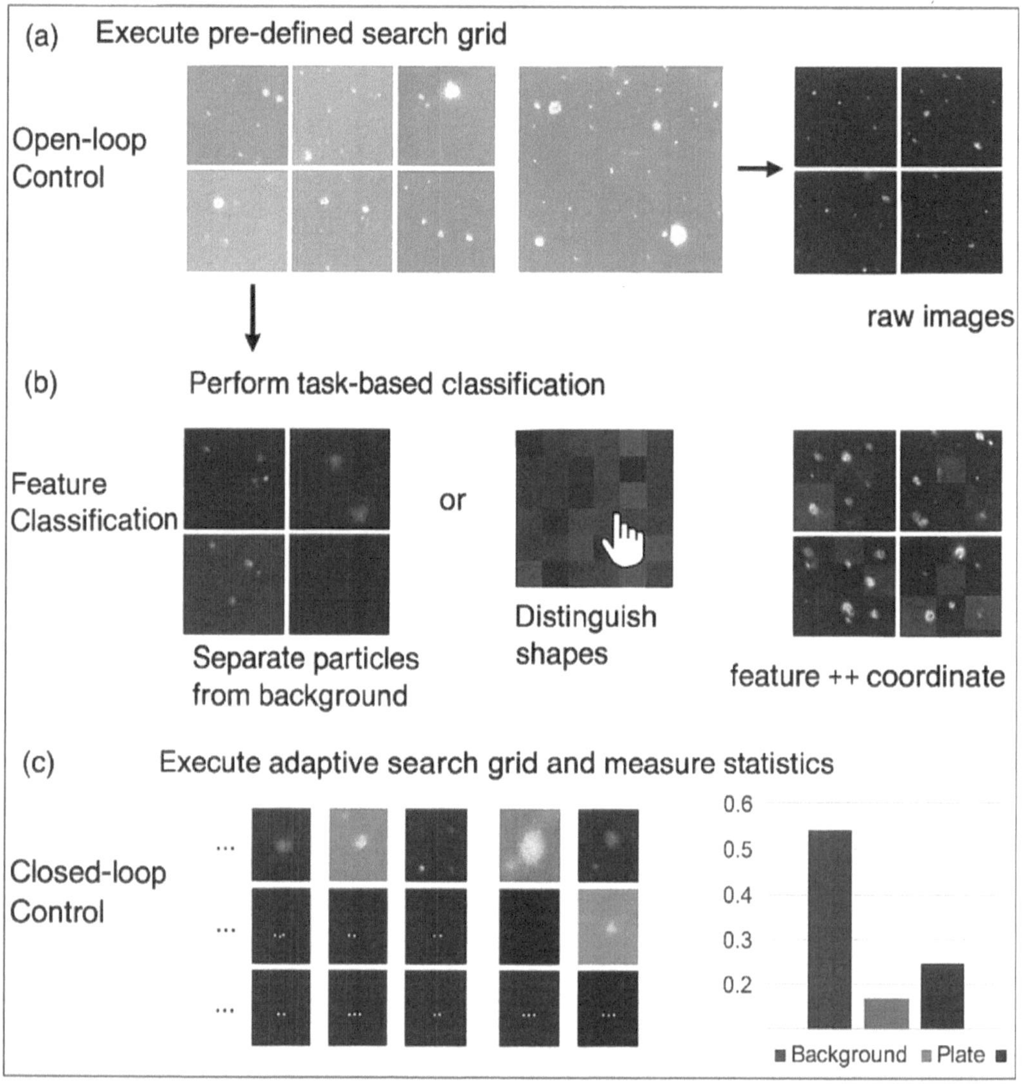

FIGURE 7.6 Control system architecture with data acquisition and processing, showing (a) Open-loop Control, (b) Feature Classification, (c) Closed-loop Control. (Adapted from Olszta et al. [86] (Open access).)

atomic forces. However, AFM is limited to measuring adhesion strength, magnetic forces, and mechanical properties. Additionally, AFM allows the examination of nanostructure properties such as frictional, nanomechanical, electrical, magnetic, and thermal characteristics. Traditional AFM faces challenges, including limited scanning areas, slower scanning speeds, thermal drift in acquired micrographs, high human interference, time-consuming operation, and the need for highly skilled operators. Compared to SEM, AFM has a restricted scanning area of only 10–20 micrometers. In contrast, SEM can scan an area of 150 ×150 micrometers in near real-time, albeit with lower resolution. The slower scanning speed of AFM leads to thermal drift in micrographs, potentially affecting the accuracy of distance measurements between topographical elements.

Through ML training, neural networks can learn various aspects of image recognition. For instance, initial layers may focus on recognizing edges, while deeper layers discern specific facial features like eyes or noses, ultimately enabling individual identification through images. This

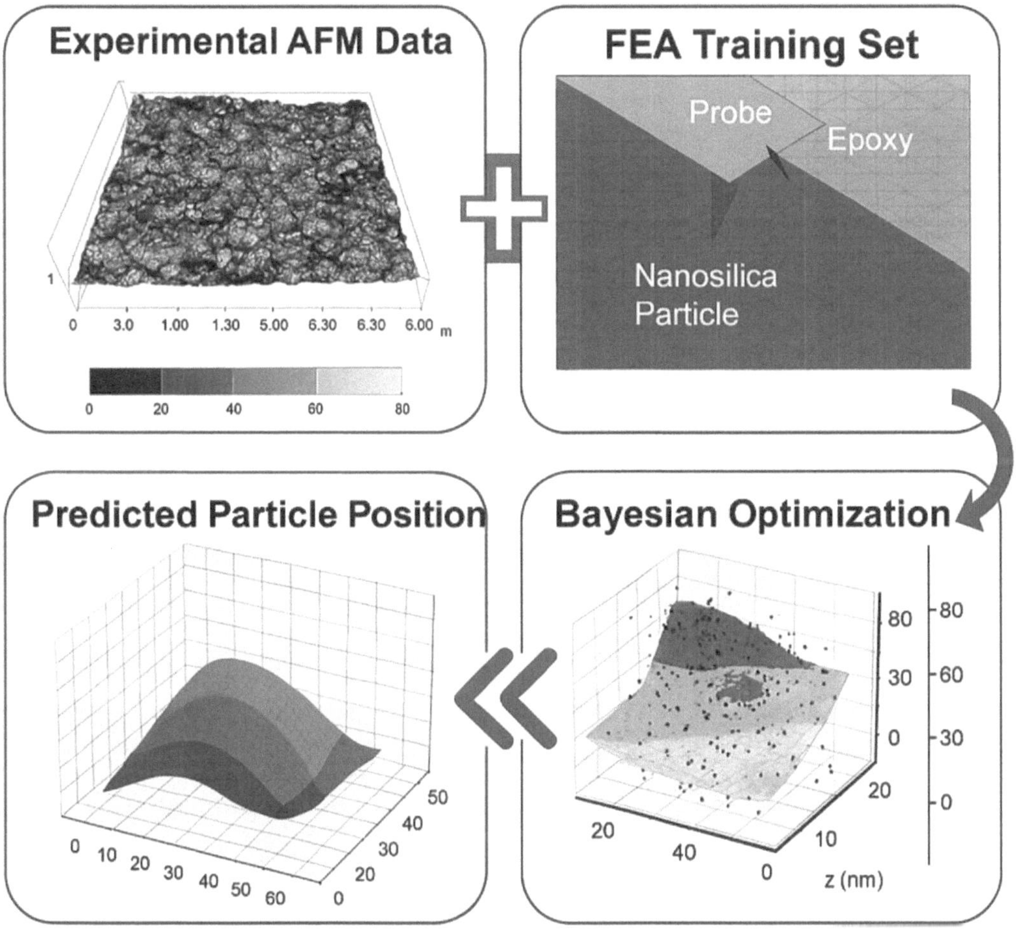

FIGURE 7.7 Schematic of integrating AFM and data science to locate nanoparticles in FRPCs. (Adapted with permission from Thiem et al. [88].)

approach can also enhance AFM imaging capabilities, improving analysis and interpretation for polymer matrix composites. Deep learning algorithms facilitate the development of autonomous AI-computer systems, reducing the need for constant human intervention [87–89]. Trained to interpret data, computers analyze and derive insights independently, efficiently detecting and classifying objects within AFM images and identifying specific nanoscale features within advanced composites. Huitian et al. utilized a deep learning algorithm to automatically recognize nanowires in AFM images, enabling the division of nanowires into abstracted polygonal line segments suitable for AFM probe manipulation [87]. Thiem et al. recently developed a method combining data science, continuum micromechanics, and experimental AFM figures to locate 142 nm nano silica particles in 3D with respect to an abraded surface, as shown in Figure 7.7 [88].

7.5 MICROSCOPY DATA ANALYSIS AND MODELING TECHNIQUES

This section thoroughly summarizes the approaches used to analyze microscope data using AI and ML algorithms. It covers advanced image analysis, pattern recognition, and predictive modeling, emphasizing their importance in extracting insights from microscopy images and understanding the samples under study.

7.5.1 Image Analysis

AI-enabled image analysis transforms the characterization and quantification of polymer matrix composites, enhancing accuracy, speed, and reproducibility. AI systems use advanced algorithms to analyze composite images, extracting valuable data like fiber orientation, distribution, and morphology. This technology enables optimization of manufacturing processes, customization of material properties, and acceleration of product development cycles, driving innovation in composite materials. Popular tools like OpenCV, TensorFlow, Keras, PyTorch, Scikit-Image, ImageJ, DeepLabCut, and CellProfiler facilitate such analysis [90–94]. Behera et al. used ImageJ to quantify fiber–matrix debonding in CFRP composites affected by hygrothermal aging, as illustrated in Figure 7.8, showcasing AI-enabled image thresholding tools' capability in measuring interfacial debonding [95].

7.5.2 Pattern Recognition

Utilizing AI-enabled pattern recognition revolutionizes the characterization of polymer matrix composites by identifying intricate patterns within materials and extracting vital information like fiber orientation and dispersion. Nair et al. employed pattern recognition for non-destructive AE data analysis in CFRP retrofitted RC beams for failure mode identification [96]. This approach allows

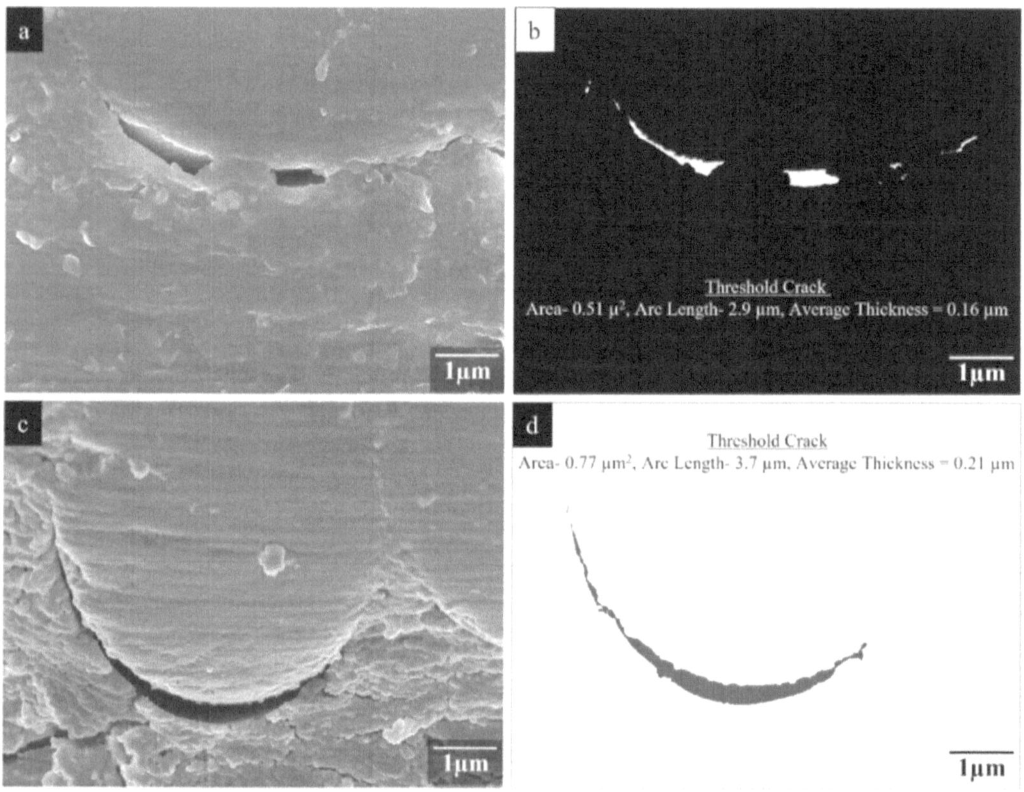

FIGURE 7.8 Application of AI-enabled ImageJ tool to quantify fiber–matrix debonding: (a, c) Raw microscopy images; (b, d) Corresponding thresholded images of (a) and (c), respectively, highlighting debonding volume. (Adapted from author's publication [90].)

researchers to optimize composite design, predict mechanical properties, and customize material compositions for specific applications, advancing polymer matrix composite technology.

7.5.3 Predictive Modeling

ML can tackle non-linear regression and classification challenges, particularly in scenarios with numerous parameters. Continuous training enhances ML's ability to discern complex relationships between input data and output findings. Cross-validations and dataset expansion further bolster ML's prediction capabilities [97]. Adeli and Yeh's seminal 1989 study [98] demonstrated the application of ANN in modeling steel beams, marking a milestone in structural engineering. Since then, researchers have employed various AI techniques like ANN, adaptive neuro-fuzzy inference systems (ANFIS), genetic expression programming (GEP), genetic algorithms (GAs), and genetic programming (GP) to address complex engineering problems [99,100]. Machello et al. proposed using random tree-based ML models to forecast the retention of tensile strength in high-temperature exposed FRPCs [101]. Kim et al. utilized an AI-based surface quality prediction model to assess CFRP milling surface finish quality, employing novel signal processing features to enhance model performance [102]. Yu et al. employed an ML-based back-propagation neural network model to predict the compressive strength of CFRP-reinforced concrete columns, reporting high prediction accuracy and robustness [103]. Due to its multifaceted benefits, efficient strength prediction via AI is crucial for various applications and sectors.

7.6 FUNDAMENTALS OF MACHINE LEARNING

ML, a branch of science, aims to create algorithms that can learn from data. These algorithms can identify relationships between different pieces of information and use that knowledge to predict new sets of data.

There are three main categories of ML problems:

- **Classification:** Involves connecting discrete labels to vectors of measurable quantities to create a computational model. With training on known data, this algorithm can forecast labels for novel measures.
- **Regression:** Focuses on connecting continuous value inputs and targets, or outputs, to create a computational model.
- **Density Estimation:** Focuses on estimating probability density functions (PDFs) without target or output values from measured data samples.

The two main categories of ML algorithms are supervised and unsupervised, as explained below.

- **Supervised Learning** involves using labeled data to train a model, where each data point has a matching value or category. After learning the correlation between inputs and outputs, the model can be applied to forecast outputs for fresh, unobserved data. Supervised learning techniques usually handle classification and regression issues.
- **Unsupervised Learning:** Focuses on unlabeled data, with the objective being to find buried structures or patterns within the data. These algorithms apply to tasks such as customer grouping and image segmentation. Unsupervised learning is used to carry out density estimates.

Extracting damage-sensitive features is crucial for implementing a successful ML paradigm for damage assessment, but defining a universal set of such features for all structural systems (e.g., aerospace, civil, oil, and gas) is practically impossible. Worden et al. [104] proposed seven axioms for SHM, three of which are particularly relevant to damage assessment in composite structures: (i) while damage location and existence can be determined in an unsupervised learning mode, damage type and severity are usually best left to supervised learning; (ii) sensors cannot directly measure damage, so feature extraction using statistical classification and signal processing is required to transform sensor data into damage information; and (iii) a measurement's vulnerability to damage is correlated with shifting operational and environmental variables, making intelligent feature extraction crucial.

7.7 MACHINE LEARNING FOR ASSESSING DAMAGE IN LAMINATED COMPOSITES

This section provides a comprehensive review of ML techniques for assessing damage in laminated composites, encompassing detection, quantification, and localization. In supervised and unsupervised settings, the classification of various ML models, including ANN, SVM, AR, Bayesian classifiers, k-means, and kNN. Additionally, we examine miscellaneous techniques developed from combinations of basic learning algorithms while other relevant ML approaches are outlined separately. The extensive mathematical specifics of the learning algorithms; instead, we will focus on providing insights on damage-sensitive features and ML techniques.

7.7.1 Artificial Neural Networks

ANNs have a long history in damage detection, with early applications exploring their potential for SHM [105]. Their success spans various industries, proving their value in fault identification for diverse components and systems. Within the realm of composite structures, ANNs continue to gain prominence. For instance, Worden and Dulieu-Barton [106] showed how a multilayer perceptron ANN could effectively detect and locate damage in composite plates using natural frequencies as the input feature. Similarly, Rizzo and Lanza di Scalea [107] leveraged a feedforward neural network to pinpoint impact damage sites in composite structures by analyzing changes in ultrasonic wave propagation features.

Yan et al. [108] integrated vibration response analysis with ANNs to provide even more precise damage detection, achieving high accuracy in pinpointing damage in a CFRP plate. When visual inspection is viable, He et al. [109] demonstrated that damage in composite structures could be reliably detected and segmented by employing a CNN directly on captured images. Addressing the complexities of classifying failure modes, Valdes-Gonzalez et al. [110] utilized AE features as inputs to an ANN, establishing a robust framework for failure mode identification in composite materials. Finally, considering the curvature of components, Chetwynd et al. [111] used multilayer perceptron neural networks combined with piezoelectric transducers for damage detection in curved carbon fiber reinforced panels.

Ghaboussi et al. [112] pioneered the application of ANN for damage detection in 1991. Since then, ANN has succeeded in various fields, including fault detection in induction motors, bearings, gears, and ship design. In composite structures, ANN has been significantly utilized for damage assessment. Researchers have predominantly used back-propagation neural networks and feed-forward back-propagation neural networks to locate delamination damages and determine their size [113,114]. Sung et al. [115] integrated ANN with the Levenberg–Marquardt algorithm to locate dynamic impact damage. Researchers have also combined X-ray computed tomography, CNN, and AE signals to detect and measure delamination in CFRP composites [116–117]. Bhat et al. [118] successfully separated noise from failure modes in CFRP specimens by using ANN to remove noise from AE signals and categorize events that belonged to various failure modes.

7.7.2 Support Vector Machine (SVM)

SVMs offer a robust approach to damage detection in composite materials due to their ability to handle high-dimensional data and identify complex non-linear relationships. A key advantage of

SVMs is their capacity to construct a hyperplane that maximizes the separation between damaged and undamaged states in a higher-dimensional feature space. This is achieved by employing kernel functions.

The application of SVMs in damage detection of composites extends across various domains. Sun et al. [119] demonstrated the effectiveness of SVMs in classifying delamination damage severity within CFRP laminates using features extracted from Lamb wave signals. Similarly, Chrysohoidis et al. [120] implemented a hybrid approach combining PCA for dimensionality reduction and SVMs for damage classification in composite structures based on AE data. SVMs have also proven valuable for real-time health monitoring of composite structures. Zhang et al. [121] proposed an SVM-based framework for online delamination detection in composite beams, utilizing strain gauge data as input. Furthermore, considering the complexities of composite structures with curvature, Xiao et al. [122] employed SVMs in conjunction with guided wave signals to achieve efficient damage detection in curved composite shells.

SVM is widely used in many sectors, including gas turbine engine defect diagnostics, AE source identification, rotating machine fault detection, and gear fault diagnosis. Researchers have extracted damage-sensitive characteristics from piezoelectric sensor responses using time-frequency-based and time-embedding approaches and then used one-class SVMs to identify and classify damage types in innovative composite laminates. To detect damage in real-time to composite helicopter rotor blades, Pawar et al. [123] created an SVM-based methodology. They used vibratory hub loads and moments, divided into matrix cracking, delamination initiation, and severe delamination. In a comparison of the effectiveness of SVM and ANN for identifying and evaluating crack damage in fiber composite panels, Farooq et al. [124] found that SVM performed better. Dib et al. [125] created a unique classifier based on one-class SVM to detect damage in glass fiber composite plates. They built feature vectors using segmented guided wave data and the Fourier transform.

7.7.3 Leveraging K-Means and K-Nearest Neighbors (KNN) for Damage Assessment in Composites

Unsupervised learning techniques like k-means clustering offer valuable insights into damage characterization within composite materials. The 'k' represents the desired number of clusters, and the algorithm groups data points based on their proximity to the calculated cluster centers. In contrast, kNN is a supervised learning algorithm for classification and clustering. In kNN, 'k' signifies the number of nearest neighbors considered during the classification of a new data point.

To analyze AE signals arising from composite damage, researchers have successfully combined k-means and kNN. Sause [126] explored the potential of AE analysis to distinguish between various failure modes in composites. Using a pattern recognition framework, k-means clustering was utilized for unsupervised grouping of AE events, followed by kNN classification for assigning damage mode labels to the clusters. Furthermore, Zhou et al. [127] developed a hybrid approach for real-time delamination monitoring in composites under cyclic loading. Their methodology involved k-means clustering for grouping AE signals based on amplitude and peak frequency features. This clustering output was then fed into a kNN classifier for the final determination of the delamination growth stage.

K-means is an unsupervised learning technique where 'K' indicates the number of clusters inside unlabeled datasets. Data points are allocated to clusters based on their separation from the cluster center. In contrast, supervised learning techniques such as kNN are employed for classification and clustering purposes. The 'K' symbol indicates the nearest neighbors used to categorize a test sample. To cluster the AE data from tensile tests on unidirectional glass/polyester composites, Godin et al. [128] used both supervised and unsupervised classifiers. As seen in flow chart Figure 7.9 of the pattern recognition study, AE signals were first clustered using k-means based on different signal parameters. These clusters were then used as labeled data for kNN classification.

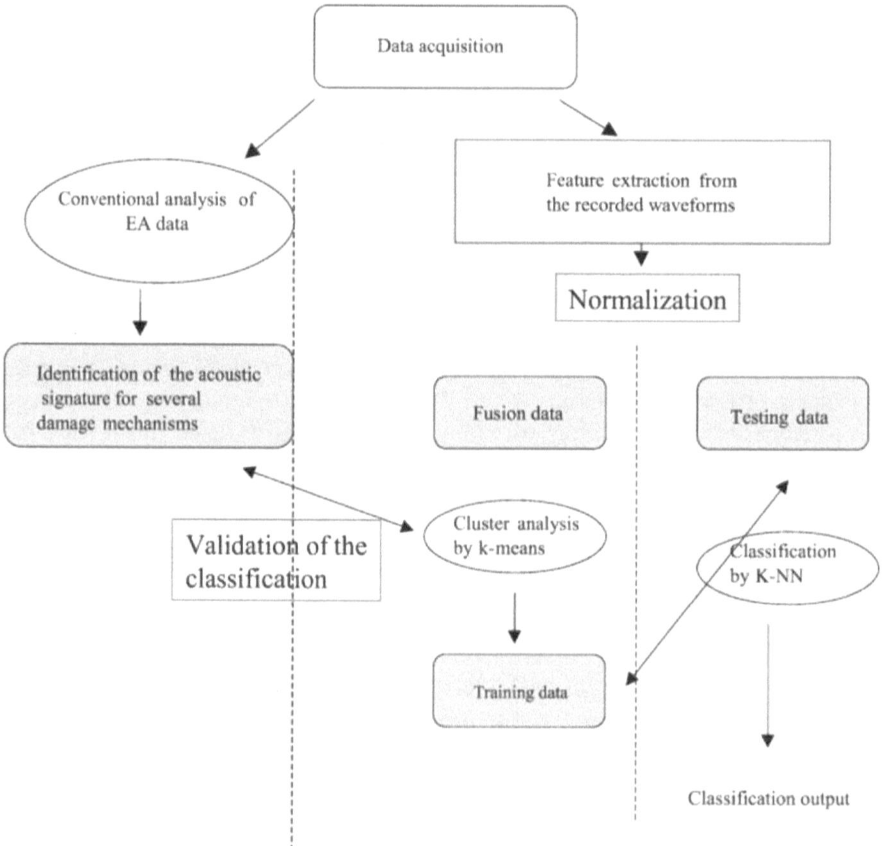

FIGURE 7.9 Flow chart of the pattern recognition analysis. (Adapted with permission from Godin et al. [128].)

The method effectively categorized AE signals associated with interfacial decohesion and matrix cracking. The primary steps of the pattern recognition analysis of the AE data are depicted in the suggested flow chart. Pashmforoush et al. [129] clustered several damage mechanisms using a GA and the k-means algorithm in a mode I delamination test of a sandwich composite. PCA was used to identify and minimize the discriminative features of the AE signals, and the integrated techniques were then used to cluster the signals into groups. The results were confirmed by microscopic studies, which showed that frequency was the best descriptor for differentiating between different damage mechanisms.

7.7.4 Principal Component Analysis (PCA)

PCA is valuable in enhancing damage detection and classification by reducing the dimensionality of structural dynamic response signals or other damage-sensitive features. This technique facilitates the visualization of feature clusters, making damage identification more intuitive. Furthermore, PCA is useful for validating sensors and distinguishing between structural degradation and the influence of environmental factors. Here's a different example demonstrating the power of PCA: for analyzing AE data from composite materials, Barile et al. [130] integrated PCA with a hierarchical clustering approach. This combination enabled effective feature dimensionality reduction and unsupervised grouping of AE events. PCA visualization revealed distinct clusters strongly correlated with damage mechanisms like matrix cracking, fiber–matrix debonding, and fiber failure. PCA is widely used to improve damage detection and classification by reducing the dimensionality of structural dynamic response signals or damage-sensitive features. This allows for the visualization of feature clusters. PCA is also useful for validating sensors and separating structural degradation from environmental influences [131].

7.7.5 Autoregressive Models

Autoregressive (AR) models offer a statistical approach to predicting future values of a variable of interest based on its historical patterns and corresponding model parameters. This makes them suitable for various damage detection scenarios. AR models have found applications in diverse domains like wind turbine monitoring, bearing fault diagnostics, and health assessment of civil structures.

Researchers have successfully leveraged AR models within the realm of composite damage detection. For instance, Liu et al. developed an approach to identify damage in laminated composite beams using a time-dependent AR model [132]. They further employed a cross-validation technique to enhance the robustness of their damage detection methodology. In a more general context, Zhu et al. presented an AR model-based framework for assessing structural damage in the time domain [133]. This methodology demonstrates the potential of AR models in a variety of SHM applications, including those related to composites.

A statistical technique called the Autoregressive (AR) model anticipates a variable of interest by considering the model's parameters and past values of the same variable. AR models have been used for damage identification in several sectors, including wind turbine damage detection, rolling bearing fault diagnosis, and civil infrastructure SHM. Using AR models, Nardi et al. [134] investigated the identification of low-velocity impact-induced delamination in innovative composite laminates. The best AR model orders were identified, and linear discriminant analysis (LDA) was used to convert AR parameters into a new subspace.

7.7.6 Bayesian Classifiers

Bayesian classifiers utilize Bayes' theorem, a robust statistical framework that allows for the quantification of trade-offs in decision-making through the calculation of costs and probabilities. These classifiers estimate the expected value of different actions and can be updated with new data to improve their performance continuously. Researchers have successfully employed various Bayesian classifier approaches when dealing with complex damage scenarios in composite materials. Li et al. [135] proposed a Bayesian compressive sensing (BCS) framework for delamination detection in composite laminates using Lamb wave signals. BCS leverages the inherent sparsity of damage signatures within the signal domain to achieve efficient damage identification. Xuan et al. [136] developed a Bayesian online changepoint detection (BO-CPD) algorithm for real-time health monitoring applications to detect damage in composite structures under varying loading conditions. This methodology utilizes online sensor data to dynamically update the damage probability distribution, enabling real-time damage assessment.

Bayesian classifiers apply Bayesian theory, quantifying trade-offs in decisions using costs and probabilities. These classifiers estimate the expected action values and are updated with fresh data. Peng et al. [137] present a probabilistic framework for determining the extent and position of delamination in carbon–carbon composites using Lamb wave-based damage detection features. Using Bayesian updating, the suggested Bayesian imaging method (BIM) creates a probability image of the delaminated region. The system allows precise delamination size, location identification, and uncertainty-bound prediction by including signal features from Lamb wave-based diagnostics into the BIM. The process's effectiveness is assessed by comparison with X-ray pictures, offering confirmation and proposing directions for further investigation. Nguyen et al. [138] used variations in the amplitudes and occurrence periods of fundamental Lamb wave mode peaks as damage indicators as they studied Bayesian networks for hole-type damage detection on an innovative composite plate. Dissipated thermal energy, temperature, and AE counts were used as damage indices by Rabiei et al. [139] to monitor multi-stage degradation in composite materials during fatigue using dynamic Bayesian networks. Addin et al. [140] evaluated Naïve-Bayes based on Lamb wave amplitude features chosen using the k-means method to classify delamination, crack, and hole damage in graphite/epoxy laminates. A Bayesian framework for estimating non-linear system characteristics from free-decay vibrations was proposed by Nichols et al. [141].

7.7.7 ANN, PCA, KNN, and LDA

Researchers have devised robust damage detection systems for composite materials by strategically combining fundamental techniques like LDA, PCA, ANN, and kNN. These ensembles offer benefits such as enhanced feature extraction, dimensionality reduction, identification of hidden patterns, and selection of crucial features. Techniques may also be combined in ensemble classifiers where multiple classifiers are integrated, often using a voting mechanism for the final classification decision.

Here are some examples of these combinations in action:

- For identifying and localizing damage in glass fiber composite beams, He et al. [142] developed a framework incorporating CNNs, wavelet transform, and the Mahalanobis–Taguchi system (MTS).
- To identify failure modes in composites, Guo et al. [143] combined AE analysis, WPT pre-processing, and a deep CNN (DCNN) architecture.
- Tibaduiza et al. [144] explored a multi-phase SHM framework for CFRP plates integrating statistical pattern recognition (SPR), multifractal analysis (MFA), and an unsupervised ML classifier.

This section examines ML frameworks created by combining fundamental tools like LDA, PCA, ANN, and kNN. Enhancing discriminative abilities, extracting features, cutting dimensions, finding hidden patterns, or choosing relevant features are possible objectives. When using ensemble classifiers, several are combined, and voting is typically used to choose the winner. Researchers have used WPT, LDA, measured amplitude ratio (MAR), unsupervised waveform clustering (UWC), ANN, Lamb wave testing with PCA, and pattern recognition to classify, characterize, and forecast damage's existence, kind, and severity [145–147]. Peak frequency content and unsupervised pattern recognition algorithms were used by Gutkin et al. [148] to investigate failures in CFRP. Godin et al. [149] classified AE signals in glass/epoxy composites by combining KSOM and k-means. Harmony search and k-means were combined by Pashmforoush et al. [129] for AE-based damage classification in glass/polyester composites. K-means and fuzzy C-means were investigated by Al-Jumaili et al. [150] for unsupervised AE classification in carbon fiber laminates. MPCA, DWT, SPE, and SOM were employed by Tibaduiza et al. [151] to detect and classify damage in CFRP plates and aircraft fuselage. PCA and ML were used by Vitola et al. [152] to assess damage in composite and aluminum constructions. Jimenez et al. [153] retrieved guided wave features using AR Walker Yule to evaluate classification techniques for delamination damage in wind turbine blades.

7.7.8 Less Common or Newly Developed Learning Algorithms

The field of damage detection in composite materials is constantly evolving, with researchers exploring novel and less-frequently used learning algorithms for enhanced detection capabilities. Here are a few examples:

- Zhang et al. [154] proposed a damage detection methodology for composite structures based on GANs to tackle the challenges of varying environmental and operational conditions. This technique generates artificial data to augment training datasets, improving the detection algorithm's robustness in the presence of noise and environmental variation.
- Chen et al. [155] presented a structural damage detection system leveraging transfer learning and CNNs. They applied a pre-trained CNN model on target image datasets from the field of interest, overcoming challenges associated with limited data availability.
- Sun et al. [156] developed a unique approach combining sparse representation with an online dictionary learning algorithm for real-time health monitoring of composites. This enabled adaptive feature extraction and damage detection even under evolving operational conditions.

Tang et al. [157] introduced a Monte Carlo method to estimate PDFs of damage-sensitive features to differentiate changes in guided wave signals caused by environmental factors from structural damage. Researchers have also used unsupervised damage classifiers like wavelet analysis from sensor responses and discordancy analysis to identify and locate delamination in composite plates [158–160].

7.8 CONCLUSION

The integration of AI techniques has significantly advanced the characterization and quantification of damage in fiber reinforced composites, spanning various NDT methods and ML algorithms. However, future research should focus on refining ML models tailored for composite damage assessment, exploring multi-modal and multiscale approaches to enhance understanding, and leveraging emerging technologies like Internet of Things (IoT) and big data analytics for real-time monitoring. Addressing these challenges holds promise for enhancing the reliability and durability of composite materials, paving the way for their widespread application in diverse industries.

REFERENCES

1. Harris B. *Fatigue in Composites*, Cambridge, Woodhead Publishing, 2003.
2. Bathias C. An engineering point of view about fatigue of polymer matrix composite materials. *Int J Fatigue* 2006; 28: 1094–9. https://doi.org/10.1016/j.ijfatigue.2006.02.008
3. Philippidis TP, Vassilopoulos AP. Fatigue of composite laminates under off-axis loading. *Int J Fatigue* 1999; 21: 253–62. https://doi.org/10.1016/S0142-1123(98)00073-5
4. Behera A, Dupare P, Thawre MM, Ballal AR. Effect of fatigue loading on stiffness degradation, energy dissipation, and matrix cracking damage of CFRP [±45]3S composite laminate. *Fatigue Fract Eng Mater Struct* 2019: 1–13. https://doi.org/10.1111/ffe.13065
5. Nayak NV. Composite materials in aerospace applications. *Int J Sci Res Publ* 2014; 4(9): 1–10.
6. Mrazova M. Advanced composite materials of the future in the aerospace industry. *INCAS Bull* 2013; 5(3): 139–50.
7. Curtis PT, Moore BB. The effects of environmental exposure on the fatigue behavior of CFRP laminates. *Composites* 1983; 14: 294–300. https://doi.org/10.1016/0010-4361(83)90019-8
8. Tsai YI, Bosze EJ, Barjasteh E, Nutt SR. Influence of hygrothermal environment on thermal and mechanical properties of carbon fiber/fiberglass hybrid composites. *Compos Sci Technol* 2009; 69: 432–7. https://doi.org/10.1016/j.compscitech.2008.11.012
9. Alessi S, Pitarresi G, Spadaro G. Effect of hydrothermal ageing on the thermal and delamination fracture behaviour of CFRP composites. *Compos Part B Eng* 2014; 67: 145–53. https://doi.org/10.1016/j.compositesb.2014.06.006
10. Takeda SI, Tsukada T, Sugimoto S, Iwahori Y. Monitoring of water absorption in CFRP laminates using embedded fiber Bragg grating sensors. *Compos Part A Appl Sci Manuf* 2014; 61: 163–71. https://doi.org/10.1016/j.compositesa.2014.02.018

11. Hussain MS, Anilchandra AR, Jagannathan N, Manjunatha CM. Fatigue life prediction of a unidirectional carbon fiber composite under off-axis spectrum loads using 3D constant life diagram. *Mater Perform Charact* 2016; 5: MPC20150048. https://doi.org/10.1520/mpc20150048
12. Liu Y, Mahadevan S. Probabilistic fatigue life prediction of multidirectional composite laminates. *Compos Struct* 2005; 69: 11–9. https://doi.org/10.1016/j.compstruct.2004.04.012
13. D'Amore A, Giorgio M, Grassia L. Modeling the residual strength of carbon fiber reinforced composites subjected to cyclic loading. *Int J Fatigue* 2015; 78: 31–7. https://doi.org/10.1016/j.ijfatigue.2015.03.012
14. Dvorak GJ. Composite materials: Inelastic behavior, damage, fatigue and fracture. *Int J Solids Struct* 2000; 37: 155–70. https://doi.org/10.1016/S0020-7683(99)00085-2
15. Kumar K, Kumar P, Deb D, Unguresan M-L, Muresan V. Artificial intelligence and machine learning based intervention in medical infrastructure: A review and future trends. *Healthcare* 2023; 11(2): 207. https://doi.org/10.3390/healthcare11020207
16. Naresh K, Shankar K, Velmurugan R, Gupta NK. Statistical analysis of the tensile strength of GFRP, CFRP and hybrid composites. *Thin-Walled Struct* 2018; 126: 150–61. https://doi.org/10.1016/j.tws.2016.12.021
17. Behera A, Thawre MM, Ballal A, Babrekar P, Vaidya P, Vijetha S, et al. Effect of Moisture Content and Fiber Orientation on the Mechanical Behavior of GFRP Composites. 2021. https://doi.org/10.1007/978-981-15-3639-7_51
18. Behera A, Dehury J, Rayaguru BP. The consequence of SiC filler content on the mechanical, thermal, and structural properties of a jute/Kevlar reinforced epoxy hybrid composite. *Silicon* 2023. https://doi.org/10.1007/s12633-023-02543-y
19. Gong Y, Zhang B, Mukhopadhyay S, Hallett SR. Experimental study on delamination migration in multidirectional laminates under mode II static and fatigue loading, with comparison to mode I. *Compos Struct* 2018; 201: 683–98.
20. Bie BX, Huang JY, Fan D, Sun T, Fezzaa K, Xiao XH, Qi ML, Luo SN. Orientation-dependent tensile deformation and damage of a T700 carbon fiber/epoxy composite: A synchrotron-based study. *Carbon* 2017; 121: 127–33.
21. Behera A, Dehury J, Thawre MM. A comparative study on laminated and randomly oriented Luffa-Kevlar Reinforced hybrid composites. *J Nat Fibers* 2019; 16. https://doi.org/10.1080/15440478.2017.1414653
22. Vishwakarma A, Behera A, Thawre MM, Ballal AR. Effect of thickness variation on static behaviour of carbon fiber reinforced polymer multidirectional laminated composite. *Mater Res Express* 2019; 6. https://doi.org/10.1088/2053-1591/ab44fa
23. Behera A, Thawre MM, Ballal A. Failure analysis of CFRP multidirectional laminates using the probabilistic Weibull distribution model under static loading. *Fibers Polym* 2019; 20. https://doi.org/10.1007/s12221-019-1194-9
24. Behera A, Thawre MM, Ballal A. Effect of matrix crack generation on Fatigue life of CFRP multidirectional laminates. *Mater Today Proc* 2018; 5. https://doi.org/10.1016/j.matpr.2018.06.374
25. Maimi P, Rodrıguez H, Blanco N, Mayugo JA. *Numerical Modelling of Matrix Cracking and Interlaminar Failure in Advanced Composite. materials* Woodhead Publication, 2015.
26. Barbero EJ, Cirnosso FA, Campo FA. Benchmark solution for degradation of elastic properties due to transverse matrix cracking in laminated composites. *Compos Struct* 2013; 98: 242–52.
27. Andreopoulos AG, Tarantili PA. Water sorption characteristics of epoxy resin–UHMPE fibers composites. *J Applied Polymer Science* 1998; 70(4): 747–55.
28. Girard H, Doitrand A, Koohbor B, Rinaldi R, Godin N, Long D, Bikard J, Trouillet-Fonti L. Numerical simulation of fiber–matrix debonding: Inverse identification of interface properties. *Eng Fract Mech* 2023; 286: 109254.
29. Chang KM, Das D, Salvati III L, Dean LM, Keshari R, Garg M, Dlott DD, Chasiotis I, Sottos NR. Durable and impact-resistant thermoset polymers for the extreme environment of low Earth orbit. *Extreme Mech Lett* 2023; 64: 102089. https://doi.org/10.1016/j.eml.2023.102089
30. Ebina M, Yoshimura A, Sakaue K, Waas AM. High fidelity simulation of low velocity impact behavior of CFRP laminate. *Compos A Appl Sci Manuf* 2018; 113: 166–79. https://doi.org/10.1016/j.compositesa.2018.07.022. 2017; 131: 153–64.

31. Paradiso A, Mendoza I, Bellafato A, Lamberson L. Failure behavior of woven fiberglass composites under combined compressive and environmental loading. *J Compos Mater* 2020; 54: 519–33.
32. Behera A, Thawre MM, Ballal A. Hygrothermal aging effect on physical and mechanical properties of carbon fiber/epoxy cross-ply composite laminate. *Mater Today Proc* 2019; 28. https://doi.org/10.1016/j.matpr.2019.12.329
33. Behera A, Vishwakarma A, Thawre MM, Ballal A. Effect of hygrothermal aging on static behavior of quasi-isotropic CFRP composite laminate. *Compos Commun* 2020; 17: 51–5. https://doi.org/10.1016/j.coco.2019.11.009
34. Behera A, Dupare P, Thawre MM, Ballal A. Effects of hygrothermal aging and fiber orientations on constant amplitude fatigue properties of CFRP multidirectional composite laminates. *Int J Fatigue* 2020; 136: 105590. https://doi.org/10.1016/j.ijfatigue.2020.105590
35. Behera A, Thawre MM, Ballal A, Manjunatha CM, Krishnakumar D, Chandankhede S. A modified piecewise linear constant life diagram for fatigue life prediction of carbon fiber/polymer multidirectional laminates. *Polym Compos* 2023; 44. https://doi.org/10.1002/pc.27176
36. Kek T, Potočnik P, Misson M, Bergant Z, Sorgente M, Govekar E, Šturm R. Characterization of biocomposites and glass fiber epoxy composites based on acoustic emission signals, deep feature extraction, and machine learning. *Sensors* 2022; 22: 6886. https://doi.org/10.3390/s22186886
37. Ellenberg A, Branco L, Krick A, Bartoli I, Kontsos A. Use of unmanned aerial vehicle for quantitative infrastructure evaluation. *J of Infrast Syst* 2015; 21(3): 04014054.
38. Rosso MM, Cucuzza R, Marano GC, Aloisio A, Cirrincione G. Review on deep learning in structural health monitoring. In Bridge Safety, Maintenance, Management, Life-Cycle, Resilience and Sustainability, CRC Press 2022: 309-315.
39. Xu H, Su X, Wang Y, Cai H, Cui K, Chen X. Automatic bridge crack detection using a convolutional neural network. *Applied Sciences* 2019; 9(14): 2867.
40. Reagan D, Sabato A, Niezrecki C. Feasibility of using digital image correlation for unmanned aerial vehicle structural health monitoring of bridges. *Structural Health Monitoring* 2018; 17(5): 1056–1072.
41. Assi L. Application of digital image correlation in structural health monitoring of bridge infrastructures: A review. *MDPI* 2021; 6(12): 176.
42. Samborski S, Korzec I. Application of the acoustic emission technique for damage identification in the fiber reinforced polymer composites. *Adv Sci Technol Res J* 2023; 17: 210–21. https://doi.org/10.12913/22998624/156602
43. Barile C, Casavola C, Pappalettera G, Vimalathithan PK. Sound of a composite failure: An acoustic emission investigation. *IOP Conf Ser Mater Sci Eng* 2022; 1214: 012006. https://doi.org/10.1088/1757-899x/1214/1/012006
44. Panasiuk K, Dudzik K. Determining the stages of deformation and destruction of composite materials in a static tensile test by acoustic emission. *Materials* 2022; 15. https://doi.org/10.3390/ma15010313
45. Muir C, Tulshibagwale N, Furst A, Swaminathan B, Almansour AS, Sevener K, et al. Quantitative benchmarking of acoustic emission machine learning frameworks for damage mechanism identification. *Integr Mater Manuf Innov* 2023; 12:70–81. https://doi.org/10.1007/s40192-023-00293-8
46. Spyrou ED, Kappatos V, Bekiaris E. Energy Value Estimation of Acoustic Emission Signal Using Advanced Machine Learning Algorithms within Composite Specimen Application. International Conference on Electrical, Computer, Communications and Mechatronics Engineering, ICECCME 2022, Institute of Electrical and Electronics Engineers Inc.; 2022. https://doi.org/10.1109/ICECCME55909.2022.9988494
47. Qiao S, Huang M, Liang Y, Zhang S zhu, Zhou W. Damage mode identification in carbon/epoxy composite via machine learning and acoustic emission. *Polym Compos* 2023; 44: 2427–40. https://doi.org/10.1002/pc.27254
48. Rubio-González C, Urquijo-Ventura MDPD, Rodríguez-González JA. Damage progression monitoring using self-sensing capability and acoustic emission on glass fiber/epoxy composites and damage classification through principal component analysis. *Compos Part B: Engg* 2023; 254: 110608. https://doi.org/10.1016/j.compositesb.2023.110608
49. Rahman MA, Rahman MM, Ashraf A. Automatic dispersion, defect, curing, and thermal characteristics determination of polymer composites using micro-scale infrared thermography and machine learning algorithm. *Sci Rep* 2023; 13. https://doi.org/10.1038/s41598-023-29270-z

50. Stawiarski A, Chwał M, Barski M, Augustyn M. The Wrinkles Characterization in GFRP composites by infrared active thermography. *Materials* 2023; 16. https://doi.org/10.3390/ma16124236
51. Chang HL, Ren HT, Wang G, Yang M, Zhu XY. Infrared defect recognition technology for composite materials. *Front Phys* 2023; 11. https://doi.org/10.3389/fphy.2023.1203762
52. Rahman MA, Becerril J, Ghosh D, Islam N, Ashraf A. Non-Destructive Infrared Thermographic Curing Analysis of Polymer Composites. *Proceedings of the ASME 2022 International Mechanical Engineering Congress and Exposition. Volume 3: Advanced Materials: Design, Processing, Characterization and Applications; Advances in Aerospace Technology*. Columbus, Ohio, USA. October 30–November 3, 2022. V003T03A035. ASME; 2022. https://doi.org/10.1115/IMECE2022-96116.
53. Sobczak M, Machynia A, Dworakowski Z, Roemer, J. Experimental Setup for Non-destructive Testing of Composite Structures Using Laser Spot Thermography. In: Puchalski, A., Łazarz, B.E., Chaari, F., Komorska, I., Zimroz, R. (eds) *Advances in Technical Diagnostics II*. ICTD. Applied Condition Monitoring 2023; 21: 75–84. https://doi.org/10.1007/978-3-031-31719-4_8
54. Spytek J, Mrowka J, Pieczonka L, Ambrozinski L. Multi-resolution non-contact damage detection in complex-shaped composite laminates using ultrasound. *NDT and E International* 2020; 116. https://doi.org/10.1016/j.ndteint.2020.102366
55. Alhammad M, Avdelidis NP, Ibarra-Castanedo C, Zolotas A, Maldague XP V. Machine learning on thermographic images for the detection and classification of damage on composites. *SPIE-Intl Soc Optical Eng*; 2022, p. 18. https://doi.org/10.1117/12.2618088
56. Phartiyal R, Agnihotri PK, Gupta N. Electrical Resistance Tomography (ERT) applied to Epoxy composites. ICD 2022 – IEEE 2022 4th International Conference on Dielectrics, Proceedings, Institute of Electrical and Electronics Engineers Inc.; 2022, p. 777–80. https://doi.org/10.1109/ICD53806.2022.9863231
57. Diaz-Escobar J, Díaz-Montiel P, Venkataraman S, Díaz-Ramrez A. Classification and characterization of damage in composite laminates using electrical resistance tomography and supervised machine learning. *Struct Control Health Monit* 2023; 2023. https://doi.org/10.1155/2023/1675867
58. Arora G, Pathak H. Nanoindentation characterization of polymer nanocomposites for elastic and viscoelastic properties: Experimental and mathematical approach. *Composites Part C: Open Access* 2021;4. https://doi.org/10.1016/j.jcomc.2020.100103
59. Arevalo SE, Ebenstein DM, Pruitt LA. A methodological framework for nanomechanical characterization of soft biomaterials and polymers. *J Mech Behav Biomed Mater* 2022; 134. https://doi.org/10.1016/j.jmbbm.2022.105384
60. Zhang Z, Bellisario D, Quadrini F, Jestin S, Ravanelli F, Castello M, et al. Nanoindentation of multifunctional smart composites. *Polymers (Basel)* 2022; 14. https://doi.org/10.3390/polym14142945
61. Jeong K, Lee H, Kwon OM, Jung J, Kwon D, Han HN. Prediction of uniaxial tensile flow using finite element-based indentation and optimized artificial neural networks. *Mater Des* 2020: 196. https://doi.org/10.1016/j.matdes.2020.109104
62. Puchi-Cabrera ES, Rossi E, Sansonetti G, Sebastiani M, Bemporad E. Machine learning aided nanoindentation: A review of the current state and future perspectives. *Curr Opin Solid State Mater Sci* 2023; 27. https://doi.org/10.1016/j.cossms.2023.101091
63. Arora G, Pathak H. Nanoindentation characterization of polymer nanocomposites for elastic and viscoelastic properties: Experimental and mathematical approach. *Composites Part C: Open Access* 2021; 4. https://doi.org/10.1016/j.jcomc.2020.100103
64. Konstantopoulos G, Semitekolos D, Koumoulos EP, Charitidis C. Carbon fiber reinforced composites: Study of modification effect on weathering-induced ageing via nanoindentation and deep learning. *Nanomaterials* 2021; 11. https://doi.org/10.3390/nano11102631
65. Gupta P, Schadler LS, Sundararaman R. Dielectric properties of polymer nanocomposite interphases from electrostatic force microscopy using machine learning. *Mater Charact* 2021; 173. https://doi.org/10.1016/j.matchar.2021.110909
66. Duan X, Guan X, Yuan H, Tang W, He J. Constitutive Modeling for the Temperature-Dependent Stress-Strain Behavior of Polymer-Bonded Composite Materials, 2022.
67. Hao P, Spronk SWF, Sevenois RDB, Van Paepegem W, Gilabert FA. Characterizing pure polymers under high speed compression for the micromechanical prediction of unidirectional composites. *Polymers (Basel)* 2023; 15. https://doi.org/10.3390/polym15051262

68. Malalli CS, Ramji BR. Mechanical characterization of natural fiber reinforced polymer composites and their application in Prosthesis: A review. *Mater Today Proc* 2022; 62: 3435–43. https://doi.org/10.1016/j.matpr.2022.04.276
69. Esmaeili H, Rizvi R. Machine learning predictions and benchmarking of non-linear mechanical behavior of polymer composites, SPIE-Intl Soc Optical Eng; 2023, p. 29. https://doi.org/10.1117/12.2658666
70. Hamidi YK, Berrado A, Altan MC. Machine learning applications in polymer composites. *AIP Conf Proc* 2020; 2205, American Institute of Physics Inc.. https://doi.org/10.1063/1.5142946
71. Sharma A, Mukhopadhyay T, Rangappa SM. Advances in computational intelligence of polymer composite materials: machine learning assisted modelling, analysis and design. *Arch Computat Methods Eng* 2022. http://doi.org/10.1007/s11831-021-09700-9
72. Riedmiller M. Advanced supervised learning in multilayer perceptrons—From back-propagation to adaptive learning algorithms. *Comput Stand Interfaces* 1994; 16: 265–78. https://doi.org/10.1016/0920-5489(94)90017-5
73. Barlow HB. Unsupervised learning. *Neural Comput* 1989; 1: 295–311. https://doi.org/10.1162/neco.1989.1.3.295
74. Hand DJ. Principles of data mining. *Drug-Saf* 2007; 30: 621–22. https://doi.org/10.2165/00002018-200730070-00010
75. Blum AL, Langley P. Selection of relevant features and examples in machine learning. *Artif Intell* 1997; 97: 245–71. https://doi.org/10.1016/S0004-3702(97)00063-5
76. Yu L, Liu H. Efficient feature selection via analysis of relevance and redundancy. *J Mach Learn Res* 2004; 5: 1205–24.
77. Quijada-Pioquinto JG, Kurkin EI, Minaev EY, Gavrilov AV. Technique of the identification, quantification, and measurement of carbon short-fibers in SEM images using the instance segmentation. Proceedings – 9th IEEE International Conference on Information Technology and Nanotechnology, ITNT 2023, Institute of Electrical and Electronics Engineers Inc.; 2023. https://doi.org/10.1109/ITNT57377.2023.10139073
78. Khan S, Rahmani H, Shah SAA. A guide for convolutional neural networks for computer vision. *Synth Lect on Comp Vision* 2018; 8(1): 1–207.
79. Padma E, Rohith DNV, Charan EVS. Mask RCNN: object detection approach using machine learning techniques. *J of Eng Sci* 2022; 13(3): 488–98.
80. Shelkovnikova TE, Egorov SF, Gulyaev PV. Application of generative adversarial neural networks for the formation of databases in scanning tunnelling microscopy. *Comp Optics* 2023; 47(2): 314–22. https://doi.org/10.18287/2412-6179-CO-1144
81. Yemelyanova MG, Smailova SS, Baklanova OE. Detection of surface defects in welded joints during visual inspections using machine vision methods. *Comp Optics* 2023; 47(1): 112–7. https://doi.org/10.18287/2412-6179-CO-1137
82. Wu L, Chung CN, Major Z, Adam L, Noels L. From SEM images to elastic responses: A stochastic multiscale analysis of UD fiber reinforced composites. *Compos Struct* 2018; 189: 206–27. https://doi.org/10.1016/j.compstruct.2018.01.051
83. Riedrich OK, Beyond the Lens: AI for Microscopy Image Analysis, www.zeiss.com/microscopy/en/resources/insights-hub/life-sciences/ai-for-microscopy-image-analysis.html
84. Guided Acquisition: Automate your microscopy -Detect rare events with ease, ZEN Guided Acquisition for Life Sciences Applications, Product flyer, www.zeiss.com/microscopy/en/products/software/zeiss-zen.html
85. Leitherer A, Yeo BC, Liebscher CH, Ghiringhelli LM. Automatic identification of crystal structures and interfaces via artificial-intelligence-based electron microscopy. *NPJ Comput Mater* 2023; 9. https://doi.org/10.1038/s41524-023-01133-1
86. Olszta M, Hopkins D, Fiedler KR, Oostrom M, Akers S, Spurgeon SR. An Automated scanning transmission electron microscope guided by sparse data analytics. *Microsc Microanal* 2022; 28: 1611–21. https://doi.org/10.1017/S1431927622012065
87. Huitian B, Sen W. Deep-learning-based nanowire detection in AFM images for automated nanomanipulation. *Nanotechnol Precis Eng* 2021; 4(1):013002. https://doi.org/10.1063/10.0003218

88. Thiem J, Cole DP, Dubey U, Srivastava A, Ashraf C, Henry TC, Bakis CE, Vashisth A. Using data science to locate nanoparticles in a polymer matrix composite. *Compos Sci Technol* 2022; 218: 109205. https://doi.org/10.1016/j.compscitech.2021.109205
89. Huang B, Li Z, Li J. An artificial intelligence atomic force microscope enabled by machine learning. *Nanoscale* 2018; 10: 21320–6. https://doi.org/10.1039/c8nr06734a
90. Aydın A, Guney S. Fault detection system for paper cup machine based on real-time image processing. *Eng Appl of Artificial Intel* 2024; 133(A): 108070. https://doi.org/10.1016/j.engappai.2024.108070
91. Luo L, Zhang B, Lei Y, Zhang G, Zhang Z, Meng B, et al. Identification of voids and interlaminar shear strengths of polymer-matrix composites by optical microscopy experiment and deep learning methodology. *Polym Adv Technol* 2021; 32: 1853–65. https://doi.org/10.1002/pat.5226
92. Botifoll M, Pinto-Huguet I, Arbiol J. Machine learning in electron microscopy for advanced nano characterization: current developments, available tools, and future outlook. *Nanoscale Horiz* 2022; 7: 1427–77. https://doi.org/10.1039/d2nh00377e
93. Rueden CT, Schindelin J, Hiner MC, DeZonia BE, Walter AE, Arena ET & Eliceiri KW. ImageJ2: ImageJ for the next generation of scientific image data. *BMC Bioinform* 2017; 18, 529.
94. Schneider CA, Rasband WS & Elicieri KW. NIH image to ImageJ: 25 years of image analysis. *Nat Methods* 2012; 9, 671–5.
95. Behera A, Bhoi NK, Thawre MM, Ballal A, Das D. Quantification of hygrothermal aging-induced interfacial debonding of carbon fiber/epoxy composites at nano-to-micrometer length scales. *J of Compos Materials* 2023; 57(30): 4637–47. https://doi.org/10.1177/00219983231213912
96. Nair A, Cai CS, Kong X. Acoustic emission pattern recognition in CFRP retrofitted RC beams for failure mode identification. *Compos B Eng* 2019; 161: 691–701. https://doi.org/10.1016/j.compositesb.2018.12.120
97. Liu X, Liu T, Feng P. Long-term performance prediction framework based on XGBoost decision tree for pultruded FRP composites exposed to water, humidity, and alkaline solution. *Compos Struct* 2022; 284: 115184.
98. Adeli H, Yeh C. Perceptron learning in engineering design. *Comput Aided CivInfrastruct Eng* 1989; 4(4): 247–56.
99. Tam L-h, Zhou A, Wu C. Nanomechanical behavior of carbon fiber/epoxy interface in hygrothermal conditioning: a molecular dynamics study. *Mater Today Commun* 2019; 19: 495–505.
100. Bas¸yigit C, Akkurt I, Kilincarslan S, Beycioglu A. Prediction of compressive strength of heavyweight concrete by ANN and FL models. *Neural Comput Appl* 2010; 19(4): 507–13.
101. Machello C, Aghabalaei Baghaei K, Bazli M, Hadigheh A, Rajabipour A, Arashpour M, et al. Tree-based machine learning approach to modelling tensile strength retention of Fibre Reinforced Polymer composites exposed to elevated temperatures. *Compos B Eng* 2024; 270. https://doi.org/10.1016/j.compositesb.2023.111132
102. Kim JW, Nam J, Kim GY, Lee SW. Artificial intelligence (AI)–based surface quality prediction model for carbon fiber reinforced plastics (CFRP) milling process. *Int J Precis Eng Manuf-Smart Techy* 2023; 1: 35–47. https://doi.org/10.57062/ijpem-st.2022.0052
103. Yu Y, Hu T. Machine learning based compressive strength prediction model for CFRP-confined columns. *KSCE J Civ Eng* 2024; 28: 315–27. https://doi.org/10.1007/s12205-023-0827-0
104. Worden K, Farrar CR, Manson G, Park G. The fundamental axioms of structural health monitoring. *Proc R Soc A: Math Phys Eng Sci* 2007; 463: 1639–64. https://doi.org/10.1098/rspa.2007.1834
105. Hajela P, Soeiro FJ. Structural damage detection based on static and modal analysis. *J Guid Control Dyn* 1990; 13: 409–15.
106. Worden K, Dulieu-Barton JM. An overview of intelligent fault detection in systems and structures. *Struct Health Monit* 2004; 3: 85–98.
107. Rizzo P, Lanza F. Ultrasonic wave propagation in progressive damage media: theoretical prediction and experimental validation using wavelets. *NDT&E Int* 2005; 38: 147–159.
108. Yan WJ, Chen WQ, Jiang ZY. Damage detection of CFRP plate based on vibration response analysis and neural network. *Proceedings of the 18th International Congress on Sound and Vibration (ICSV 18)* 2007: 3585–3591.
109. He Y, Lu Y, Jiang Z. Damage identification for composite structures using a convolutional neural network based on visualized structural connections. *Struct Health Monit* 2022; 21: 474–87.

110. Valdes-Gonzalez GC, Dorronsoro JR. Neural network for online, in-service structural damage identification. *Proceedings of the 4th European Workshop on Structural Health Monitoring* 2008.
111. Chetwynd D, Mustapha F, Worden K, Rongong JA, Pierce SG, Dulieu-barton JM. Damage localisation in a stiffened composite panel. *Strain* 2008; 44: 298–307. https://doi.org/10.1111/j.1475-1305.2007.00371.x
112. Ghaboussi J, Garrett Jr J, Wu X. Knowledge-based modelling of material behavior with neural networks. *J Eng Mech* 1991; 117(1): 132–53.
113. Islam AS, Craig KC. Damage detection in composite structures using piezoelectric materials. *Smart Mater Struct* 1994; 3: 318–28.
114. Okafor AC, Chandrashekhara K, Jiang YP. Delamination prediction in composite beams with built-in piezoelectric devices using modal analysis and neural network. *Smart Mater Struct* 1996; 5: 338–47.
115. Sung DU, Oh JH, Kim CG, Hong CS. Impact Monitoring of Smart Composite Laminates Using Neural Network and Wavelet Analysis. *J Intell Mater Syst Struct* 2000; 11(3): 180–90. https://doi.org/10.1106/N5E7-M37Y-3MAR-2KFH
116. Sammons D, Winfree WP, Burke E, Ji S. Segmenting delaminations in carbon fiber reinforced polymer composite CT using convolutional neural networks. *AIP Conf Proc*, vol. 1706, American Institute of Physics Inc.; 2016. https://doi.org/10.1063/1.4940585
117. Bar HN, Bhat MR, Murthy CRL. Identification of failure modes in GFRP using PVDF sensors: ANN approach. *Compos Struct* 2004; 65: 231–7. https://doi.org/10.1016/j.compstruct.2003.10.019
118. Bhat C, Bhat MR, Murthy CRL. Acoustic emission characterization of failure modes in composites with ANN. *Compos Struct* 2003; 61: 213–20. https://doi.org/10.1016/S0263-8223(03)00068-0
119. Sun Z, Chang CC, Qian C. Classification of Lamb wave damage patterns utilizing support vector machine and differential evolution algorithm. *Proceedings of the SPIE* 2016; 9803.
120. Chrysohoidis NA, Karalis GD, Tsioukas VE. Damage assessment of composite structures using acoustic emission monitoring and a deep learning classification model. *European Conference on Acoustic Emission Testing* (EWGAE) 2021
121. Zhang Z, Pan J, Luo W, In-situ structural damage monitoring of a composite beam based on support vector machine. *J Vibroeng* 2017; 19(6): 4065–76.
122. Xiao F, Hulsey M, Balasubramaniam K. Lamb wave-based damage detection in anisotropic curved panels using a support vector machine. *Procedia Eng* 2017; 173: 1476–83.
123. Pawar PM, Jung SN. Support vector machine based online composite helicopter rotor blade damage detection system. *J Intell Mater Syst Struct* 2008; 19: 1217–28. https://doi.org/10.1177/1045389X07084713
124. Farooq M, Zheng H, Nagabhushana A, Roy S, Burkett S, Barkey M, et al. Damage detection and identification in smart structures using SVM and ANN. *Smart Sens Phenom Tech Net Syst Integr* 2012; 8346, SPIE: 83461O. https://doi.org/10.1117/12.915189
125. Dib G, Karpenko O, Koricho E, Khomenko A, Haq M, Udpa L. Ensembles of novelty detection classifiers for structural health monitoring using guided waves. *Smart Mater Struct* 2018; 27. https://doi.org/10.1088/1361-665X/aa973f
126. Sause MGR. *In Situ Monitoring of Fiber-Reinforced Composites: Theory, Basic Concepts, Methods, and Applications*. Springer, 2016.
127. Zhou Y, Jia Y, Liu W. A novel intelligent method for real-time delamination detection in composite structures based on k-means clustering and probabilistic neural network. *Struct Health Monit* 2022; 21(1): 208–22.
128. Godin N, Huguet S, Gaertner R, Salmon L. Clustering of acoustic emission signals collected during tensile tests on unidirectional glass/polyester composite using supervised and unsupervised classifiers. *NDT and E International* 2004; 37: 253–64. https://doi.org/10.1016/j.ndteint.2003.09.010
129. Pashmforoush F, Khamedi R, Fotouhi M, Hajikhani M, Ahmadi M. Damage classification of sandwich composites using acoustic emission technique, and k-means genetic algorithm. *J Nondestruct Eval* 2014; 33: 481–92. https://doi.org/10.1007/s10921-014-0243-y
130. Barile C, Casavola C, Pappalettera G. Analysis of damage phenomena in composite materials through acoustic emission technique and unsupervised algorithms. *Struct Health Monit* 2020; 19(6): 1671–80.
131. Oskouei AR, Heidary H, Ahmadi M, Farajpur M. Unsupervised acoustic emission data clustering for the analysis of damage mechanisms in glass/polyester composites. *Mater Des* 2012; 37: 416–22. https://doi.org/10.1016/j.matdes.2012.01.018

132. Liu M, Zhang Z, Zhou S. Damage detection of laminated composite beams based on time-dependent autoregressive model and cross-validation technique. *Compos Struct* 2022; 286: 115322.
133. Zhu J, Wang Y, Qing X. An autoregressive model-based approach for time-domain structural damage detection. *J Vibroeng* 2020; 22(4): 709–21.
134. Nardi D, Lampani L, Pasquali M, Gaudenzi P. Detection of low-velocity impact-induced delaminations in composite laminates using Auto-Regressive models. *Compos Struct* 2016; 151: 108–13. https://doi.org/10.1016/j.compstruct.2016.02.005
135. Li W, Sun C, Zhou K. Compressive sensing for damage detection in composite laminates based on Bayesian framework. *Smart Mater Struct* 2019; 28(8): 085013.
136. Xuan H, Wang W, Ghobakhloo A. Bayesian online changepoint detection for damage identification in composite structures under varying loading conditions. *Struct Health Monit* 2022; 21(1): 142–58.
137. Peng T, Saxena A, Goebel K, Xiang Y, Sankararaman S, Liu Y. A novel Bayesian imaging method for probabilistic delamination detection of composite materials. *Smart Mater Struct* 2013; 22. https://doi.org/10.1088/0964-1726/22/12/125019
138. Nguyen M, Wang X, Su Z, Ye L. Damage identification for composite structures with a Bayesian network. *Proceedings of the 2004 Intelligent Sensors, Sensor Networks and Information Processing Conference*, Melbourne, Australia, 2004: 307–11. https://doi.org/10.1109/ISSNIP.2004.1417480
139. Rabiei E, Droguett L, Modarres M. Damage monitoring and prognostics in composites via dynamic Bayesian networks. *Annual Reliability and Maintainability Symposium (RAMS)*, Orlando, FL, USA, 2017: 1–7. https://doi.org/10.1109/RAM.2017.7889668
140. Addin O, Sapuan SM, Mahdi E, Othman M. A Naïve-Bayes classifier for damage detection in engineering materials. *Mater Des* 2007; 28: 2379–86. https://doi.org/10.1016/j.matdes.2006.07.018
141. Nichols JM, Link WA, Murphy KD, Olson CC. A Bayesian approach to identifying structural nonlinearity using free-decay response: Application to damage detection in composites. *J Sound Vib* 2010; 329: 2995–3007. https://doi.org/10.1016/j.jsv.2010.02.004
142. He Y, Li M, Meng Z. Damage identification of glass fiber composite beams with random excitation based on CNN, wavelet analysis and MTS. *Compos Struct* 2022; 282: 115103.
143. Guo, F., Li, W., Jiang, P., Chen, F., & Liu, Y. Deep learning approach for damage classification based on acoustic emission data in composite materials. *Materials* 2022; *15*(12): 4270.
144. Tibaduiza, DA, Mujica, LE, Rodellar, J., Güemes, A. Structural damage detection using principal component analysis and damage indices. *J Intell Mater Syst Struct* 2016; 27: 233–48.
145. McCrory JP, Al-Jumaili SK, Crivelli D, Pearson MR, Eaton MJ, Featherston CA, et al. Damage classification in carbon fibre composites using acoustic emission: A comparison of three techniques. *Compos B Eng* 2015; 68: 424–30. https://doi.org/10.1016/j.compositesb.2014.08.046
146. Kessler SS, Spearing SM, Soutis C. Damage detection in composite materials using Lamb wave methods. *Smart Mater Struct* 2002; 11: 269.
147. Oliveira RD, Marques AT. Health monitoring of FRP using acoustic emission and artificial neural networks. *Comput Struct* 2008; 86: 367–73. https://doi.org/10.1016/j.compstruc.2007.02.015
148. Gutkin R, Green CJ, Vangrattanachai S, Pinho ST, Robinson P, Curtis PT. On acoustic emission for failure investigation in CFRP: Pattern recognition and peak frequency analyses. *Mech Syst Signal Process* 2011; 25 :1393–407. https://doi.org/10.1016/j.ymssp.2010.11.014
149. Godin N, Huguet S, Gaertner R. Integration of the Kohonen's self-organising map and k-means algorithm for the segmentation of the AE data collected during tensile tests on cross-ply composites. *NDT and E Int* 2005; 38: 299–309. https://doi.org/10.1016/j.ndteint.2004.09.006
150. Al-Jumaili SK, Holford KM, Eaton MJ, McCrory JP, Pearson MR, Pullin R. Classification of acoustic emission data from buckling test of carbon fibre panel using unsupervised clustering techniques. *Struct Health Monit* 2015; 14(3): 241–51. https://doi.org/10.1177/1475921714564640
151. Tibaduiza DA, Torres-Arredondo MA, Mujica LE, Rodellar J, Fritzen CP. A study of two unsupervised data driven statistical methodologies for detecting and classifying damages in structural health monitoring. *Mech Syst Signal Process* 2013; 41: 467–84. https://doi.org/10.1016/j.ymssp.2013.05.020
152. Vitola J, Pozo F, Tibaduiza DA, Anaya M. Distributed piezoelectric sensor system for damage identification in structures subjected to temperature changes. *Sensors (Switzerland)* 2017; 17. https://doi.org/10.3390/s17061252
153. Jimenez AA, Munoz CQG, Marquez FPG. Machine Learning for Wind Turbine Blades Maintenance Management. *Energies* 2018; 11(1): 13. https://doi.org/10.3390/en11010013

154. Zhang W, Li R., Zhang H. Damage detection of composite structures under varying temperature and noise based on generative adversarial networks. *Compos Struct* 2022; 284: 115182.
155. Chen Z, He J, Azarmi F. Transfer learning for structural damage assessment of composite plates from limited datasets. *Adv Eng Softw* 2022; 166: 103055.
156. Sun H, Wang Z, Wang X. Online dictionary learning for sparse representation-based damage detection of composite structures. *Compos Struct* 2022; 279: 114765.
157. Tang J, Yan G. Monitoring, and detection of structural damage under complex environment using guided waves, Singapore International NDT Conference & Exhibition, 2013.
158. Sohn H, Park G, Wait JR, Limback NP, Farrar CR. Wavelet-based active sensing for delamination detection in composite structures. *Smart Mater Struct* 2004; 13: 153–60. https://doi.org/10.1088/0964-1726/13/1/017
159. Sultan MTH, Worden K, Pierce SG, Hickey D, Staszewski WJ, Dulieu-Barton JM, et al. On impact damage detection and quantification for CFRP laminates using structural response data only. *Mech Syst Signal Process* 2011; 25: 3135–52. https://doi.org/10.1016/j.ymssp.2011.05.014
160. Fotouhi M, Heidary H, Ahmadi M, Pashmforoush F. Characterization of composite materials damage under quasi-static three-point bending test using wavelet and fuzzy C-means clustering. *J Compos Mater* 2012; 46(15): 1795–808. https://doi.org/10.1177/0021998311425968

8 Molecular Dynamics Simulation of Mechanical Behavior of Penta-Graphene

Sumit Sharma, Sanjeev Kumar, and Shahram Ajori

8.1 OVERVIEW

The fascinating atomic configuration of pentagon-only carbon in penta-graphene (PG), a novel two-dimensional (2D) metastable carbon allotrope, is reminiscent of the Cairo pentagonal tiling. PG has excellent dynamic and mechanical stability, allowing it to survive high temperatures up to 1000 K, as shown by extensive theoretical studies. PG exhibits an outstanding negative Poisson's ratio and extremely high ideal strength in terms of mechanical properties, both of which are higher than those of graphene. PG possesses an inherent quasi-direct band gap with a significantly large magnitude, approximately measuring 3.25 eV. This band gap is comparable to the band gaps observed in materials such as ZnO and GaN. It is worth noting that this characteristic sets PG apart from graphene, which typically requires functionalization to induce the opening of a band gap. Additionally, PG can be separated from T12-carbon. PG rolled into nanotubes with pentagon-based structures when it was compressed. The inherent semiconducting characteristics of these nanotubes are dependent on their chirality. PG can also be stacked in a variety of ways to create three-dimensional (3D) twin T12-carbon structures that are stable and have even wider band gaps than the original T12-carbon structure. PG and its derivatives have a lot of potential because of their adaptability for a variety of uses in the disciplines of nanoelectronics and nanomechanics. The creation of innovative nanoscale materials and devices is made possible by PG's exceptional mechanical and electrical characteristics [1–4]. A broad variety of carbon-based materials in various dimensions have been found thanks to the adaptability of carbon atoms in the formation of numerous allotropes with distinctive electrical configurations. In particular, the 2D hexagonal lattice of carbon atoms known as graphene has attracted a lot of interest in the field of materials. The exceptional features of graphene make it a promising material for use in energy storage, photodetectors, supercapacitors, spin filters, and optoelectronics. Graphene's semi-metallic structure, which is characterized by a lack of a band gap, is one of its limitations, limiting its direct application in nanoelectronic devices and nanotechnology. As a result, scientists have been actively investigating ways to modify the graphene band gap using both theoretical and practical methods. The investigation of alternative materials with inherent band gaps and a wide range of outstanding qualities is being done concurrently. Modern materials research aims to develop a strong relationship between structure and property, which significantly depends on the arrangement, make-up, and form of the constituent parts. The goal of the research is to produce a new class of innovative materials with superior characteristics and enhanced device performance by logically manipulating these structural building pieces. As a result, substantial work has gone into finding geometric building blocks that are different from the typical hexagonal shape that can be used to create these unique materials. In conclusion, there has been a lot of study done in the hunt for carbon-based materials with tunable band gaps and unique features. This research entails

 DOI: 10.1201/9781003664611-9

looking into many kinds of materials other than graphene and analyzing how structural components affect the general characteristics of these materials. These activities have the potential to open new possibilities for the creation of cutting-edge materials and technology [5].

8.2 GRAPHENE

PG is an extremely thin sheet made of carbon atoms bound together in an sp^2 arrangement. It is a single layer that can either be suspended from or be bonded to a non-carbon substrate. This layer is typically only one atom thick. PG has a width that can range from a few nanometers to a microscale. Monolayer or single-layer PG, the purest form of PG, has been shown to be favorable for high-frequency circuit applications. However, as more layers are added, unique characteristics appear. As a case study, tri-layer graphene has three layers, while bi-layer graphene has two. These layered arrangements display distinctive qualities and have the potential for specialized applications.

Due to its exceptional mechanical, electrical, thermal, and optical capabilities, graphene is frequently employed in nanocomposites, sensors, energy storage systems, and optical apparatus, among others. It is common knowledge that graphene's mechanical properties are crucial to its practical applications. Numerous testing techniques have so far been used to experimentally determine key crucial graphene mechanical properties. For instance, using a technique called nanoindentation, Lee et al. [6] determined that monolayer graphene has inherent breaking strengths of 130 GPa and 1 TPa, respectively, and that its Young's modulus is 1 TPa used an in-situ tensile testing technique using a scanning electron microscope (SEM) to determine that monolayer graphene has a tensile strength of 56–60 GPa (SEM). They also discovered that graphene exhibits typical brittle cracks and that its maximum tensile strain can be as high as 5%. So according to studies by the Lou group, who also accomplished a pre-crack and the first-ever "in-situ" tensions of suspending graphene layers, graphene's critical strain energy release rate and critical stress intensity factor are each 15.9 J/m^2 and 4 MPa $m^{1/2}$, correspondingly. Additionally, by the use of nonlinear finite computation and "in-situ" SEM tensile analysis, the Filleter group found that graphene-oxide had a greater fracture toughness than pure graphene [7].

8.3 PENTA-GRAPHENE

A kind of carbon is called PG. It possesses a pentagon framework with both sp^2- and sp^3-like carbon atoms and appears to be quasi-two-dimensional. PGs, in contrast to "Graphene", have a non-zero band gap that makes optoelectronic applications possible. It is hypothesized that PG, also known as the Cairo pentagon's spade, is a 2D metastable graphite substance made entirely of carbon pentagons. The new carbon polymorphism is robust mechanically and dynamically, and it can tolerate temperatures up to 1000 K, according to current theoretical research. Due to its distinctive atomic structure, PG is much superior to graphene and has an incredibly high ideal strength and a peculiarly negative Poisson's ratio. PG, unlike graphene, shares similarities with materials like ZnO and GaN due to its inherent quasi-direct band gap, which can reach up to 3.25 eV. However, to introduce a band gap, PG requires functionalization. This functionalization is necessary to create an energy difference within the materials. PG, in contrast with graphene, has an inherent quasi-direct band gap of up to 3.25 eV, for which comparable those of "ZnO and GaN". However, it needs to be functionalized to establish an energy deficit. Significantly, T12-carbon is capable of exfoliating PG. It can be rolled up to make semiconductor nanomaterials based on polyhedra, irrespective of chirality. Steady 3D twin T12-carbon nanostructures with a band gap even greater than those of T12-carbons can be produced by laminating materials in various ways. PG as well as its derivatives are expected to be used in a wide variety of nanomechanics and nanoelectronics applications because of their versatility [8–10].

8.4 MOLECULAR DYNAMICS

To study microscopic events, the powerful and economical method of molecular dynamics (MD) simulation is often used. MD offers important insights into complex systems that would be challenging or impossible to examine experimentally by modeling the motion and interactions of molecules and atoms based on Newton's equations of motion. For straightforward classical mechanics examples, manual analysis is sufficient; however, real-world systems require several interacting atoms, making analytical computations impossible. Computer simulations are a useful tool for in-depth microscopic study in such circumstances. There are several crucial steps in the MD process. The system of interest is first represented by initializing the molecules and their surroundings. Then, a thorough examination is done of molecule-to-molecule interactions as well as interactions between molecules and their surroundings. In this analysis, forces, energy, and other pertinent attributes are evaluated. Finally, to enable the simulation to advance over time, the system's structure is periodically modified based on these interactions and traits. Researchers can better comprehend molecular behavior, investigate complex phenomena, and forecast system responses under varied conditions by utilizing MD simulations. Additionally, when compared to conventional experimental methods, MD simulations have a far lower environmental impact, making them a more environmentally responsible choice. Overall, MD simulations are essential for expanding our understanding of science and for the creation of novel products, medicines, and technology [11].

8.4.1 Molecular Dynamics Process

Newton's formula for motion, which is given by, can be utilized to represent the motion of a particle:

$$m_i \frac{d^2xi}{dt^2} = f_i \tag{8.1}$$

$$f_i = m_i a_i \tag{8.2}$$

where m_i is the molecular mass, a_i is the particle's acceleration, and f_i is the force acting on the particle. The energy potential gradient may be utilized as well to determine the force as follows:

$$f_i = -\frac{\partial V}{\partial x_i} \tag{8.3}$$

From Equations (8.1) and (8.3), we get,

$$-\frac{\partial V}{\partial x_i} = m_i \frac{d^2xi}{dt^2} \tag{8.4}$$

where V stands for the entire system's energy potential. There are several techniques to evaluate a particle's state at a specific time in an MD simulation, as listed below:

8.4.2 Verlet Method

When the velocity terms are not necessary to estimate a molecule's position at the following time step, the Verlet technique is used. Instead of immediately calculating the velocity, this integration approach makes use of the particle's prior two locations and acceleration. The Verlet approach can approximatively predict the track of the molecule in MD simulations using this data:

$$x(t+\delta t) = x(t) + v(t)\,\delta t + \frac{1}{2}a(t)\,\delta t^2 \tag{8.5}$$

$$x(t-\delta t) = x(t) - v(t)\,\delta t + \frac{1}{2}a(t)\,\delta t^2 \tag{8.6}$$

From Equations (8.5) and (8.6), we get,

$$x(t+\delta t) = 2x(t) - x(t-\delta t) + a(t)\,\delta t^2 \tag{8.7}$$

where x indicates the molecular position, v the velocity, and a is *the* acceleration.

8.4.3 Velocity Verlet Method

In some circumstances, it becomes vital to keep the system's temperature constant. In such instances, the particle's position at the following time step is calculated using a method that considers both the positions and velocities of the molecules. The first step of the procedure is to specify the initial positions and speeds of each molecule, after which the forces that are acting on them are assessed. The new positions of the molecules can be estimated using the relevant equations utilizing the information that is obtained:

$$x(t+\delta t) = x(t) + v(t)\,\delta t + \frac{1}{2}a(t)\,\delta t^2 \tag{8.8}$$

Furthermore, one can determine the molecules' new velocities through,

$$v(t+\delta t) = v(t) + \frac{1}{2}\left[a(t) + a(t+\delta t)\right]\delta t \tag{8.9}$$

The trajectory of each particle can be determined by repeatedly iterating these actions at regular intervals. Until the appropriate number of stages have been performed, this process is repeated. As a result, the system's state may be anticipated given certain conditions, such as constant pressure and temperature. This iterative method enables the study of diverse phenomena at the molecular level and permits simulation of the system's behavior through time, giving insights into its dynamics.

8.4.4 Leapfrog Method

The procedure entails leapfrogging evaluations of the particle locations, forces, and velocities in discrete time steps. To estimate the particle's new position, it determines the particle's velocity at $\left(t - \frac{1}{2}\delta t\right)$ to half of the time step. The particle's position by using the following relations, one can forecast the particle's novel location at time t and its velocity at a prior time step are both used in this method. These calculations can be applied repeatedly to simulate and analyze the particle behaviors and trajectories:

$$x(t+\delta t) = x(t) + v\left(t + \frac{1}{2}\delta t\right)\delta t \tag{8.10}$$

$$v\left(t + \frac{1}{2}\delta t\right) = v\left(t - \frac{1}{2}\delta t\right) + a(t)\,\delta t \tag{8.11}$$

For calculate the velocities at time t, this estimate can be used:

$$v(t) = \frac{1}{2}\left[v\left(t - \frac{1}{2}\delta t\right) + v\left(t + \frac{1}{2}\delta t\right)\right] \tag{8.12}$$

8.4.5 Forcefields

In order to determine the interacting potential of a system, forcefields are utilized. Forcefields serve the purpose of accurately representing the potential energy surface of various groups of molecules. They are numerical equations designed to depict particle interactions. The primary role of forcefields is to incorporate both bonded aspects (such as covalent bonds) and non-bonded interactions, including van der Waals forces and electrostatic forces:

$$Etotal = Ebonded + Enon-bonded \tag{8.13}$$

where the following summations provide the components of the covalent and noncovalent contributions:

$$Ebonded = Ebond + Eangle + Edihedral \tag{8.14}$$

The energy of interactions between non-bonded atoms is considered for

$$Enon-bonded = Eelectrostatic + Evan\ der\ Waals + Ehydrogen-bond \tag{8.15}$$

$$Etotal = \sum_{i=1}^{Nbond} Ebond + \sum_{i=1}^{Nangle} Eangle + \sum_{i=1}^{Ndihed} Edihed + \sum_{i=1}^{Nnb} Enb \tag{8.16}$$

Any simulation's forcefield selection is essential since it affects the particle system's potential energy. COMPASS (Condensed-phase Optimized Molecular Potentials for Atomistic Simulation Studies) is a forcefield option that is well known and frequently used in MD simulations. It distinguishes out as an initial ab initio forcefield capable of precisely predicting properties for numerous inorganic and organic substances in both condensed and gas phases.

To account for numerous system interactions, the forcefield COMPASS uses a variety of terminology. Bond (b), angle (θ), dihedral (ϕ), out-of-plane angle (χ), and cross-terms are some of these. Two non-bonded functions are also included in it: a Coulombic function to deal with electrostatic interactions and a 9-6 Lennard-Jones potential to deal with van der Waals interactions:

$$E_{total} = E_b + E_\theta + E_\varphi + E_{b,\theta} + E_{\theta,\varphi} + E_{\theta,\theta'} + E_{\theta,\theta',\varphi} + E_q + E_{vdW} \tag{8.17}$$

In which,

$$E_b = \sum_b \left[k_2 (b - b_0)^2 + k_3 (b - b_0)^3 + k_4 (b - b_0)^4 \right] \tag{8.18}$$

$$E_\theta = \sum_\theta \left[k_2 (\theta - \theta_0)^2 + k_3 (\theta - \theta_0)^3 + k_4 (\theta - \theta_0)^4 \right] \tag{8.19}$$

$$E_\phi = \sum_\phi \left[k_1 (1 - \cos 1\phi) + k_2 (1 - \cos 2\phi) + k_3 (1 - \cos 3\phi) \right] \tag{8.20}$$

$$E\chi = \sum_{ç} k_2 \chi^2 \tag{8.21}$$

$$E_{b,b'} = \Sigma k (b - b_0)(b' - b'_0) \tag{8.22}$$

$$E_{b,\theta} = \sum_{b,\theta} k (b - b_0)(\theta - \theta_0) \tag{8.23}$$

$$E_{b,\phi} = \sum_{b,\phi} (b - b_0)\left[k_1 \cos 1\varphi + k_2 \cos 2\varphi + k_3 \cos 3\varphi \right] \tag{8.24}$$

$$E_{\theta,\phi} = \sum_{\theta,\phi} (\theta - \theta_0)\left[k_1 \cos 1\varphi + k_2 \cos 2\varphi + k_3 \cos 3\varphi \right] \tag{8.25}$$

$$E_{\theta,\theta'} = \sum_{\theta,\theta'} k (\theta - \theta_0)(\theta' - \theta'_0) \tag{8.26}$$

$$E_{\theta,\theta',Æ} = \sum_{\theta,\theta',\varphi} k (\theta - \theta\theta_0)(\theta' - \theta'_0) \cos Æ \tag{8.27}$$

$$E_q = \sum_{ij} \frac{q_i q_j}{r_{ij}} \tag{8.28}$$

$$E_{vdW} = \sum_{ij} \epsilon_{ij} \left[2\left(\frac{r_{ij}^{\,0}}{r_{ij}} \right)^9 - 3\left(\frac{r_{ij}^{\,0}}{r_{ij}} \right)^6 \right] \tag{8.29}$$

k_1, k_2, k_3, and k_4 are experimentally determined force constants.
Let's define the following variables:
b: Bond length after stretching.

θ: Bond angle after bending.
b_0: Equilibrium bond length.
θ_0: Equilibrium bond angle.
χ: Out-of-plane inversion angle.
ϕ: Bond torsion angle.
E, b: Energy due to the interaction between bond stretching.
E, θ: Energy due to the interaction between bond stretching and bending.
E, ϕ: Energy due to the interaction between bond stretching and torsion.
E, θ, ϕ: Energy due to the interaction between bond stretching, bending, and torsion.
E, θ, θ': Energy due to the interaction between bond bending and torsion.
E, θ, θ', ϕ: Energy due to the interaction between bond bending, torsion, and bending.

Additionally, we have

q_i, q_j: atomic charges on the atoms/molecules.
ϵ_{ij}: Well depth or bond dissociation energy.
r_{ij}: Separation between atoms/molecules.
r_0: Distance in which the interaction energy between two atoms is zero.

In summary, the equation can be written as follows:

$$E = E, b + E, \theta + E, \phi + E, \theta, \phi + E, \theta, \theta' + E, \theta, \theta', \phi$$

8.5 MATERIALS AND METHODS

An extensive computational study has demonstrated that the carbon polymorph is both dynamically and mechanically stable, exhibiting remarkable thermal stability up to 1000 K. Notably, PG exhibits unique properties such as a negative Poisson's ratio and exceptionally high ideal strength, surpassing that of graphene. PG has a 2D structure composed of carbon pentagons, resembling the Cairo pentagonal tiling (Figure 8.1). The capacity of PG to separate from T12-carbons, a similar carbon structure, is a fascinating characteristic. These pentagon-based nanotubes, when rolled into nanotubes, display semiconducting qualities regardless of their chirality. Additionally, stable 3D twin structures of T12-carbon can be created by stacking PG in various configurations; these twin structures have even wider band gaps than T12-carbon itself. PG and its derivatives have outstanding adaptability, which points to a variety of possible uses in nanotechnology and nanomechanics. The creation of cutting-edge nanomechanical systems and nanoelectronic devices is made possible by these special features.

The initial stage of the creation of this research required validating earlier, closely related studies. The study carried out by Ajori and Eftekharfar [12] was used as a reference in order to achieve this. As shown in Figure 8.2, their study used MD simulation to evaluate the mechanical characteristics of a PG sheet. To ensure consistency and reliability, the results were cross-examined using a methodology that was similar.

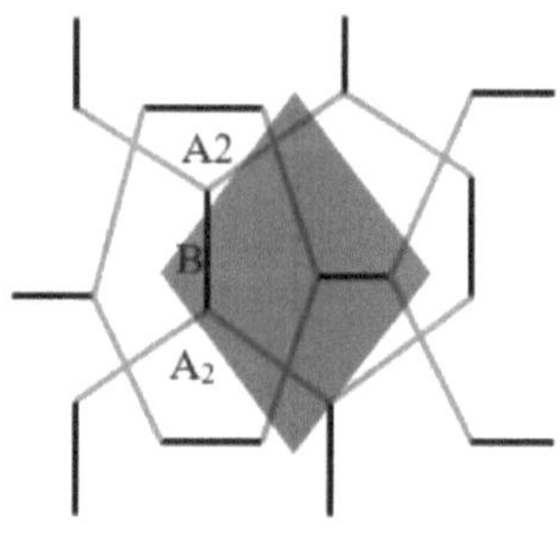

FIGURE 8.1 Lattice structure of penta-graphene.

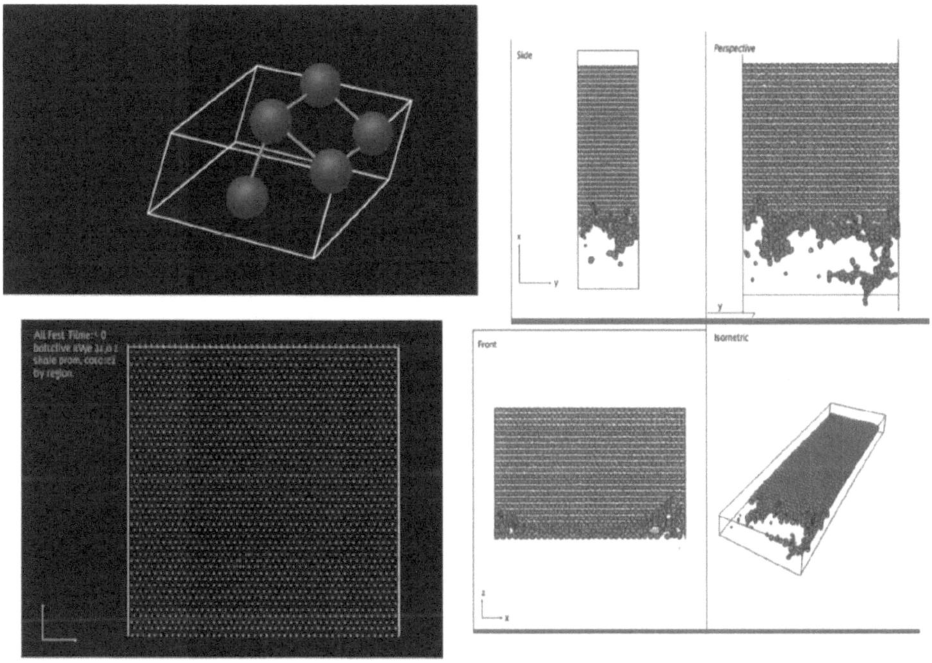

FIGURE 8.2 Validation work (a) Unit cell: Penta-graphene (b) Penta-graphene sheet (30 × 30) (c) PG sheet prepared using MD simulation.

The findings from their research demonstrated that the presence of the PG sheet led to an augmentation in the elastic constants of the PG sheet. The validation results exhibited a close similarity, with a deviation of less than 10%, primarily attributed to a few unidentified parameters that were assumed. Figure 8.3 presents a comprehensive comparison between the outcomes obtained in their study and the results derived from this validation work. Figure 8.4 illustrates a graphical depiction comparing the values of the elastic constants obtained in the study conducted by Ajori and Eftekharfar [12] with the results obtained in the current work.

Following an extensive review of the literature and confirmation of the procedure during the validation work, the primary study got underway using the same methodology as the previous author. In Figure 8.5, the workflow for this research is shown. Two software applications, BIOVIA Material Studio and LAMMP, were used to carry out the theoretical calculations for the MD simulations. Additionally, AVITO software was used to visualize the results. The software used for the purpose of this research has consecutive steps, which are represented visually in Figure 8.6.

8.5.1 Modeling

The following in-order stages were used to conduct the modeling process:

8.5.1.1 Creation of Pristine Graphene Sheet to Study Volume Fraction

To build the PG sheet, graphite structure has been imported from the ceramics section of the material studio software. To obtain the single layer of the PG sheet, graphite has been trimmed to the required size. The graphene sheet (GE) used for the simulation, as per the study done by Ajori and Eftekharfar [12], is shown in Figure 8.7.

Parameters that were taken for the creation of PG structure have been shown in Tables 8.1 and 8.2.

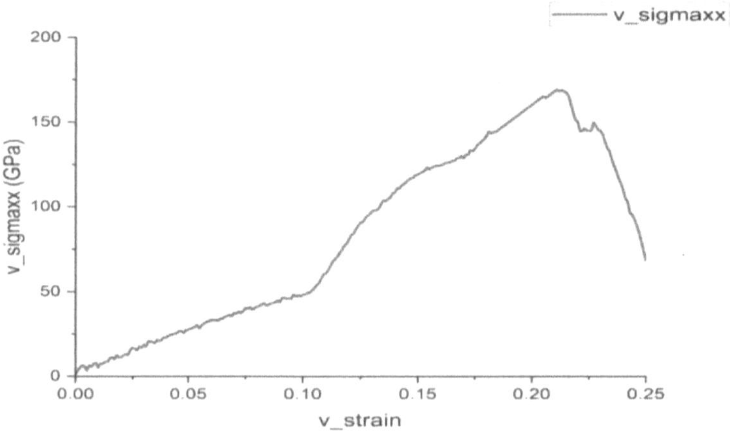

FIGURE 8.3 Validation work.

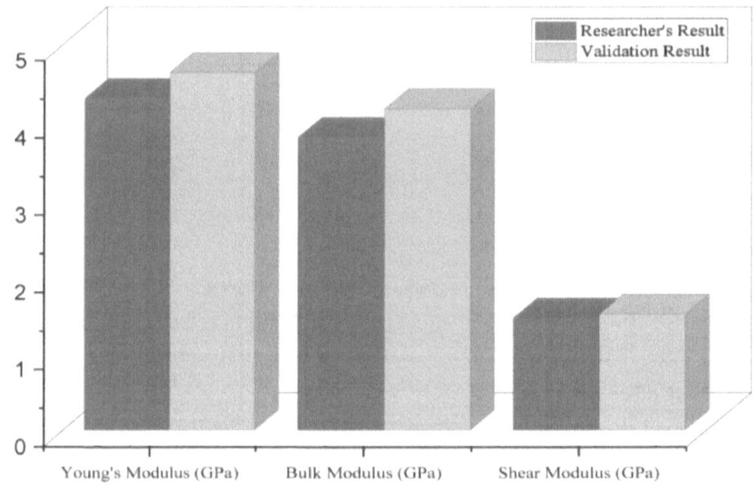

FIGURE 8.4 Research validation.

8.5.1.2 Creation of PDV (Particle Data Visualization)

Define the output: Create the desired output in the data file or input script for LAMMPS. To do this, we have defined the values and have stored in the Particle Data Visualization (PDV) file, such as particle locations, velocities, forces, or any other pertinent characteristics. To generate the necessary data, we might need to utilize the proper LAMMPS commands, such as dump or compute.

8.5.1.3 Create Data File for LAMMPS

The data file contains details regarding the system's shape, types of atoms, atomic locations, and other relevant data. These are the key points:

i. Define the system: Establish the dimensions, form, and make-up of your simulated system. Indicate the quantity, nature, and any other pertinent characteristics of the atoms. Data from experiments, theoretical calculations, or other sources can be used to determine this information.

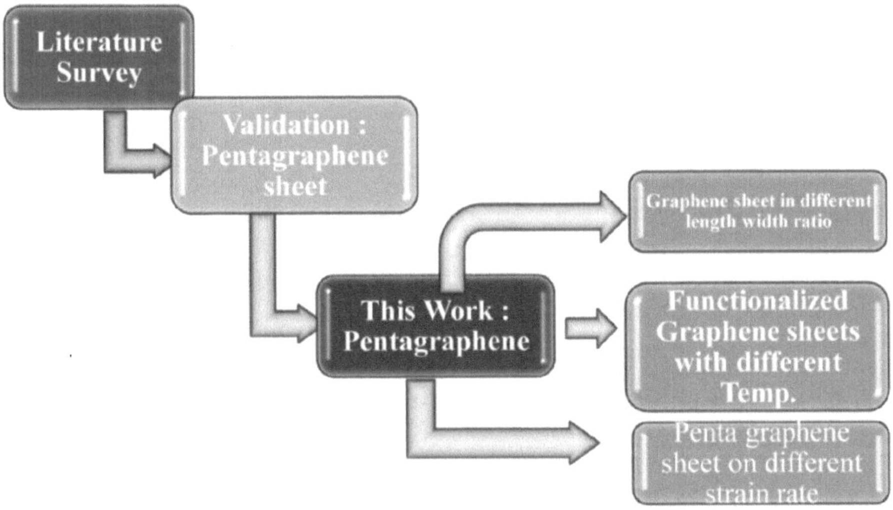

FIGURE 8.5 A workflow for the processes done for this study.

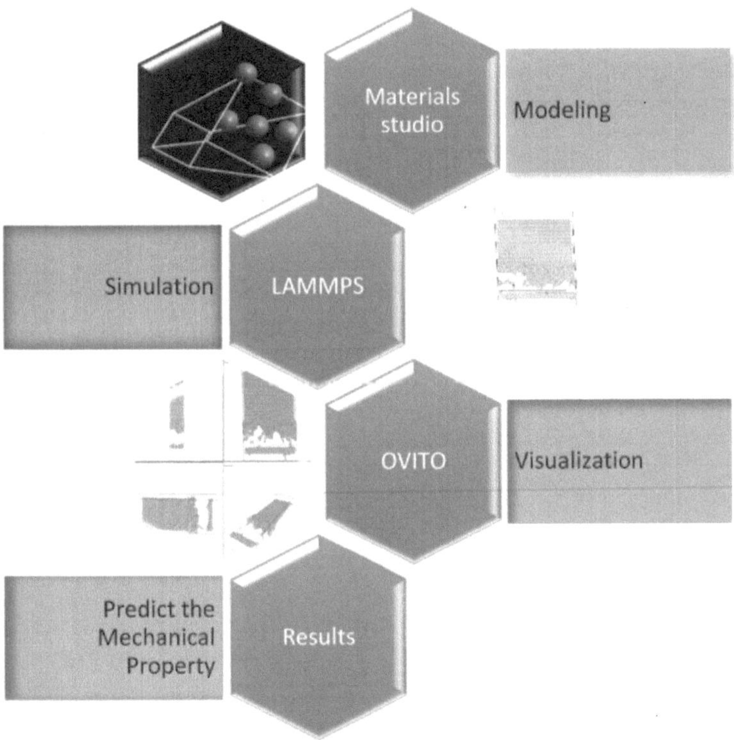

FIGURE 8.6 Molecular dynamics methodology.

ii. Format of data: Put the data in an arrangement that is easy for LAMMPS to read and understand. The data file format normally consists of various parts, each of which contains certain system data. Atom kinds, atomic coordinates, atomic masses, bond topology, and any other necessary information are frequently included in these parts.

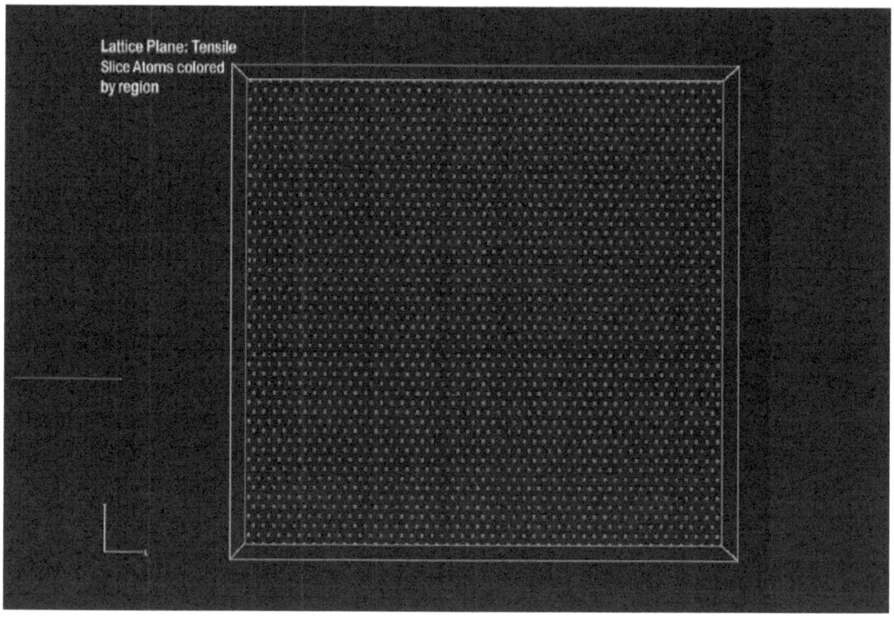

FIGURE 8.7 Penta-graphene monolayer sheet.

TABLE 8.1
Coordinates for the Unit Cell

Coordinates:	3.64271 0.0 0.0 0.0	3.64271 0.0	0.0 0.0 20.0
Cartesian coordinates: -			
C	1.7733055131952025	1.7732970803128694	−1.1466000000000000E−004
C	0.42510952801533969	2.2463562120224903	0.60658372000000005
C	1.3001489098448005	0.42513151175324854	−0.60684369999999999
C	3.1215303668577570	1.3001484945797204	0.60658533999999997
C	2.2463754382552139	3.1215341042659093	−0.60678213999999997
C	−4.8040026695845867E−002	−4.8054710459620564E−002	−1.1410000000000000E−004

TABLE 8.2
Parameters for Creation of Unit Cell

Symmetry	Optimized Lattice Constants (Å)	Bond Length (Å)	Bond Angle
P-421m	a = b = 3.64 (3.64)	C1–C2 = 1.55 (1.55)	C2–C1–C2= 133.5° (134.2°)
space group		C2–C2 = 1.35 (1.35)	C1–C2–C1= 112.2°
113			C2–C2–C1 = 113.2°

iii. Create a new file using a text editor or a scripting language, then write the prepared data into it.

iv. The data file is saved: The data file should be saved with a suitable name and file extension, such as ".data".

TABLE 8.3
Process Parameters

Parameters	Values
Module	Amorphous cell
Cell size	35 × 20 × 6
Forcefield	Airebo
Quality	Fine
Packing Density	0.5 g/cm^3
Runtime	100,000
Time step	0.5 fs
Strain rate	2.5 × 10^9/s, 2.5 × 10^7/s, 2.5 × 10^8
Temperature (K)	300, 500, 700, 900,

8.5.1.4 Import LMP Application from LAMMPS

Material Properties, Biomolecular Simulations, Nanoscale Systems, Material Design, Fluid Dynamics, Reaction Kinetics to determine we can use LAMMPS. LAMMPS is extensively used because of its adaptability, extensibility, and effective parallelization capabilities to investigate and understand the behavior of complex systems at the atomic level.

For the simulation LMP APPLICATION is needed.

8.5.1.5 Potential File (A Forcefield Parameter File)

The forces and energies between atoms or molecules are defined by the potential energy function (PEF), which provides the parameters that are needed. The format of the potential file in LAMMPS typically depends on the forcefield being used. In this simulation, the authors have used "AIREBO" potential. It is a forcefield that simulates the interactions of carbon-based molecules, such as hydrocarbons, in MD simulation. The covalent and non-covalent interactions are both included in the AIREBO potential, which enables it to capture the structural, energetic, and dynamic characteristics of carbon-based systems. It takes into account torsion rotation, non-bonded interactions between atoms, bonding stretches, and bond bending. A potential file or parameter file, which usually includes data on bond lengths, bond angles, torsion angles, force constants, and other factors required for the computation of interatomic forces and energy, is usually given with the potential parameters.

8.5.1.6 In File

An input script file containing all the commands and settings required for defining the simulation setup, specifying the forcefield parameters, setting up the simulation box, defining the time integration parameters, controlling the output, and analyzing the results of the simulation is typically created before a LAMMPS simulation is run. You can add more commands and settings to the input script to suit the needs of your simulation. For comprehensive details on the various commands, options, and their usage, refer to the LAMMPS manual along with appropriate sources. The main parameters used in the process have been shown in Table 8.3 (Figures 8.8–8.10).

8.5.2 Molecular Dynamics Simulations

Using the Forcite module, we ran dynamics simulations that let us watch how atoms moved in response to calculate the forces. Based on modified classical equations of motion, these simulations consider how temperature and pressure affect the systems. We chose a thermodynamic ensemble and entered relevant information, such as the simulation temperature and time step, to carry out these

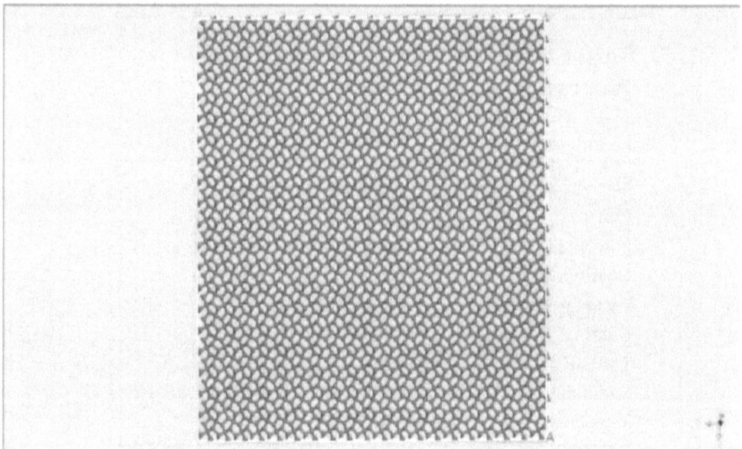

FIGURE 8.8 Monolayer penta-graphene.

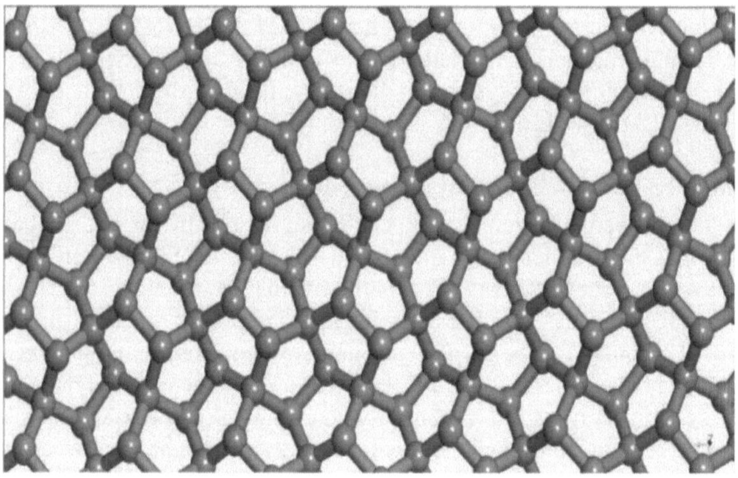

FIGURE 8.9 Pristine penta-graphene sheet.

simulations. To minimize the internal stress of the composite, we used various canonical ensembles, such as NVT (number of particles (N), volume (V), and temperature (T), which are constant) and isothermal–isobaric ensembles, such as NPT (number of particles (N), pressure (P), and temperature (T) are fixed). These ensembles allowed us to regulate the temperature and pressure during the simulations, which resulted in a more precise characterization of the system's behaviors. Using this method, we were able to investigate the dynamics of the composite material and watch how each atom responded to the applied forces. We obtained an understanding of the material's structural changes and its reaction to external environments by considering the impacts of temperature and pressure.

8.5.2.1 Mechanical Properties

PG's mechanical characteristics were also studied as part of the Forcite module. In this study, three specific mechanical properties have been investigated:

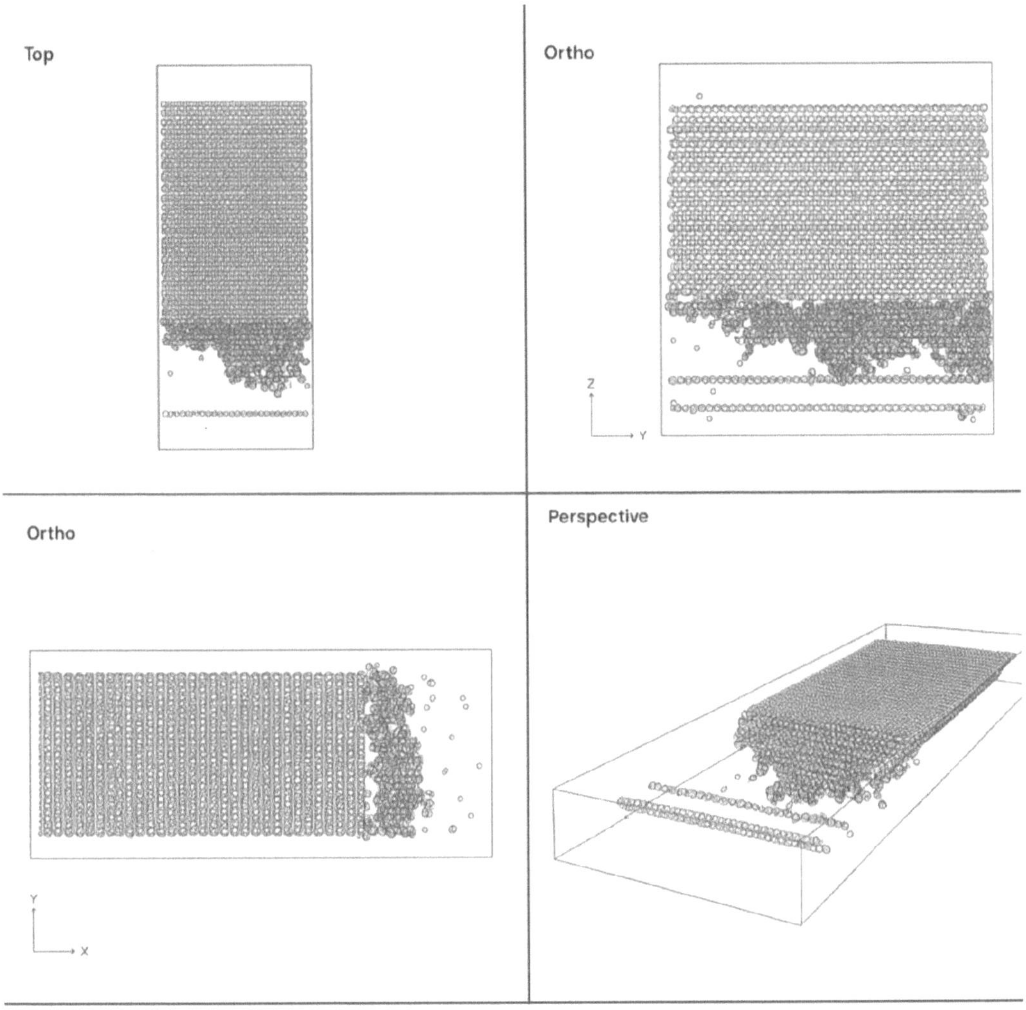

FIGURE 8.10 After applying tensile force, functionalized penta-graphene sheets.

i. Elastic Constants

The stresses used in the constant strain approach have a maximum strain amplitude of 0.003 and were applied in the Cartesian Y direction. Then, internal relaxation was achieved by optimizing the structure. The virial stress idea was then used to calculate the stress in each direction. The 35 × 20 stiffness matrix (C_{ij}) and compliance matrix (S_{ij}) can be obtained using the following relations to characterize the stress–strain behavior:

$$\sigma_i = C_{ij}\varepsilon_j \tag{8.30}$$

$$\varepsilon_i = S_{ij}\sigma_j \tag{8.31}$$

ii. Young's modulus (E)

Young's modulus is a mechanical property that represents the stiffness of an elastic material when a force is applied along its length. It is calculated as the ratio of the longitudinal stress (σ) to the longitudinal strain (ε):

$$E_i = \frac{\sigma_i}{\varepsilon_i} \tag{8.32}$$

Only in the region where the stress is inversely proportional to the strain and the material behaves elastically is Young's modulus relevant and useful. When the external force is eliminated in this elastic zone, the material resumes its original shape. However, as the applied stresses continue to rise, the material can flow, undergo permanent deformation, or even fracture beyond the elastic limit.

iii. Bulk modulus (B)

A fundamental elastic constant known as the bulk modulus measures a material's capacity to compress. It is calculated by dividing the fractional volume compression that the material underwent by the ratio of the pressure increase. In other words, it gauges a material's resistance to volume changes brought on by applied pressure. The bulk modulus sheds important light on how materials react when compressed or compressed. Understanding how a material will behave and deform in response to outside forces or pressure changes is useful. The study of fluid mechanics, acoustics, and materials science are a few engineering applications that make use of the bulk modulus as well:

$$K = \frac{dP}{\left(-\frac{dV}{V}\right)} \tag{8.33}$$

Voigt–Reuss–Hill approximation was used to get a mean of the bulk modulus:

$$\boldsymbol{B}_H = \frac{\left[\boldsymbol{B}_V + \boldsymbol{B}_R\right]}{2} \tag{8.34}$$

iv. Shear Modulus (S)

The key property of a material that describes its resistance to shearing deformation is the shear modulus. Rapid transmission of shear energy is made possible by a material's high resistance to shearing. The ratio of shear stress (τ) to shear strain (γ) is known as the shear modulus. In plainer language, it measures how well a material can withstand being bent or distorted by shear pressures. A stiffer material with a higher shear modulus may resist greater shearing forces without significantly deforming

$$S = \frac{\tau}{\gamma} \tag{8.35}$$

Voigt–Reuss–Hill approximation was used to get a mean of the bulk modulus.

$$S_H = \frac{S_V + S_R}{2} \tag{8.36}$$

v. Stress–Strain Behavior

A PG sheet undergoes internal pressures causing it to deform when it is stressed. Force applied on the sheet per unit area is known as the stress, and the deformation that results in relation to the original size of the sheet is known as the strain. There are essential characteristics that can be used to define the stress–strain behavior of a PG sheet;

Elastic Region:
The PG sheet shows elastic behavior under low tension, which results in a reversible deformation. Sheet regains its natural shape after the stress is released without sustaining any long-lasting distortion. According to Hooke's law, the relation between stress and strain in this area is linear in nature.

Plastic Deformation:
PG sheet experiences a plastic deformation when it has passed the yield point, which means its deformation occurs more as the stress increases. Sheets go through localized structural modifications as the connection between stress and strain becomes nonlinear.

Yield Point:
PG sheet may reach a yield point as the stress rises, at which point plastic deformation starts to occur. Even after the force is released, the material continues to have ongoing deformation at this location. The material's elastic behaviors are limited by the yield point.

Failure Point:
Failure or fracture point of the PG sheet is reached. Stress level is the point at which the sheet can no longer sustain deformations and snaps. Failure point shows the material's maximum strength.

The following procedures have been taken in order to use LAMMPS software for predicting the stress–strain behavior:

a. The strain–stress curve of the PG sheet throughout the tensile process was obtained using the constant-strain method.
b. To obtain the stress–strain curve, a code is used that is available under the MD option of LAMMPS software. With the OVITO Materials Visualizer, we provide all data to MD and documents, allowing us to develop, modify, and debug. The code before running them on the command prompt.
c. The stress–strain code calculates strain by applying stresses in one direction while applying none in the others.
d. The output is a study table that contains the values of strains for different applied and measured stresses and is designed to be run from the Command prompt (LAMMP).
e. Stress–strain curves have been created for the various PG volume fractions using these data.

vi. Creep behavior

Studying the creep behavior regarding engineering materials, particularly regarding time and temperature, is necessary to determine their long-term durability and efficacy. When a material is subjected to a steady load or stress over an extended period without going beyond its yield strength, a process known as creep develops. It causes an irreversible, gradual, and permanent deformation of the material. Creep behaviors typically depend on temperatures, which are frequently close to the melting point of the material. When

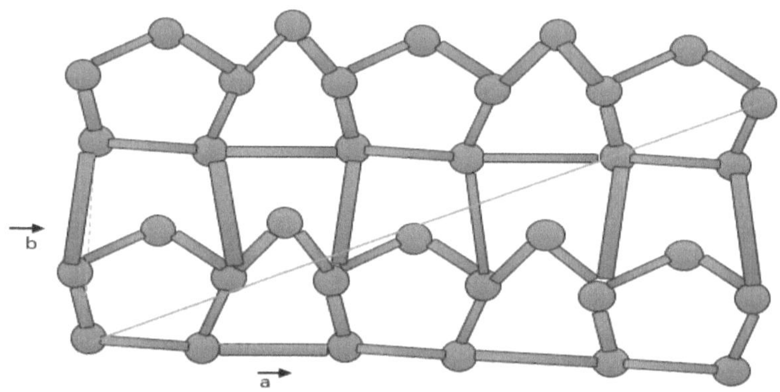

FIGURE 8.11 Creep behavior (stretched PG sheet).

developing PG for applications such as hydrostatic pressure containers, sealings, joints, and fittings that face consistent loads, it is essential to understand creep behaviors. PG experiences creep because of the substance's viscoelastic deformation of the material. Researchers have investigated adding nanoparticles to materials, especially PG compounds, to improve the creep resistance and stress–strain behaviors. Utilizing the stress–strain code imported into LAMMP software, the uniaxial tensile stress method is used to analyze the creep behavior of PG. All the PG crystals are initially subjected to a load equal to 60% of the yield strength determined from the behavior of the material. The load is then raised gradually and incrementally until it is below the yield strength. The same stress is then applied repeatedly over a long period of time within a simulation. Researchers hope to learn more about the creep resistance displayed by PG sheets by evaluating various loads. For determining PG's applicability for diverse applications and optimizing its mechanical properties for long-term viability under continual loads or strains, creep behavior in PG must be studied (Figure 8.11).

8.6 RESULTS AND DISCUSSION

This section discusses the outcomes of the PG sheet's MD simulation. The effects of different length-to-width (L/W) ratios of PG on the mechanical properties of pristine PG have been comparatively studied. The impact of functionalization of PG sheets has also been discussed in this section. The section has been divided into two parts, one to study the effect of different temperatures as well as different strain rates of PG sheets and the other for the effect of the L/W ratio of PG sheets.

8.6.1 Effect of Temperature

The characteristics and behavior of PG are significantly influenced by temperature. To study the effect of PG models have been created based on the temperature of PG sheet (i.e., 300 K, 500 K, 700 K, and 900 K). The following mechanical properties have been predicted using MD simulations.

The variation in stress with temperature (T) is shown in Figure 8.12. The curves in each set are generally found to be fairly similar before failure. Thus, it can be inferred that, at a considerable distance from the failure point, the relationship between stress and engineering strain remains relatively unaffected by variations in temperature. Young's modulus shares this trait in that it varies only slightly as temperature rises. On the other hand, as the temperature rises, PG's ultimate stress, ultimate strain, and toughness tend to decrease. The sole deviation from this pattern is seen around 900 K, where the material exhibits ductile behavior. The yield point is shown on each graph at this temperature. The existence of a hardening phase at low temperatures is clear when examining the stress–strain behavior of PG at various temperatures. The cutoff function, angle shifts between covalent bonds, and bond elongation are the three causes of this phenomenon. The cutoff function is a mathematical function that acts independently of physical factors like temperature and is a crucial component of the PEF. It is crucial for the atomistic simulation of nanostructures in many different ways. The stress–strain relationship is affected by this function because it controls how close atoms are to one another, but it is unaffected by temperature. Angle changes between covalent bonds and bond elongation are the two main elements that must be taken into account in addition to the tension process in order to more accurately explain the hardness that is observed at lower temperatures. There are two main phases that precede bond rupture. The smooth area of the strain–stress graph reflects the variations in bond angles that occur during the initial loading steps. These angle shifts, however, have a limit because atoms can't go too near to one another. The second stage officially begins at this point, and the hardening segment is now visible. Bond elongation becomes prominent at this stage, resulting in considerable changes to bond length and stronger covalent bonding. As a result, the stress is far greater than it was during the initial loading stage. The difference in fracture temperature compared to perfect PG is largely caused by lower temperatures, which severely restrict atom movement. Other parameters, besides Young's modulus and Poisson's ratio, converge to comparable values as the temperature rises sufficiently. A hardening segment is present at low temperatures, according to the stress–strain behaviors of PPG at various temperatures. Bond elongation, angle moves between covalent bonds, and the cutoff function are some of the causes of this hardening. These elements, together with the constrained atomic movement at low temperatures, produce a different stress–strain response in contrast to high temperatures or perfect PG (Table 8.4).

8.6.2 Stress–Strain Behavior

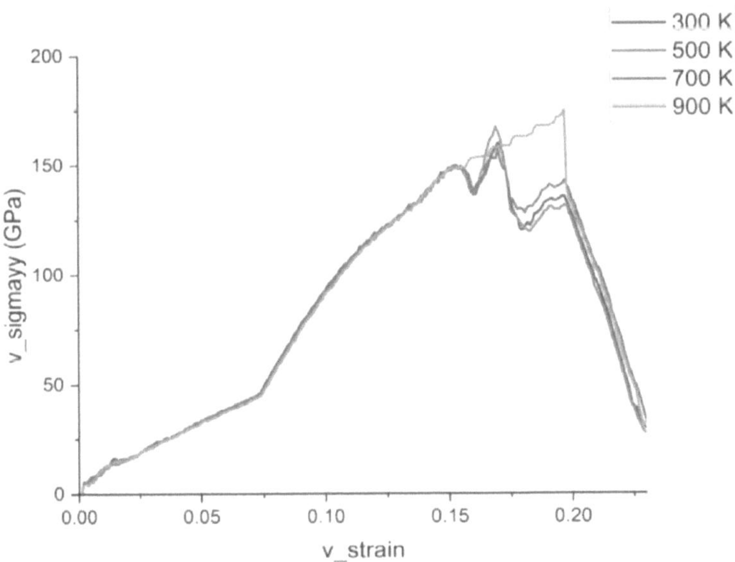

FIGURE 8.12 Effect of temperature on stress–strain behavior of penta-graphene.

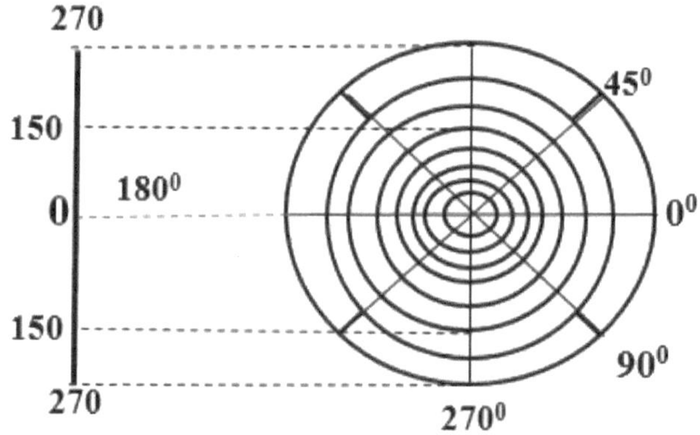

FIGURE 8.13 Young's modulus (N/m).

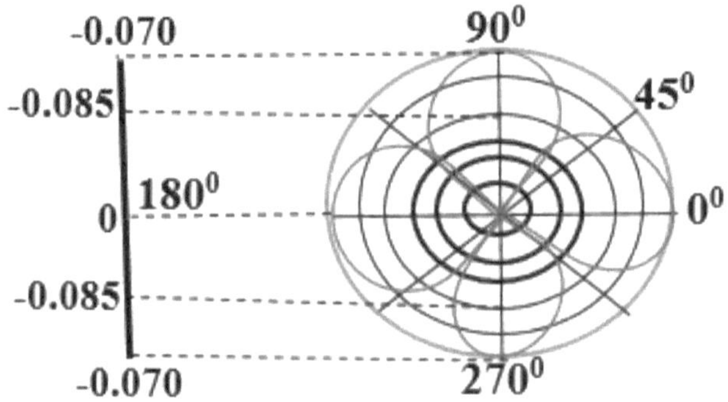

FIGURE 8.14 Poisson's ratio.

TABLE 8.4
Results of Ultimate Tensile Strength, Fracture Strain, and Elastic Modulus for Different Temperatures

Temperature (K)	Number of Atoms	UTS	FS	Elastic Modulus
300	1014	149.2614	0.17	114.85
500	1014	167.384	0.169	139.744
700	1014	157.019	0.17	161.922
900	1014	148.0017	0.21	539.174

8.6.3 Effect of Strain Rate

The characteristics and behavior of PG are significantly influenced by different strain rates. To study the effect, PG models have been created based on the different strain rates of PG sheets, i.e., (1.0×10^{10}, 1.0×10^{9}, 1.0×10^{8}). The following mechanical properties have been predicted using MD simulations. The results show that the applied strain rate has a considerable impact on the mechanical response and failure behavior of PG. For accurate PG performance prediction in real-world applications, it is vital to understand this strain rate dependence. PG sheets' mechanical characteristics, such as stiffness, strength, and toughness, depend on the strain rate. PG shows increased stiffness and strength at lower strain rates, allowing it to bear greater external loads. PG, on the other hand, exhibits decreased stiffness and strength at greater strain rates, resulting in lower load-carrying capacity. The mechanical behavior of PG sheets is significantly affected by the applied strain rate. Lower strain rates result in a more ductile response with increased plasticity, while higher strain rates result in more brittle behavior with reduced plastic deformation (Figures 8.13 and 8.14).

8.6.4 Impact of Varying Length and Width Ratio of Penta-Graphene Sheets

The impact of changing the length and width of PG sheets on their mechanical characteristics has been examined in this work using MD simulations. The findings show that the ultimate tensile strength (UTS) and failure behavior depend on the L/W ratio, even though the mechanical parameters are initially equal for sheets of varied dimensions. Notably, an L/W ratio of 2 results in the maximum UTS. These findings provide useful information for modifying the characteristics of PG. PG sheets exhibit identical mechanical characteristics, such as linear elastic deformation and yielding, up to a specific strain, independent of the sheet dimensions, as shown in Figure 8.15. This behavior shows that the length and width variations have no impact on the initial structural integrity. A significant mechanical characteristic known as the UTS was found to be dependent on the L/W ratio of PG sheets. Remarkably, the greatest UTS was identified at a 2 L/W ratio. This ratio allowed sheets to resist tensile deformation better, which allowed them to withstand greater stress before failure (Tables 8.5 and 8.6).

8.6.5 Structure's Strength and the L/W Ratio

Because PG sheets of various sizes exhibit comparable behavior before failure, it may be that interatomic bonding and lattice structure, rather than sheet size, have a greater impact in the overall

TABLE 8.5
Results of Ultimate Tensile Strength, Fracture Strain, and Elastic Modulus for Different Strain Rates

Strain Rates	Number of Atoms	UTS	FS	Elastic Modulus
0.01/s	1014	136.7451	0.2	540.67
0.001/s	1014	144.579	0.18	543.59
0.0001/s	1014	147.375	0.18	531.181

TABLE 8.6
Results of Ultimate Tensile Strength, Fracture Strain, and Elastic Modulus for Different Length and Width Ratios

Length and Width Ratio	Number of Atoms	UTS	FS	Elastic Modulus
0.1	360	76.06919	0.23	232.331
0.2	480	85.47744	0.19	316.302
0.3	720	106.4125	0.19	387.283
0.75	2400	127.5868	0.21	435.44
01	4800	146.0228	0.21	524.381
1.5	3600	141.9959	0.21	499.297
1.75	4200	138.6344	O.21	493.375
02	5600	149.26134	0.21	532.376
2.5	6000	144.731	O.21	524.7
03	7200	144.44131	0.21	517.361

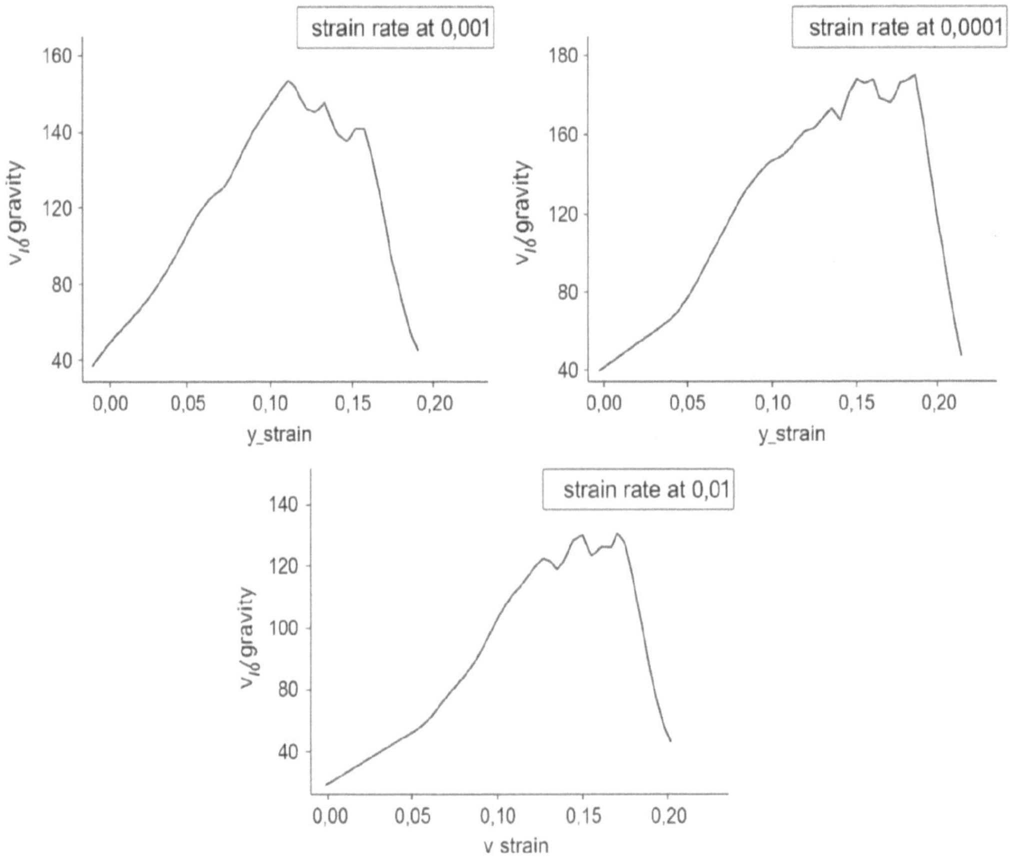

FIGURE 8.15 Stress–strain behavior at different strain rates.

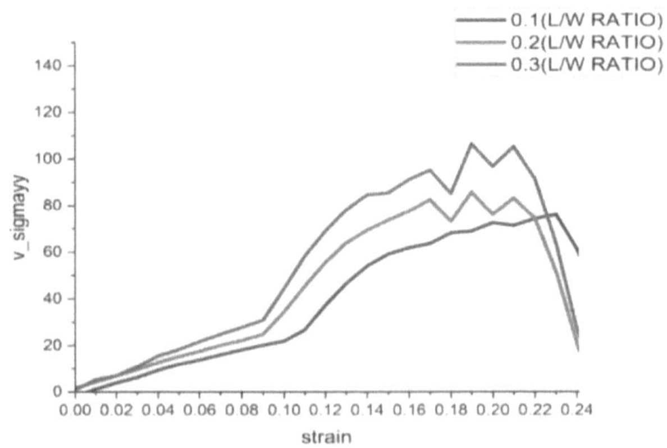

FIGURE 8.16 Stress–strain behavior at different aspect ratios.

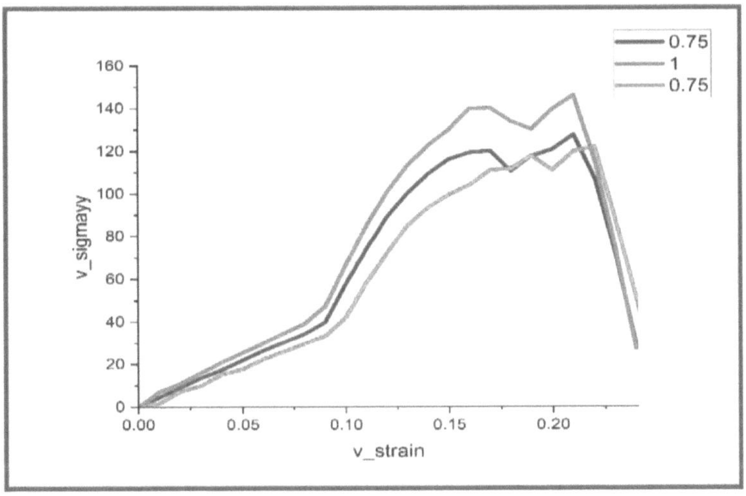

FIGURE 8.17 Stress–strain behavior at different aspect ratios.

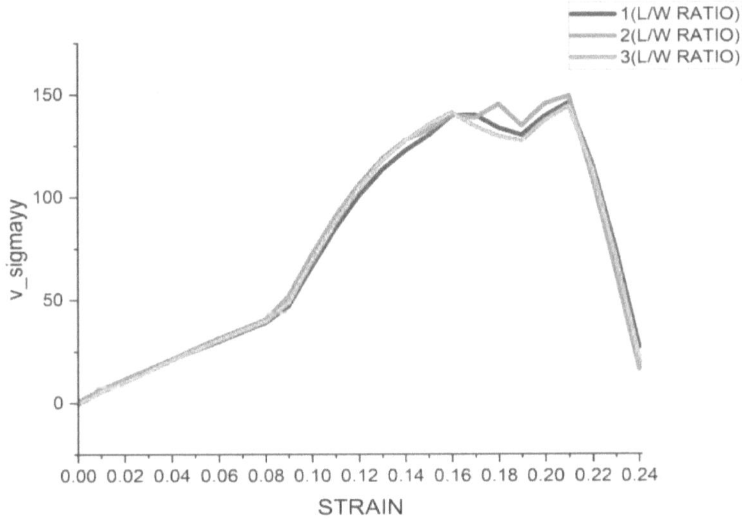

FIGURE 8.18 Stress–strain behavior at different aspect ratios.

structural stability. That it is showing the maximum UTS results from an L/W ratio of 2 indicates a connection between sheet geometry and mechanical strength. The creation of PG-based materials with higher mechanical characteristics can be influenced by this realization. PG sheets' mechanical characteristics, particularly their UTS and failure behavior, are impacted by changes in their length and width (Figures 8.16–8.19).

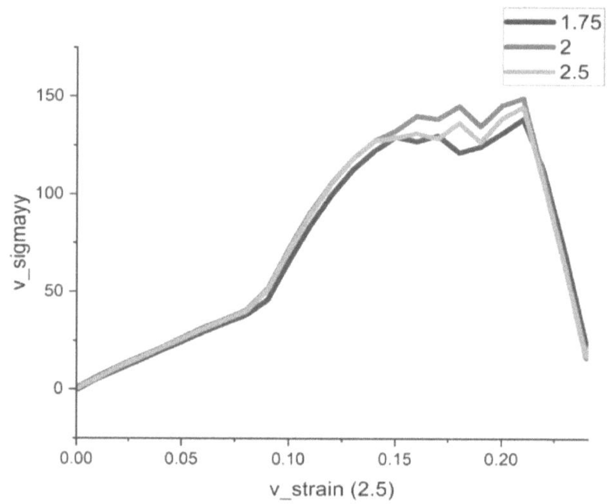

FIGURE 8.19 Stress–strain behavior at different aspect ratios.

REFERENCES

1. Zhang, S., Zhou, J., Wang, Q., Chen, X., Kawazoe, Y., & Jena, P. (2015). Penta-graphene: A new carbon allotrope. *Proceedings of the National Academy of Sciences of the United States of America*, 112(8), 2372–2377. https://doi.org/10.1073/PNAS.1416591112
2. Miranda, W. D. S. A., Frazão, N. F., Moreira, E., & Azevedo, D. L. (2022). Penta-belt: A new carbon nanobelt. *Journal of Molecular Structure*, 1263. Elsevier B.V. https://doi.org/10.1016/j.molstruc.2022.133055
3. Zhang, B., Sun, J. Y., Ruan, M. Y., & Gao, P. X. (2020). Tailoring two-dimensional nanomaterials by structural engineering for chemical and biological sensing. *Sensors and Actuators Reports*, 2(1), 100024. https://doi.org/10.1016/J.SNR.2020.100024
4. Zhang, Y. Y., Pei, Q. X., Cheng, Y., Zhang, Y. W., & Zhang, X. (2017). Thermal conductivity of penta-graphene: The role of chemical functionalization. *Computational Materials Science*, 137, 195–200. https://doi.org/10.1016/j.commatsci.2017.05.042
5. Nazir, M. A., Hassan, A., Shen, Y., & Wang, Q. (2022). Research progress on penta-graphene and its related materials: Properties and applications. *Nano Today*, 44, 101501. https://doi.org/10.1016/J.NANTOD.2022.101501
6. Lee, M. Q. (2017). Mechanical properties of penta-graphene, hydrogenated penta-graphene, and penta-CN2 sheets. *Computational Materials Science*, 136, 181–190. https://doi.org/10.1016/j.commatsci.2017.05.004
7. Niu, M., Miao, L., Xu, Z., Zhao, Y., Hao, W., Li, J., Li, J., Sui, C., & Wang, C. (2022). Tunable tensile mechanical properties of bilayer graphene through inter-layer rotation. *Diamond and Related Materials*, 130, 109522. https://doi.org/10.1016/j.diamond.2022.109522
8. Feng, Y. T., Li, H. B., & Shao, Z. G. (2021). F2 adsorption on penta-graphene: A first-principles study. *Surfaces and Interfaces*, 27. https://doi.org/10.1016/j.surfin.2021.101569
9. Hassan, A., Guo, Y., Wang, Q., Kawazoe, Y., & Jena, P. (2019). Interfacial properties of penta-graphene-metal contacts. *Journal of Applied Physics*, 125(6). https://doi.org/10.1063/1.5085414
10. Torabi Rad, M., & Foroutan, M. (2022). Wettability of penta-graphene: A molecular dynamics simulation approach. *Journal of Physical Chemistry C*, 126(3). https://doi.org/10.1021/acs.jpcc.1c06826

11. Sharma, S., Kumar, P., Chandra, R., Singh, S. P., Mandal, A., & Dondapati, R. S. (2019). Overview of BIOVIA materials studio, LAMMPS, and GROMACS. In *Molecular Dynamics Simulation of Nanocomposites using BIOVIA Materials Studio, Lammps and Gromacs*. https://doi.org/10.1016/B978-0-12-816954-4.00002-4
12. Ajori, S., & Eftekharfar, A. R. (2022). Mechanical properties and fracture analysis of defective penta-graphene under temperature variation: Insight from molecular dynamics. *Diamond and Related Materials*, 124. https://doi.org/10.1016/j.diamond.2022.108956

9 Blast Performance Assessment of Metallic Plate and Composite Laminate

A Comparative Numerical Study

Murlidhar Patel, Deepak Kumar, and Shivdayal Patel

9.1 INTRODUCTION

The attacks with explosives carried out by terrorists pose a serious threat that has immediate and far-reaching effects [1]. Deaths and injuries account for most of the catastrophic human expenditure. In addition to their physical effects, these attacks leave communities with long-lasting scars of fear and worry. Disruption in society is widespread, impacting everyday routines and vital services. There has been significant economic damage, necessitating money for compensation and restoration. The design of protective structures is essential for reducing the harmful consequences of terrorist blasts [2,3]. One way to lessen the effects of explosions is to use materials that are resistant to blasts, provide redundancy to the structure, and create barriers. The capacity of a structure to resist shock waves and fragmentation is increased by specialized features, including reinforced structural parts made of high-strength materials [4]. AL-6XN steel's superior corrosion resistance, high strength, ductility, and toughness make it an excellent option for blast-resistant applications [5]. This austenitic stainless steel alloy resists explosive blast assaults by terrorists by withstanding dynamic stresses and retaining mechanical characteristics at high temperatures, hence providing structural integrity. Its resistance to fragmentation improves occupant safety, and its weldability makes it easier to construct sturdy structures. Because of its minimal maintenance requirements and capacity to absorb and disperse energy during explosions, AL-6XN steel is a dependable material for creating blast-resistant structures. The outstanding characteristics of the AL-6XN steel, as compared to the other grades of steel, highlight how well it protects buildings and their people from the damaging effects of explosive blasts [6]. The utilization of fiber-reinforced polymer composites, such as carbon fiber-reinforced polymer (CFRP), basalt fiber-reinforced polymer (BFRP), and Kevlar fiber-reinforced polymer (KFRP) composites, is essential in mitigating the damaging consequences of terrorist explosive blast assaults on structural elements [7]. Because of their exceptional strength-to-weight ratio, these composites allow for lightweight construction while maintaining the strength required to withstand blast pressures. Their capacity to absorb and disperse energy reduces structural damage during explosions [8]. Researchers are continuously trying to precisely manage how composite materials respond to blast forces relative to their specialized design options, which maximize structural performance. These materials also have a lesser chance of fragmentation, which improves occupant safety, and they are resistant to corrosion, which lowers maintenance costs [9–11]. Fiber-reinforced composites, with their adaptability and simplicity of installation, have become indispensable portions of the building of blast-resistant architecture [12].

Many researchers previously determined the performance of various metallic plates and composite laminates under air blast loadings. The influence of boundary conditions on the deformation behavior of an isometric plate against blast loadings was investigated by Mohammadzadeh et al.

 DOI: 10.1201/9781003664611-10

[13]. They reported that the fixed plate provides higher blast resistance than the simply supported plate. The effect of steel plate curvature on the blast performance was predicted by Veeredhi et al. [14]. They noted that the increase in the curvature of the plate increased the blast resistance of the plate. Yuan et al. [15] experimentally, numerically, and analytically determined the AA6061 plates' dynamic response under confined blast loadings. They found that the peak overpressure on the interface and saturated impulse don't show a linear relationship with stand-off distance (SoD). Neuberger et al. [16] experimentally and numerically analyzed the blast performance of a scaled circular steel plate and found a good correlation between experimental and numerical results. Tian et al. [17] investigated the blast resistance behavior of the circular steel plate under repeated blast loadings. They noted that the mid-point deflection of the plate increases with an increase in the number of blast loadings applied and verified their newly proposed method to solve repeated blast loading problems. Gan et al. [18] studied the effect of explosive charge shapes on the blast performance of steel plates. They found that the cubical-shaped explosive significantly deforms the plate as compared to the spherical-shaped explosive blast. Zhang et al. [19] determined the effect of various shapes of pre-formed holes on the steel plates' blast resistance under confined conditions and also compared their findings with the same mass of solid steel plate's blast resistance. They pointed out that solid plates deform higher than plates with pre-formed holes under similar conditions of blast loading. The effects of uncoated, back face polyurea-coated, and both-face polyurea-coated AA6063 plate blast mitigation capabilities were discovered by Zhu et al. [20]. They noted that the back face-coated Al plate provides the highest deformation resistance against blast loadings. In the case of combined blast and fragment loadings, the aluminum plate coated on both sides showed high resistance to deformation [21]. Mohotti et al. [22] reported the effectiveness of the polyurea coating on the blast performance of the steel plate. They found that the polyurea coating on the steel plate enhances the blast resistance of the steel plate. The different configurations of polyurea and steel used in plates' blast performance were discovered by Hou et al. [23]. They observed that the back face polyurea-coated steel plates have the lowest peak deflection under blast loading as compared to other configurations. Zhang et al. [24] evaluated the blast resistance of the polyurea-coated steel plates and the uncoated steel plates of equal masses under confined conditions. They found that the polyurea-coated steel plate provides inferior blast resistance to the same mass of steel plate. A reduction in the thickness of the polyurea coating on a steel plate while maintaining the total mass of the plate's constant improved the deflection resistance of the plate under confined blast loads. Ackland et al. [25] also confirm that the steel plate provides better protection against blast loading in comparison with the same mass of polyurea-coated steel plate.

In comparison to fiber-reinforced composites (i.e., CFRP, BFRP, or KFRP), AL-6XN steel has lower specific strength and stiffness, is heavier, and has poorer fatigue resistance. However, it does have strong mechanical properties and resistance to corrosion. Better weight-to-strength ratios, improved fatigue resistance, and increased design flexibility are offered by fiber-reinforced composite materials, whereas steel manufacture is intricate and expensive. In addition to being more easily repaired, composites perform well in extreme conditions. The decision between AL-6XN steel and composites ultimately comes down to factors like manufacturing complexity, corrosion resistance, weight considerations, and mechanical properties. Composites are frequently chosen for applications where design flexibility, fatigue resistance, and weight reduction are crucial [26–28]. Langdon and Cantwell [29] reported that the natural fiber-reinforced polymer composite can't provide better blast resistance than the CFRP and glass fiber-reinforced polymer (GFRP) composites. They also reported that the woven fiber-reinforced polymer composite has good deformation resistance under blast loading [30]. Zhu and Khanna [31] also designed a protective structure of transparent GFRP composite laminate that can sustain high-intensity blast loads. Li et al. [32] analyzed the effect of fragment impact loadings, blast loading, and combined blast and fragment loading on the damage behaviors of GFRP laminate. They observed that fiber tensile failure as well as delamination are the major damage problems in the composite laminate for both the impact loading and the combined blast and fragment loading cases. For purely blast loading cases, delamination is the main damage problem in the laminate. Zhao et al. [33] reported that the combined blast and fragment loading increased the deflection and severe deformation in the composite laminates. Süsler et al. [7] investigated the various types of fiber-reinforced composite laminates' blast performance. They revealed that the CFRP laminate has higher deflection resistance characteristics than the GFRP, KFRP, and BFRP composite laminates. Gargano et al. [8] experimentally and numerically evaluated that the overpressure on the CFRP laminate increased with a decrease in SoD for the same mass of explosive detonation. Results in the deformation of the CFRP laminate increased. Fontaine et al. [34] experimentally discovered the blast responses of long-term sea water-exposed CFRP laminate and non-sea water-exposed CFRP laminate. They noted that the CFRP laminate's long-term exposure to sea water has marginally lower blast resistance than the non-exposed CFRP laminates. Lee et al. [35] also analyzed the blast performance of CFRP laminate by using different techniques, and they suggested that equivalent air blast overpressure is a good technique for predicting the blast performance of the laminate. Wanchoo et al. [36] studied the CFRP laminates of rubber-modified and non-rubber-modified resins and their blast performances. They reported that the rubber-modified resin used in the composite laminate has improved the deformation resistance. Liu et al. [37] experimentally investigated the effect of the position of the polyurea coating on the blast mitigation performance of the CFRP laminate and also compared their performance with that of uncoated CFRP. But they failed to keep all the laminates' masses constant. They observed that the CFRP laminates' deflection and deformation resistance improved with polyurea coating and found that the coating on the back face of the laminate provided a more significant enhancement in blast resistance. Reda Taha et al. [38] evaluated the blast performance of the carbon nanotube-added CFRP composite laminates. They noted that the addition of carbon nanotubes enhanced the blast resistance of the CFRP laminate. A similar response for carbon nanotube-added CFRP composite laminate was also observed by Altunc et al. [39]. Zhou et al. [40] experimentally demonstrated the GFRP and Al-incorporated fiber metal laminates' (FMLs') blast resistance under different conditions of blast loading with the same mass of explosive detonation. They found that the FMLs were highly deformed under fully confined blast conditions as compared to fully vented blast conditions. Xu et al. [41] also analyzed the blast resistance of GFRP and Al-incorporated FMLs. They observed that the perforated area of the FML increased with an increase in the mass of the explosive for blast loading.

The summary of this literature survey depicts that the steel plates have superior blast mitigation characteristics in comparison with the same mass of various configurations of polyurea-coated steel plates. In the case of composite laminates, the blast resistance characteristics of CFRP are incomparable with those of other composite laminates. However, this chapter deals with an effective comparison of the blast performances of the steel plate and equivalent CFRP composite laminates. For this effective comparative analysis, a series of numerical simulations have been conducted to analyze their center point deflection response, kinetic energy profile, energy absorption characteristics, and damage behavior under the various air blast loadings.

9.2 COMPUTATIONAL MODELING

9.2.1 Modeling of Air Blast Loading

In air blast loading, a rapid and large amount of energy is released due to an explosive charge's explosion, resulting in a blast shock wave created by the compression of surrounding air. This generated shock wave expanded with very high velocity and struck the object [42]. Due to this interaction of the shock wave with the object, the variations in pressure profile on the object's surface specific point are represented in Figure 9.1. After the detonation of an explosive, a generated shock wave arrives at the target at time (t_{ar}) and suddenly increases the pressure on the target surface from ambient pressure (P_a) to peak overpressure (P_o). This increased pressure gradually and

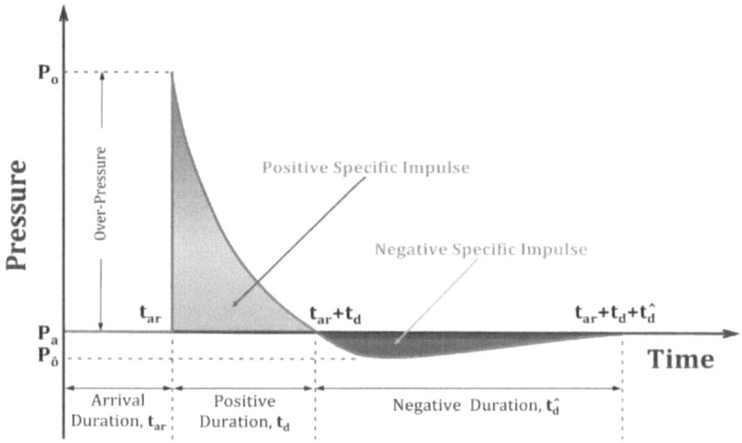

FIGURE 9.1 A variation in pressure profile on a specific point of target under blast loading.

exponentially decreases to ambient pressure (P_a) at time (t_d), and this pressure on the target further decays to peak negative pressure ($P_{\hat{o}}$) and again finally reaches ambient pressure (P_a) after time ($t_{\hat{d}}$). However, this pressure variation on the target is divided into two phases: the first phase is called the positive pressure phase (from t_{ar} to t_{ar} + t_d), and the second phase is the negative pressure phase (t_{ar} + t_d to t_{ar} + t_d + $_{t\hat{d}}$). The area under the curve of the positive pressure phase indicates the positive specific impulse (I^p) applied to the target by the shock wave. Similarly, the area under the curve of the negative pressure phase represents the negative specific impulse (I^n) applied to the target by the shock wave [43]. The target is mainly damaged during the positive pressure phase, and the duration of the negative pressure phase is too long as compared to the positive pressure phase [44]. Hence, the study of this phase is mostly considered by the researchers to determine the damage behavior of the protective structures under blast loadings. The pressure changes in free-field blast loading are characterized by the modified Friedlander equation as represented in Equation (9.1), and the positive specific impulse is calculated by using Equation (9.2) given as follows [5,45]:

$$P(t) = \left(P_o - P_a\right)\left(1 - \frac{t - t_{ar}}{t_d}\right)e^{-(t-t_{ar})\varnothing/t_d} \tag{9.1}$$

$$I^p = \int_{t_{ar}}^{t_{ar}+t_d} P(t)\,dt \tag{9.2}$$

where the symbol ∅ denotes the time decay constant.

In the present chapter, the in-built Conventional Weapons Effects Program (CONWEP) module was used for the calculation of these pressures and specific impulses.

9.2.2 Specifications of Geometry, Meshing, and Boundary Conditions

In this chapter, to study the blast mitigation behavior of both the steel plate and CFRP composite laminate of equal area densities, only their quarter parts were designed using Abaqus CAE. The designed plate and laminates are symmetrical about two planes (x–z and y–z planes), which permit the design of quarter models. Because the mirror pattern facility available in Abaqus can provide full model results. The study of quarter parts helps to save the time required for computation and

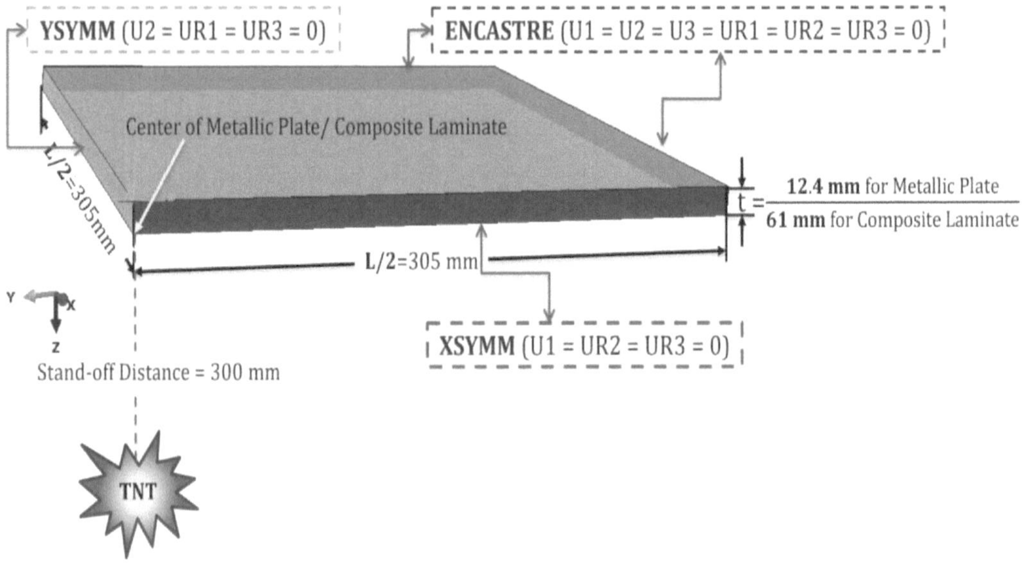

FIGURE 9.2 Details of target plate's specifications with applied loading and boundary conditions.

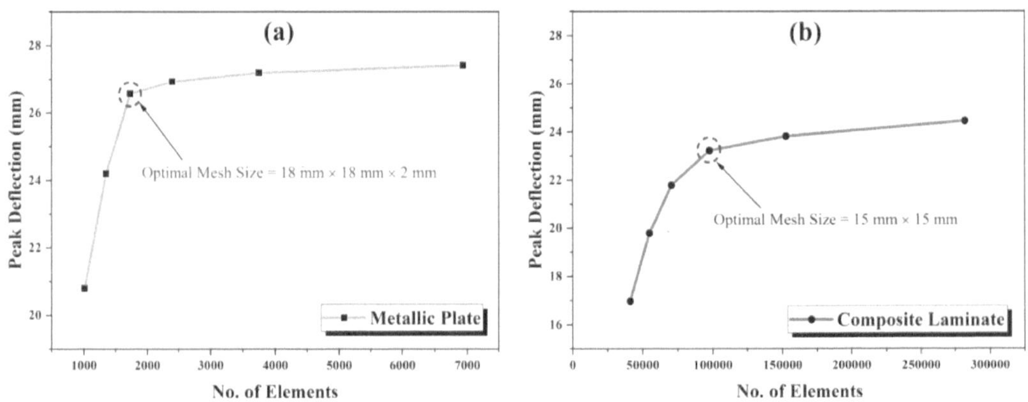

FIGURE 9.3 Mesh convergence study for (a) metallic plate and (b) composite laminate.

minimize the storage requirements. The in-plane dimensions of both the metallic plate and composite laminate are 305 mm × 305 mm. To maintain the masses of the metallic plate and composite laminate constant at 9 kg, their thicknesses were set at 12.4 mm and 61 mm, respectively. These dimensions are represented in Figure 9.2.

The two side faces of these metallic and composite plates were subjected to fixed (ENCASTRE) boundary conditions, and the other two adjacent side faces were subjected to symmetry (XSYMM and YSYMM) boundary conditions, as demonstrated in Figure 9.2. The top in-plane surfaces of the metallic plate and composite laminates were defined as target surfaces for air blast loading by using the CONWEP module. The TNT explosive charges of different masses (1 kg–3 kg) were applied at a 300 mm SoD in a positive z-direction from the center of the top surface, as shown in Figure 9.2. The mesh convergence analyses were performed for both the metallic plate and composite laminate, and obtained results are displayed in Figure 9.3. After performing mesh sensitivity analysis, the element size and element type for the metallic plate were decided to be 18 mm × 18 mm × 2 mm and C3D8R, respectively. Similarly, for each composite lamina, the mesh size and element type are used

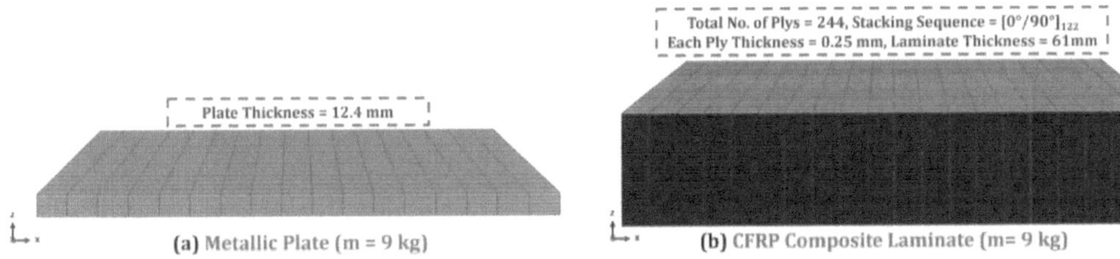

FIGURE 9.4 Finite element models of the designed (a) metallic plate and (b) composite laminate.

as 15 mm × 15 mm and S4R, respectively. The C3D8R elements are preferred in this study because of their reduced integration approach, which lessens shear locking and improves stability over the C3D8 elements in finite element analysis, especially in thin structures or bending-dominated situations. For blast simulations where precise modeling of mechanical behavior and stability is crucial, this element is the preferable option because of its enhanced bending (transverse deflection) behavior accuracy, compatibility with sophisticated material models, and typically acceptable computational cost increase. The modeled CFRP composite laminate has 244 laminas, and they are arranged with a stacking sequence of $[0°/90°]_{122}$. Each lamina is 0.25 mm thick. All these details are mentioned in Figure 9.4 with finite element models of both metallic plates and composite laminates of equal masses. The adjacent laminas are connected by using the zero-thickness cohesive contact available in Abaqus.

9.2.3 Modeling and Characteristics of Materials

Due to the high strength, excellent corrosion resistance, and high ductility of AL-6XN superaustenitic stainless steel, it has been chosen for this study. The CFRP composite laminate has the highest strength-to-weight ratio as compared to the other fiber-reinforced composites. However, an equivalent composite laminate was designed for an effective comparison. In air blast loadings, the structures are deformed with a high strain rate. Therefore, to analyze the deformation behavior of the metallic plate, a rate-dependent Johnson–Cook (J–C) model was used. This material model has the ability to provide a satisfactory result. Because the J–C model considers the combined effect of strain, strain rate, and temperature. The constitutive equation of this model is given by the following equation [46,47]:

$$\sigma_1 = \left\{A + B\left(\varepsilon_1\right)^n\right\}\left\{1 + C ln\left(\frac{\dot{\varepsilon}_1}{\dot{\varepsilon}_0}\right)\right\}\left\{1 - \left(\frac{T - T_0}{T_1 - T_0}\right)^m\right\} \tag{9.3}$$

where σ_1 stands for flow stress, $\dot{\varepsilon}_1$ is the equivalent plastic strain rate, ε_1 is the equivalent plastic strain, and T indicate the current temperature. The description of other parameters and mechanical properties of AL-6XN steel used for material assignment to the metallic plate are catalogued in Table 9.1.

In composite laminates, various types of failures or damage occur when they are subjected to air blast loadings. These damages may be delamination, fiber breakage, matrix cracking, or a combination of them. One important kind of failure in composite laminates is delamination, also called inter-laminar damage, in which the material separates into layers at their interfaces, losing structural integrity. This separation can happen between the laminate's adjacent plies or between the laminate and any bound structure (such as substrate surfaces or adhesive layers). Delamination is especially dangerous since it can spread quickly under stress and lower the laminate's ability to support loads.

TABLE 9.1
Material Properties of AL-6XN Steel with J–C Parameters

Symbol	Parameter	Value
ρ	Density (kg/m^3)	7850
E	Young's modulus (GPa)	161
υ	Poisson's ratio	0.35
A	Yield stress (MPa)	400
B	Strain hardening constant (MPa)	1500
n	Strain hardening coefficient	0.4
C	Strengthening coefficient of strain rate	0.045
m	Thermal softening coefficient	1.2
$\dot{\mu}_0$	Reference strain rate	0.001
T_0	Room temperature (K)	293
T_1	Melting temperature (K)	1800

Source: [6].

When compressive or tensile forces inside a composite laminate are too great for the fibers to withstand the air blast loads, the resulting fiber-breaking damage takes place. These high-strength fibers are mainly responsible for the strength of composite materials; the matrix material is usually a polymer resin. The performance and integrity of the laminate may suffer from broken fibers caused by applied loads greater than the fibers' strength. When the resin matrix material encircling the reinforcing fibers microcracks under blast loading, it causes matrix cracking in composite laminates. This increases the likelihood of delamination, reduces load-bearing capacity, and degrades stiffness. The structural integrity of the laminate may be affected by these fractures, which might start at stress concentrations and spread throughout the matrix. The fiber breakage and matrix cracking come under the intra-laminar damage of the composite laminates.

To determine the intra-laminar damage behavior of the CFRP composite laminate, Hashin [49] and Puck–Schurmann [50] damage initiation criteria were used. To anticipate matrix failure under tension and fiber failure under both tension and compression, the damage initiation criteria provided by Hashin [49] were used. These criteria are in-built in Abaqus. To accurately anticipate matrix failure under compression, the user-defined function VUMAT of Abaqus/Explicit was used to implement Puck–Schurmann's [50] criterion. As compared to Hashin's matrix compression failure [49] criterion, Puck–Schurmann's [50] matrix compression failure criterion was chosen for the study due to its ability to predict matrix compression failure more accurately [51]. These damage initiation criteria are given by the following equations:

Fiber tensile failure ($\sigma_{11} \geq 0$):

$$\left(\frac{\sigma_{11}}{X_1}\right)^2 + \left(\frac{\sigma_{12}}{S_{12}}\right)^2 = 1 \tag{9.4}$$

Fiber compressive failure ($\sigma_{11} < 0$):

$$\left(\frac{\sigma_{11}}{X_2}\right)^2 = 1 \tag{9.5}$$

Matrix tensile failure ($\sigma_{22} \geq 0$):

$$\left(\frac{\sigma_{22}}{Y_1}\right)^2 + \left(\frac{\sigma_{12}}{S_{12}}\right)^2 = 1 \tag{9.6}$$

Matrix compressive failure ($\sigma_{22} < 0$):

$$\left(\frac{\sigma_t^n}{S_{23}^A + \mu_t^n \sigma_n^n}\right)^2 + \left(\frac{\sigma_l^n}{S_{12} + \mu_l^n \sigma_n^n}\right)^2 = 1 \tag{9.7}$$

The stresses operating in this case are σ_{11}, σ_{22}, and σ_{12}, respectively, along the longitudinal, transverse, and shear axes. The composite's tensile and compressive strengths, along the fiber direction, are denoted by X_1 and X_2, respectively. The transverse strengths under tension and compression are denoted by Y_1 and Y_2, respectively. The shear strength characteristics in the longitudinal and transverse directions are denoted by S_{12} and S_{23}, respectively. The Puck–Schurmann [50] criteria state that in unidirectional fiber-reinforced composite laminates, the matrix fails in shear mode at a fracture plane θ = 53° from the direction of the compression stress [52]. The friction coefficients are expressed here as μ_i^n (i = l, n, t) based on the Mohr–Coulomb failure theory. Equation (9.8) states that the stress tensor σ_{ij} (i, j=1, 2, 3) is rotated to determine σ_i^n (i = l, n, t) in the fracture plane using the $\left[T(\theta)\right]$ (transformation matrix). Equation (9.9) is used to get the transverse shear strength in the fracture plane (S_{23}^A):

$$\sigma_{\mathrm{i}}^n = \left[T(\theta)\right]\sigma_{\mathrm{ij}}[T(\theta)]^T \tag{9.8}$$

$$S_{23}^A = \frac{Y_2}{2}\left(\frac{1-\sin(2\theta-90)}{\cos(2\theta-90)}\right) \tag{9.9}$$

The details of the used damage evolution criteria and inter-laminar damage criteria are described in the literature [10,53]. The material properties assigned to each CFRP composite lamina are listed in Table 9.2. The cohesive properties used to join the composite plies are taken from the literature [48].

9.2.4 Numerical Approach Validation

Validation studies are essential for verifying the precision and reliability of numerical models in numerical analysis through the comparison of computational model results with experimental data or proven analytical solutions. These investigations guarantee that the numerical simulations faithfully replicate the behavior of practical methods, hence establishing trust in their prediction power. Therefore, in this study, the obtained numerical results for peak overpressure on the AL-6XN steel plate and its radial deflection at 1 kg–3 kg of TNT blast for a 0.1 m SoD are compared with the literature [5] experimental results of the AL-6XN steel plate of the same configuration. These comparisons of the peak overpressure and radial deflection are shown in Figure 9.5(a) and (b), respectively. To verify the composite laminate model's applicability for the analysis of its air blast performance, the adopted numerical approaches for CFRP were validated by comparing the obtained results for CFRP composite laminates at 100 g explosive blast at 1 m SoD with both experimental and numerical results of the literature [8]. The comparison of the profile of CFRP composite laminate center point deflection for the present numerical study and literature [8] experimental and numerical cases

TABLE 9.2
Material Properties of CFRP Composite

Parameter	Symbol	Value
Density (kg/m^3)	ρ	1600
Young's modulus (GPa)	E_1	153
	E_2	10.3
	E_3	10.3
Poisson's ratio	υ_{12}	0.3
	υ_{23}	0.4
	υ_{13}	0.3
Shear modulus (GPa)	G_{12}	6
	G_{23}	3.7
	G_{13}	6
Longitudinal tensile strength (MPa)	X_1	2537
Longitudinal compressive strength (MPa)	X_2	1580
Transverse tensile strength (MPa)	Y_1	82
Transverse compressive strength (MPa)	Y_2	236
Longitudinal shear strength (MPa)	S_{12}	90
Transverse shear strength (MPa)	S_{23}	40
Longitudinal tensile fracture energy (kJ/m^2)	G_X^1	91.6
Longitudinal compressive fracture energy (kJ/m^2)	G_X^2	79.9
Transverse tensile fracture energy (kJ/m^2)	G_Y^1	0.22
Transverse compressive fracture energy (kJ/m^2)	G_Y^2	1.1

Source: [48].

is shown in Figure 9.5(c). Its deformation contour comparison is also represented in Figure 9.5(d). Figure 9.5(a) depicts that the maximum percentage error in the numerically obtained overpressure in the present study is only 3.6% as compared to the literature [5] results. The obtained peak deflection of CFRP composite laminate also has only 2.3% error with respect to literature experimental [8] results, as depicted in Figure 9.5(c). Figure 9.5(b) and (d) also shows good correlations between the present study's steel plate numerical deflection profile and composite laminate deformation contours, respectively, to literatures [5] and [8] experimental results. Due to the small percentage of errors involved in the present numerical models of metallic plate and composite laminate, they prove the authenticity of the present chapter's numerical analysis under various applied blast loads.

9.3 RESULTS AND DISCUSSION

9.3.1 Variations in Overpressures

The changes in peak pressures over the target surface in the radially outward direction of the designed plates at 1 kg–3 kg of TNT blast loads for a 300 mm SoD are shown in Figure 9.6(a). The time-dependent pressure and impulse variation at the center of the plate at various applied air blast loads are represented in Figure 9.6(b). The intensity of pressure applied to the target by the blast shock wave is inversely proportional to the shock wave travelling distance required to reach the target surface-specific point and is directly proportional to the mass of explosive charge used for air blast loading. Therefore, Figure 9.6(a) depicts that the peak pressure on the target surface of the plate for a fixed mass of spherical TNT blast gradually decreases from the target surface center point to the radial outward direction. Because the spherically propagated shock wave first

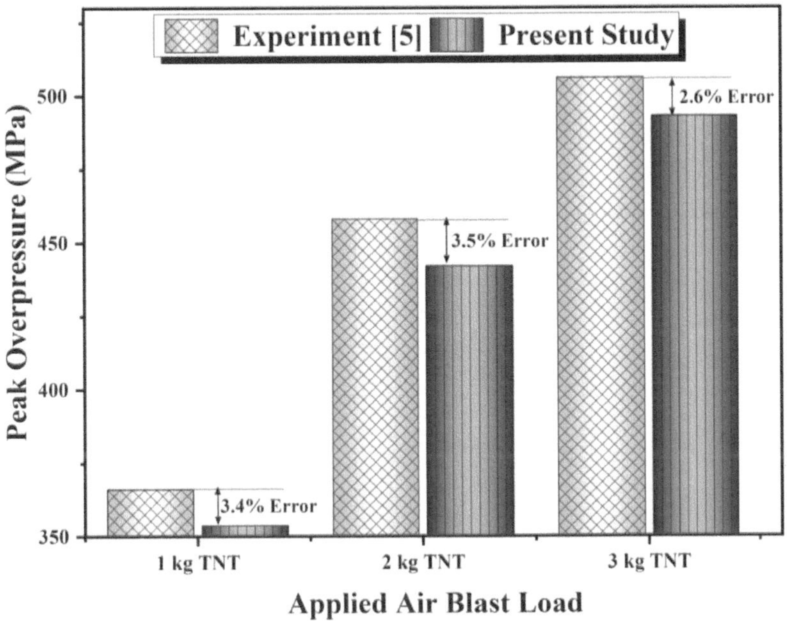

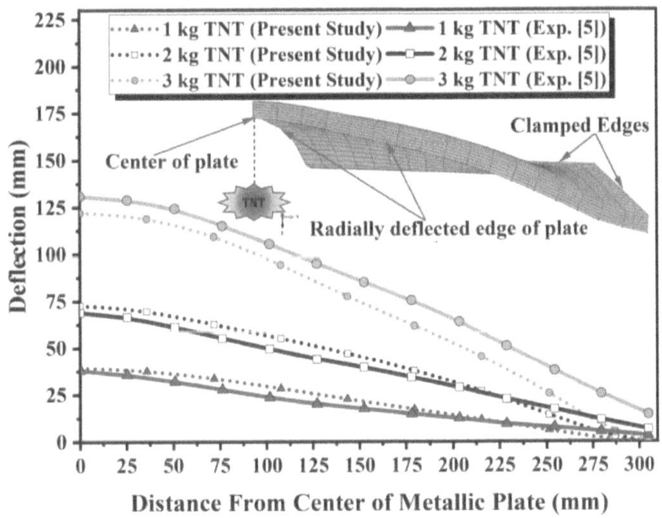

FIGURE 9.5 Validation of (a) peak overpressure on AL-6XN steel plate (b) redial deflection of AL-6XN steel plate (c) CFRP composite laminate center deflection and (d) CFRP composite laminate deformations.

interacts with the center of the target plate due to its smallest travelling distance from the explosive charge center as per the applied loading conditions, and this travelling distance of the shock wave increases with an increase in radial distance from the center of the plate, however, the intensity of the experienced peak pressure by radial elements gradually decreases [54]. A high mass of TNT blasts releases a high amount of energy, which causes the pressure on each element of the target surface to be the highest for the 3 kg TNT blast, followed by 2 kg and 1 kg TNT blasts, as shown in Figure 9.6(a). Figure 9.6(b) demonstrates that the arrival duration of the shock wave to the target's center point decreases with high mass of TNT blast and impacts the highest peak pressure at 3 kg TNT, followed by 2 kg and 1 kg TNT blasts. It is also observed that the exponential decay duration

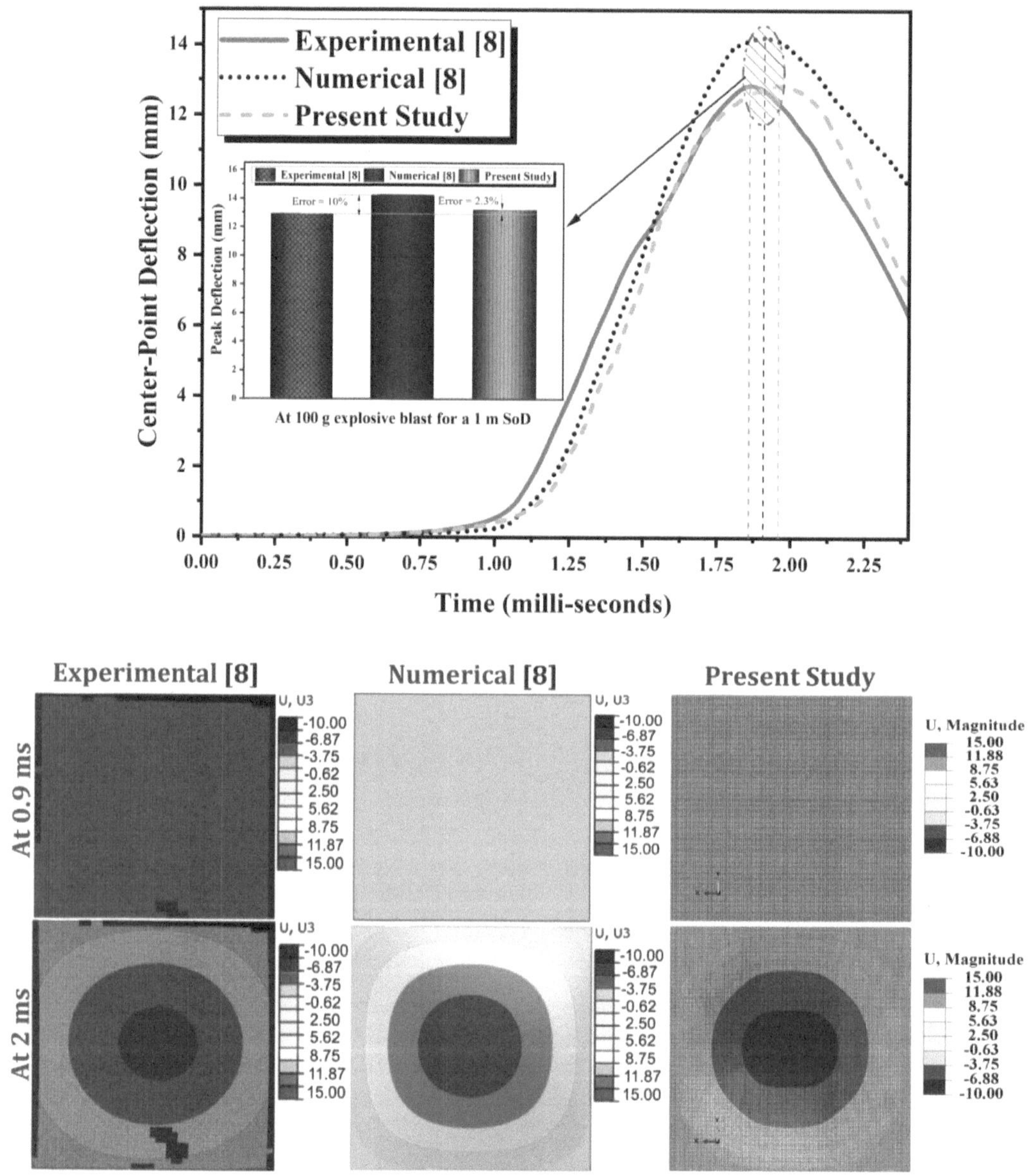

FIGURE 9.5 (Continued)

of the pressure is the highest for 3 kg of TNT blast, followed by 2 kg and 1 kg of TNT blast. Due to this shorter arrival duration and higher decay duration, with a higher peak pressure at a higher mass of TNT blast, the impacted impulse is also higher, as can be clearly seen in Figure 9.6(b). The observed peak pressures applied on the target plate at 1 kg–3 kg of TNT blast for a 300 mm SoD are 81.8 MPa, 117.9 MPa, and 144.5 MPa, respectively. These results in the experienced specific impulse recorded at the center of the target plate are 3.17 kPa s, 5.71 kPa s, and 8.22 kPa s, respectively. For 1 kg and 2 kg of TNT blasts, 43.4% and 18.4%, respectively, reduction in peak pressure and 61.4% and 30.5%, respectively, decrement in impulse on the target plate were observed with regard to 3 kg of TNT blast. These percentage reductions in pressure and impulse are also indicated in Figure 9.6(b).

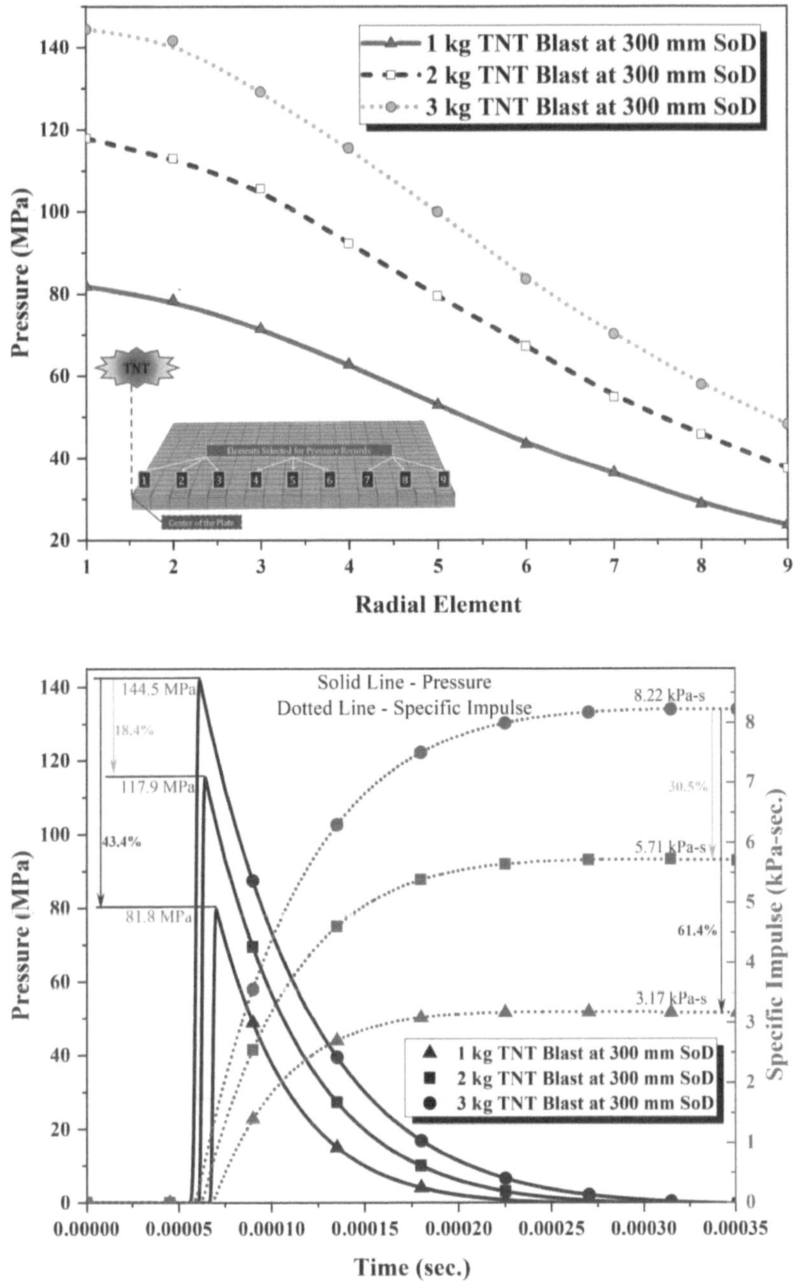

FIGURE 9.6 (a) Changes in peak pressure in the radial direction and (b) time-dependent pressure and specific impulse variation at the center.

9.3.2 Comparison of Deflection Behavior

Due to the highest pressure and the highest impulse impacted on the center of the target plates at each blast load, the center point deflection responses of both metallic plates and composite laminates are used for a significant comparison of their deflection behavior. These comparisons of time-dependent deflection responses of both metallic plates and composite laminates at various air blast loadings

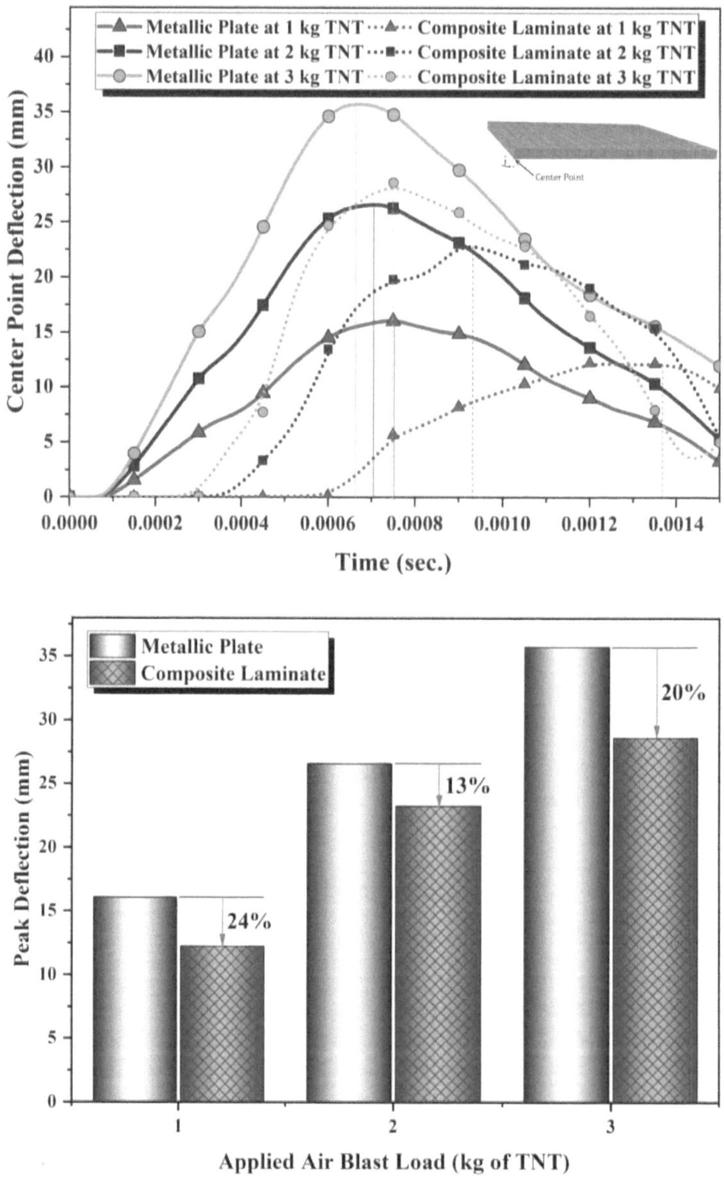

FIGURE 9.7 Comparison of (a) time-dependent center point deflection and (b) peak deflection of metallic and composite laminates under air blast loads.

are shown in Figure 9.7(a). From this figure, it is observed that both metallic plate and composite laminate center points deflected more quickly at 3 kg TNT blast loads and got peak deflection earlier than at 2 kg and 1 kg TNT blast loads. Similarly, a higher peak deflection and quicker response are obtained at 2 kg of TNT blast as compared to 1 kg of TNT blast. This happens due to the shorter arrival duration and higher intensity of the shock wave at the higher mass of the TNT blast. After getting peak deflection at the center of the metallic plate and composite laminate, their center point deflections are decreasing due to their spring-back effect. The response of metallic plates is faster than that of composite laminates due to their different mechanisms of deformation. These deformation mechanisms are compared in Section 9.3.3. The experienced peak deflections of the designed

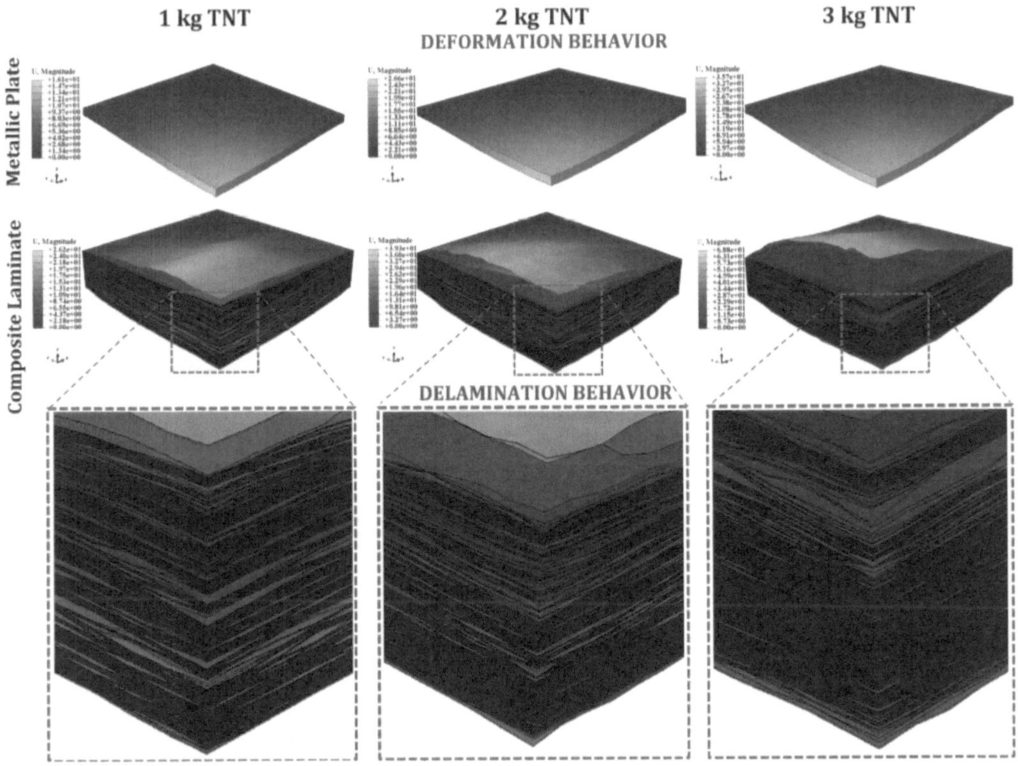

FIGURE 9.8 Comparison of the deformation of metallic plates and composite laminates.

metallic plate and composite laminate at all applied blast loads are compared in Figure 9.7(b). From this figure, it is noted that the center point peak deflections of the composite laminate are smaller than those of the metallic plate for every blast load applied. For the blast loads of 1 kg–3 kg TNT, the composite laminate shows a reduction in peak deflection by 24%, 13%, and 20%, respectively, as compared to the metallic plate for a fixed 300 mm SoD. This result indicates that the composite laminate has a higher capability to mitigate the adverse effects of the blast event.

9.3.3 Comparison of Deformation and Damage Behavior

The deformed metallic plates and composite laminates at various blast loads are displayed in Figure 9.8. All these deformed figures were recorded after 1.5 ms of blast. The fiber and matrix damages in the CFRP composite laminates at this instant are shown in Figure 9.9. From Figure 9.8, it is observed that the central regions of both the metallic plate and composite laminate are highly deformed under all applied blast loads, and the deformation in both the metallic plate and composite laminate increases with an increase in the mass of TNT used for blast loads.

These types of deformation behavior are found in the designed plates due to the pressure and impulse variations on the plate, which also vary in this pattern, as discussed in Section 9.3.1. It is observed that the composite laminate deformation is more uniform and global as compared to the same mass of steel plate. This is due to the composite laminate having an excellent ability to distribute the impacted blast loads over the whole in-plane area of the laminate. When the blast loads are applied to the AL-6XN steel plate, it experiences a high-intensity blast pressure and deforms with a series of damage mechanisms, starting with elastic deformation and ending with plastic deformation. Finally, the plate is stretched and bent in the out-of-plane direction, as represented in

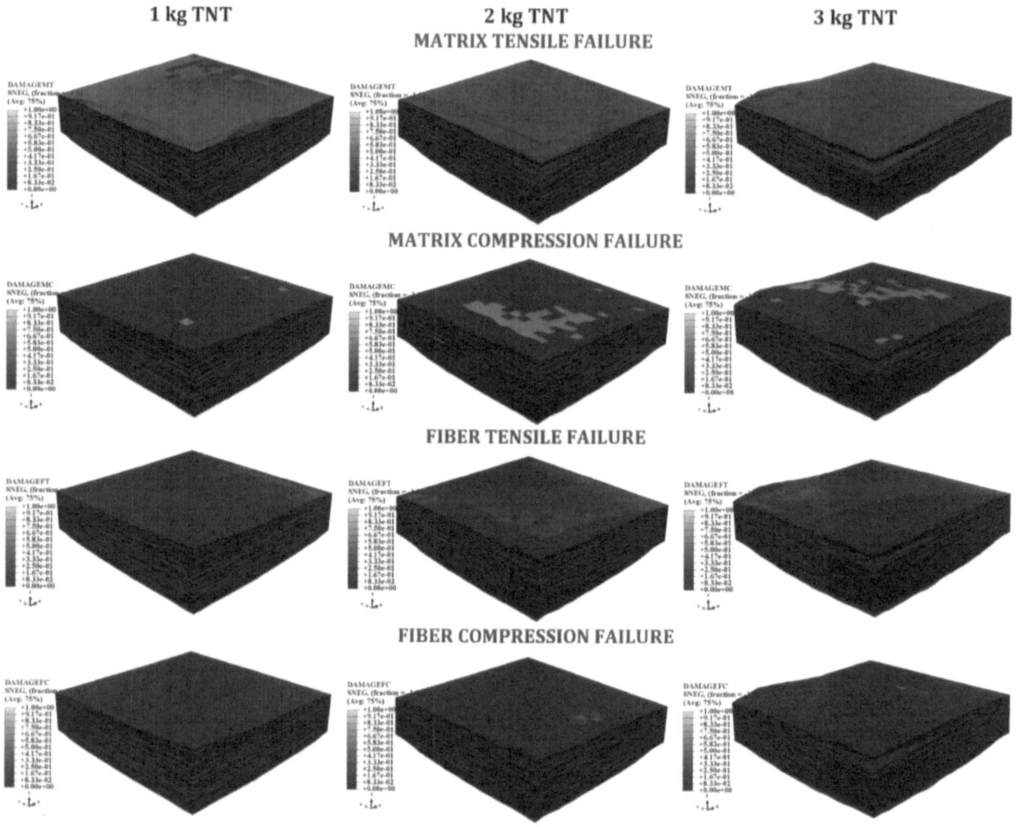

FIGURE 9.9 Damage behaviors of CFRP composite laminates under air blast loads.

Figure 9.8. In the case of composite laminates, when the blast wave impacts the laminate, it absorbs the blast wave's kinetic energy. Results in the composite laminate first undergoing elastic deformation; thereafter, localized plastic deformation occurs in areas where stress is concentrated. Shear forces at interfaces between laminas cause damage to initiate, which can result in matrix cracking, delamination, and fiber–matrix debonding. The progressive delamination, fiber breakage, and matrix cracking result from these damage initiation processes propagating as the blast loading increases. The delamination failures of the CFRP composite laminates are clearly shown in Figure 9.8. The matrix and fiber failures after damage propagation in the composite laminates are represented in Figure 9.9. From Figures 9.8 and 9.9, it is noted that delamination failure is the major mechanism of failure in the CFRP composite laminate under blast loads, followed by matrix tensile failure, fiber tensile failure, matrix compression failure, and fiber compression failure. All these failures in the composite laminate increase with an increase in blast loads.

9.3.4 Comparison of Kinetic Energy and Energy Absorption

The changes in the average kinetic energies and energy absorption of both the metallic plate and the CFRP composite laminate with respect to time are shown in Figure 9.10(a) and (b), respectively. From Figure 9.10(a), it is observed that the peak average kinetic energy of the composite laminate is higher than that of the metallic plate at each blast load. This happens due to the global delamination between the laminas, which increases the average kinetic energy of the composite laminate as compared to the metallic plate's kinetic energy. In the case of metallic plate kinetic energies,

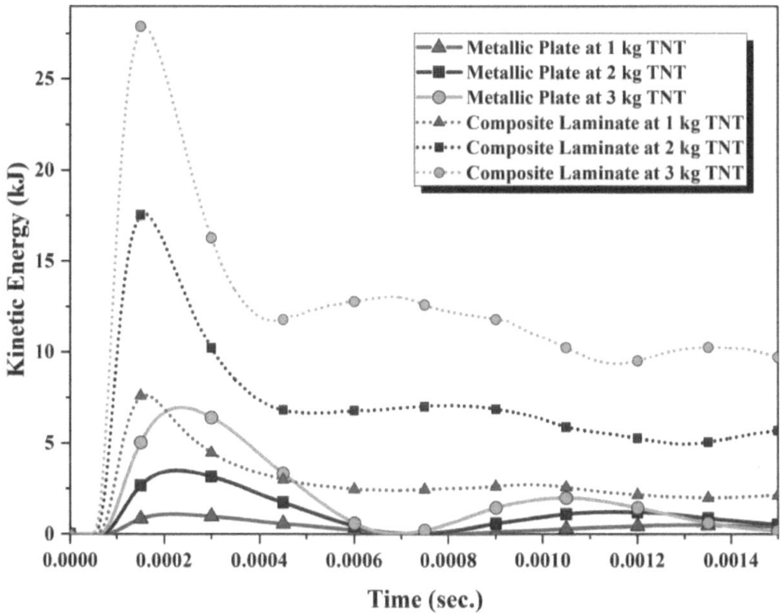

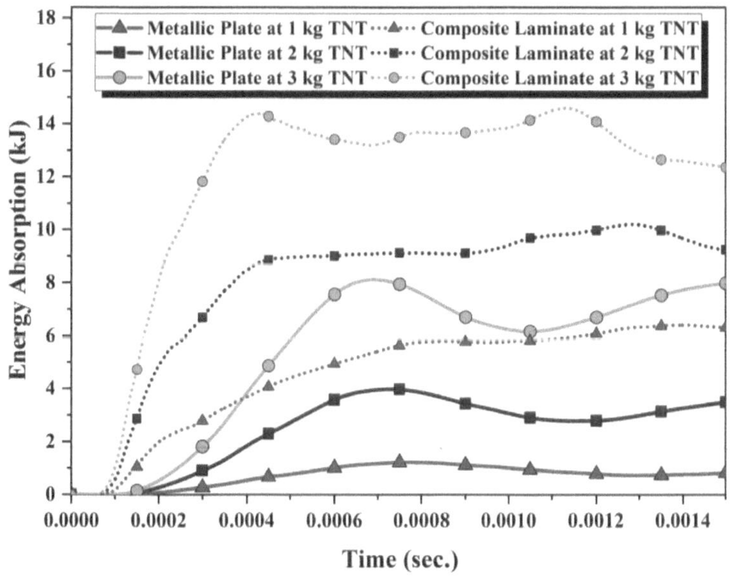

FIGURE 9.10 Time-dependent variation in (a) kinetic energy and (b) internal energy of metallic plates and composite laminates at various air blast loadings.

the average kinetic energy of the metallic plate reached zero near 0.7 ms and 1.5 ms. The first diminished kinetic energy indicates that the metallic plate deflected its maximum in a downward direction (negative z-direction). The second peak observed in the kinetic energy curves of the metallic plate between 0.7 ms and 1.5 ms indicates the spring-back effect of the metallic plate. But the value of the second peak of the kinetic energy is much smaller than the first peak; this result indicates that the spring-back effect in the metallic plate is very small. Hence, the deformation in the plate after 1.5 ms is considered permanent. But in the case of composite laminate, the value of kinetic energies is not zero; this is due to the continued delamination in the composite laminate, which requires more time to get a permanent deformation.

From Figure 9.10(b), it is noted that the composite laminate has higher energy-absorbing capability at each blast load as compared to the metallic plate. This enhancement in absorbed energy is due to the more global deformation and uniform load distribution capacity of the composite laminate as compared to metallic plates. The delamination of the laminas in the composite laminate also helps to improve the energy absorption of the laminate. The energy absorption of both the composite laminate and metallic plate increases with an increase in the mass of TNT used for blast loading. This happens because their deformation also increases with an increase in the mass of TNT for blast loading [55]. The combination of these higher energy absorption capabilities and smaller peak deflections of the CFRP laminates makes composite laminates, as compared to metallic plates, highly desirable for applications requiring enhanced blast resistance, such as aerospace, defense, and infrastructure protection.

9.3.5 Applications of the Outperforming CFRP Composite Laminate

From the analysis of Sections 9.3.2–9.3.4, it is discovered that the CFRP composite laminate outperforms the equivalent metallic plate for every circumstance of blast loading. Due to their advantageous properties, the CFRP composite laminates have versatile applications in many fields to withstand blast loadings. They can play a vital role for aircraft structures in the aerospace industry because they offer robustness against possible hits and strong air blast loads during flight. CFRP laminates can act as important constituents of body armor, vehicle armor, and bunker armor systems in defense and military applications, protecting people and equipment from explosive threats. Additionally, CFRP laminates can be used in infrastructure to strengthen buildings and retrofit existing structures, reducing the impacts of air blasts and improving overall safety, and in marine vessels to improve structural integrity and resist air blast loadings in maritime applications. The use of CFRP laminates is advantageous to the automotive industry as well since they can sustain air blast loads seen in a variety of driving circumstances owing to their strength-to-weight ratio in armored vehicles and high-performance racing automobiles. The CFRP composite laminates are essential for enhancing safety and robustness against air blast loadings in all these applications.

9.4 CONCLUSIONS

The present chapter deals with the blast mitigation characterization of metallic plates and composite laminates of equal areal densities. The summary of this comparative numerical analysis is drawn as follows:

- The CFRP composite laminate showed a decrease in peak deflection of the center point by 13%–24% as compared to the high-strength AL-6XN steel plate for blast loads of 1 kg–3 kg TNT.
- The metallic plate fails only due to being deformed by out-of-plane stretching, and the composite laminate mainly fails due to delamination, followed by matrix tensile cracking, fiber tensile breakage, matrix compressive cracking, and fiber compressive breakage.
- Due to delamination failure and globalized deformation of the composite laminate, its maximum kinetic energy and energy absorption capability are higher than the equivalent metallic plate for similar blast loading conditions.
- An increase in the mass of TNT for blast loads impacted increased pressure and impulse on both the metallic plate and composite laminate, which increased deflection, deformation, kinetic energy, and energy absorption by them.

- Due to the enhanced blast mitigation capability of the CFRP composite laminate in terms of smaller peak deflection and higher energy absorption capacity than the metallic plate, it may be preferred for making various parts of military vehicles, body armor, blast-resistant doors and windows, aircraft fuselages and wings, etc.
- In the future, the addition of various nano-fillers (carbon nanotubes, alumina, silicon carbide, etc.) can further improve the blast performance of the CFRP composite laminate.

REFERENCES

[1] Angelides, S. C., Morison, C., Burgan, B. A., Kyprianou, C., Rigby, S. E., and Tyas, A., 2023, "A Methodology for Predicting Far-Field Blast Loading on Structures," *Structures*, **58**(December), p. 105619.

[2] Chen, D., Jing, L., and Yang, F., 2019, "Optimal Design of Sandwich Panels with Layered-Gradient Aluminum Foam Cores under Air-Blast Loading," *Compos. Part B Eng.*, **166**(October 2018), pp. 169–186.

[3] Imbalzano, G., Linforth, S., Ngo, T. D., Lee, P. V. S., and Tran, P., 2018, "Blast Resistance of Auxetic and Honeycomb Sandwich Panels: Comparisons and Parametric Designs," *Compos. Struct.*, **183**(1), pp. 242–261.

[4] Li, J., Qin, Q., and Zhang, J., 2021, "Internal Blast Resistance of Sandwich Cylinder with Lattice Cores," *Int. J. Mech. Sci.*, **191**(September 2020), p. 106107.

[5] Dharmasena, K. P., Wadley, H. N. G., Xue, Z., and Hutchinson, J. W., 2008, "Mechanical Response of Metallic Honeycomb Sandwich Panel Structures to High-Intensity Dynamic Loading," *Int. J. Impact Eng.*, **35**(9), pp. 1063–1074.

[6] Patel, M., and Patel, S., 2024, "Design and Analysis of Lightweight Stiffened Honeycomb Metallic Sandwich Panels under High-Intensity Air Blast," *Proc. Inst. Mech. Eng. Part L J. Mater. Des. Appl.*, **238**(1), pp. 22–38.

[7] Süsler, S., Kurtaran, H., Türkmen, H. S., Kazancı, Z., and Lopresto, V., 2020, "An Experimentally Validated Numerical Method for Investigating the Air Blast Response of Basalt Composite Plates," *Mech. Adv. Mater. Struct.*, **27**(6), pp. 441–454.

[8] Gargano, A., Das, R., and Mouritz, A. P., 2019, "Finite Element Modelling of the Explosive Blast Response of Carbon Fibre-Polymer Laminates," *Compos. Part B Eng.*, **177**(September), p. 107412.

[9] Gaur, B., Patel, M., and Patel, S., 2023, "Strain Rate Effect on CRALL under High-Velocity Impact by Different Projectiles," *J. Brazilian Soc. Mech. Sci. Eng.*, **45**(2), p. 103.

[10] Patel, M., and Patel, S., 2024, "Blast Mitigation Analysis of Novel Designed Sandwich Structures Using Novel Approaches," *Mech. Adv. Mater. Struct.*, **31**(25), pp. 7195–7217.

[11] Patel, M., Patel, S., and Ahmad, S., 2023, "A Novel Design of Blast Proof Sandwich Structure with Hybrid Skin," *Mater. Today Proc.*, **87**(1), pp. 221–227.

[12] Patel, M., and Patel, S., 2022, "Novel Design of Honeycomb Hybrid Sandwich Structures under Air-Blast," *J. Sandw. Struct. Mater.*, **24**(8), pp. 2105–2123.

[13] Mohammadzadeh, B., Kang, J., and Im, S., 2020, "Blast Loaded Plates: Simplified Analytical Nonlinear Dynamic Approach," *Structures*, **28**, pp. 2034–2046.

[14] Veeredhi, L. S. B., Ramana Rao, N. V., and Veeredhi, V. R., 2018, "Effect of Plate Curvature on Blast Response of Structural Steel Plates," *J. Inst. Eng. Ser. A*, **99**(3), pp. 463–471.

[15] Yuan, Y., Li, X., Zhang, C., Tan, P. J., and Chen, P., 2022, "Impulse Saturation in Metal Plates under Confined Blasts," *Int. J. Impact Eng.*, **168**(June), p. 104308.

[16] Neuberger, A., Peles, S., and Rittel, D., 2007, "Scaling the Response of Circular Plates Subjected to Large and Close-Range Spherical Explosions. Part I: Air-Blast Loading," *Int. J. Impact Eng.*, **34**(5), pp. 859–873.

[17] Tian, W., Zhai, H., Liu, Y., Guo, Q., and Shi, X., 2023, "A Response Prediction Method for Clamped Circular Plates Subjected to Repeated Blast Loading," *Int. J. Impact Eng.*, **180**(April), p. 104684.

[18] Gan, L., Zong, Z., Chen, Z., Wu, T., Lin, J., and Li, M., 2023, "Differences in Responses of Square Steel Plates Exposed to Blast Loads Generated by Cubic and Spherical Explosives," *Thin-Walled Struct.*, **182**(October 2022), p. 110332.

[19] Zhang, C., Tan, P. J., and Yuan, Y., 2022, "Confined Blast Loading of Steel Plates with and without Pre-Formed Holes," *Int. J. Impact Eng.*, **163**(September 2021), p. 104183.
[20] Zhu, H., Wang, X., Wang, Y., Wu, G., Cao, Y., and Ji, C., 2023, "Dynamic Responses of Aluminum Alloy Plates Coated with Polyurea Elastomer Subjected to Repeated Blast Loads: Experimental and Numerical Investigation," *Thin-Walled Struct.*, **189**(April), p. 110912.
[21] Zhang, L., Ji, C., Zhao, C., Wang, Y., and Wang, X., 2023, "Non-Monotonous Strengthening Effects of Polyurea Layer on Polyurea-Aluminum Composite Plate Subjected to Blast of Charge with Prefabricated Fragments," *Thin-Walled Struct.*, **192**(August), p. 111194.
[22] Mohotti, D., Fernando, P. L. N. L. N., Weerasinghe, D., and Remennikov, A., 2021, "Evaluation of Effectiveness of Polymer Coatings in Reducing Blast-Induced Deformation of Steel Plates," *Def. Technol.*, **17**(6), pp. 1895–1904.
[23] Hou, H., Chen, C., Cheng, Y., Zhang, P., Tian, X., Liu, T., and Wang, J., 2019, "Effect of Structural Configuration on Air Blast Resistance of Polyurea-Coated Composite Steel Plates: Experimental Studies," *Mater. Des.*, **182**(November), p. 108049.
[24] Zhang, C., Yuan, Y., Su, X., Tan, P. J., Zhang, Q., and Chen, P., 2023, "Large Inelastic Response of Polyurea-Coated Steel Plates to Confined Blast Loading," *Thin-Walled Struct.*, **183**(December 2022), p. 110454.
[25] Ackland, K., Anderson, C., and Ngo, T. D., 2013, "Deformation of Polyurea-Coated Steel Plates under Localised Blast Loading," *Int. J. Impact Eng.*, **51**, pp. 13–22.
[26] Patel, M., 2025, "A Comprehensive Review of Functionally Graded Materials and Their Ballistic Impact Performance: Current Status and Future Challenges," *Next Materials*, 8(July), p.100704.
[27] Chauhan, S., Nagargoje, A.R., Patel, M. and Roy, A., 2025, "Optimal Replacement of Glass Fabric with Carbon Fabric in Epoxy Matrix Composite," *Mechanics of Advanced Composite Structures*, 12(3), pp.507-522.
[28] Patel, M., Sonkar, L. and Patel, S., 2025, "Assessment of Ply Stacking Sequence Effect on Damage Behavior of CFRP Composite Laminate Under Low-Velocity Impacts," *Advances in Materials Science and Engineering*, 2025(1), p.4349535.
[29] Langdon, G. S., and Cantwell, W. J., 2020, "The Blast Response of Composite and Fiber-Metal Laminate Materials," *Polymer Composites in the Aerospace Industry*, Elsevier, pp. 415–439.
[30] Langdon, G. S., and Cantwell, W. J., 2015, "The Blast Response of Composite and Fibre-Metal Laminate Materials Used in Aerospace Applications," *Polymer Composites in the Aerospace Industry*, Woodhead Publishing, pp. 371–391.
[31] Zhu, H., and Khanna, S. K., 2016, "Dynamic Response of a Novel Laminated Glass Panel Using a Transparent Glass Fiber-Reinforced Composite Interlayer under Blast Loading," *Int. J. Impact Eng.*, **89**, pp. 14–24.
[32] Li, J., Huang, C., Ma, T., Huang, X., Li, W., and Liu, M., 2019, "Numerical Investigation of Composite Laminate Subjected to Combined Loadings with Blast and Fragments," *Compos. Struct.*, **214**(February), pp. 335–347.
[33] Zhao, C. Z., Qiang, L. S., Zhang, R., Zhang, Q. C., Zhong, J. Y., Zhao, Z. Y., and Lu, T. J., 2023, "Dynamic Response of UHMWPE Plates under Combined Shock and Fragment Loading," *Def. Technol.*, 27(September), pp. 9–23.
[34] Fontaine, D., LeBlanc, J., and Shukla, A., 2021, "Blast Response of Carbon-Fiber/Epoxy Laminates Subjected to Long-Term Seawater Exposure at Sea Floor Depth Pressures," *Compos. Part B Eng.*, **215**(November 2020), p. 108647.
[35] Lee, J., Lacy, T. E., and Pittman, C. U., 2021, "Lightning Mechanical Damage Prediction in Carbon/Epoxy Laminates Using Equivalent Air Blast Overpressure," *Compos. Part B Eng.*, **212**(August 2020), p. 108649.
[36] Wanchoo, P., Chaudhary, B., Li, H. W. X., Matos, H., and Shukla, A., 2023, "Blast Failure and Energy Analysis of Rubber-Modified Carbon-Fiber Vinyl-Ester Composite Laminates," *Mech. Mater.*, **183**(January), p. 104685.
[37] Liu, Q., Guo, B., Chen, P., Zhai, H., Guo, Y., and Tang, S., 2022, "Experimental Investigation Blast Resistance of CFRP/Polyurea Composite Plates under Blast Loading," *Thin-Walled Struct.*, **181**(September), p. 110149.
[38] Reda Taha, M. M., Colak-Altunc, A. B., and Al-Haik, M., 2009, "A Multi-Objective Optimization Approach for Design of Blast-Resistant Composite Laminates Using Carbon Nanotubes," *Compos. Part B Eng.*, **40**(6), pp. 522–529.

[39] Altunc, A. B., Kim, J. J., Al-Haik, M., and Reda Taha, M. M., 2011, "Reliability-Based Design of Blast-Resistant Composite Laminates Incorporating Carbon Nanotubes," *Compos. Struct.*, **93**(8), pp. 2042–2048.

[40] Zhou, H., Zheng, C., Lu, A., Zhu, Z., Kong, X., and Wu, W., 2023, "An Experimental Study of the Effects of Degrees of Confinement on the Response of Thermoplastic Fibre–Metal Laminates Subjected to Blast Loading," *Thin-Walled Struct.*, **192**(May), p. 111125.

[41] Xu, M., Yang, Y., Lei, H., Wang, P., Li, X., Zhang, Z., and Fang, D., 2020, "Dynamic Response of Fiber Metal Laminates Subjected to Localized High Impulse Blast Loading," *Compos. Struct.*, **243**(December 2019), p. 112216.

[42] Patel, M., and Patel, S., 2024, "Investigation of Air-Blast Mitigation by Single-V and Multi-V Shape Face Sheets Used Honeycomb Sandwich Panels," *Mater. Today Proc.* 102, pp. 386–392.

[43] Patel, M., Patel, S., Ahmad, S., and Soares, C. G., 2023, "Numerical Investigation of Air-Blast Performance of Cross-Filled Honeycomb Sandwich Panel," *Proceedings of the ASME 2023 42nd International Conference on Ocean, Offshore and Arctic Engineering. Volume 2: Structures, Safety, and Reliability*, American Society of Mechanical Engineers, Melbourne, Australia, p. V002T02A005.

[44] Patel, M., and Patel, S., 2023, "Influence of Honeycomb Core Height on the Blast Mitigation Performance of Sandwich Panel," *Mater. Today Proc.*, **74**(4), pp. 611–620.

[45] Patel, M., and Patel, S., 2023, "Numerical Analysis on Hexagonal Honeycomb Sandwich Structure Under Air-Blast Loading," *Lecture Notes in Mechanical Engineering*, V.K. Gupta, C. Amarnath, P. Tandon, and M.Z. Ansari, eds., Springer Nature Singapore, pp. 725–734.

[46] Patel, M., and Patel, S., 2024, "Dynamic Response Optimization of the Multistage Sandwich Structures Imperiled to Explosive Loading," *Proc. Inst. Mech. Eng. Part L J. Mater. Des. Appl.*, 238(10), pp. 1999–2017.

[47] Patel, M., and Patel, S., 2025, "Determination of high-strength polymer composite reinforcement effect on the armor-grade aluminum plates' explosion resistance—A computational perspective," *Polym. Compos.*, 46(4), pp. 3770-3790.

[48] Shi, Y., Swait, T., and Soutis, C., 2012, "Modelling Damage Evolution in Composite Laminates Subjected to Low Velocity Impact," *Compos. Struct.*, **94**(9), pp. 2902–2913.

[49] Hashin, Z., 1980, "Failure Criteria for Unidirectional Fiber Composites," *J. Appl. Mech.*, **47**(2), pp. 329–334.

[50] Puck, A., Schu, H., and Schurmann, H., 2002, "Failure Analysis of FRP Laminates by Means of Physically Based Phenomenological Models," *Compos. Sci. Technol.*, **62**, pp. 1633–1662.

[51] Patel, S., and Patel, M., 2022, "The Efficient Design of Hybrid and Metallic Sandwich Structures under Air Blast Loading," *J. Sandw. Struct. Mater.*, **24**(3), pp. 1706–1725.

[52] Singh, H., Namala, K. K., and Mahajan, P., 2015, "A Damage Evolution Study of E-Glass/Epoxy Composite under Low Velocity Impact," *Compos. Part B Eng.*, **76**, pp. 235–248.

[53] Patel, M., Patel, S., and Ahmad, S., 2023, "Blast Analysis of Efficient Honeycomb Sandwich Structures with CFRP/Steel FML Skins," *Int. J. Impact Eng.*, **178**(April), p. 104609.

[54] Patel, M., and Patel, S., 2023, "Impact of Stand-Off Distance on the Dynamic Response of Honeycomb Sandwich Panel in Air Blast," *Mechanika 2023. Proceedings of the 27th International Scientific Conference*, KTU, pp. 83–87.

[55] Patel, M., and Patel, S., 2023, "Effect of Honeycomb Cell Size on the Air-Blast Performance of Sandwich Panels," *Mater. Today Proc.*, **78**(4), pp. 792–797.

10 Critical Review on Impact of Nanofillers in of Biopolymer Composites

Sasmita Kar, Sarojrani Pattnaik, and Mihir Kumar Sutar

10.1 INTRODUCTION

Natural wood-based biopolymer composites are universally accepted due to their environmental sustainability, low cost, lower density, easy availability, lightweight, and superior strength. Now in many research fields, it has the utmost attraction in comparison to man-made polymers due to its ideal load-carrying capacity. The inherent hollow structure and chemical constituents of biofibers are typically more responsible for stress distribution and intensify the strength and modulus simultaneously. Though it has a lot of good features, it still has a lot of limitations like sensitivity to light or heat, inhomogeneous, poor fiber-matrix adhesive bonding, hydrophilic, and heterogeneous nature [1]. The characteristics of green polymers are also enhanced by incorporating organic or inorganic fillers, and by various surface treatments. The presence of various chemical constituents such as cellulose, hemicelluloses, lignin, and pectin and the crystallinity of biofibers drastically influence water intake behavior. However, fibers extracted from different portions of the plants are also responsible for the rate of water absorption (WA) due to the varying chemical constituents, physical structure, and dimensions of fibers. Excluding the above factors there are also several other factors that affect the polymer composites for moisture intakes such as different chemical, and physical treatments along with time durations, types of yarn, number of fibers, density, diameter, and twisting of the yarn, space in the yarn or woven fabric, architecture of fabrics, permeability of each fiber, fabrication method, form (short or long), size, and distribution of fibers. In addition to this, void contents, interfacial adhesion between fiber, and matrix, type of polymers, crystallinity of polymers, other physical, and chemical properties of polymers, weight percentage of natural fibers, and thickness of polymer composites significantly impart the moisture gain [2]. This leads to the following failures in composites such as swelling of fibers, debonding of fiber, and matrix, develops interfacial shear stress, and finally diminishes the structural integrity, and strength as well [3]. The moisture uptake can be optimized by adopting several techniques such as physical and chemical treatment of green fibers; various coating of fibers, using nano- and microfillers, hybridizing with synthetic fibers, and also in few cases by adopting advanced techniques of manufacturing of composites.

The current review represents the inclusion of various nanoparticles in elevating the strength by reducing the hydrophilic nature of the biocomposites. A characteristically favorable point is that currently most of the waste products collected from farming, food, and textile industries are economically reprocessed by some chemical and physical treatments. The biopolymer polylactic acid (PLA) which is a biomass-based products naturally obtained protein-based polymers like, soya, rubber, maize, gums cellulose, and starch is numerously used in most of the products of industrial, medical, textile, packaging, renewable energy, and agricultural application [4]. It is highly

DOI: 10.1201/9781003664611-11

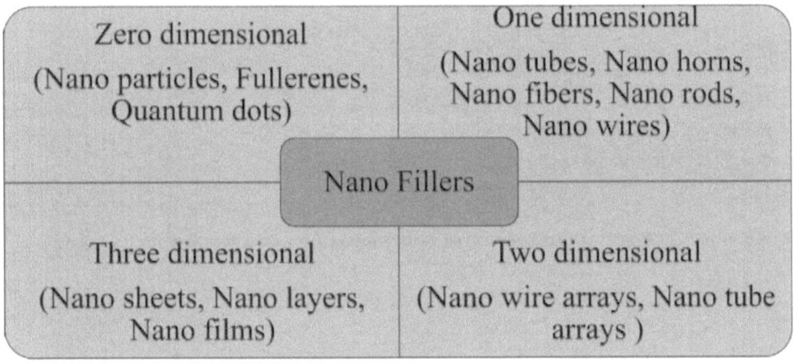

FIGURE 10.1 Dimensional classification of nanofillers.

appreciable because of its lightweight, least manufacturing cost, and superior strength. Green fiber-reinforced biopolymer composites have superior strength and modulus but not exactly higher than synthetic polymer composites. So, to magnify the strength as well as other characteristics like thermal properties, chemical resistance, corrosion resistance, and antimicrobial properties, nano- or microfillers are added up to certain wt.% to achieve the optimum value. Meanwhile, nanoparticles also help to modify and improve the miscibility tendency as well as morphology in nanopolymer composites.

The dimension of the nanoparticle plays a critical role in obtaining novel characterized composites. The nanofillers are currently categorized in various dimensional forms, having unique features for property enhancement in polymer composites as shown in Figure 10.1. It has different shapes such as spherical, cylindrical, conical, spiral, hollow, thin film, and irregular.

The size effect and surface-to-volume ratio of the fillers critically affect the characteristics of composites but both in reverse. The reduction in size shows better dispersion, and greater surface area stimulates the fiber, and matrix interfacial bonding, which critically reduces the tendency [5]. The inadequate stress distribution within the fiber and matrix results in weak bonds mostly due to the chemical constituents of the green fibers, and its presence retains moisture within it. The involvement of various types of fillers, and particulates up to the nanosized excellently enhance the overall functional features of the composites [6].

There is a large variation of nanomaterials, and each of its impacts is elaborately discussed in the ongoing review. Though many review papers are highly focused on the nanofiller addition in the development of synthetic and biodegradable polymer composites, very few have been least attentive toward the trend or kinetics of WA in nanofiller-reinforced biopolymer composites.

10.2 KINETICS OF WA

WA is a very critical phenomenon, and that causes the debonding or delamination following various stages of degradation such as the presence of voids or growth of microcracks due to fiber swelling, water particles dispersed through the matrix, and finally reached the interface of fiber, and matrix, water feasible substances drain from the fiber and finally debonding starts as shown in Figure 10.2. As the fibers started swelling after moisture intake, gradually dimension increases, and due to internal stress cracks developed around the matrix surrounding the fiber, and water particle move due to capillary action around the interface. Gradually the water-soluble substances present in the fiber drained out, the fiber became weak, and debonding started. The major load-carrying components became fragile to take further load, and finally, the strength of the composite degraded [3].

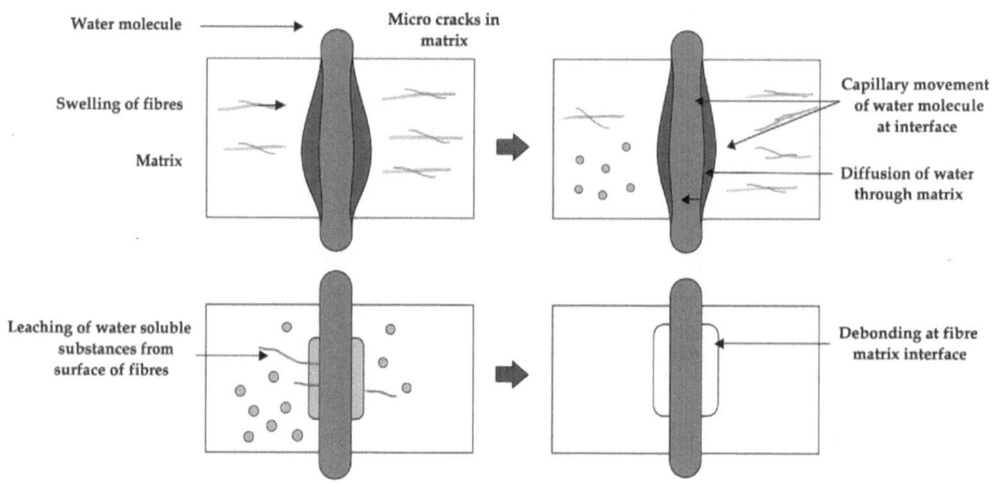

FIGURE 10.2 Structural degradation of fibers at various stages due to WA [3].

10.2.1 Water Absorption

WA phenomena are remarkably very important in the overall strength improvement of the composite materials. The distribution of water may be limited within the matrix or sometimes beyond the matrix toward the interphase and passes through the voids and microcracks through the capillary action. WA is always a time-dependent function, and with an increase in time, WA at specific it will increase, and achieve equilibrium at a certain interval of time, which is represented in the following equation form [7].

$$M_a(\%) = \frac{W_f - W_i}{W_i} \times 100 \tag{10.1}$$

where M_a represents the WA at a specific time, W_f represents the final weight of the specimen collected after the time interval of immersion, and W_i is the initial or dry weight of the samples.

The coefficient of the thickness swelling in composites is equally important which can be represented as in the following equation form [3].

$$T(\%) = \frac{T_f - T_i}{T_f} \times 100 \tag{10.2}$$

where T represents thickness swelling coefficient of composites. T_i and T_f represent thickness before swelling and final thickness after swelling. At various atmospheric conditions, the dispersion mechanism of WA is investigated by various researchers using Fickian dispersion at normal temperature, pressure, and non-Fickian dispersion principle comparatively at higher temperatures. The water dispersion technique is quite complex because the curve shown in Figure 10.3 is linearly increased for some time and becomes stable for a certain duration then gradually increases. The linear curve depends on the initial water intake due to fiber, matrix type, or voids, and then it becomes saturated and stable for some time. After that water intake suddenly increases due to the passage through the

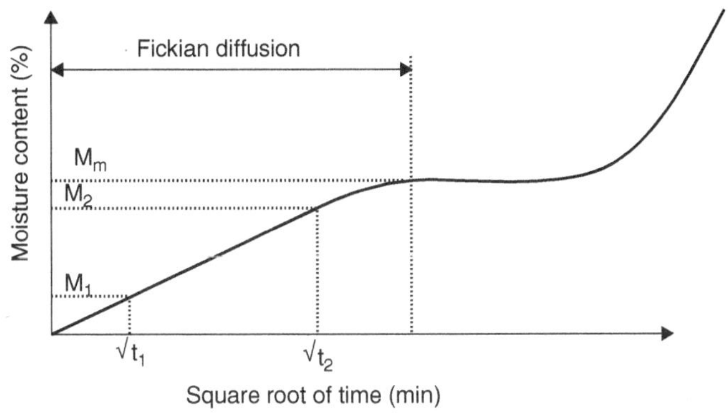

FIGURE 10.3 General responses of polymer composites [7].

TABLE 10.1
Different Types of Moisture Dispersion Techniques in Polymer Composites

Types of Dispersion		Dispersion Index (n)	Time Factor (t)	Dispersion Techniques
Type 1	Low Fickian	$n < 0.5$	$t^{-0.5}$	Water dispersion is very slow in comparison to the motion of polymer sections.
	Normal Fickian	$n = 0.5$	$t^{0.5}$	
Type 2	Dispersion	$n = 1.0$	Independent of time	Dispersion process is much more energetic than the distraction process.
	Superior dispersion	$n > 1.0$	t^{n-1}	
Type 3	Non-Fickian	$0.5 < n < 1.0$	t^{n-1}	Motion of water molecules are along with polymer section, which lies in between type 1 and type 2 dispersion.

Source: [2].

cracks, and weak fiber matrix bonding mostly in the case of fiber-reinforced polymer composites, and the dispersion coefficient is studied in the Fickian dispersion model.

Moisture dispersion coefficient (D) is represented by Equation 10.3.

$$D = \pi\left[\left(\frac{h}{4Mm}\right)\left(\frac{M_2 - M_1}{\sqrt{t_2} - \sqrt{t_1}}\right)^2\right] \tag{10.3}$$

where h = specimen thickness, M_m = maximum content of moisture, M_1, and M_2 are the masses of the specimen at time t_1 and t_2, respectively. Different types of moisture dispersion are represented in Table 10.1.

10.2.2 Contact Angle

The angle of contact in a composite surface with water particles immensely affects the rate of WA. In the case of a higher contact angle, the surface area of the liquid is less, and it shows minimum spread on the surface, which finally results in optimum WA but just the reverse occurs in the case of a lower contact angle [8]. The contact angle significantly increases by adding nanofiller in composites and makes the structure more compact in between filler, and matrix bonding. The glossiness of the surface due to the addition of nanofiller restricts the passage of moisture.

10.3 NANOFILLERS

The nanofillers are predominantly categorized in different ways such as carbon-based organic, and inorganic fillers. Inorganic fillers are mostly nature-derived, and the classification of the nanofillers is represented in Figure 10.4.

10.3.1 Nanoclay

Nanoclay is easily accessible and extracted from the bentonite in the form of montmorillonite (MMT) to be used in several polymer composites. Halloysite nanotube (HNT) is also one type of nanoclay that provides superior strength, and other properties. Ramakrishnan et al. [9] enormously studied the influence of Cloisite 20-A-type nanoclay dispersed in epoxy with a variation of wt.% on jute fiber-reinforced composites. The mechanical and physical properties of WA are studied by considering the factors like different concentrations of NaOH treatment, variation in effective fiber length, and weight fractions. It is found that though 30 mm fiber length, and 20 wt.% fraction results in superior overall characteristics but 5 wt.% nanoclay, and NaOH treatment gave lower and higher mechanical strengths. It is also observed that with an increase in fiber length, the WA gradually increases. Kord et al. [10] performed the experimental analysis for 50 wt.% of beech wood powder-reinforced polypropylene (PP) by adding nanoclay filler in different concentrations. The fungal decay resistance and mechanical properties are analyzed properly, where it is found that a 5% concentration of nanoclay shows maximum hardness and the least weight loss. Simultaneously impact strength and moisture gain are drastically reduced after the addition of nanoclay up to 3% concentration but bend strength, and its modulus gradually increases.

The enhancement in mechanical, thermal, and physical behavior of silicate layered reinforced vinyl ester polymer composites, and the addition of nanoclay up to 5 wt.% concentration is studied by Alateyah et al. [11]. It is observed that WA is 13 times more without clay composites in comparison to the 5 wt.% clay included composites. However, water-immersed composites result in weak hardness but the maximum layer of silicate retains the strength as well as hardness up to a certain extent. Najafi et al. [12] illustrated the tensile strength, impact strength, and WA behavior of reed-flour-reinforced PP composites including nanoclay, manufactured by injection molding method. Tensile strength improved due to nanoclay but impact strength, and WA gradually reduced with an increase in wt.% of nanoclay addition. The physical and thermal properties of woven kenaf fabric with nonwoven bamboo fiber-mat-reinforced epoxy hybrid composites are prepared by h and layup process. Three different forms of nanoclay such as MMT, HNT, and organic refined MMT

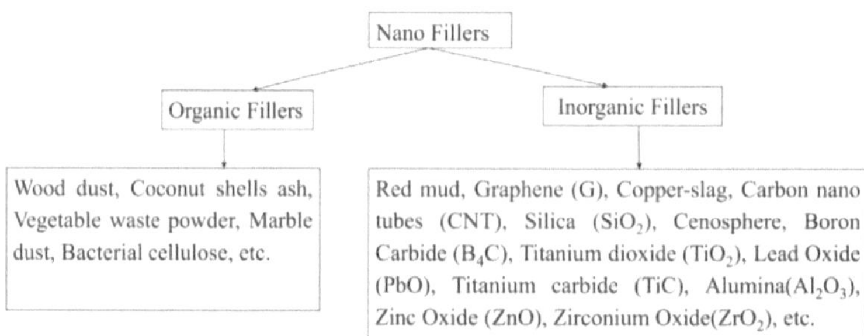

FIGURE 10.4 A schematic of classification of nanofillers.

(OMMT) fillers of 1 wt.% added carefully in hybrid composites. Chee et al. [13] summarized that OMMT addition in bamboo (B)- and kenaf (K)-reinforced composites revealed quite better physical as well as thermal expansion as compared to other fillers. It is also intensely observed that the more void content more is the WA as shown in Figure 10.5.

Alhuthali et al. [14] prepared the composite by reinforcing the sheets of recycled cellulosic fibers in vinyl ester known as eco-composites, and another along with Cloisite 30B nanoclay of varying weight percentage in vacuum conditions. At 5 wt.% inclusion of nanoclay optimizes WA but reduces the impact, and fracture toughness, and shows brittle characteristics, whereas from 1 to 3 wt.% revealed improvement in impact as well as fracture toughness.

Alhuthali and Low [15] enumerated the WA, and mechanical behavior of HNT reinforced vinyl ester resin (VER) composites using a large stirring speed. Finally, it is found that 5% HNT attributes maximum strength, toughness, thermal stability, and most preferable water resistance. The amount of nanoparticles plays an exceptional role such as the concentration of the filler the lower the cross-linking capacity of the polymer and matrix which at last reduce the bond strength. Menezes et al. [16] focused on the cross-linking potential of organic refined nanoclay along with the mechanical characteristics of hybrid composites used in the dental application and also compared it with the broadly used silica nanofillers. From the experimental result, it is concluded that up to 2.5% organic refined nanoclay is uniformly well dispersed but more than that clustered together without proper cross-linking. So 2.5% organic refined clay is permissible for dental use whereas it is observed that the early used microsilica shows gradual improvement in properties. Opelt and Coelho [17] critically searched the mechanism of effective reinforcement of sodium, and organically modified nanoclay on polymer composites along with toughening techniques. The effective stage of dispersion of nanoclay enhances the efficiency of polymer composites which is clearly known as its transition state that depends on volume or weight fraction of nanomaterials. Haq et al. [18] examined the mechanical, and hygro-thermal aspects of bio-based unsaturated polymer fused carefully in soybean oil, and reinforced with industrial hemp fiber, and nanoclay in varying proportions. Bio-based matrix composites mostly enhance the toughness compromising the stiffness but by adding 1.5% nanoclay filler both achieved synergistic effects including the improvement in hygro-thermal behavior. Simultaneously it is also noticed that multifunctional features of biocomposites are drastically developed by adding optimum nanoclay with pure bioresins. Kord [19] also experimentally

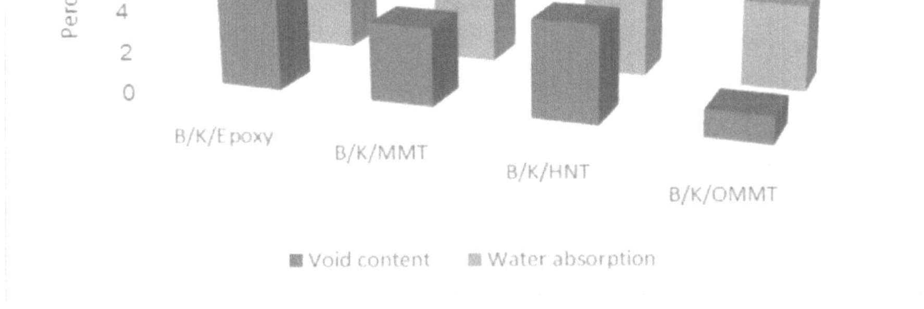

FIGURE 10.5 Relative comparison of void content with percentage of WA [13].

found that with an increase in nanoclay content in hemp fiber added to PP composites water intake, thickness swelling characteristic linearly reduced. The optimum WA of a few nanoclay-filled biocomposites is tabulated in Table 10.2.

10.3.2 Silica (SiO_2)

Anbukarasi et al. [20] explored the supremacy of nano-SiO_2 sphere inclusion on overall characteristics development of luffa, and coir fiber supported nanocomposites. It is validated that nano-SiO_2 sphere up to 3% volume declared quite improved overall mechanical strength, and modulus, and also shows efficient reduction in WA to 0.14%. Nanofillers have noble characteristics and immense potential that without increasing the density of the composites can successfully improve the other characteristics. Nourbakhsh et al. [21] explored the impact of mineral filler such as nano-SiO_2 on mechanical as well as physical behavior in varying wt.% filled in biopolymer composites consisting of rice husk, and beech bark separately, which are the cellulosic waste materials. It is remarkably observed that lower the filler content within 3 wt.% results in higher mechanical strength, and optimal WA due to the size effect it develops well cross-linking, and superior interfacial bonding. But the composites having higher hollow-cellulosic fiber content absorb maximum water that is in the case of beech bark fiber containing higher cellulose, and lower lignin as compared to rice husk. Nano-SiO_2 (NDS) is also successfully introduced in green cellulosic waste baggage (B) fiber reinforced in polyethylene (PE) integrated with maleic anhydride by Hosseini et al. [22]. The increase in fiber and filler content dramatically elevates the thickness swelling and WA simultaneously but impact strength shows a reverse trend, and other mechanical properties are moderately improved as shown in Figure 10.6.

Zuo et al. [23] tried to enhance the attributes of PLA by adding natural bamboo fiber with nano-SiO_2 for suitable interfacial bonding development. The exceptional result is observed, where mechanical strength as well as WA increased dramatically with 1.5 wt.% SiO_2 addition. The well-defined crystal lattice is developed due to effective cross-link bonding between fiber, and polymer by proper diffusion of SiO_2. The strength of the material's surface wettability also determines the rate of WA by observing the surface contact angle between the outer droplet surface, and the solid surface. The hydrophobic nature is enhanced with a larger contact angle that is depicted in Figure 10.7(a).

The individually nanocoated fibers show a tendency to enhance the overall properties. Bayart et al. [24] analyzed the organic nano-SiO_2 coating effect on nanocoated flax fiber reinforced with PLA, noncoated flax fiber with PLA, and only pure PLA on mechanical, and WA behavior as shown in Figure 10.8(b). WA of coated fiber is minimal in comparison to noncoated one; also it highly improves the tensile strength, and modulus due to the homogeneous dispersion of nano-SiO_2. The trend of humidity uptake w.r.t. is shown in Figure 10.8(b). Liu et al. [8] enumerated the significance of both nano-SiO_2, and ZnO as additives up to 3, and 4 wt.% in xylan film, a component of plant-hemicellulose-reinforced polyvinyl alcohol (PVA) composites. A total of 3 wt.% ZnO and SiO_2 dramatically reduce the water, and oxygen absorption capacity but simultaneously enhance the tensile strength. The trend of ZnO- and SiO_2-coated film composites is shown in Figure 10.8.

10.3.3 Titanium Dioxide (TiO_2)

TiO_2 is a ceramic basically white in color like SiO_2 which also takes a vital role in optimizing the WA behavior including the permeability tendency of any other gases that reduce the strength of the composites ultimately.

Nayak et al. [25] analyzed to upgrade the residual strength, thermal, and MA behavior of glass fiber polymer composites filled with organic nano-TiO_2 filler. It is remarked that 0.1 wt.% of filler minimizes the WA about 9%, and residual flexural, and interlaminar shear strength is enhanced close to 19 and 18%, respectively. The experimental result also shows quite good conformation with Weibull parametric distributions. Shaili et al. [26] studied the physical, thermal,

TABLE 10.2
WA Trend in Different wt.% of Nanoclay Biocomposites

Type of Fiber	Form of Fiber	Nanoclay	Type of Matrix	Fabrication Process	Optimum WA in wt.% of Nanofiller (%)	References
Jute	Short fiber	Nanoclay	Epoxy	H and layup	5	[9]
Beech wood flour	Powder form	Nanoclay	Polypropylene (PP)	Melt compounding and injection molding	3	[10]
Silicate	Layered form	Nanoclay	Vinyl ester matrix	Melt intercalation and compression molding	5	[11]
Reed fiber	Powder form	Nanoclay	Maleated PP	Injection molding	4	[12]
Bamboo and kenaf	Nonwoven and woven mat form	Nanoclay (OMMT, HNT, MMT)	Epoxy	Hand layup	1	[13]
Cellulose fiber	Sheet	Nanoclay	Vinyl ester matrix	Vacuum layup	5	[14]
Without fiber	–	1,3,5 wt.% of HNT	VER	Vacuum layup	5	[15]
Boron-alumina silicate glass	Spherical form of 4 μm radius	Organo-modified clay	Bis-GMA (35%), UDMA (40%), and TEGDMA (25%)	Light curing	2.5	[16]
Raw hemp	Chopped (25 mm)	Nanoclay	Unsaturated polyester resin (UPE) and epoxidized methyl soyate (EMS)	Compression molding	1.5	[18]
Hemp	Short fiber (100 mm)	Nanoclay	PP and Maleic anhydride–grafted PPs	Injection molding	5	[19]

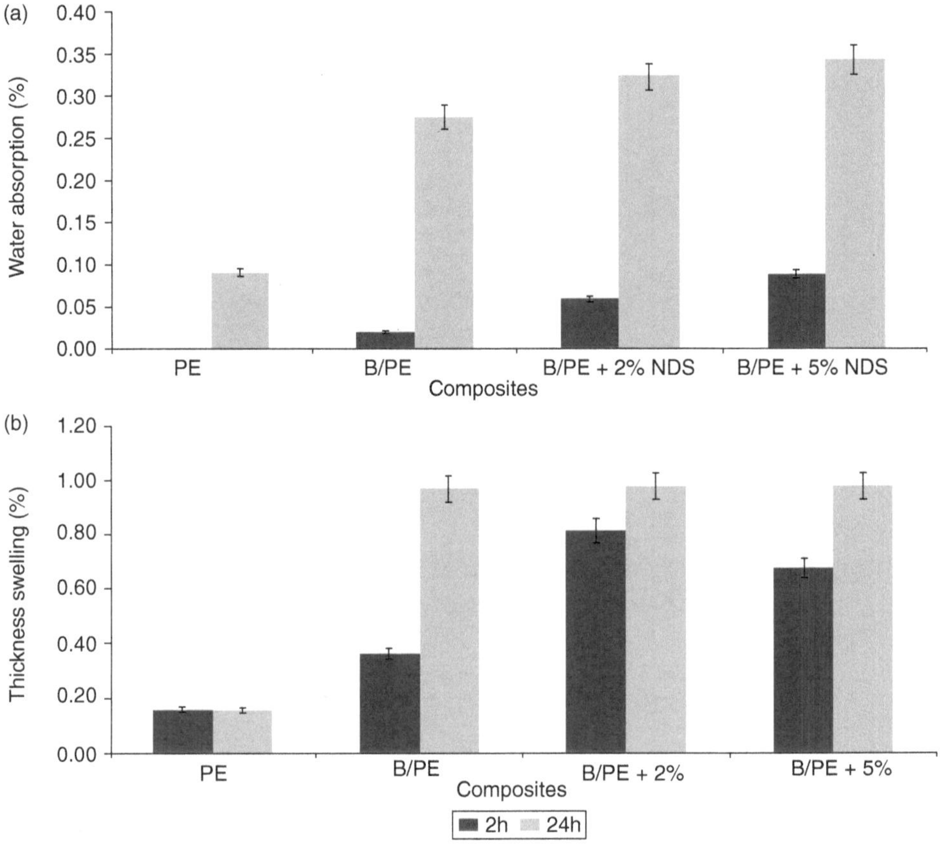

FIGURE 10.6 (a) Comparison of WA, and (b) thickness swelling with different wt.% of nano-SiO_2 [22].

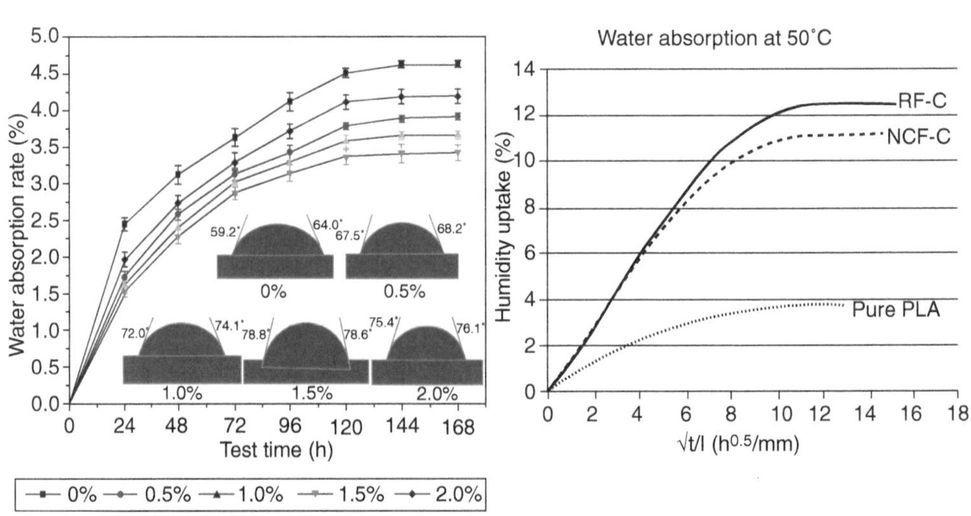

FIGURE 10.7 (a) WA with a contact angle of varying wt.% of SiO_2 nanofiller [23]. (b) Humidity uptake (%) of pure PLA, and coated materials [24].

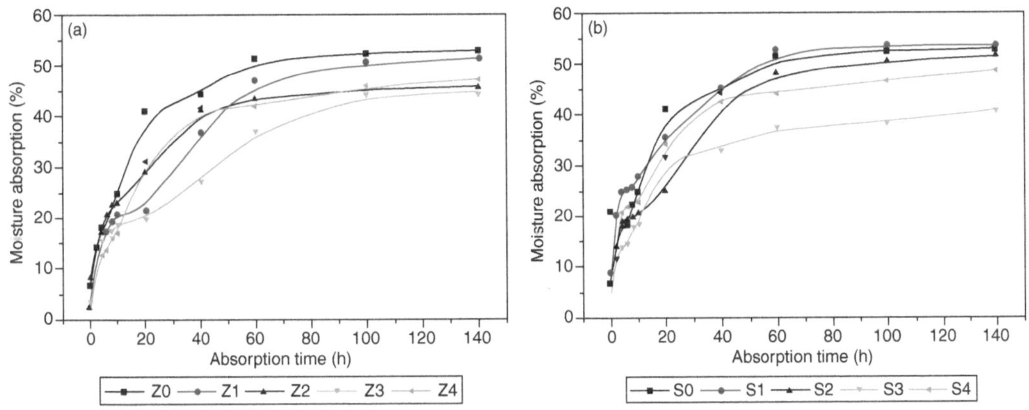

FIGURE 10.8 WA trend in composite film of (a) ZnO-coated (Z0–Z4) samples, and (b) SiO_2-coated (S0–S4) samples [8].

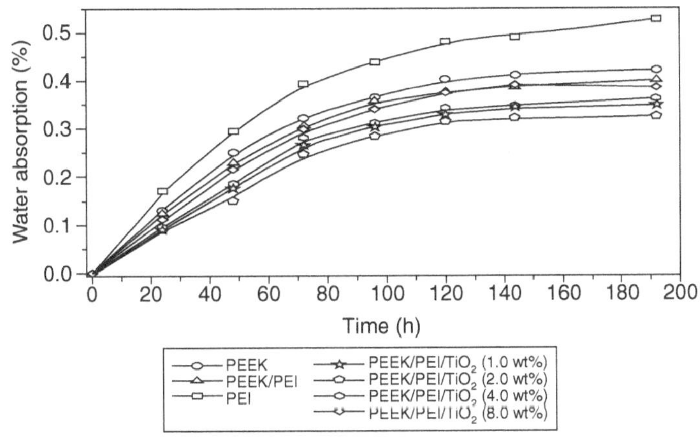

FIGURE 10.9 Water diffusion behavior of PEI, PEEK, and PEEK/PEI/TiO_2 composites [27].

and functional characteristics of well-dispersible soybean polysaccharides added with nano-TiO_2 in varying proportions. The moisture and oxygen permeability reduced drastically at 5 wt.% addition of filler than others. Diez-Pascual and Diez-Vicente [27] manufactured the biodegradable nanopolymer composites by sonication, subsequently followed by melt blending polyether ether ketone (PEEK) and polyether amide (PEI) with reinforcement of nano-TiO_2 additive, and studied various characteristics for prolong load-carrying application. It is observed that 4 wt.% nano-TiO_2 remarkably improved the overall mechanical properties as well as optimized the WA due to an increase of consistent nanoparticle distribution in the matrix. The kinetics influenced by TiO_2 is represented in Figure 10.9.

Currently, the functional properties of bio-fungicide chitosan which is collected from the shells of crabs, prawns, etc. are very popular in industrial fields (Figure 10.9). Chang et al. [28] briefly enumerated the significance of nano-TiO_2, and glycerol monolaurate (GML) used as additives in chitosan matrix, which is mostly used for food packaging applications. It is noticed that the addition of nano-TiO_2 exceedingly reduces the WA capacity along with oxygen, and carbon dioxide (CO_2) permeability, which may cause huge hazards in many packaging industries. Nano-TiO_2 is very effectively used in the manufacturing of artificial organs like bone, and tissues in biomedical engineering fields. Khan et al. [29] carefully investigated the significance of nano-TiO_2, and nano-hydroxyapatite (n-HAp) introducing in arabinoxylan (AX) a pure biomaterial used in biomedical tissue, and bone development. Consequently, it is observed that with the increment in nanoceramic TiO_2 optimization of the swelling became 27.5% as these are less hydrophobic in nature, and also enhance the cross-linking between the polymers. The high cross-linking feature develops strong intermolecular or interlaminar bonding. The impact of varying weight percentages of nano-TiO_2 filler in flax-epoxy-reinforced composites, manufactured by compression molding process is meticulously analyzed by Prasad et al. [30]. It is observed that 0.7 wt.% of nano-TiO_2 results in better performance in the improvement of mechanical characteristics as well as a 31.66% reduction in water dispersion coefficient due to perfect distribution by mechanical stirring followed by sonication. Wang et al. [31] used nano-TiO_2 in pure form, and different treated forms with coupling agents such as trimythoxyvinylsilane (A171) and (3-aminopropyl) triethoxysilane (KH550). The wheat straw fiber is reinforced with PP, and uniformly dispersed with pure and reformed nano-TiO_2 by using a hot extrusion process. It is remarked that KH550 reformed TiO_2 accomplished better mechanical as well as physical properties such as WA.

10.3.4 Graphene

Abdullah et al. [32] conducted the water permeability test for nanographene particles added in composites of coconut shell fiber reinforced with unsaturated polyester in varying proportions by h, and layup process. It is consequently ensured that with the increase in nanographene addition at 1.5 wt.%, the behavior gradually increases in comparison to 1 wt. %, and at 0.5 wt.%, this may be due to improper distribution, and aggregation of nanoparticles. The clustering of fillers generally forms voids and reduces interlaminar or intermolecular bondage. Whereas Ashok, and Kalaichelvan [33] also utilized graphene nanofiller in different wt.% for epoxy composites reinforced with luffa fiber, and noticed that 2 wt.% filler results in better mechanical as well as physical characteristics. It is also declared that water permeability is reduced at 2 wt.% addition of nanofiller, and the coefficient of dispersion drops to 38% in comparison to without filled composites.

10.3.5 Nano-marble Dust

Marble dust (MD) is considered most economic environmental waste, and collected from various marble industries, and functionally used in composite manufacturing. Kar et al. [34] analyzed the physical, mechanical, and thermal characteristics of nano-MD particles added in sisal, and jute fabric hybrid composites. It is finally obtained that nano-MD particles drastically diminish the water intake, and thickness swelling features in comparison to nonfilled composites. It also enhances the strength through better cross-linking characteristics and interlaminar bonding between fiber and epoxy. Thangaraj et al. [35] manufactured the short sisal, and banana fiber within 1–2 cm in length-reinforced hybrid PP composites through injection molding method by introducing 50–60 nm MD particles. It is observed that water intake capacity enhanced by adding nano-marble dust particles may be due to its higher density.

10.3.6 Other Nanofillers

Ashok et al. [36] investigated the influence of nano-lead oxide (PbO) incorporated in luffa-epoxy composites by the process of compression molding. From the physical test, it is analyzed that WA was reduced effectively by adding nano-PbO in comparison to the specimens having only luffa fibers. Figure 10.10 represents the tendency of WA in varying wt.% of PbO incorporated in composites, and observed that 1.25 wt.% PbO addition results in optimum WA.

The physical property of nano-titanium carbide (TiC), microparticles of coir-filled hybrid basalt, and innegra-fabric-reinforced composites in biosynthetic epoxy are studied by Mohit et al. [37]. TiC

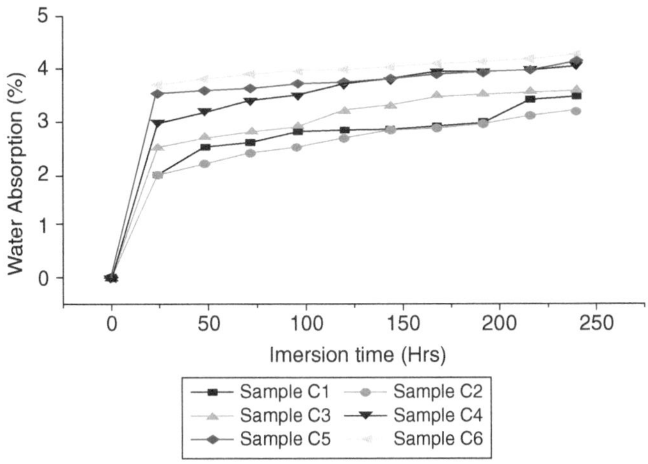

FIGURE 10.10 WA trend of pure luffa and PbO-reinforced composite samples (C1–C6) [36].

positively influences the water intake as well as the porosity of the composites, but the higher contentment of microcoir affects it negatively. The nature-extracted haritaki nanopowder is used as filler in epoxy composites reinforced with tindora tendrils plant fiber, and manufactured by using both h and layup processes along with compression molding process. Nanofiller imparts less moisture intake due to large surface area, and bonding in comparison to the unfilled composites but represents a similar trend without any specific difference because of the intrinsic WA tendency of haritaki filler [38]. The optimum WA due to the wt.% of different nanofillers are listed in Table 10.3.

10.4 APPLICATION OF NANOPARTICLES

Currently, nanoparticles have wide applications in various fields such as packaging, the automobile industry, the construction field, the manufacturing of cosmetic products, electrical, and electronic appliances, medicines, and biomedical fields, sports, and ballistic equipment. Natural fibers such as flax, hemp, jute, bamboo, and kenaf are mostly considered engineered fibers, basically used for interior, and exterior components of automobiles such as seat backs, bumpers, door panels, dashboards, gas tanks, heads, and trunk liners, automobile sunroof frames, and package trays [39,40]. German-made automotive companies such as BMW, Mercedes, Ford, Audi, and Volkswagen mostly concentrate on cellulosic fiber composites for manufacturing their outer and inner components. Hybrid flax and sisal-fiber-reinforced epoxy composites are used as door panels in E grade model of Mercedes Benz, whereas rubber and coir fiber are considered to be more suitable for seat manufacturing in A-grade model of the same [41]. Kenaf fiber is used in the Mondeo model of Ford; sisal/flax hybrid fiber reinforcement in polyurethane is utilized as a door panel in Audi, and similarly flax fiber in PP is used by Volkswagen for the Chevrolet 2000 model. Apart from automobile applications, natural polymer composites are also utilized in electrical and electronic equipment such as printed circuit boards, solar panels, organic solar cells, capacitors, insulators, and gears. The sports industry has equally utilized the biofibers in bicycle frames, baseball bats, tennis rackets, helmets, hockey sticks, sports gear, bows and arrows used in archery, racing bikes, etc. [42]. The most conventional construction industry also has wide applications of nanofillers due to their strength and durability. Nanopolymers are also being used in medical science for healing wounded or burned skins, and in dentistry for dental brackets, dental filling materials, and various dental repairing unique liquid epoxy nanocomposites are used [43]. Nanotalc, extracted from corn starch, is outstandingly employed in the packaging industry [44].

TABLE 10.3
Optimum WA in Different wt.% of Nanofiller Biocomposites

Type of Fiber	Form of Fiber	Nano Filler	Type of Matrix	Fabrication Process	Optimum WA in wt.% of Nanofiller (%)	References
E-glass	Long	SiO_2, Al_2O_3, Si_3N_4	Vinyl ester resin	Pultrusion	1, 1, 5	[7]
Beech wood xylan	Powder form	ZnO, SiO_2	PVA resin	Ultrasonic dispersion and solution casting	3, 4	[8]
Luffa and coir	Chopped (4 mm)	SiO_2 nanosphere	Epoxy	Injection molding	3	[20]
Rice husk and beech bark	Short fiber	SiO_2 (0–4%)	Polypropylene (PP)	Injection molding	3	[21]
Bagasse	Flour form (40 mesh)	SiO_2 powder	High-density polyethylene (HDPE)	Injection molding	5	[22]
Bamboo	Short	SiO_2	Polylactic acid (PLA)	Compression molding	1.5	[23]
Flax	Woven fabric	SiO_2	PLA	Hot pressing	5	[24]
Coconut shell fiber	Long form	Graphene	Unsaturated polyester (UPE)	Hand layup	1.5	[32]
Sisal, and jute	Plain woven fabric	Marble dust	Epoxy	Hand layup	10	[34]
Banana, and sisal	Short (1–2 mm)	Marble dust (50–60 nm)	PP	Injection molding	5	[35]
Luffa	Mat form	PbO (20 nm)	Epoxy resin	Compression molding	1.25	[36]
Basalt-innegra	Woven fabric	TiC (10 nm), and coir (250 μm)	Epoxy resin	Hand layup	5	[37]

Finally, and above all nanofiller with natural fiber composite is still in a developing stage for further future application due to its sustainability. In concern to sustainability phenomena incorporation of nanoparticles or nanofibers and natural polymer composites drastically reduce the environmental impact. Subsequently, the above-stated sustainable bio-nanocomposite materials significantly improve the mechanical, thermal, and acoustic behavior in comparison to conventional materials.

10.5 CONCLUSION

The nanofiller incorporated in biopolymer-reinforced composites plays a promising role in the synergetic improvement of the overall behavior of the same. In biopolymer composites, the most negative tendency is the WA, which drastically affects the strength of the composites. The rate of moisture intake significantly depends on the void content during manufacturing, improper bonding between fiber, and matrix, type of fiber, size of fiber, and its corresponding chemical constituents, and structures, type of matrix, type of coupling agents, and also the type of treatment recommended for green fibers. From the above review, it is finally obtained that nanoparticle size, amount, specific surface characteristics, and density have a special effect on the optimization of WA in biopolymer composites. The coated natural fiber significantly reduces the moisture intake in comparison to the normally distributed nanofillers within the matrix. As per the survey among all the nanofillers, MD nanofillers are used up to a maximum of 10% to minimize the WA in polymer composites. It may be further explored more considering other different kinds of nanofillers in different composites, and various application fields, as very few types are explained in this study. The nanofillers collected from different biowaste products also can be further studied to minimize the major water diffusion problem.

REFERENCES

1. Rajeshkumar, G., Seshadri, S.A., Ramakrishnan, S., Sanjay, M.R., Siengchin, S., and Nagaraja, K.C., 2021. A comprehensive review on natural fiber/nano-clay reinforced hybrid polymeric composites: Materials, and technologies. *Polymer Composites*, *42*(8), pp.3687–3701.
2. Al-Maharma, A.Y., and Al-Huniti, N., 2019. Critical review of the parameters affecting the effectiveness of treatments used for natural composites. *Journal of Composites Science*, *3*(1), p.27.
3. Wang, Q., Chen, T., Wang, X., Zheng, Y., Zheng, J., Song, G., and Liu, S., 2023. Recent progress on aging of plant fiber reinforced polymer composites. *Polymers*, *15*(20), p.4121.
4. Raj, N. and Selvakumar, S., 2024. A Review on biopolymers as an eco-friendly and sustainable. In Proceedings of the Indian Geotechnical Conference 2022 Volume 4: Geotechnics: Learning, Evaluation, Analysis and Practice (GEOLEAP) (Vol. 4, p. 355). Springer Nature..
5. Hosseini, S.B., 2017. A review: Nanomaterials as a filler in natural fiber reinforced composites. *Journal of Natural Fibers*, *14*(3), pp.311–325.
6. Mishra, T., Mandal, P., Rout, A.K., and Sahoo, D., 2022. A state-of-the-art review on potential applications of natural fiber-reinforced polymer composite filled with inorganic nanoparticle. *Composites Part C: Open Access*, *9*, p.100298.
7. Won, J.P., Yoon, Y.N., Hong, B.T., Choi, T.J., and Lee, S.J., 2012. Durability characteristics of nano-GFRP composite reinforcing bars for concrete structures in moist and alkaline environments. *Composite Structures*, *94*(3), pp.1236–1242.
8. Liu, X., Chen, X., Ren, J., Chang, M., He, B., and Zhang, C., 2019. Effects of nano-ZnO, and nano-SiO_2 particles on properties of PVA/xylan composite films. *International Journal of Biological Macromolecules*, *132*, pp.978–986.
9. Ramakrishnan, S., Krishnamurthy, K., Rajasekar, R., and Rajeshkumar, G., 2019. An experimental study on the effect of nano-clay addition on mechanical, and water absorption behaviour of jute fibre reinforced epoxy composites. *Journal of Industrial Textiles*, *49*(5), pp.597–620.
10. Kord, B., Jari, E., Najafi, A., and Tazakorrezaie, V., 2014. Effect of nanoclay on the decay resistance, and physicomechanical properties of natural fiber-reinforced plastic composites against white-rot fungi (*Trametes versicolor*). *Journal of Thermoplastic Composite Materials*, *27*(8), pp.1085–1096.
11. Alateyah, A.I., Dhakal, H.N., and Zhang, Z.Y., 2014. Water absorption behavior, mechanical, and thermal properties of vinyl ester matrix nanocomposites based on layered silicate. *Polymer-Plastics Technology, and Engineering*, *53*(4), pp.327–343.
12. Najafi, A., Kord, B., Abdi, A., and Ranaee, S., 2012. The impact of the nature of nanoclay on physical and mechanical properties of polypropylene/reed flour nanocomposites. *Journal of Thermoplastic Composite Materials*, *25*(6), pp.717–727.
13. Chee, S.S., Jawaid, M., Sultan, M.T.H., Alothman, O.Y., and Abdullah, L.C., 2020. Effects of nanoclay on physical and dimensional stability of Bamboo/Kenaf/nanoclay reinforced epoxy hybrid nanocomposites. *Journal of Materials Research, and Technology*, *9*(3), pp.5871–5880.
14. Alhuthali, A., Low, I.M., and Dong, C., 2012. Characterization of the water absorption, mechanical, and thermal properties of recycled cellulose fibre reinforced vinyl-ester eco-nanocomposites. *Composites Part B: Engineering*, *43*(7), pp.2772–2781.
15. Alhuthali, A., and Low, I.M., 2013. Water absorption, mechanical, and thermal properties of halloysite nanotube reinforced vinyl-ester nanocomposites. *Journal of Materials Science*, *48*, pp.4260–4273.
16. Menezes, L.R., Silva, E.O.D., Rocha, A.C.D.S., Oliveira, D.C.R.S.D., and Campos, P.R.D.B., 2018. The applicability of organomodified nanoclays as new fillers for mechanical reinforcement of dental composites. *Journal of Composite Materials*, *52*(7), pp.963–970.

17. Opelt, C.V., and Coelho, L.A., 2016. Reinforcement and toughening mechanisms in polymer nanocomposites–reinforcement effectiveness, and nanoclay nanocomposites. *Materials Chemistry, and Physics*, *169*, pp.179–185.
18. Haq, M., Burgueño, R., Mohanty, A.K., and Misra, M., 2008. Hybrid bio-based composites from blends of unsaturated polyester, and soybean oil reinforced with nanoclay, and natural fibers. *Composites Science, and Technology*, *68*(15–16), pp.3344–3351.
19. Kord, B., 2012. Effect of nanoparticles loading on properties of polymeric composite based on hemp fiber/polypropylene. *Journal of Thermoplastic Composite Materials*, *25*(7), pp.793–806.
20. Anbukarasi, K., Hussain, S.I., Roseline, A.A., and Kalaiselvam, S., 2019. Effect of SiO_2 nanospheres on mechanical, thermal, and water absorption behaviour of luffa-coir/epoxy hybrid composites. *Materials Research Express*, *6*(12), p.125618.
21. Nourbakhsh, A., Baghlani, F.F., and Ashori, A., 2011. Nano-SiO_2 filled rice husk/polypropylene composites: Physico-mechanical properties. *Industrial Crops, and Products*, *33*(1), pp.183–187.
22. Hosseini, S.B., Hedjazi, S., Jamalirad, L., and Sukhtesaraie, A., 2014. Effect of nano-SiO_2 on physical and mechanical properties of fiber reinforced composites (FRCs). *Journal of the Indian Academy of Wood Science*, *11*, pp.116–121.
23. Zuo, Y., Chen, K., Li, P., He, X., Li, W., and Wu, Y., 2020. Effect of nano-SiO_2 on the compatibility interface, and properties of polylactic acid-grafted-bamboo fiber/polylactic acid composite. *International Journal of Biological Macromolecules*, *157*, pp.177–186.
24. Bayart, M., Gauvin, F., Foruzanmehr, M.R., Elkoun, S., and Robert, M., 2017. Mechanical and characterization of PLA composites reinforced with nano-coated flax fibers. *Fibers, and Polymers*, *18*, pp.1288–1295.
25. Nayak, R.K., Mahato, K.K., and Ray, B.C., 2016. Water absorption behavior, mechanical, and thermal properties of nano TiO_2 enhanced glass fiber reinforced polymer composites. *Composites Part A: Applied Science, and Manufacturing*, *90*, pp.736–747.
26. Shaili, T., Abdorreza, M.N., and Fariborz, N., 2015. Functional, thermal, and antimicrobial properties of soluble soybean polysaccharide biocomposites reinforced by nano TiO_2. *Carbohydrate Polymers*, *134*, pp.726–731.
27. Díez-Pascual, A.M., and Díez-Vicente, A.L., 2015. Nano-TiO_2 reinforced PEEK/PEI blends as biomaterials for load-bearing implant applications. *ACS Applied Materials, and Interfaces*, *7*(9), pp.5561–5573.
28. Chang, X., Hou, Y., Liu, Q., Hu, Z., Xie, Q., Shan, Y., Li, G., and Ding, S., 2021. Physicochemical, and antimicrobial properties of chitosan composite films incorporated with glycerol monolaurate, and nano-TiO_2. *Food Hydrocolloids*, *119*, p.106846.
29. Khan, M.U.A., Haider, S., Shah, S.A., Abd Razak, S.I., Hassan, S.A., Kadir, M.R.A., and Haider, A., 2020. Arabinoxylan-co-AA/HAp/TiO_2 nanocomposite scaffold a potential material for bone tissue engineering: An in vitro study. *International Journal of Biological Macromolecules*, *151*, pp.584–594.
30. Prasad, V., Joseph, M.A., and Sekar, K., 2018. Investigation of mechanical, thermal, and water absorption properties of flax fibre reinforced epoxy composite with nano TiO_2 addition. *Composites Part A: Applied Science, and Manufacturing*, *115*, pp.360–370.
31. Wang, D., Xuan, L., Han, G., Wong, A.H., Wang, Q., and Cheng, W., 2020. Preparation, and characterization of foamed wheat straw fiber/polypropylene composites based on modified nano-TiO_2 particles. *Composites Part A: Applied Science, and Manufacturing*, *128*, p.105674.
32. Abdullah, A., Bakar, M.A., Mohd, S.H., and Masri, M.N., 2021, December. Effect of graphene nanoplatelets on water absorption properties of coconut shell reinforced unsaturated polyester composites. *Journal of Physics: Conference Series*, *2129*(1), p.012072. IOP Publishing.
33. Ashok, K.G., and Kalaichelvan, K., 2020. Mechanical, ballistic impact, and water absorption behavior of luffa/graphene reinforced epoxy composites. *Polymer Composites*, *41*(11), pp.4716–4726.
34. Kar, S., Mishra, S., Pattnaik, S., and Sutar, M.K., 2023. Impact of nano-marble dust on characteristic development of sisal-jute hybrid composites. *Proceedings of the Institution of Mechanical Engineers, Part C: Journal of Mechanical Engineering Science*, *20*, p.09544062231154083.
35. Thangaraj, S., Vigneshwaran, M., Muthuraj, M., Shenbagaraj, R., and Girimurugan, R., 2023. Distinctive study on banana/sisal fiber hybrid composites filled with nano marble dust particles. *Materials Today: Proceedings*.

36. Ashok, K.G., Kalaichelvan, K., and Damodaran, A., 2022. Effect of nano fillers on mechanical properties of luffa fiber epoxy composites. *Journal of Natural Fibers*, *19*(4), pp.1472–1489.
37. Mohit, H., Srisuk, R., Sanjay, M.R., Siengchin, S., Khan, A., Marwani, H.M., Dzudzevic-Cancar, H., and Asiri, A.M., 2021. Nanoparticles addition in coir-basalt-innegra fibers reinforced bio-synthetic epoxy composites. *Journal of Polymers, and the Environment*, *29*, pp.1–13.
38. Ramesh, M., Deepa, C., Niranjana, K., Rajeshkumar, L., Bhoopathi, R., and Balaji, D., 2022. Influence of Haritaki (*Terminalia chebula*) nano-powder on thermo-mechanical, water absorption, and morphological properties of Tindora (*Coccinia grandis*) tendrils fiber reinforced epoxy composites. *Journal of Natural Fibers*, *19*(13), pp.6452–6468.
39. Mohanty, A.K., Khan, M.A., and Hinrichsen, G., 2000. Surface modification of jute, and its influence on performance of biodegradable jute-fabric/Biopol composites. *Composites Science, and Technology*, *60*(7), pp.1115–1124.
40. Davoodi, M.M., Sapuan, S.M., Ahmad, D., Aidy, A., Khalina, A., and Jonoobi, M., 2011. Concept selection of car bumper beam with developed hybrid bio-composite material. *Materials& Design*, *32*(10), pp.4857–4865.
41. Shinoj, S., Visvanathan, R., Panigrahi, S., and Kochubabu, M., 2011. Oil palm fiber (OPF), and its composites: A review. *Industrial Crops and Products*, *33*(1), pp.7–22.
42. Mishra, T., Mandal, P., Rout, A.K., and Sahoo, D., 2022. A state-of-the-art review on potential applications of natural fiber-reinforced polymer composite filled with inorganic nanoparticle. *Composites Part C: Open Access*, p.100298.
43. Saba, N., Md Tahir, P., and Jawaid, M., 2014. A review on potentiality of nano filler/natural fiber filled polymer hybrid composites. *Polymers*, *6*(8), pp.2247–2273.
44. López, O.V., Castillo, L.A., García, M.A., Villar, M.A., and Barbosa, S.E., 2015. Food packaging bags based on thermoplastic corn starch reinforced with talc nanoparticles. *Food Hydrocolloids*, *43*, pp.18–24.

11 The Various Types of Severe Plastic Deformation Processes to Develop Ultrafine Grain Structure in Materials with Enhanced Mechanical Properties

Deepak Kumar, Murlidhar Patel, R. Seetharam, and K. Ponappa

11.1 INTRODUCTION

During the Bronze-to-Iron Age transition, severe plastic deformation (SPD) power was recognized. The earliest SPD method presumably employed for machinery for processing was repetitive hammering and folding, often called repeated smashing and twisting [1–3]. Iron has a higher melting point than bronze, which makes casting it difficult. As a result, metal forming gained attention for processing tools made of iron. In order to create wrought iron with increased toughness, the iron ingot that was taken out of the furnace had to be repeatedly heated, folded, and hammered in order to eradicate carbon and compress slag. This labour-intensive process was mostly used to create basic yet tactical equipment like guns; nevertheless, competent smiths were also able to create intricate, multi-piece instruments that are currently on display in some museums. The making of patterned Damascus steel swords has historically been linked to the process of repetitive hammering and folding [4]. It is still unclear if the name "Damascus" alludes to the ancient capital of Syria or to the damask pattern, which is the patterned form of these swords. High carbon wootz steels, also known as Fuladh in Syria, watered Pulad in Persia, Bulat in Russia, and Bintie in China, were mostly exported from Sri Lanka and India to Persia and other surrounding nations, where they were forged to create the Damascus steel swords [3,5,6]. These swords developed an uneven structure with low- and high-carbon areas as a result of constant forging and folding, which gave them a patterned appearance. In the past century, it has been demonstrated that the enormous carbon content and diffusion of carbide elements in the morphology of these swords were the primary causes of their remarkable strength, which had been a mystery for the eighteenth century. While some scholars have cast doubt on the existence of carbon nanotubes in these swords, several investigations conducted this century have revealed that extreme forging may enhance the strength of swords due to the progress of carbon nanotubes. Many ancient cultures, including Egypt, Iran, China, Russia, Korea, Japan, Turkey, the Philippines, and Indonesia, eventually began to use the repetitive beating procedures to create decorated swords. The repetitive forging and hammering process was presumably known to European cultures, and Vikings frequently employed it to make their weapons [3,5,6].

 DOI: 10.1201/9781003664611-12

The early Viking swords and the Damascus steel swords differed in that the former were made from decorative forging, or joining various pieces of steel and iron together, while the latter were primarily made with a single piece of steel that contains high carbon content, through numerous forging and folding passes [7].

On the other hand, it is thought that the renowned Ulfberht Viking swords, like the Damascus swords, were fashioned from a single piece of foreign steel. The figure displays a few swords that appear to have been made via pattern forging or repetitive hammering and folding: (a) Persian Shamshir [8], (b) Indonesian Kris [9], (c) Chinese sword [10], (d) Japanese Katana [11], (f) Korean sword [12], (g) Isonakami Jingu Shrine [13], (h) and Indian blade [9].

SPD becomes progressively more popular as an effective technique for controlling micro- and nanostructures [14,15]. Thus, it results in improved mechanical and functional properties in a wide range of materials like metals, alloys, polymers, intermetallics, and ceramics. Using a range of materials, from single- to multi-phase materials, and creating UFG materials on different scales without the requirement for alloying elements are two crucial advantages of SPD methods [16,17]. In the last few years, scientists have concentrated on using several SPD approaches for controlling the material's grain size. It is evident from the Hall–Petch relation that lower grain size significantly augments the inherent characteristics of metals and alloys. Moreover, the potential for developing superplastic at high temperatures can be achieved by the reduction in grain size. High-level hydrostatic stress is typically used in the SPD method without affecting the cross-section, resulting in ultra fine-grained structure and high-angle grain boundaries. Over the last few decades, there has been a significant increase in the bulk ultrafine-grain structure developed through SPD. The objective is to achieve superior mechanical properties such as elevated strength, good ductility, superior super plasticity, less coefficient of friction, high wear and corrosion resistance, high cycle fatigue life [16,18,19]. The materials created by SPD technology have the potential to be used in both basic research and practical applications because they are void of porosity and contamination. Even though SPD technology has an abundance of possibilities in terms of microstructure refinement, it has not yet effectively generated a particular desired, uniformly refined structure or preferred material attributes. Furthermore, the present SPD techniques have several limitations, such as the need for sophisticated equipment, expensive processing fees, a convoluted procedure, and challenges with industrialization and automation for large-scale manufacturing [20–22]. These days, SPD methods are moving from the field of lab-based research into the manufacturing of different UFG materials on a commercial basis. A prerequisite for distinguishing the SPD process from other metal-forming techniques is that the workpiece cross-sectional dimensions should not change. This makes it possible to repeatedly push and build up extremely high stresses. Large-scale samples can still be processed using recently developed traditional metal-forming techniques that apply strong plastic deformation.

It is widely accepted that microstructural features, particularly grain size, determine the mechanical characteristics of polycrystalline materials at ambient temperature and their high-temperature superplastic behaviour. The well-known Hall–Petch relation states that materials possess more strength at room temperature when the grain size decreases [23–25]. At high temperatures, the potential to create superplastics is also enhanced by a reduction in particle size. Grain size is one of the most crucial aspects of the microstructure of a polycrystalline material. Materials with a typical grain size of greater than 10 μm are referred to be coarse-grained materials in the industry. Grain sizes of fine-grained metals treated by industrial thermo-mechanical processing range from around 1 to 10 μm. The material having a grain size less than 1 μm, referred to as ultrafine grained, or less than 100 nm, referred to as nanograined, is difficult to manufacture using standard thermo-mechanical and metal-forming techniques [26,27] The main causes are lack of high-angle grain boundaries, limited levels of hydrostatic compressive stresses, and limited plastic strain resulting from restrictions in the decreased cross-section.

In many research labs throughout the world today, addressing metallic materials by applying extreme plastic deformation has gained significant importance. The definitions of the pertinent phrases utilized in this field of research and a recent article that outlined the fundamentals of SPD processing were utilized to develop the terminology used in this study. Being able to refine grains to sizes not possible with traditional thermo-mechanical treatments is the most alluring aspect of SPD processing. Therefore, SPD processing can produce grains with diameters between submicrometres and nanometres, while traditional methods lead to grain sizes of several micrometres [28]. The SPD procedure leads to grains that are referred as UFG. UFG solids are described as "bulk solids with uniformly mixed microstructures having large angles of misorientation and grain size of less than 1 micrometre". The broad aspects of SPD, the science and art of producing advanced bulk nanostructured materials using SPD techniques, are covered in this chapter as a result of collaborative efforts. This chapter aims to cover the most noteworthy discoveries that can advance the science, technology, and commercialization of the SPD research in the future. Scientists working in the field review both historical development and recent trends, and some terminological aspects are discussed.

11.2 SEVERE PLASTIC DEFORMATION PROCESSES

The SPD methodologies are the focus of this study, since they are one of the most intriguing procedures developed for grain refining. Due to their capability to develop significant grain refinement in completely dense, bulk-scale materials, thereby holding potential structural applications hence these procedures are highly popular. The submicrometre [29] (100–1000 nm) and nanoscale (<100 nm) categories contain the attainable grain sizes. Such materials that have undergone SPD processing are sometimes referred to as NanoSPD materials; nevertheless, only the latter may be considered, by traditional definition, to be nanostructured. In this chapter a quick overview of the basic SPD approach is provided, highlighting the key distinctions between the various NanoSPD procedures as well as their common characteristics. Langford, Cohen, and Landel [30] and Cohen [29] showed in the 1960s that the microstructure of Fe–0.003%C subjugated to large stresses through wire drawing leads to fine grains of 200–500 nm range and UFG structure [31]. This creates a distinct boundary that separates NanoSPD materials from more traditional materials that include subgrain features generated by ordinary metal-forming processes like cold rolling. In spite of the difference, all efforts paved the way for massive strain deformation-based microstructure refinement, where inducing plastic stress in the material is a difficult and precisely demanding process. It necessitates a significant investment in tool design, which must be both suited for processing materials without damaging the material and robust enough to withstand repeated high stresses during material formation. Considering the distinctive characteristics of SPD methodology is that large strains are applied short of appreciably altering the materials overall dimensions. This is made possible by unique tool designs that restrict material flow and raise the hydrostatic pressure to a high level. This hydrostatic pressure's existence provides a hint for reaching the high stresses necessary for remarkable grain refining. Once hydrostatic pressure is high, many crystalline materials that are normally brittle, such as tungsten oxide, B_2O_3 glasses, and amorphous materials, gain a significant amount of ductility and may be stretched to significant stresses without breaking [32]. These days, there are several SPD method variations that either explicitly or implicitly make use of this common property of high hydrostatic pressure in order to fabricate a wide range of UFG materials.

11.2.1 Equal Channel Angular Pressing (ECAP)

The most advanced SPD processing method available today is equal channel angular pressing (ECAP), often more accurately called equal channel angular extrusion in some publications. When a sample is forced through a die travels across the plane where the two channels converge, a simple shear strain is created. Because the sample cross-sectional dimensions don't change, it may be repeatedly pressed, which causes extremely enormous stresses to build up. Figure 11.1 provides an illustration of the ECAP procedure with a 90° channel angle [33,34]. For instance, the comparable (von Mises) equivalent strain (Equation 11.1) of 1.15 is injected every pass in ECAP when the channels are 90° apart [35,36]. The transverse, flow, and longitudinal planes within the as-pressed billet are represented by the planes X, Y, and Z in Figure 11.1, respectively. It is feasible to create many ECAP variations that include rotating billet along the pressing axis in between the passes; these variations often provide varied microstructure. One important question in ECAP is how much strain is applied with each individual pass through the die. The grain size of a metal is often considerably reduced during ECAP processing, which should result in significant strengthening.

$$\varepsilon_{equivalent} = N\frac{2}{\sqrt{3}}\cot(\varnothing) \qquad \text{Equation 11.1 [34]}$$

N is the number of ECAP passes.

It is therefore appropriate to remember that V. Segal [37] developed the primary benefits and principles of ECAP in a number of early publications. These principles included the extrusion process, the calculation of the optimal process parameters which includes frictional balance, tool geometry, strain path, and its effectiveness in grain refinement. Additionally, he showed that ECAP works best when the following conditions are met: (i) there is very little friction between the die walls and the billet; (ii) There are roughly 90° of angle between each channel; (iii) The entire sharp outer corner is occupied, resulting in narrow shear zone. The first criterion promoted the creation of dies with lower friction by introducing new, efficient lubricants and surface hardness of the channel walls, movable walls, etc. [38,39]. The relevance of back-pressure in the processing of materials with superior mechanical characteristics and a uniform microstructure was realized as a result of the third criterion. Segal demonstrated how a die's perceived simplicity of design might affect its performance. Specifically, he emphasized against using dies that have a corner arc since this might result in a widely dispersed plastic zone that resembles a fan. This is the same as creating artificially high friction, which disperses shear and results in a substantial heterogeneity of strain. Sadly, several subsequent investigations that used a "simplified" die design with rounded outer corners ignored this crucial warning, paying a heavy price in the form of significant variability of the deformed structure. On the other hand, samples with a homogeneous microstructure across the billet might be created by adhering to Segal's theory. Segal's concepts allowed for the construction of industry-scale ECAP facilities for processing of billets as big as 50×50 sq. m area and 600 mm in length. Standard laboratory-scale ECAP rigs are capable of handling large-size materials also [40,41].

11.2.2 High Pressure Torsion (HPT)

The term "high pressure torsion" (HPT) refers to a processing method that originated from Bridgman's anvils and combines high (GPa range) pressure with torsional straining [42,43]. Many researchers now recognize this technique as the most effective method for refining grains; however, one drawback of this method is that it can only process small coin-shaped samples, usually 10–15 mm in diameter and 1 mm in thickness. The samples produced by HPT are mainly used for research due to size restrictions. A thin disk that is positioned between large anvils, subjected to pressure P, and then torsionally stressed by rotating one of the anvils is often used in HPT method. This scenario is schematically shown in Figure. 11.2. The majority of the data that is now accessible relates to the usage of thin disks, although there are limited experimental results available for tiny cylindrical samples treated by HPT. The strain (Equation 11.2) fluctuates throughout the disk, reaching a maximum at the perimeter and zero at the disk core, which might be problematic for HPT

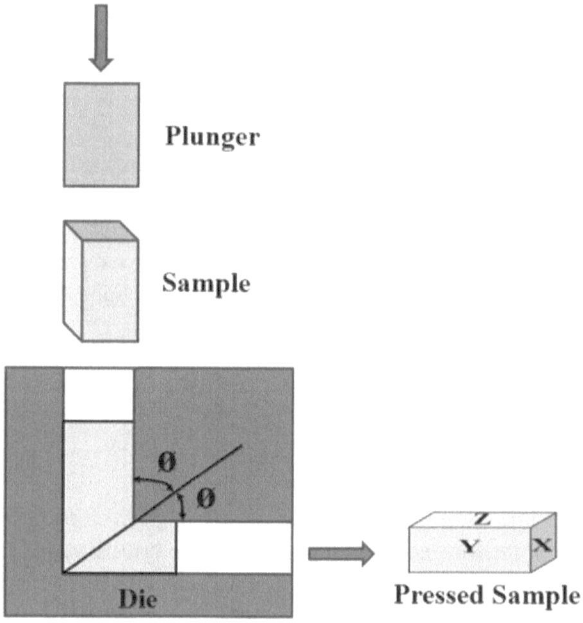

FIGURE 11.1 Schematic illustration of equal channel angular pressing.

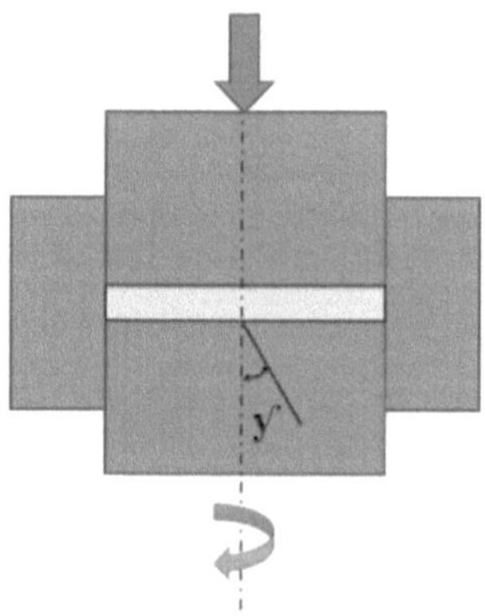

FIGURE 11.2 Schematic illustration of high pressure torsion.

processing [44–46]. This suggests that the structures created by HPT processing might be intrinsically nonhomogeneous, but preliminary tests revealed that if a high enough pressure was applied and the torsional strain was maintained for a large enough number of revolutions, a reasonable degree of homogeneity could be achieved throughout the disk's surface. The non-uniform deformation is a significant problem with most of the SPD process, also in HPT. For example, assuming the workpiece geometry remains unchanged during HPT straining, the shear strain at the rotation axis should be 0 and increase uniform in the circular path. This indicates that the material in close proximity to the sample's rotation axis ought to remain unaltered [47].

Several microstructural interpretations, microhardness experiments demonstrate a rather regular grains dissemination, as long as the anvil's stress and number of revolutions are vast enough that do not support anvil. The misalignment of the anvils' axes makes it nearly difficult to achieve an optimum HPT deformation, which is how Vorhauer and Pippan [48] explained the disparity. Strain gradient plasticity modelling has provided an appropriate interpretation of this structural progression towards homogeneity. The question of whether the through-thickness of HPT disks parallel to the rotation axis is homogeneous has garnered attention recently [49].

$$\varepsilon_{equivalent} = N\frac{2}{\sqrt{3}}\frac{\pi R}{t} \quad \text{Equation 11.2 [50]}$$

R is the radius from the axis, t is the thickness of the sample, and N is the number of revolutions.

11.2.3 Accumulative Roll Bonding (ARB)

Figure 11.3 provides a schematic illustration of the accumulative roll bonding (ARB) process. In the 1990s, Saito et al. [51] developed the ARB process as a means of establishing an SPD process suitable for the continuous manufacture of bulky materials with UFG structures. He proposed ARB as a method to get over the main drawbacks of the HPT and ECAP, particularly the latter's small workpiece size and the former's low productivity. The fact that ARB uses a traditional rolling facility is its biggest technological benefit [51,52]. The common SPD procedures handled very small specimens in an intermittent manner. Rolling deformation is the most efficient method of working with metal that ARB use for continually manufacturing plates, sheets, bars, and pipes.

For instance, in the first phase of ARB, 4 mm plate thickness is rolled with a 50% reduction in thickness. The 2 mm thick rolled plate is divided in half, stacked to provide the initial dimension prior to the initial rolling, and then rolled again by reducing the thickness by 50%. Essentially, these processes may be repeated indefinitely, allowing SPD to be applied to produce plates with UFG or nanostructures that are 2 mm thick [51,53]. Rolling in ARB is a two-step process that produces one-body solid materials by deforming the material and sticking it together. This process is called roll-bonding and is used to create clad sheets. The contact areas of the sheets are thoroughly degreased and wire-brushed to ensure optimal bonding. Rolling, cutting, brushing, and stacking are the steps in this process that are continued until the sheet eventually experiences a significant amount of strain. ARB was successfully utilized to process a variety of materials, including interstitial-free steel, pure aluminium, AA5083, laminated structures and composites based on aluminium and magnesium [54]. Regardless of the kind of metal or alloy produced, the UFG structure created during repeated rolling is not equal-sized but instead has pancake-like grains, much like in conventional rolling.

Moreover, the sample may be dynamically heated as it is being rolled, although this needs to be done at a temperature that prevents recrystallization. In an ARB method, the equivalent strain (eN) (Equation 11.3) following N cycles is expressed as eN = 0.80 N [51]. Rather from being three-dimensionally equiaxed, the UFG structure that is ARB-generated is really pancake-like and

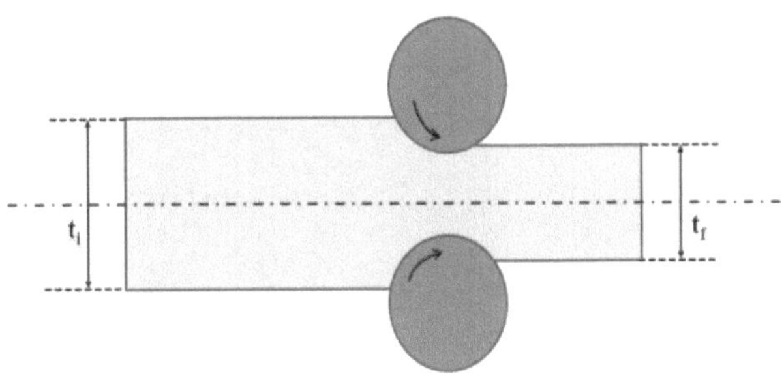

FIGURE 11.3 Schematic illustration of accumulative roll bonding.

prolonged laterally. Whichever material (including metals and alloys) is used, the microstructural characteristic will remain the same. This technology is used to fabricate metal matrix-based composite having coated mixed powder and subsequently proceed to accumulating roll bonding technique [55,56].

Moreover, mechanical mixed, and layer-by-layer structures made of irregular and nanosized materials can be processed using ARB. It is considered that nonhomogeneous materials exhibit exceptional mechanical performance; ARB was additionally used to construct heterostructured materials. One of the few SPD processes still available for the continuous manufacturing of big, heavy items is ARB. The establishment of massive equipment restricts large-scale metal productions, making it challenging to acclimatize SPD at large scale like those in the steel industry. On the other hand, the idea of ARB has previously been used in small-scale enterprises to produce thin stainless steel strips that are about 0.1 mm thick [57,58]. Successful scaling up of the ARB method has been accomplished even in academic laboratories. Future advancements in essential investigation and the industrial fabrication of nanosized materials are anticipated from ARB.

$$\varepsilon_{equivalent} = N\frac{2}{\sqrt{3}}\ \ln[\frac{t_i}{t_f} \qquad \text{Equation 11.3 [59]}$$

t_i is the initial thickness of the sample, t_f is the final thickness of the specimen, and N is the number of the passes.

11.2.4 Multi-Dimensional Forging (MDF)

Since ancient times, Forging has drawn interest from metal shaping in a number of ways. Scientists also found that this concept was a helpful way to increase plastic strain. Bridgman invented a double-stage forging process that permits substantial plastic deformation of materials. As an SPD methodology, the multi-dimensional forging (MDF) method, in its current form, has been in use at least since the 1990s [60]. A number of research groups successfully produced UFG materials using this approach in the 1990s and early this century. Dynamic recrystallization in alloys and metals is associated with the MDF process. The operating principle of MDF is shown in Figure 11.4

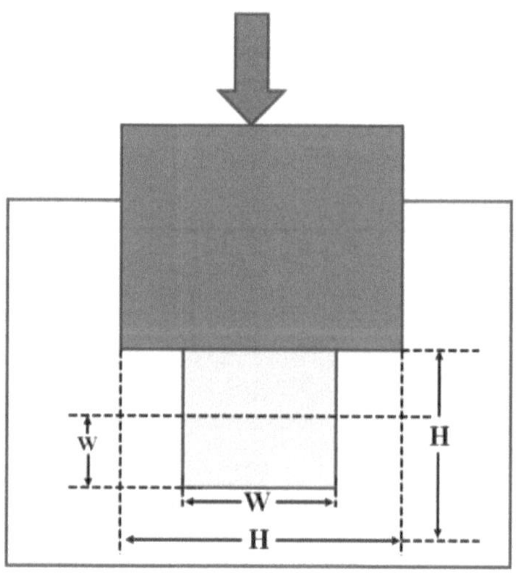

FIGURE 11.4 Schematic illustration of multi-dimensional forging.

in a clear and concise manner. It is predicated on the suppositions of several forging operations, including setting and pulling, with variations in the directions of load application. Compared to ECAP and HPT, the homogeneity of the MDF-induced strain is lower [61,62]. However, brittle materials can be made nanostructured using this method since low tool-specific loads and high processing temperatures are required at the beginning of the operation. By choosing the appropriate temperature-strain rate deformation regimes, the required grain refinement can be achieved. Large-scale materials with nanocrystalline structures can be produced using this process, which is normally applied at temperatures between 0.1 and 0.5 T_m, where T_m stands for the absolute melting temperature [18,63].

The sample is repeatedly forged along three orthogonal axes during the MDF process, with a 90° rotation in between passes. The simplest MDF approach is forging without the use of any specific dies; however, it can also be done using a die under constraint (cyclic closed die forging). The aggregate of the strains attained in every cycle in MDF produces a significant plastic strain ($\varepsilon = \ln(hi/hf)$), where hi is the initial height of the billet and hf is the final height of the billet. The key benefit of MDF is that it can be used with big hydrolytic presses that are currently in use in the industry. This makes the process appealing for use in commercial settings where large bulk samples need to be produced. Elongated strains at the surfaces and less hydrostatic pressure are the main drawbacks of the MDF process, which cause fractures to form and restrict the to and fro movement that can be performed at room temperatures [64,65]. The heterogeneity of the microstructure in MDF, as in other forging processes, is another drawback. In order to get over these restrictions, the MDF method is used at elevated temperatures, which can result into macro grains but also sufficient formability and homogeneity. In spite of all of these shortcomings, MDF is effectively used to the processing of Ti, Mg alloys, steels, and other materials, in addition to the grain refining of easily deformed materials like copper and aluminium.

$$\varepsilon_{equivalent} = N\frac{2}{\sqrt{3}}\ln\left[\frac{H}{W}\right]$$ Equation 11.4 [54]

11.2.5 Twist Extrusion (TE)

Another variation of a straightforward shear deformation process that was first presented by Beygelzimer et al. approximately ten years ago is twist extrusion (TE). A prismatic billet is extruded via a "twist die" in the TE processing method as shown in Figure 11.5 [32,66]. TE enrich capability, which is a benefit, but it has the same general issue of irregular deformation like HPT along the extrusion axis. The idea behind the TE method was first used to refine grain sizes in the late 2000s.

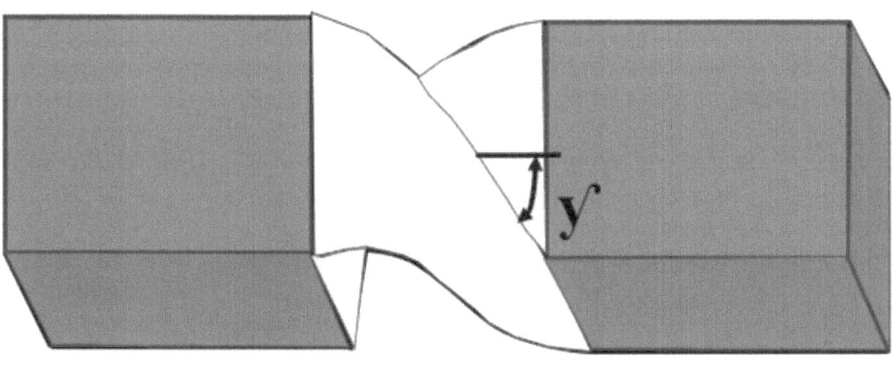

FIGURE 11.5 Schematic illustration of twist extrusion.

A workpiece needs to be forced through an extrusion die configuration during this intense plastic deformation process. Along its longitudinal axis, it must twist through a predetermined necessary angle while maintaining its cross-sectional dimensions and shape. As a result, after each twist-extruded pass, sample pushed during extrusion die and regains its original size and shape [67]. For this reason, it is easy to repeat the previous stages over material to achieve outstanding grain refinement. With the exception of circular geometries, a wide range of cross-sectional shapes can be severely plastically distorted employing this technique. The expression of maximum and minimum strain is given in Equations 11.5 and 11.6. In practice, the plastic strain is not evenly distributed throughout the sample cross-section, which is similar to the HPT approach. However, when length increases away from the axis, the amount of plastic strain increases as well, leading to significantly superior grain refinement in the material's farther-off portions. The resulting microstructural heterogeneities produce non-uniform mechanical characteristics, with the weakest area throughout the sample being the centre of the axis's cross-section. It is projected that, at least for pure metals, increasing the number of TE passes can improve the microstructural heterogeneity previously noted. The characteristics of material earlier TE die, inside the die, and after it exits the die; sample has undergone deformation due to twisting [68].

$$\varepsilon_{equivalent}^{mini.} \approx 0.4 + 0.1\tan(\gamma) \qquad \text{Equation 11.5 [65]}$$

$$\varepsilon_{equivalent}^{max.} \approx N\frac{2}{\sqrt{3}}\tan(\gamma) \qquad \text{Equation 11.6 [65]}$$

11.3 MICROSTRUCTURAL FEATURE AND FORMATION MECHANISM OF UFG

Large strain is introduced into a material component by the SPD procedures under typically high pressures, which inevitably results in notable microstructural changes. During the SPD process, the most well-known microstructural change is grain refinement to submicron or nanoscale range. It is generally known that at large strains, this kind of grain refining eventually saturates to a steady-state level [69]. In the realm of NanoSPD, comprehending to fine grains and the physical events that underlie grain refinement saturation remain crucial. Along with the production of UFG with a high percentage of high-angle grain borders, other distinctive microstructural properties include precipitate formation/dissolution and grain boundary segregation, in addition to several types of defects like twinning, dislocations, stacking faults, and vacancies [70,71]. As previously stated, these morphological characteristics are not exclusive to metallic materials that have undergone SPD processing; several oxide ceramics exhibit comparable characteristics. This section covers the key microstructural characteristics of SPD-processed materials that result in better mechanical and functional performance. SPD has made significant progress in the study of metallic materials with UFG structures, which have micron sized grain size and have been the subject of intense global research in recent decades. SPD was discovered to be able to realize UFG structures in the majority of metals and alloys. It is still unclear, nevertheless, why and how SPD processing results in the formation of UFG structures [72]. The primary method of refining grain in traditional metallurgy is recrystallization, which occurs during annealing following plastic deformation.

However, without annealing, UFG structures are already visible during deformation of materials that have undergone SPD processing. Other slip systems are triggered in polycrystalline materials' grains in addition to the principal slip system having large shear stress because of limitations from nearby grains. Different local crystal rotations result from the activation of various slip mechanisms in various grain regions. As a result, neighbouring areas become increasingly misoriented, giving rise to geographically required borders (GNBs) [72]. Within each domain that is split by GNBs, a large number of dislocations are also stored. These dislocations tend to create low-energy topologies known as incidental dislocation boundaries (IDBs).

GNBs and IDBs split the initial grains as plastic deformation increases, and finally, during SPD, finer subdivision structures at the nanoscale are created [71,72]. High densities of dislocations are present in the ultrafine lamellar structures, which are extended along the primary deformation direction. This suggests that a type of severely distorted microstructures is also present in the UFG structures generated by SPD. Both positive and negative effects on the mechanical characteristics can result from the distinctive UFG structures created by SPD. Additionally, it has recently been discovered that in some metals and alloys, completely recrystallized UFG structures maintaining micron sized grain sizes at the time of annealing process are possible. Grain diameters in both alloys range from several hundred nanometres, exhibiting fully recrystallized structures. Additionally, it has been made clear that such fully recrystallized structures possess great combination of ductility and strength.

11.4 MECHANICAL PROPERTIES

The most studied characteristics of materials that have been substantially distorted are their mechanical properties. Since the beginning of ancient metal making, it has been widely understood that deforestation strains the hardening process. At least as early as the eighteenth century, scientists have described strain hardening. Bulffingeri, a German–Russian scientist, postulated a power–law relationship between strength and strain that is comparable known as Hollomon equation. Hollomon [73] investigated the shear strength using torque measurements throughout the HPT process in the 1930s and found that the strength eventually saturates to a steady state and that the power–law stress–strain connection is invalid at large strains [74]. Additionally, he stated that surprising strain softening, as opposed to predicted strain hardening, is seen in pure lead. Subsequent measurements of the hardness and tensile characteristics by other scientists verified the existence of steady state at high stresses. These days, measuring microhardness in metals and alloys that have undergone SPD processing is a rapid mechanical characterization test to check for steady-state occurrence and the degree of deformation. There are three primary hardness-strain behaviours that have been documented for pure metals; however, the degree of purity and processing temperature have an impact on these behaviours: (i) Hardness diminishes with increasing strain (despite grain refinement) and saturates at large stresses for low melting temperature metals such indium, lead, tin, and zinc [75,76].

The ultrahigh pure UFG aluminium steady-state hardness is likewise less than coarse grain material. (ii) The hardness rises initially and then falls to an optimum value above the hardness level for coarse-grained materials after reaching a maximum for metals with modest melting temperatures, such as pure magnesium and pure aluminium. (iii) The hardness of metals with high melting temperatures usually rises until it saturates at high strains. The majority of metals and alloys that have undergone SPD processing exhibit the latter hardness-strain behaviour. Additionally hardness measures, bending, compression, and tensile testing on highly deformed materials reveal that these materials typically have higher strengths because of the presence of significant fractions of certain defects and also because fine grains leads to validation of Hall–Petch relations. In addition to their high strength, these materials' plasticity is often within an acceptable range, despite several efforts to increase their uniform ductility or even attain super plasticity at ambient temperature [77]. Other mechanical characteristics of these materials that have been studied recently include fracture toughness, wear resistance, fatigue resistance, and creep resistance. Some of the most important mechanical properties are presented in this section.

11.4.1 Strength and Ductility

The two main grain-size-dependent properties of a material that control almost every aspect of its mechanical response are strength and ductility. Because these qualities have been consistently asserted in the works as the ones that gained by grain refinement. The two most significant "secondary" characteristics, toughness and fatigue resistance, are directly influenced by ductility and strength. A structural material that combines high strength, adequate ductility, and high fracture toughness is excellent. Nanomaterials exhibit tremendous strength, but because of their weak strain hardening capabilities, constant elongation, or ductility. Metals extensively sheared by ECAP can exhibit a desirable mix of elevated strength and ductility, according to a 1981 study by Segal et al. [34]. Later, Valiev et al. observed a similar problem as a contradiction of strength and ductility. In spite of these studies, SPD-processed nanomaterials typically exhibit limited uniform elongation. Significant efforts have been made recently to strengthen nanomaterials without diminishing their ductility through the use of stacking faults, nanotwins, bimodal structures, precipitation control, grain boundary engineering, heterostructure development, lattice softening, gradient structure generation, and other techniques. Steels have been effectively strengthened and made more durable through the process of grain refining [78]. It should be highlighted that pure materials and diluted solutions, which are most receptive to grain refinement, possess tremendous ability to replace certain standard alloys due to the additive nature by which the several mechanisms elevate the strength. The blend of flow stress and limited work-hardening ability of materials fabricated by SPD is the primary cause of this loss of ductility. In certain instances, the materials treated with SPD may even have more ductility than their traditionally strain-hardened counterparts. In addition to maintaining ductility, this strength increase was also accompanied by a 1.5–2 increase in tensile elongation to rupture [79]. It is clear that SPD processing does not usually result in uniform elongation, but it can significantly resist the plastic flow of the material during post necking.

High-strength metallic materials typically exhibit low work hardening that results in poor ductility, premature failure, and plastic instability at a premature stage of plastic deformation. In these alloys, the phase stability of the γ phase plays a crucial role in both the microstructural development during the SPD process and the deformation behaviour in the mechanical testing that follows. During mechanical tests following SPD processing, strain-induced transformation and deformation twinning are expected to occur in the untransformed γ phase due to its strong plastic deformation capability. In lattice-softened alloys, simultaneous activation of these deformation mechanisms during plastic deformation is a typical feature; nevertheless, the precise mechanism behind the impact of phase stability on the deformation behaviour and ductility remains unclear. Despite their tremendous strength, the poor uniform elongation (i.e. ductility) of nanocrystalline and UFG metallic materials severely limits their uses. Since surface modification is essential to enhancing the mechanical characteristics of metallic materials, most material failures originate there. The enhanced strain hardening capacity of the gradient-structured materials and the coarse-grained matrix are responsible for the improved ductility. An additional method to attain a concurrent enhancement in both ductility and strength can be taken from the traditional "coarse grain" realm: conventional metallurgy has acknowledged that enhanced ductility is linked to "delayed necking" [80]. This can be accomplished when dislocations are blocked and unable to adequately accommodate induced plastic forces. In such cases, alternative processes of deformation, such as phase transitions or twinning, can be activated. Strength and ductility were simultaneously enhanced, which was explained by increased dislocation buildup and by resisting the dislocation movement. For other alloys, a similar approach based on uniformly dispersed nanoscale precipitates stabilizing the microstructure has previously used. It may be used in a variety of UFG and nanosized components, providing opportunities for extensive industrial uses.

11.4.2 Fracture Toughness

Applications for nanostructured materials treated with SPD may be found in the energy, biological, or mechanical fields. The resistance to crack formation, or fracture toughness, of these nanostructured materials is a key factor in determining their dependability in use. The structural integrity of engineering materials is always under risk due to flaws and crack-like imperfections. In materials science, this is well understood, and it has resulted in damage-tolerant design, particularly for structural elements that are crucial to safety. Nevertheless, fracture toughness is quite important for purposes other than component design [81]. Sufficient fracture toughness is also necessary throughout the production process of components built from SPD-processed materials so that failure at the time of machining can be prevented. The same problem occurs at the time of functioning, when materials must maintain a certain level of mechanical integrity throughout manufacture or for subsequent use even when they are not primarily subjected to strong mechanical loads. Finally, during SPD processing, fracture resistance even becomes a problem. The development of fractures, which frequently characterizes a material's deformability in a particular SPD process, is influenced by the material's resistance to crack propagation, or fracture toughness, in addition to the process's typical stress state. Thus, in the realm of NanoSPD, fracture behaviour is usually very important. The study of fracture behaviour is very new in the NanoSPD community compared to other literatures. Only in the past 15 years, a number of research teams begun to look closely at the fracture behaviour of materials prepared through SPD. The limits of the well-established fracture mechanics principles are a fundamental reason for the small size of this research topic [82]. For instance, there are minimum specimen dimensions required by linear elastic fracture mechanics. Materials handled by HPT can yield better coherent understandings about the fracture of UFG materials. Although it may seem at first that this SPD methodology is not very appropriate for these kinds of investigations, the HPT method has some unbeatable benefits over other procedures when it comes to fracture mechanics-related topics. The ability of HPT's high hydrostatic compressive stress to inhibit the development of pores or cracks at the time of deformation is one of its primary benefits. Furthermore, it has been demonstrated that materials treated by other SPD methods that exhibit a high potential for usage in industry in the future also exhibit general trends similar to those discovered in the HPT-processed materials. Numerous investigations have found that there can be noticeable variations in the fracture toughness of materials that have undergone selective pressure processing. It is primarily possible to classify SPD-processed materials into ductile and brittle groups based on how the coarse-grained counterparts behave. Nevertheless, no grain-size dependent patterns encompassing a vast grain size regime can be identified, unlike the yield strength of materials, where such trends are well documented. This may have to do with the fact that, when grain sizes get very small, the fracture characteristics frequently exhibit a highly anisotropic behaviour and a fracture type transition. The degree of anisotropy is evidently dependent on the major fracture type. It is more prominent in the case of the brittle type because there is only one route for crack propagation, primarily along the elongated microstructure (Figure 11.6(a) and (b)) where intercrystalline fracture is present.

On the other hand, micro-ductile fracture is more common for the ductile kind, similar to fracture growth. Recognizing the non-crystalline fracture process is still lacking. Thus far, the intercrystalline fracture may have intrinsic causes in addition to being the only result of grain boundary embrittlement. However, the fracture anisotropy in materials treated with SPD also has advantageous properties [83]. It makes it possible to produce remarkable combinations of strength and fracture toughness in certain testing directions that are challenging to do with traditional high-strength materials.

11.4.3 Creep Behaviour

The time-dependent aspect of plastic deformation is known as creep, and it typically happens at a constant temperature with constant load and/or tension. While creep can happen at any temperature, it becomes more significant at higher temperatures (often above 0.4Tm, or the absolute melting

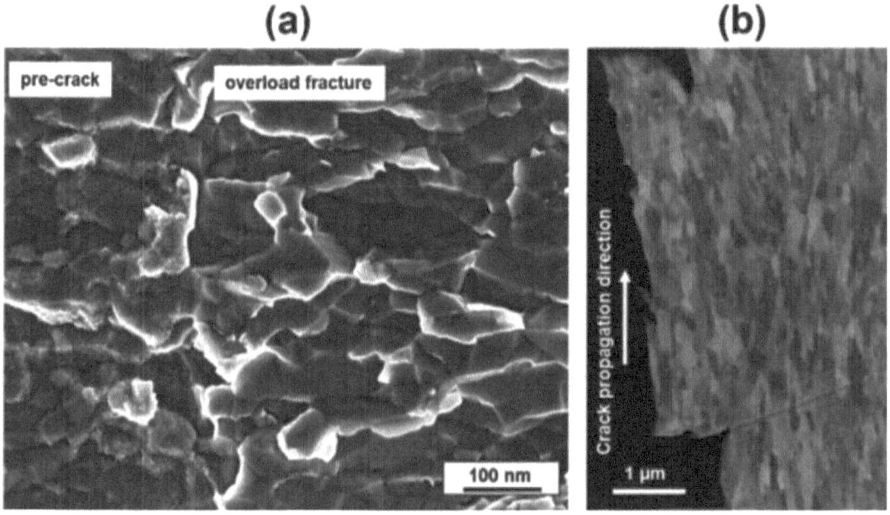

FIGURE 11.6 Fracture features of metals exemplified with iron through SPD: (a) intercrystalline crack propagation in severely deformed metals and (b) crack along the elongated grain structure [84].

point), when diffusional processes happen very fast. It was not until the previous year when high-temperature creep in metals was first studied, but following World War II, a great deal of theoretical and experimental research was conducted on the subject [85]. Explanations of the empirically observed creep behaviour, numerous dislocation prototypes and concepts of creep have been put forth. Currently, it is well acknowledged that while materials with typical grain sizes can occasionally be both strong and ductile, they rarely are. However, the numerous published findings regarding creep behaviour in connection to mean grain size vary greatly and are sometimes at odds with one another. Applying SPD during processing has become a recognized method for creating large-scale UFG materials. The huge internal interface extension of UFG materials provides a means of studying previously unfeasible locations through new research. While there has been some success in understanding creep behaviour in UFG materials, there are still many unanswered questions regarding this intricate phenomenon [86]. It should be emphasized that while grain growth happens readily during creep experiments, it can be challenging to identify novel creep behaviour in UFG materials.

The key aspect of the creep studies conducted on UFG-developed materials having uneven microstructures is that the operating mechanisms and creep behaviour are comparable to those observed in conservative coarse-grained materials that have not undergone SPD processing (Figure 11.7). Grain size, the percentage of high-angle grain boundaries, testing methods, creep settings, and SPD processing pathways all have an impact on the variation in creep resistance among UFG-developed and large-sized grain materials [87,88].

When evaluated during similar creep settings, the UFG-developed materials show a loss in creep life compared to the equivalent coarse-grained materials. In contrast, a notable increase in creep fracture plasticity is observed with application of SPD.

11.4.4 Fatigue

In addition to strength and ductility under monotonic stress, the cyclic properties of UFG materials must be considered for potential engineering applications. Enhancing fatigue characteristics is difficult and crucial at the same time. These characteristics were thoroughly investigated in connection with UFG metals, and this subject is covered in multiple thorough reviews [89]. The plastic strain

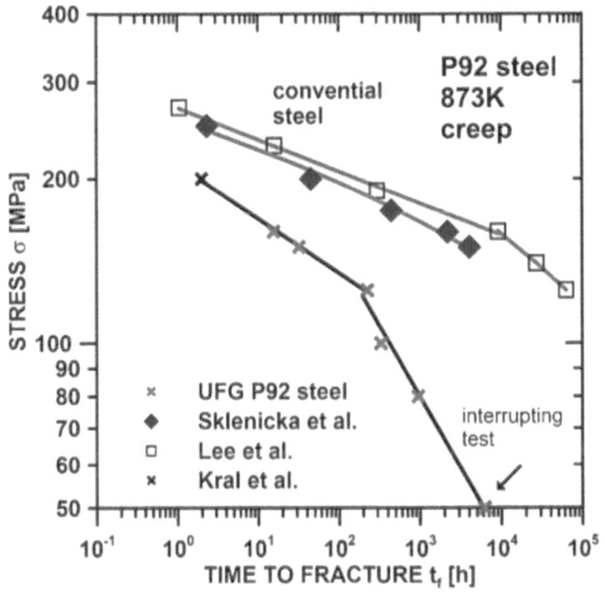

FIGURE 11.7 Creep curves for aluminium before and after ECAP process with multiple passes [86].

component predominates in the total applied strain at large strain amplitudes that correspond to short fatigue lives, and ductility is the primary determinant of fatigue life. The method of reducing fatigue can be established by the following relationship: the endurance limit increases with increasing strength under constant load. The fracture strength determines the fatigue life, and at large fatigue existences, the strain due to elastic amplitude is more important than the plastic one. As a result, the limit of endurance enhance as tensile strength upsurges. This is by no means a rigid "law", and the material and grain refinement method employed will determine how well the strategy for increasing fatigue characteristics by increasing the ultimate tensile strength works. Grain boundaries and their unique role in the characteristics of UFG materials is a much-discussed topic in the literature [90]. It is obvious that many events cannot be explained by the interfacial energy and increased diffusion between grains. Grain boundaries, for example, are important in SPD metal fatigue. On the one hand, compared to coarse-grained structures, fine-grained structures typically result in high fatigue lives. However, it was found that the grain boundaries have a role in the relatively poor stability of the UFG structure, as well as the tendency for fatigue fracture formation and recovery and coarsening of the grain during cycling. Moreover, grain boundaries contribute to crack initiation and propagation, as well as the commonly observed shear banding. Therefore, grain boundaries seem to be the most important structural component [91]. The fatigue behaviour of UFG metals is easier to characterize than that of common polycrystals and single crystals, despite the intricacy of the nanocrystalline structures. The main reason for this simplification is the absence of dislocation patterning within the UFG structures, as the usual parameters of the substantial components of the structure, both grains and cells, are lower than the typical length range of dislocation structures that self-organization during cyclic loading would induce.

11.5 CONCLUSION AND FUTURE SCOPE

In this chapter, we discussed the various processes that explain the behaviour of the material during SPD, provided an overview of the current approaches, and provided a brief history of SPD techniques as a way of generating UFG materials. The objective of this chapter is to provide the reader with an overview of the topic, insights into the mechanics underlying SPD, and information on the remarkable material properties produced by SPD processing. By carefully choosing the materials and processing method, one can acquire increased qualities for a variety of applications as substantial number of materials can be prepared using SPD techniques. The majority of the materials investigated for SPD processing are metals, alloys, and compounds that are currently practically employed in the coarse-grained form. To fully benefit from SPD, chemical compositions must be optimized for the SPD processing conditions, even if SPD permits the creation of novel nanostructural characteristics.

The materials prepared using SPD have been potentially promising for use in biomedical applications, as life expectancy rises and sophisticated biomaterials become more necessary. Furthermore, the NanoSPD field will contribute to energy-related issues in the future given the global demands to achieve carbon neutrality and reduce CO_2 emissions, especially given the promising energy-related properties reported for the SPD-processed materials. Such applications still require the development of SPD techniques at the industrial scale, along with combinations of materials science, chemistry/physics, and theoretical computations.

An appealing new SPD processing application should revitalize established SPD techniques. Specifically, it seems that TE, HPT, and other more recent techniques are appropriate for creating a variety of spiral designs that are advantageous for ductility and strength. We think this sector will have an interesting future since it provides a great platform for the development of unique materials and processes.

REFERENCES

[1] Durand-Charre, Madeleine. *Microstructure of steels and cast irons*. Springer Science & Business Media, vol. 41, 2004.

[2] J. de Lapérouse, Metallurgy: Early metallurgy in mesopotamia, *Encycl. Hist. Sci. Technol. Med. Non-Western Cult.* (2014) 1624–1634.

[3] M. Durand-Charre, Damascus and pattern-welded steels, in: *Damascus and Pattern-Welded Steels*, EDP Sciences, n.d.

[4] M. Reibold, P. Paufler, A.A. Levin, W. Kochmann, N. Pätzke, D.C. Meyer, Carbon nanotubes in an ancient Damascus sabre, *Nature.* 444 (2006) 286.

[5] M. Sachse, P. Knighton, *Damascus Steel: Myth, History, Technology, Applications*, Verlag Stahleisen, 2008.

[6] J.D. Verhoeven, The mystery of Damascus blades, *Sci. Am.* 284 (2001) 74–79.

[7] K. Hjardar, V. Vike, *Vikings at War*, Casemate Publishers, 2016.

[8] O.D. Sherby, J. Wadsworth, Damascus steels, *Sci. Am.* 252 (1985) 112–121.

[9] J. Wadsworth, Archeometallurgy related to swords, *Mater. Charact.* 99 (2015) 1–7.

[10] J.T. Wang, Historic retrospection and present status of severe plastic deformation in China, *Mater. Sci. Forum*, 503–504 (2006) 363–370.

[11] M. Weber, S. Banaschak, M.A. Rothschild, Sharp force trauma with two katana swords: Identifying the murder weapon by comparing tool marks on the skull bone, *Int. J. Legal Med.* 135 (2021) 313–322.

[12] V.B. Ginzburg, R. Ballas, *Flat Rolling Fundamentals*, CRC Press, 2000.

[13] Deibert MC, inventor; Monsanto Research Corp, assignee. Laminar electrode including hydrophobic and hydrophilic layfrs; method of making; fuel cell therewith; and method of using fuel cell. United States patent US 3,457,113. 1969 Jul 22.

[14] R.Z. Valiev, R.K. Islamgaliev, I. V Alexandrov, Bulk nanostructured materials from severe plastic deformation, *Prog. Mater. Sci.* 45 (2000) 103–189.

[15] R.Z. Valiev, I.V Alexandrov, Nanostructured materials from severe plastic deformation, *Nanostruct. Mater.* 12 (1999) 35–40.

[16] R.Z. Valiev, Y. Estrin, Z. Horita, T.G. Langdon, M.J. Zechetbauer, Y.T. Zhu, Producing bulk ultrafine-grained materials by severe plastic deformation, *Jom.* 58 (2006) 33–39.

[17] T.G. Langdon, The processing of ultrafine-grained materials through the application of severe plastic deformation, *J. Mater. Sci.* 42 (2007) 3388–3397.
[18] A.K. Ghosh, W. Huang, Severe deformation based process for grain subdivision and resulting microstructures, in: *Investigations and Applications of Severe Plastic Deformation*, Springer, 2000: pp. 29–36.
[19] D. Kumar, R. Seetharam, K. Ponappa, A review on microstructures, mechanical properties and processing of high entropy alloys reinforced composite materials, *J. Alloys Compd.* 972 (2023) 172732.
[20] T.C. Lowe, R.Z. Valiev, X. Li, B.R. Ewing, Commercialization of bulk nanostructured metals and alloys, *MRS Bull.* 46 (2021) 265–272.
[21] D. Kumar, S.K. Rajak, R. Seetharam, H. Singh, Effects of hBN and Y_2O_3 addition on mechanical and tribological behavior of SiC ceramic matrix composites prepared by spark plasma sintering, *Silicon.* 15 (2023) 2297–2311.
[22] D. Kumar, S.K. Rajak, R. Seetharam, H. Singh, Effect of Y_2O_3 additive on mechanical and wear behaviour of hBN reinforced SiC ceramic matrix composite, *Mater. Today Proc.* 87, Part 1 (2023) 193–196.
[23] S. Yoshida, T. Ikeuchi, T. Bhattacharjee, Y. Bai, A. Shibata, N. Tsuji, Effect of elemental combination on friction stress and Hall-Petch relationship in face-centered cubic high/medium entropy alloys, *Acta Mater.* 171 (2019) 201–215.
[24] M. Patel, S.K. Sahu, M.K. Singh, Fabrication and investigation of mechanical properties of SiC particulate reinforced AA5052 metal matrix composite, *J. Mod. Mater.* 7 (2020) 26–36.
[25] M. Patel, S.K. Sahu, M.K. Singh, Mechanical, tribological and corrosion behaviour of aluminium alloys and particulate reinforced aluminium or aluminium alloy metal matrix composites-A review, *I-Manager's J. Mater. Sci.* 8 (2020) 40.
[26] M. Noda, K. Funami, Y. Suwahara, Effects of constraint and strain path on evolution of ultrafine grained microstructure by multi-axial alternative forging, *Mater. Sci. Forum.* 475–479 (2005) 3471–3474.
[27] N.K. Bhoi, H. Singh, S. Pratap, Developments in the aluminum metal matrix composites reinforced by micro/nano particles – A review, *J. Compos. Mater.* 54 (2020) 813–833.
[28] R. Pippan, S. Scheriau, A. Taylor, M. Hafok, A. Hohenwarter, A. Bachmaier, Saturation of fragmentation during severe plastic deformation, *Annu. Rev. Mater. Res.* 40 (2010) 319–343.
[29] G. Langford, M. Cohen, Strain hardening of iron by severe plastic deformation, *ASM Trans Quart.* 62 (1969) 623–638.
[30] R.F. Landel, L.E. Nielsen, *Mechanical Properties of Polymers and Composites*, CRC Press, 1993.
[31] R.Z. Valiev, Y.H. Estrin, Z. Horita, T.G. Langdon, M.J. Zehetbauer, Y.T. Zhu, Producing bulk ultrafine-grained materials by severe plastic deformation, *Journal of Minerals, Metals & Materials Society* 58 (2003) 33–39.
[32] Y.T. Zhu, T.G. Langdon, R.S. Mishra, S.L. Semiatin, M.J. Saran, T.C. Lowe, *Ultrafine Grained Materials*, *The Minerals, Metals & Materials Society*,. 2002 p. 43.
[33] V.M. Segal, Materials processing by simple shear. *Materials Science and Engineering: A* 197, no. 2 (1995): 157–164.
[34] V.M. Segal, Equal channel angular extrusion: from macromechanics to structure formation. *Materials Science and Engineering: A* 271, no. 1–2 (1999): 322–333.
[35] R.Z. Valiev, N.A. Krasilnikov, N.K. Tsenev, Plastic deformation of alloys with submicron-grained structure, *Mater. Sci. Eng. A.* 137 (1991) 35–40.
[36] Y.T. Zhu, Ultrafine grained materials III: Proceedings of symposium: held during the 2004 TMS Annual Meeting in Charlotte, North Carolina, USA, March 14-18, 2004, (No Title) (2004).
[37] V. Segal, Modes and processes of severe plastic deformation (SPD), *Materials (Basel).* 11 (2018) 1175.
[38] P. Frint, M. Hockauf, T. Halle, G. Strehl, T. Lampke, M.F.X. Wagner, Microstructural features and mechanical properties after industrial scale ECAP of an Al 6060 alloy, *Mater. Sci. Forum.* 667–669 (2011) 1153–1158.
[39] H. Ueno, K. Kakihata, Y. Kaneko, S. Hashimoto, A. Vinogradov, Enhanced fatigue properties of nanostructured austenitic SUS 316L stainless steel, *Acta Mater.* 59 (2011) 7060–7069.
[40] I.J. Beyerlein, L.S. Tóth, Texture evolution in equal-channel angular extrusion, *Prog. Mater. Sci.* 54 (2009) 427–510.

[41] W. Skrotzki, N. Scheerbaum, C.-G. Oertel, H.-G. Brokmeier, S. Suwas, L.S. Toth, Recrystallization of high-purity aluminium during equal channel angular pressing, *Acta Mater.* 55 (2007) 2211–2218.

[42] V.K. Truong, R. Lapovok, Y.S. Estrin, S. Rundell, J.Y. Wang, C.J. Fluke, R.J. Crawford, E.P. Ivanova, The influence of nano-scale surface roughness on bacterial adhesion to ultrafine-grained titanium, *Biomaterials.* 31 (2010) 3674–3683.

[43] Y. Estrin, E.P. Ivanova, A. Michalska, V.K. Truong, R. Lapovok, R. Boyd, Accelerated stem cell attachment to ultrafine grained titanium, *Acta Biomater.* 7 (2011) 900–906.

[44] Y. Kusadome, K. Ikeda, Y. Nakamori, S. Orimo, Z. Horita, Hydrogen storage capability of $MgNi_2$ processed by high pressure torsion, *Scr. Mater.* 57 (2007) 751–753.

[45] K. Edalati, A. Yamamoto, Z. Horita, T. Ishihara, High-pressure torsion of pure magnesium: Evolution of mechanical properties, microstructures and hydrogen storage capacity with equivalent strain, *Scr. Mater.* 64 (2011) 880–883.

[46] G.F. Lima, A.M. Jorge, D.R. Leiva, C.S. Kiminami, C. Bolfarini, W.J. Botta, Severe plastic deformation of Mg-Fe powders to produce bulk hydrides, *J. Phys. Conf. Ser.* 144 (2009) 12015.

[47] Y. Estrin, E. Rabkin, R.J. Hellmig, M. Kazakevich, A. Zi, Severe plastic deformation by solid state infiltration, *Mater. Sci. Eng. A.* 410 (2005) 165–168.

[48] A. Vorhauer, R. Pippan, On the homogeneity of deformation by high pressure torsion, *Scr. Mater.* 51 (2004) 921–925.

[49] Á. Révész, Z. Kánya, T. Verebélyi, P.J. Szabó, A.P. Zhilyaev, T. Spassov, The effect of high-pressure torsion on the microstructure and hydrogen absorption kinetics of ball-milled $Mg_{70}Ni_{30}$, *J. Alloys Compd.* 504 (2010) 83–88.

[50] P.W. Bridgman, Reflections on rupture, *J. Appl. Phys.* 9 (1938) 517.

[51] Y. Saito, H. Utsunomiya, N. Tsuji, T. Sakai, Novel ultra-high straining process for bulk materials—Development of the accumulative roll-bonding (ARB) process, *Acta Mater.* 47 (1999) 579–583.

[52] Y. Saito, N. Tsuji, H. Utsunomiya, T. Sakai, R.G. Hong, Ultra-fine grained bulk aluminum produced by accumulative roll-bonding (ARB) process, *Scr. Mater.* 39 (1998) 1221–1227.

[53] T. Hausöl, V. Maier, C.W. Schmidt, M. Winkler, H.W. Höppel, M. Göken, Tailoring materials properties by accumulative roll bonding, *Adv. Eng. Mater.* 12 (2010) 740–746.

[54] S.-H. Lee, T. Sakai, Y. Saito, H. Utsunomiya, N. Tsuji, Strengthening of sheath-rolled aluminum based MMC by the ARB process, *Mater. Trans. JIM.* 40 (1999) 1422–1428.

[55] P.S. Sahoo, A. Meher, M.M. Mahapatra, P.R. Vundavilli, Understanding the fabrication of ultrafine grains through severe plastic deformation techniques: An overview, *JOM.* 74 (2022) 3887–3909.

[56] P. Hidalgo-Manrique, C.M. Cepeda-Jiménez, O.A. Ruano, F. Carreño, Effect of warm accumulative roll bonding on the evolution of microstructure, texture and creep properties in the 7075 aluminium alloy, *Mater. Sci. Eng. A.* 556 (2012) 287–294.

[57] N. Tsuji, New routes for fabricating ultrafine-grained microstructures in bulky steels without very-high strains, *Adv. Eng. Mater.* 12 (2010) 701–707.

[58] I. Daigo, Y. Matsuno, Y. Adachi, Substance flow analysis of chromium and nickel in the material flow of stainless steel in Japan, *Resour. Conserv. Recycl.* 54 (2010) 851–863.

[59] D. Geist, C. Rentenberger, H.P. Karnthaler, Extreme structural inhomogeneities in high-pressure torsion samples along the axial direction, *Acta Mater.* 59 (2011) 4578–4586.

[60] O.R. Valiakhmetov, R.M. Galeev, G.A. Salishchev, Mechanical properties of VT8 Ti-alloy of submicrocrystalline structure(Mekhanicheskie svoistva titanovogo splava VT8 s submikrokristallicheskoi strukturoi), *Fiz. Met. i Metalloved.* (1990) 204–206.

[61] G.E. Meyer, H.J. Henning, *Beryllium Wrought Products*, Defense Metals Information Center, Battelle Memorial Institute, 1970.

[62] G. Salishchev, R. Zaripova, R. Galeev, O. Valiakhmetov, Nanocrystalline structure formation during severe plastic deformation in metals and their deformation behaviour, *Nanostructured Mater.* 6 (1995) 913–916.

[63] A. Belyakov, T. Sakai, H. Miura, Microstructure and deformation behaviour of submicrocrystalline 304 stainless steel produced by severe plastic deformation, *Mater. Sci. Eng. A.* 319 (2001) 867–871.

[64] O. Sitdikov, T. Sakai, A. Goloborodko, H. Miura, R. Kaibyshev, Effect of pass strain on grain refinement in 7475 Al alloy during hot multidirectional forging, *Mater. Trans.* 45 (2004) 2232–2238.

[65] M. Sheik Hassan, S. Sharma, B. Kumar, A review of severe plastic deformation, *Int. Ref. J. Eng. Sci.* 6 (2017) 66–85.

[66] Y. Beygelzimer, V.N. Varyukhin, S.G. Synkov, A.N. Sapronov, V.G. Synkov, New schemes of large plastic deformations accumulating with using of hydroextrusion, *Phys. Technol. High Press.* 9 (1999) 109–111.

[67] Y.Y. Beygelzimer, D. V Orlov, High pressure effects in chemistry, *Biol Mater Sci.* 208 (2002) 311.

[68] D. Orlov, Y. Beygelzimer, S. Synkov, V. Varyukhin, N. Tsuji, Z. Horita, Plastic flow, structure and mechanical properties in pure Al deformed by twist extrusion, *Mater. Sci. Eng. A.* 519 (2009) 105–111.

[69] F.J. Humphreys, M. Hatherly, *Recrystallization and Related Annealing Phenomena*, Elsevier, 2012.

[70] N. Tsuji, N. Kamikawa, B.L. Li, Grain size saturation during severe plastic deformation, *Mater. Sci. Forum*, 539–543 (2007) 2837–2842.

[71] N. Hansen, D.J. Jensen, Development of microstructure in FCC metals during cold work, *Philos. Trans. R. Soc. London. Ser. A Math. Phys. Eng. Sci.* 357 (1999) 1447–1469.

[72] N. Hansen, R.F. Mehl, A. Medalist, New discoveries in deformed metals, *Metall. Mater. Trans. A.* 32 (2001) 2917–2935.

[73] J.H. Hollomon, Tensile deformation, *Aime Trans.* 12 (1945) 1–22.

[74] S. Erbel, *Mechanizm zmian własności metali poddanych wielkim odkształceniom*, Wydawnictwa Politechniki Warszawskiej, 1976.

[75] P.W. Bridgman, Effects of high shearing stress combined with high hydrostatic pressure, *Phys. Rev.* 48 (1935) 825.

[76] P.W. Bridgman, Shearing phenomena at high pressures, particularly in inorganic compounds." In *Proceedings of the American Academy of Arts and Sciences*, vol. 71, no. 9, pp. 387-460. American Academy of Arts & Sciences, 1937.

[77] J.G. Sevillano, P. Van Houtte, E. Aernoudt, Large strain work hardening and textures, *Prog. Mater. Sci.* 25 (1980) 69–134.

[78] W.W. Jian, G.M. Cheng, W.Z. Xu, H. Yuan, M.H. Tsai, Q.D. Wang, C.C. Koch, Y.T. Zhu, S.N. Mathaudhu, Ultrastrong Mg alloy via nano-spaced stacking faults, *Mater. Res. Lett.* 1 (2013) 61–66.

[79] Y. Wang, M. Chen, F. Zhou, E. Ma, High tensile ductility in a nanostructured metal, *Nature.* 419 (2002) 912–915.

[80] Z. Horita, K. Ohashi, T. Fujita, K. Kaneko, T.G. Langdon, Achieving high strength and high ductility in precipitation-hardened alloys, *Adv. Mater.* 17 (2005) 1599–1602.

[81] G. Serra, L. Morais, C.N. Elias, I.P. Semenova, R. Valiev, G. Salimgareeva, M. Pithon, R. Lacerda, Nanostructured severe plastic deformation processed titanium for orthodontic mini-implants, *Mater. Sci. Eng. C.* 33 (2013) 4197–4202.

[82] P. Hübner, R. Kiessling, H. Biermann, A. Vinogradov, Fracture behaviour of ultrafine-grained materials under static and cyclic loading, *Int. J. Mater. Res.* 97 (2022) 1566–1570.

[83] I.P. Semenova, J.M. Modina, A. V Polyakov, G. V Klevtsov, N.A. Klevtsova, I.N. Pigaleva, R.Z. Valiev, T.G. Langdon, Fracture toughness at cryogenic temperatures of ultrafine-grained Ti-6Al-4V alloy processed by ECAP, *Mater. Sci. Eng. A.* 716 (2018) 260–267.

[84] A. Hohenwarter, R. Pippan, Fracture toughness evaluation of ultrafine-grained nickel, *Scr. Mater.* 64 (2011) 982–985.

[85] E.N.D.C. Andrade, On the viscous flow in metals, and allied phenomena, *Proc. R. Soc. London. Ser. A, Contain. Pap. a Math. Phys. Character.* 84 (1910) 1–12.

[86] V. Sklenicka, J. Dvorak, P. Kral, M. Svoboda, M. Kvapilova, V. Kopylov, S. Nikulin, S. Dobatkin, Creep behavior of a zirconium alloy processed by equal-channel angular pressing, *Acta Phys. Pol. A.* 122 (2012) 485–489.

[87] R.Z. Valiev, T.G. Langdon, Achieving exceptional grain refinement through severe plastic deformation: New approaches for improving the processing technology, *Metall. Mater. Trans. A.* 42 (2011) 2942–2951.

[88] G.P. Grabovetskaya, K. V Ivanov, Y.R. Kolobov, Creep features of nanostructured materials produced by severe plastic deformation, *Ann. Chim. Sci. Des Matériaux*, (2002) 89–98.

[89] J.A. Schwarz, C.I. Contescu, K. Putyera, *Dekker Encyclopedia of Nanoscience and Nanotechnology*, CRC Press, 2004.

[90] R. Valiev, Principles of producing bulk nanostructured metals with unique properties by severe plastic deformation techniques, in: ISOPE Int. Ocean Polar Eng. Conf., ISOPE, 2007: p. ISOPE-I.

[91] H. Höppel, H. Mughrabi, A. Vinogradov, Fatigue properties of bulk nanostructured materials, in: *Bulk Nanostructured Materials*, Wiley online library, 2009: pp. 481–500.

12 Aluminum-Based Composites with TiB_2 and Multicomponent Alloy Reinforcements

Processing, Microstructure, and Property Enhancement

Ritukesh Sharma, Avanish Kumar, and Partha Sarathi De

12.1 INTRODUCTION

Commercial lightweight aluminum-based particulate metal matrix composites (PMMCs) are the focus of the research community on account of their outstanding properties [1–3]. Some of the advantages of aluminum-based PMMCs are as follows:

a. Increased strength and elastic modulus compared to the metal matrix.
b. Enhanced high-temperature mechanical properties and microstructural stability compared to the metal matrix.
c. Comparatively easy fabrication and low cost compared to other varieties of composites.

However, the above advantages are contingent upon a strong interfacial bonding between matrix and particulate and thermal stability of particulate compared to the matrix. Similarly, corrosion resistance of aluminum PMMC requires galvanic potential compatibility between the interface and particulate. Consequently, a proper understanding of the interaction between matrix and particulate is essential to optimize the mechanical and corrosion properties.

The objective of this chapter is to provide a comprehensive overview of the significant potential of Al–TiB_2 composites, while also addressing the primary challenges that hinder their widespread adoption. First, it delves into the remarkable properties and advantages of Al–TiB_2 composites [1–3], emphasizing their suitability for engineering applications where lightweight, yet durable materials are crucial like in aerospace and automobile applications.

However, despite their promising characteristics, the chapter cites the key challenges that impede the broader utilization of Al–TiB_2 composite. These challenges may include issues related to increased reinforcement, processing techniques, scalability of production, and potential performance limitations under certain conditions. Understanding and addressing these obstacles are essential for maximizing the benefits of the composite in practical applications.

Moreover, the chapter also explores the potential strategies to mitigate these limitations, focusing particularly on the role of multicomponent alloys as reinforcements in aluminum matrices. By incorporating various alloying elements into the composite material, researchers aim to enhance its mechanical properties by tailoring its microstructure, and thereby improving the overall performance characteristics. This discussion likely involves an examination of different alloy compositions, processing methods, and performance evaluations to assess the effectiveness of multicomponent

DOI: 10.1201/9781003664611-13

alloys in overcoming the challenges associated with Al–TiB_2 composites. Through systematic experiments and analysis, researchers aim to identify the appropriate chemistry of the alloy and optimization of processing parameters that can enhance the performance and viability of these composites.

Thus, this chapter serves to provide a comprehensive understanding of the potential benefits and challenges associated with Al–TiB_2 composites, while also exploring innovative approaches to address these challenges through the integration of multicomponent alloys. The next section describes in detail the factors which make composites a suitable material for engineering applications.

12.2 INTERACTION BETWEEN REINFORCEMENT AND MATRIX

In the micromechanics approach, the strengthening of the reinforcements occurs due to (a) coefficient of thermal expansion (CTE) mismatch strengthening, (b) Orowan strengthening, and (c) load-bearing effect of the reinforcing particles [3]. During strengthening, the particles are assumed to be spherical and uniformly distributed in the matrix [4]. These effects are discussed in detail in the subsequent subsections.

12.2.1 CTE Mismatch Strengthening

As a composite is cooled down from the processing to the ambient temperatures, a residual plastic strain is generated between the reinforcement and the matrix due to a mismatch in CTE. This increases dislocations around the particles raising the yield strength of the matrix [5]. Such yield stress increment due to CTE mismatch is given as [3]

$$\Delta\sigma_{CTE} = BG_m b\sqrt{\rho^{CTE}} \tag{12.1}$$

where B is a constant (~1.25), G_m is the shear modulus of the matrix, b is Burgers vector of the dislocation, and ρ^{CTE} is the dislocation density due to the difference in CTE. The G_m and ρ^{CTE} can be estimated as [6].

$$G_m = \frac{E_m}{2(1+\upsilon)} \tag{12.2}$$

$$\rho^{CTE} = \frac{\Lambda\Delta\alpha\Delta TV_p}{bd_p(1-V_p)} \tag{12.3}$$

where E_m is Young's modulus of the matrix, v is the Poisson's ratio of the matrix, Λ is a constant (~12), $\Delta\alpha$ is the CTE difference between matrix and reinforcement, ΔT is the difference between processing and test temperatures, d_p is the average particle size, and V_p is the volume fraction of the reinforcement particles.

12.2.2 Orowan Strengthening

The Orowan strengthening occurs due to interaction between dislocations and reinforcing particles (smaller than one micron) within the matrix [4]. The general form of the strengthening mechanism is given by the Orowan–Ashby equation [7].

$$\Delta\sigma_{Orowan} = \frac{0.4MG_m b \ln\left(\sqrt{\frac{2}{3}}\frac{d_p}{b}\right)}{\pi\mho\sqrt{1-\nu}} \tag{12.4}$$

$$\mho = \sqrt{\frac{2}{3}}d_p\left(\sqrt{\frac{\pi}{4V_p}}-1\right) \tag{12.5}$$

where M is the mean orientation factor and υ is the inter-particle spacing. The mean orientation factor represents the ease with which dislocations can move and thus affects the composite's strength and ductility.

12.2.3 Load Bearing Strengthening

A proper interface bonding between the matrix and the reinforcement allows an efficient transfer of load from the matrix to the particles [8]. This results in an increase in yield strength ($\Delta\sigma_{ys}$) in the fabricated composite [8]. The increase of the $\Delta\sigma_{ys}$ induced by effective load transfer can be calculated as [9].

$$\Delta\sigma_{load} = 0.5V_p\sigma_m \tag{12.6}$$

where σ_m represents the yield strength of the matrix/base material.

However, synergistic effects due to interaction between reinforcement and matrix do exist. Thus, particulate reinforcements present can cause microstructural evolution like grain size refinement and particulate-assisted precipitation behavior leading to further strengthening of the composites [10,11].

12.2.4 Grain Size Strengthening

The Hall–Petch strengthening occurs due to the refinement of grain size [12]. With a decrease in grain size, the yield strength increases as follows [13]

$$\Delta\sigma_y = k\left(\frac{1}{D_{MMC}^{0.5}} - \frac{1}{D_M^{0.5}}\right) \tag{12.7}$$

where D_{MMC} and D_M represent the average grain size of the processed composite and matrix while k represents the Hall–Petch coefficient [14].

12.2.5 Particulate-Assisted Precipitation Strengthening

In heat-treatable aluminum alloys, the addition of reinforcing particles influences precipitation behaviors [15,16]. Thus, with the addition of particulates, a large amount of matrix/particle interfaces are created, which affects the dissolution–precipitation kinetics during heat treatment [17,18]. Generally, the reinforcing particles accelerate the precipitating kinetics in the matrix due to the additional dislocation density generated within the matrix [4]. Such an increase in dislocation density promotes dislocation-assisted diffusion of the alloying elements from the adjacent dislocation-lean areas of the matrix [4]. This results in solute enrichment in the interfacial regions, which favors nucleation and growth during precipitation [4].

The overall yield strength of a composite is estimated by the arithmetic summation of individual strengthening mechanism contributions and is presented as [4,7]

$$\sigma_c = \sigma_m + \Delta\sigma_{CTE} + \Delta\sigma_{Orowan} + \Delta\sigma_{load} + \Delta\sigma_y \tag{12.8}$$

The assumption is that the different mechanisms are unaffected by each other and independently contribute to the yield strength of the composite. In the next subsection, the effect of reinforcement on corrosion behavior is discussed.

12.2.6 Factors Affecting Corrosion Resistance

Any exacerbation of corrosion in metal matrix composite is mainly on account of the galvanic corrosion between the reinforcement and the matrix [19]. The galvanic corrosion becomes significant for conductive reinforcements and/or due to the formation of conductive phases at the reinforcement/matrix interface [20]. Therefore, expectedly, graphite or silicon carbide (SiC) reinforced aluminum PMMC shows a poor corrosion resistance [21,22]. The aluminum matrix corrodes preferentially at an accelerated rate when galvanically coupled to a relatively inert reinforcement of adequate conductivity. Further, other factors, such as (a) segregation of alloying elements, (b) formation of the interphase between reinforcement and matrix, and (c) impurities introduced into the microstructure during processing, may also affect the corrosion behavior [20]. In fact, in Al–Graphite and Al–SiC composites, the formation of Al_4C_3 degrades the corrosion resistance of the composites [20]. Thus, a variety of factors in aluminum PMMC creates pathways for corrosion causing a breakdown of the passive film on the matrix [23].

Another possible corrosion process is the pitting phenomena. Pure aluminum under deaerated and open circuit conditions is resistant to pitting [24]. However, the presence of nobler precipitates as in alloys results in flaws on the surface oxide film, causing pitting corrosion in aerated solutions [23]. At each initiated pit site, an enhanced local anodic current density occurs with a very low oxygen concentration due to limited access into the pit [25]. The high concentration of aluminum cations within each pit encourages chloride ion migration to maintain a charge balance [25]. Further precipitation of $Al(OH)_3$ occurs once the solution inside the pit is adequately concentrated with aluminum cations [25] which lower the pH of the pit creating a solid corrosion product [25]. Thus, a low pH and a high chloride ion concentration accelerate the anodic dissolution reaction facilitating the aluminum corrosion process [25]. In the next section, a brief introduction to titanium diboride, a common reinforcement for aluminum MMCs is provided.

12.3 TITANIUM DIBORIDE: AN INTRODUCTION

Titanium diboride (TiB_2) is a ceramic material with a high melting point (2790°C), hardness (960 HV), Young's modulus (530 GPa), and good corrosion resistance [26]. Such a combination of excellent properties has made TiB_2 a well-accepted material for erosive, abrasive, corrosive, and high-temperature environment applications [17,27,28]. However, its use is most prevalent in applications like cutting tools, crucibles, wear-resistant coating, and impact-resistant purposes [17,27,28]. Notably, TiB_2 particles are non-reactive with molten aluminum, preventing the formation of brittle reaction products [29,30]. Regarding physical properties, the Hamaker constant[1] of TiB_2 is very close to that of molten aluminum, which accounts for its good wettability when compared to other standard ceramic reinforcements [26]. Thus, TiB_2 particle acts as an attractive strengthening

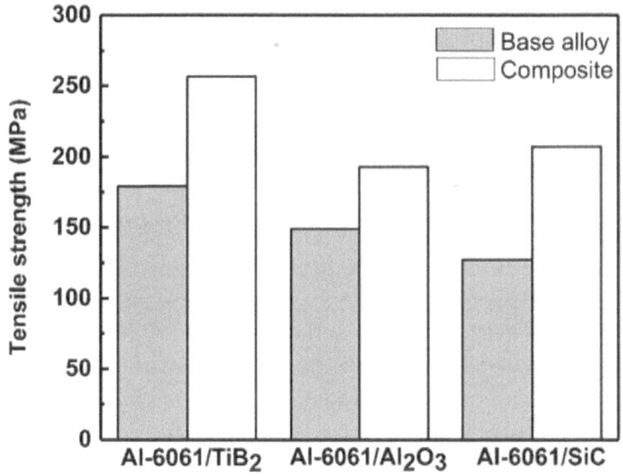

FIGURE 12.1 Tensile strength of base matrix and MMCs fabricated through stir casting. (Data from Refs. [31–33].)

reinforcement for aluminum-based composites (Figure 12.1) [31–33]. However, the amount of TiB_2 reinforcement in the aluminum matrix affects the composite property in a complex fashion.

As reported in the literature, in as-processed composite with high TiB_2 concentrations, the TiB_2 particles generally agglomerate and appear as clusters. Such segregation and clustering of TiB_2 particles occur during solidification, leading to inter-granular distribution and is attributed to [34–36].

i. High interfacial energy between aluminum and TiB_2.
ii. Lower interface velocity than the critical velocity (V_c) leading to rejection of TiB_2 particles toward the boundaries where [19]

$$V_c^2 = \frac{4\zeta T\gamma a_0 \alpha(1-£)^3}{9\pi\eta^2 r^3(1-3£)} \tag{12.9}$$

where η is the viscosity of the liquid, T is temperature, γ is the free energy of the solid/liquid interface, r is the radius of the particle, a_0 is the distance between the particle and solid/liquid interface, ζ is the curvature of the solid/liquid interface, and £ is the ratio of particle radius to interface radius.

iii. The longer cooling time of the melt causes the solidification front to sustain longer.

Furthermore, a monotonic increase in the TiB_2 fraction of in-situ composites is associated with the ensuing limitations.

i. **Formation of Al_3Ti :** The formation reaction for the Al_3Ti phase is an exothermic process and is favored at low temperatures where TiB_2 particles react with the aluminum melt to form a solid solution of (Ti, Al) B_2 [37]. The solid solution combines with titanium to form Al_3Ti particles, where TiB_2 particles act as a substrate for nucleation [37]. Consequently, a high TiB_2 fraction in the composite is susceptible to the formation of excess Al_3Ti content. These Al_3Ti intermetallics formed act as preferential sites for crack initiation and propagation [38]. Therefore, the elimination of Al_3Ti is regarded as a primary concern during the manufacture of Al–TiB_2 composites with consistent properties.

ii. **Loss of ductility :** A high TiB_2 particle content in the Al–TiB_2 composite translates to an increase in the strength of the composite. This is attributed to the efficient distribution of load between the matrix and reinforcing particles due to their good bonding [39]. Furthermore, the reinforcing particles act as strong impenetrable obstacles, and dislocation movement requires a sharp change in curvature of the dislocation line [39]. Thus, higher stress is required for the dislocation to move through the matrix [39]. However, an increased fraction of reinforcement leads to a simultaneous decrease in ductility due to reduced dislocation pile-up spacing [39]. Thus, Lu et al. found that Al–TiB_2 composite with a reinforcement percentage higher than 10% leads to brittle behavior [39]. Consequently, methods to overcome the above limitations of Al–TiB_2 composites are required. Further, randomization of such particle distribution to improve the corrosion characteristics requires a careful balance of wettability and chemical affinity [40,41]. The following section presents the effect of TiB_2 content on the grain size, strength, and corrosion resistance in an aluminum matrix-based composite.

12.3.1 Effect of Titanium Diboride on Grain Size

In the Al–TiB_2 composite, TiB_2 particles act as sites for heterogeneous nucleation, where nucleation occurs on the basal (0001) faces of the hexagonal surface [42,43]. During solidification, aluminum grains start growing from mold walls, particle impurities, and the grain refiner particles added to the melt [44]. A higher amount of grain refiner particles causes it to supersede other sites for nucleation [44]. For heterogeneous nucleation, a low surface-solid interfacial energy is possible when the product solid lattice structure matches the nuclei's surface [44]. However, in the case of solid aluminum and TiB_2, a lattice mismatch exists making the TiB_2 particles an unlikely nucleation site [44]. Nonetheless, X-ray diffraction of TiB_2 particles reveals a thin layer of a metastable Al_3Ti phase on the TiB_2 particles whose lattice structure matches with aluminum [45,46]. This Al_3Ti layer facilitates nucleation making the TiB_2 particles potent nucleating sites. Moreover, the TiB_2 particles act as obstacles against grain growth which increases with an increase in particle content (Figure 12.2) [4,10]. Consequently, an overall decrease in grain size with an increase in TiB_2 content occurs.

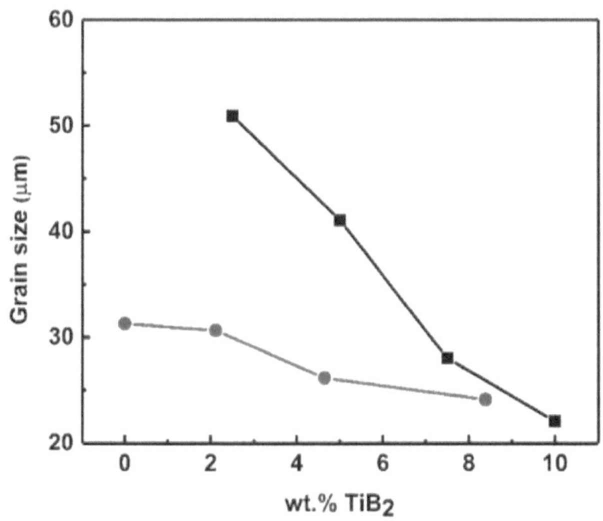

FIGURE 12.2 Variation of grain size of aluminum composite as a function of TiB_2 reinforcement content. (Data from Refs. [4,10].)

12.3.2 Effect of Titanium Diboride on Strength and Ductility

Properties such as ultimate tensile strength, yield strength, percentage elongation, and elastic modulus of the Al–TiB_2 composites depend on the amount and clustering behavior of TiB_2 during their fabrication and subsequent processing [47]. Generally, significant improvements in the mechanical properties have been reported with the increase in wt.% of TiB_2 as compared to the base alloy (Figure 12.3) [17,18,39,48]. The improvement in strength is partially attributed to the efficient transfer of load from the matrix to the TiB_2 particles due to good bonding. Besides this, the difference in thermal expansion between the aluminum matrix and the TiB_2 particles during composite processing further strengthens the matrix. Various other strengthening mechanisms such as dislocation strengthening and Orowan strengthening further contribute to the improved strength of such composites after processing. An additional grain boundary strengthening also occurs due to the refinement of grains by the TiB_2 particles within the matrix [17,18,39,48].

In Figure 12.4, the variation in elongation with TiB_2 content is presented [17,18,39,48]. Generally, the percentage elongation decreased with an increase in TiB_2 content due to decreased pile-up spacing [17,18,39,48]. However, the elastic modulus of composites improved (Figure 12.5) compared to the base alloy due to the enhanced nucleation of TiB_2 particles in the matrix alloy [17,18,39,48].

On the other hand, the clustering of TiB_2 particles has detrimental effects on the properties of the composites [49,50]. An accepted way to homogenize the particle distribution is through the addition of reactive agents like magnesium, zirconium, and ceria (CeO_2) [34,51,52]. On the other hand, conventional secondary processing techniques like rolling, extrusion, and equal channel angular pressing are also used for uniform dispersion of the particles [53,54]. The technique of friction stir welding/processing (FSW/P) also holds exceptional promise in homogenizing the particle distribution locally under solid-state conditions. The large local plastic deformation from the retreating to the advancing side during FSW/P results in a more uniform distribution of TiB_2 particles [55]. Concurrent, dynamic recrystallization during FSW/P further refines the grain size [36].

12.3.3 Effect of Titanium Diboride on Corrosion Behavior

A comparative study of the corrosion characteristics of different aluminum alloys in NaCl solution indicates that commercially pure aluminum possesses the best corrosion resistance [56,57].

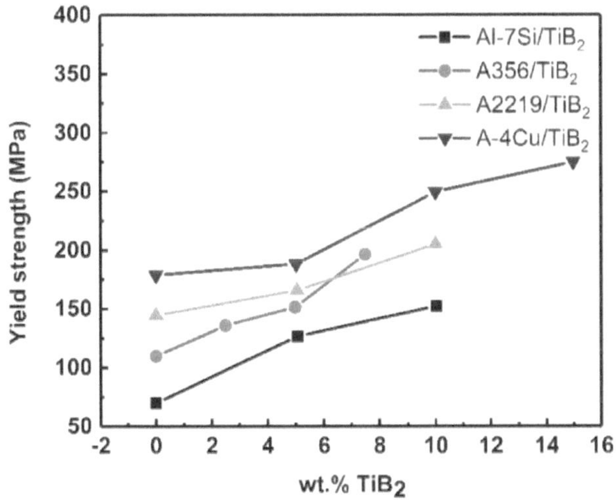

FIGURE 12.3 Variation of yield strength of aluminum composite as a function of TiB_2 reinforcement content. (Data from Refs. [17,18,39,48].)

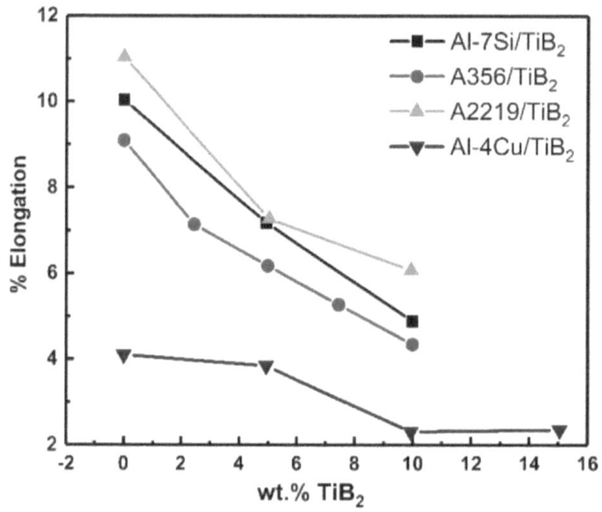

FIGURE 12.4 Variation of percentage elongation of aluminum composite as a function of TiB_2 reinforcement content. (Data from Refs. [17,18,39,48].)

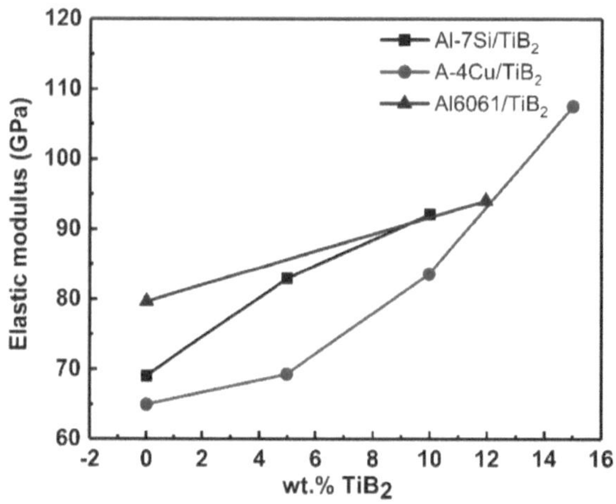

FIGURE 12.5 Effect of TiB_2 content on elastic modulus. (Data from Refs. [28,56,57,78].)

Intermetallics and second phases in aluminum alloys invariably deteriorate the corrosion resistance with matrix/reinforcement interface and areas adjoining grain boundaries acting as preferential sites for galvanic corrosion [58]. The galvanic corrosion further worsens when the reinforcement phase is impure, causing decrease in the resistivity [20]. Most of the reinforcements used are conductive and promote galvanic corrosion. However, Sun et al. recognized that TiB_2 as reinforcement improves the corrosion resistance due to the reduced galvanic coupling effect between the noble reinforcement and active aluminum matrix [59]. More recently, Poria et al. reported that the corrosion resistance of base alloy increased with the addition of 1 wt.% TiB_2 but reduced at 5.5 wt.% (Figure 12.6) [60]. Again, in-situ Al–TiB_2 composites produced by the stir casting method exhibited a higher corrosion rate in 3.5 wt.% NaCl, 0.5 mol L^{-1} H_2SO_4, and/or 0.5 mol L^{-1} NaOH electrolytic solutions, as compared to the base metal [61].

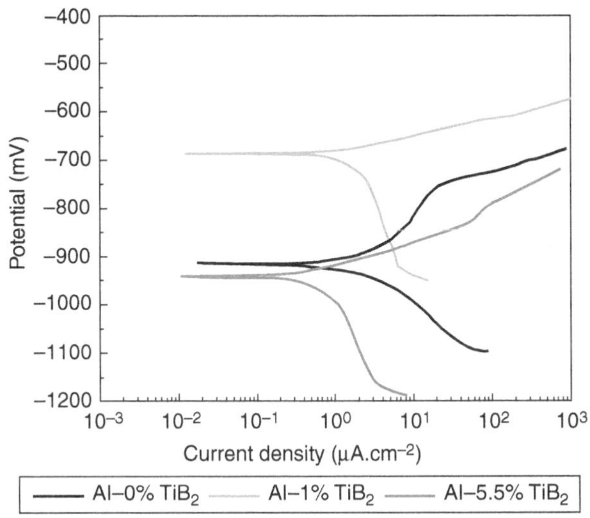

FIGURE 12.6 Tafel plot for base alloy and different wt.% of Al–TiB_2 composites. (Data from Refs. [59,60].)

More recently, literature on secondary processing techniques like FSW/P in AA2219 alloy reported an improved corrosion resistance due to the dissolution of the secondary phases [62]. Contrarily, FSW of 5 wt.% TiB_2 in AA6005 matrix is stated to deteriorate the corrosion resistance [63]. This was attributed to the absence of a stable Al_2O_3 passive film in the welded zone of the composite [63]. Thus, the effect of FSW/P on corrosion is principally due to its influence on second-phase distribution (*i.e.*, localization or randomization of the phases) and its ability to dissolve/re-precipitate the second phases [41].

In this context, multicomponent alloy particles as primary or secondary reinforcements demonstrated significantly improved interface bonding and strengthening in Al–MMC composites [8,64–66]. For example, AlCoCrFeNi particles (~10.5 μm) added to the aluminum matrix resulted in high strength and good plasticity [8]. Likewise, AlMgLiCuZn particles (~1.5 μm) in Mg-based composites enhanced the strength compared to traditional micron-size reinforcements [64]. Similarly, MoTaNaVW (3 vol.%) reinforced aluminum displayed enhanced strength without altering its corrosion resistance [65]. Again, $CoNiFeAl_{0.4}Ti_{0.6}Cr_{0.5}$ in Al–SiC composites exhibit approximately 20% enhancement in strength [66].

However, suitable multicomponent alloy selection for the aluminum matrix is essential and needs to fulfill the following criteria [8,65,66]:

i. The mechanical property of the multicomponent alloy particle supplements the softer matrix.
ii. Strong interface bonding by ensuring similar crystal structure and lattice parameter with matrix.
iii. Electrochemical similarity to minimize interfacial reaction products.
iv. Ease of processing.

Thus, considering the superior properties offered by High/Medium Entropy Alloys (H/MEAs) over a wide range of temperatures, a significant interest in investigating the properties of MMCs with H/MEAs as reinforcements is ongoing. In the next section, a brief introduction to H/MEAs is presented.

12.4 HIGH/MEDIUM ENTROPY ALLOYS: AN INTRODUCTION

Multicomponent H/MEAs are a new alloying concept where 4–5 or more elements in equimolar or near equimolar proportion are mixed to form a single-phase solid solution with unique mechanical and corrosion properties [67,68]. However, the latest studies reveal that single-phase formation is apparently an exception and rather multiple-phase formations are common. Thus, attempts have been made to predict the phase formation in H/MEAs based on parameters like [69]

$$\text{Atomic size difference } (\delta) = \sqrt{\sum_{i=1}^{N} c_i \left(1 - \frac{r_i}{\sum_{i=1}^{N} c_i r_i}\right)^2}$$

$$\text{Enthalpy of mixing } (\Delta H_{mix}) = \sum_{i=1,\, i\neq j}^{N} 4\Delta H_{mix}^{AB} c_i c_j$$

$$\text{Entropy of mixing } (\Delta S_{mix}) = -R\sum_{i=1}^{N} c_i \ln c_i$$

where c_i and r_i represent the molar content and atomic radius of the ith component, ΔH_{mix}^{AB} enthalpy for binary systems A–B and R is the universal gas constant. The conditions for a stable single-phase solid solution suggested as $\delta < 4\%$, -10 kJ $\Delta mol^{-1} < \Delta H_{mix} < 5$ kJ·mol^{-1} and $\Delta S_{mix} > 13.38$ J·K^{-1}mol^{-1} [70]. An additional parameter defined as [71]

$$\text{Valence Electron Concentration (VEC)} = \sum_{i=1}^{n} c_i n_i$$

is also used where n_i is the average number of itinerant electrons per atom. A VEC ≥ 8 is suggested to favor an FCC structure, while VEC ≤ 7 facilitates a BCC structure [70]. For a 7 < VEC < 8, an FCC + BCC structure is stable [70]. The unique properties of H/MEAs are ascribed to four physical effects and are described in the following subsections.

12.4.1 High Entropy Effect

The entropy effect in H/MEAs refers primarily to configurational entropy. Thus, in an equimolar quinary random solid solution, the configurational entropy is 1.61 R, where R is 8.31 kJ·mol^{-1} [72]. Such high configurational entropy in alloys lowers the Gibbs free energy of mixing facilitating solid solution formation. However, at low temperatures, the entropy effect reduces, promoting the transformation of the random solid solutions to intermetallics [73]. As an example, the single BCC phase HfNbTaTiZr alloy precipitates the HCP phase rich in Hf and Ta after annealing at 600 °C and 800 °C [74]. Thus, configurational entropy alone is insufficient to suppress the formation of multiple phases. In fact, thermodynamic data analysis of multicomponent alloys suggests that additionally mixing enthalpy (-10 kJ·mol^{-1}<ΔH_{mix}< 5 kJ·mol^{-1}) and non-configurational entropy play a significant role in single-phase stability [70,72].

12.4.2 Sluggish Diffusion Effect

Metastable phases and solid solutions are quite common in H/MEAs. This is primarily due to the slow diffusion in H/MEAs compared to conventional alloys and is strongly correlated to their exceptional strength [75,76] and thermal stability [77,78]. Tsai et al., in their investigation of CoCrFeMnNi alloy, ascribed such sluggish diffusion to the higher normalized activation energy than in conventional alloys due to the greater variety of atoms surrounding each lattice site of these alloys [79]. Thus, Liu et al., in their work on cold-rolled and annealed CoCrFeMnNi alloy, found a significantly high value of activation energy (~321 kJ·mol^{-1}) and slow coarsening kinetics on the grain growth behavior, which is in agreement with the sluggish diffusion effect [80]. Such sluggish diffusion is advantageous in controlling the microstructure and properties on account of increased recrystallization temperature, slower grain growth, reduced particle coarsening rate, and increased creep resistance [81].

12.4.3 Lattice Distortion Effect

Severe lattice distortion occurs in H/MEAs due to the different sizes of atoms that constitute the crystal lattices [82]. The displacement at each lattice site depends on the atom occupying that site and the atomic types surrounding it. The distortion effect is small in dilute solutions (see Figure 12.7), as the minority atom position is constrained by the majority surrounding atoms. On the other hand, owing to high solute concentration in H/MEAs, the lattice is highly strained, and the atoms are displaced from their usual lattice positions, and the distortion effect is significant. Such lattice distortion results in the decreased intensity of the X-ray diffraction peaks, increased hardness, and reduced electrical and thermal conductivity [81,83].

12.4.4 Cocktail Effect

This effect is used to describe a synergistic influence where the result is unpredictable [84,85]. According to this effect, exceptional material properties often result from unexpected interactions [81]. Thus, in H/MEAs having single or multiple phases, the bulk properties can be considered as overall contributions from size, shape, distribution, boundaries, and properties of each phase [81]. As a whole, the cocktail effect ranges from atomic-scale multicomponent composite effect to microscale multi-phase composite effect. The cocktail effect was demonstrated by Yeh et al. where an increase in aluminum content beyond a critical value in Al_xCoCrCuFeAl leads to a phase transformation from an FCC structure to a BCC structure (see Figure 12.8) [86].

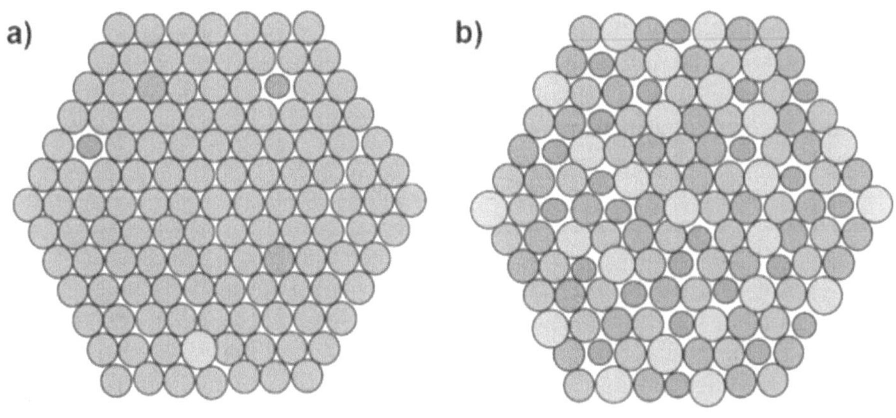

FIGURE 12.7 The effect of atom size difference on atom positions in (a) a dilute solution (b) a complex, concentrated solution. The variability in atom positions in (b) contributes to excess configurational entropy to decrease the free energy of the system. (Adapted from [82].)

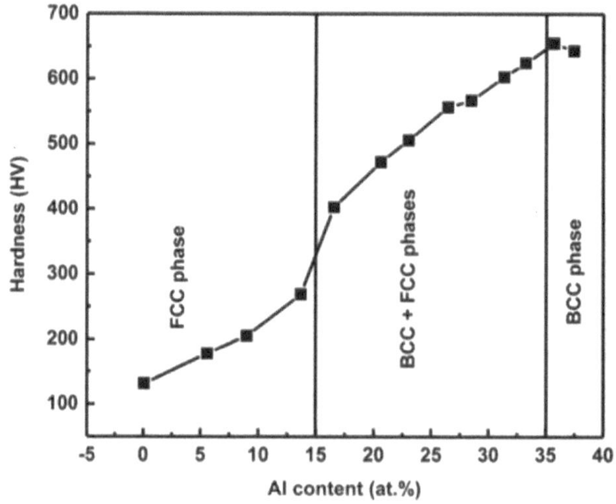

FIGURE 12.8 Cocktail effect introduced by the interaction of constituent elements in Al_xCoCrCuFeAl alloy. (Adapted from [86].)

However, most of the H/MEAs use transition elements which incur a cost in terms of the density of the alloys. The research on composites using H/MEAs as a strengthening phase suggests new possibilities for achieving tailor-made properties. The next section presents a detailed discussion on research and development of H/MEA reinforced composites.

12.4.5 High/Medium Entropy Reinforced Composites

The H/MEAs possess enhanced thermal, mechanical, and corrosion-resistant properties [87–90]. Thus, the use of H/MEAs as reinforcements to produce composites with a high strength-to-weight ratio and good ductility is receiving increased attention [91]. There are different processes by which composites are manufactured and fabrication processes have a crucial effect on their overall performance. The following subsections discuss the effect of fabrication routes on MMC microstructure.

12.4.5.1 Processing and Microstructure

Currently, several MMC fabrication methods like powder metallurgy, casting techniques, in situ routes, and chemical synthesis are in vogue [92–95]. However, compared to other routes, the melt stirring process is commercially more significant. This is principally due to advantages such as better matrix–particle bonding, easy control of matrix structure, applicability to large quantity production, and excellent productivity for near-net-shaped components [93,96]. However, poor wettability and heterogeneous distribution of reinforcement particles, especially when the density difference between reinforcement particles and the molten matrix is significant, are some of the problems associated with stir casting [26]. The problem is significant for aluminum MMCs as the low density of molten aluminum causes the strengthening particles to sink due to the density differences between the reinforcement and the matrix [26]. As discussed earlier, transition metals constitute most of H/MEAs and are significantly denser compared to aluminum. Thus, a significant challenge exists to obtain a uniform dispersion of reinforcing H/MEA particles in the aluminum matrix. In the following subsections, the advantages and disadvantages of different MMC fabrication routes are discussed.

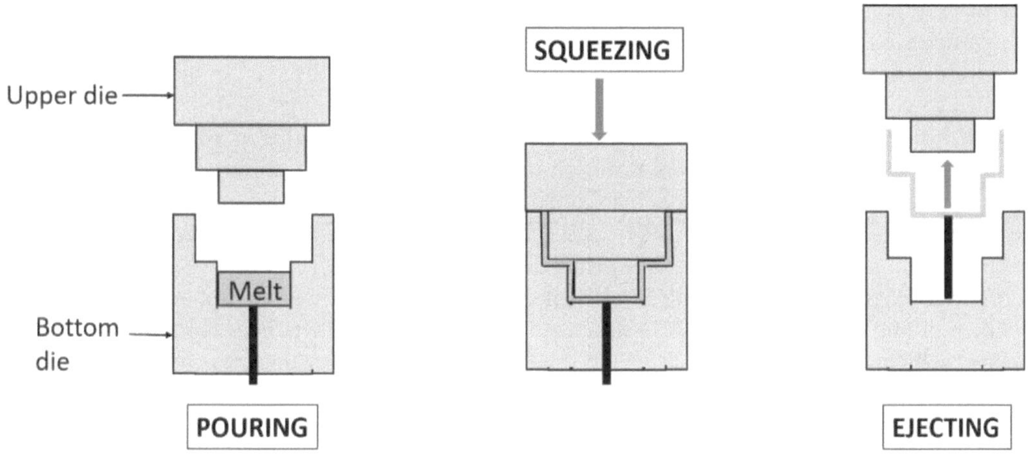

FIGURE 12.9 Schematic of squeeze casting technique.

12.4.5.1.1 Liquid Route–Casting

Literature on the use of casting routes to fabricate H/MEA reinforced composite is sparse. This is mainly due to the difficulty in controlling the dispersion and second-phase formation during the processing. As an exception, Kumar et al. fabricated the AA2024-(5–15%) Al–20Cu–10Mg composite by stir casting, homogenized at 100 °C for 24 h, where the average particle size was ~125 μm [91]. Interface dissolution time increased with an increase in weight fraction which resulted in a decrease in particle size from 8 to 4 μm with densification increasing from 2.82 to 2.96 g·cm^{-3} [91]. Lu et al. successfully fabricated $CoNiFeAl_{0.4}Ti_{0.6}Cr_{0.5}$ alloy by mechanical alloying and prepared a 5 vol.% HEA–40 vol.% SiC–AA7075 composite by squeeze casting (Figure 12.9) [66]. Both the SiC and H/MEA particles were distributed uniformly and were well-retained in the aluminum matrix [66]. Significantly, the H/MEA particles remained nanocrystalline after casting and subsequent heat treatment because of their good thermal stability [66]. Additionally, the aluminum grains were refined to ~1–3 μm, which is far smaller than expected in traditional casting and was attributed to the high-volume fraction of SiC particles limiting the grain growth [66]. Similarly, solution treatment followed by artificial aging of $Co_{0.5}NiFeCrTi_{0.5}$–SiC–AA7075 composite revealed that the amount of the second-phase dissolving into the aluminum matrix initially increased with an increase in duration of the solution treatment [97]. However, no obvious change in the amount of the second phase after 4 h was observed since only the Al_7Cu_2Fe phase was left in the matrix [97]. The interface between the Al/SiC in the composites did not show significant changes with increasing solution time, whereas gradual growth of the Al/HEA interface was observed [97]. The peak-aging time of composites during artificial aging gradually shortened with an increase in solution treatment time indicating that a longer time accelerated the precipitation kinetics [97]. Further, composites solution treated for 2 and 4 h displayed a larger supersaturation of Mg and Zn atoms along with a higher density of precipitates than that of the 1 h solution-treated composites [97]. Thus, a non-reactive Al/HEA interface in the composites can be obtained during casting processes whose effect on the mechanical properties is discussed later in this chapter.

12.4.5.1.2 Solid Route–Powder Metallurgy

In recent years, ball milling of multiple elemental powders in a high-energy ball mill to produce H/MEA particles is common [98,99]. Subsequently, these H/MEA particles are used to reinforce a metallic matrix by sintering at high temperatures. The main advantage of this route is that a high-density composite is obtained despite the high melting point elements and ceramic particles used.

As an example, Chen et al. prepared Cu-(10–20 wt.%) AlCoNiCrFe composite by ball milling the particles at 300 rpm for 60 h and then sintered the Cu and HEA particle mixture at 800 °C and 70 MPa for 30 min [98]. No apparent interfacial reaction was observed in the composite, and an average particle size of approximately 20 nm for the composite suggested an absence of grain growth during sintering [98]. However, Liu et al. observed that pre-heating of a 10 wt.% HEA composite at 850 °C for 6 h promoted the diffusion of the aluminum, cobalt, chromium, iron, and nickel from the reinforcement to the copper matrix and vice-versa, resulting in a core-shell structure [99]. However, excess pre-heating beyond 6 h causes disappearance of the core-shell structure and subsequent formation of Cu-rich and CoCrFe-rich FCC structures [99].

Further, extending the work with 5 vol.% of the HEA in an aluminum matrix at a sintering temperature of 560 °C generated an FCC-structured Al-rich transition layer between the HEA and the matrix, with an average thickness of ~4 μm [100]. The transition layer thickness increased up to 15 μm at 600 °C with columnar grains near the HEA particles and a fine equiaxed matrix [100]. More recently, Lu et al. reinforced an aluminum matrix with 7.5, 15, and 30 vol.% $CoNiFeCrAl_{0.6}Ti_{0.4}$ HEA particles by combining ball milling, hot pressing, and hot extrusion [101]. It was reported that the average particle size of the nanocrystalline HEA significantly decreased after milling and was subsequently distributed uniformly after extrusion [101]. Importantly, the HEA particles refined the aluminum matrix through pinning effects on grain boundary movement in the MMC [101]. Similar grain refinement occurred in hot-extruded Al–$FeNiCrCoAl_3$ composite with an increase in the milling time [102]. Significantly, during the hot extrusion elemental diffusion resulted in fine needle-like Al_3Fe, Al_6Fe, and nano-scaled Al_2Cu phases, which improved the compressive strength of the composite [102]. Another study on microwave sintered Al–$FeCoNi_{1.5}CrCu$ showed that at 480 °C, the sintered composite was compact with an even distribution of particles [103]. However, increased sintering temperature up to 520 °C resulted in forming an AlNi solid solution phase with a BCC structure in the HEA [103]. The HEA's phase changed from FCC to FCC+BCC, with additional Al_5Cr_8, Al_5Co_2, and Al_2Cu precipitates [103].

Similar work on Mg-based MMCs showed a similar array of complex behavior. The Mg composite contained 2.5–7 wt.%. AlMgLiCuZn HEA particles developed by Tun et al. (microwave-assisted powder metallurgy method followed by hot extrusion) showed a significant reduction in grain size [64]. Such grain size reduction indicated the occurrence of dynamic recrystallization due to HEA particles during the hot extrusion process [104]. During extrusion, incompatibility between the non-deformable hard particles and the plastically deformable matrix resulted in strain gradients around the particles [64]. This resulted in regions of high dislocation density known as deformation zones and was an ideal site for recrystallization by particle-stimulated nucleation (PSN) [64]. However, such PSN effect was specific for reinforcements of size greater than 1 μm [105,106]. Similar behavior was observed in the hot-extruded composites [64,101,102], where reinforcement particles of size greater than 1 μm acted as nucleation sites for recrystallization [107]. However, in the Mg composites, the clustering tendency increased with a progressive increase in the amount of reinforcement, negating the nucleation and grain boundary pinning tendency of the HEA particles [105]. Again, $Al_{65}Cu_{16.5}Ti_{18.5}$–$Al_{0.6}CoCrFeNi$ composites prepared by mechanical mixing and spark plasma sintering (SPS) comprised an amorphous/nanocrystalline matrix, with interdiffusion (ID) surrounding the BCC+FCC HEA particles [108]. A schematic representation of the ID layer is shown in Figure 12.10. The ID layer formed due to enhanced diffusion-reaction under the effect of localized high temperature consisted of submicron grains with a size of 200–400 nm [108]. A similar ID layer was also observed in aluminum and NiAl-based composites reinforced with 10–18 wt.% CoCrFeMnNi [109,110]. It was suggested that the composition of the ID layer was related to the diffusion coefficients and diffusion mode of the elements and was affected by the enhanced localized high temperatures [109]. In NiAl-based composites, Al_2O_3 formed during the sintering process within the HEA phase further limited the grain growth of the composite [110]. Fu et al. introduced the TiB_2–TiNiFeCrCoAl composite and investigated its properties at different reinforcement

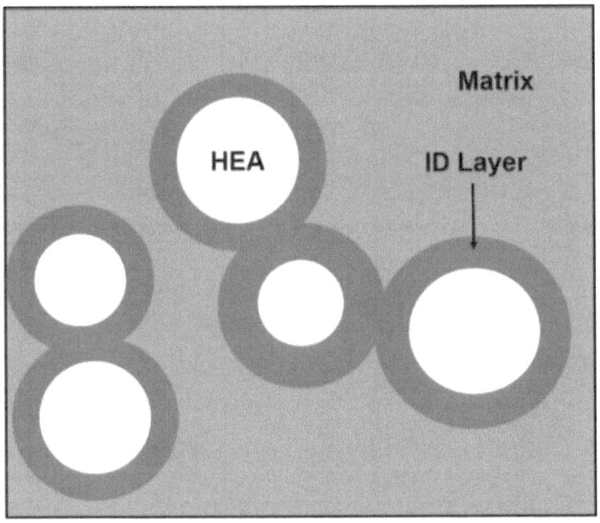

FIGURE 12.10 Schematic of the inter-diffusion layer in HEA reinforced composite. (Adapted from [108].)

percentages [111,112]. An increase in the content of HEA and ball milling time removed the oxide impurities in TiB_2 and enhanced the sintered density. Further, the solubility of TiB_2 in the HEA and the solid–liquid surface energy decreased and inhibited grain growth [111,112]. Prabhakaran et al. successfully fabricated equiatomic CrMnFeNiCu HEA through mechanical alloying and used it as a reinforcement with an aluminum matrix [113]. It was found that the densities of the composites were reduced up to 3.5% compared to the matrix material [113]. Constant load during compacting was insufficient to compress the HEA materials which failed to make a good bonding with the matrix material, hence less density was achieved [113]. Moreover, reinforcement particles generated voids which increased the porosity of the composites [113]. However, an increase in sintering temperature increased the density of the composites, reducing the porosity by ~5–10% [113]. Further, in (10–30 vol.%) CuZrNiAlTiW reinforced aluminum composite fabricated by SPS, the large concentration gradient between aluminum matrix and HEA particles induced the formation of in-situ NiAl-rich B2 and WAl_{12} phases layers around the HEA particles [114]. Moreover, the core-shell structure formed in the 10 vol.% HEA reinforcement case effectively coordinated the deformation between the HEA and the aluminum matrix [114]. On the other hand, Satyanarayana et al. developed a Tungsten heavy alloy composite using CoCrFeMnNi as the reinforcement phase without any crack formation by conventional, microwave, and SPS techniques [115]. It was found that a high heating rate and shorter holding time during sintering resulted in a significant reduction of tungsten grain size and reduced the volume fraction of the Cr–Mn rich oxide phase within the composite [115]. Apart from the traditional HEA reinforcements, refractory HEAs like MoTaNbVW were also used in variable amounts for reinforcing the aluminum matrix at a sintering temperature of 550 °C [65]. The composites exhibited low porosity, good homogeneity, few defects, and good distribution of the reinforcing phases within the matrix [65]. Further, the absence of secondary intermetallic phases at the interface between matrix/reinforcement indicated good bonding with no signs of reactivity [65].

12.4.5.1.3 Additive Manufacturing

Additive manufacturing to fabricate H/MEA reinforced composites is also prevalent where FSP for consolidation was used (Figure 12.11) [116,117]. Thus, friction deposition experiments using direct-drive rotary friction welding machines are used for depositing the reinforcement on the substrate. Rods made of matrix material were drilled to produce holes of suitable diameter using electric discharge machining. These holes were filled with high entropy alloy particles. Subsequently,

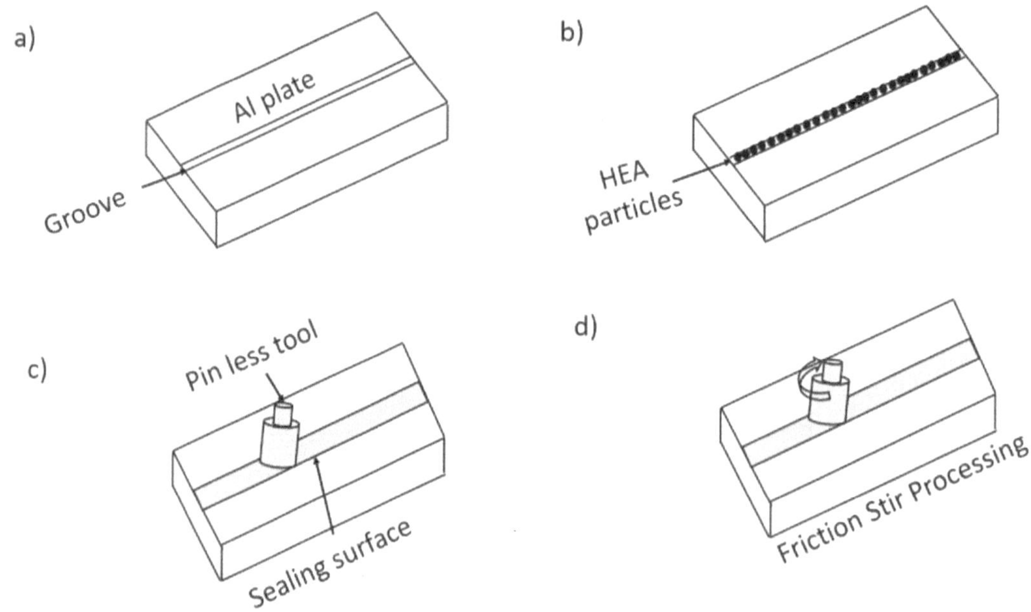

FIGURE 12.11 Schematic diagram of fabricating composite by friction stir processing. (Adapted from [117].)

FSP was utilized for the friction-induced deposition of the H/MEA particles on the substrate, where the process was repeated several times for multi-layer deposition. Thus, Karthik et al. successfully deposited multiple layers of 12 vol.% CoCrFeNi reinforcement in an aluminum matrix by FSP [116]. The FCC phase HEA particles in the composite were fine with an average size of ~5 μm and an inter-particle spacing of ~8 μm, suggesting fragmentation of the particles during processing [116]. Similarly, Gao et al. fabricated FeCoNiCrAl HEA-reinforced AA5083 composites [117]. The distribution of the HEA particles became more homogeneous with increasing processing passes along with subsequent fragmentation due to tool stirring [117]. Similarly, Yang et al. synthesized 10 vol.% AlCoCrFeNi–Al composites by Submerged FSP where a random distribution of HEA particles along with finer grains was obtained [8]. This was attributed to dynamic recrystallization, PSN, and shortened thermal cycles due to water cooling [8]. However, as in the case of powder-based processes, additively manufactured composites also displayed a typical core and shell particle/matrix arrangement. In the subsequent section, the properties exhibited by the MMCs obtained through different processing routes are discussed.

12.4.6 Properties of HEA Reinforced PMMC

12.4.6.1 Mechanical Properties of Cast MMCs

The main aim of MMCs is to produce high specific-strength materials. The principal strengthening methods in H/MEA reinforced composites are solid solution strengthening, grain boundary strengthening, and dislocation strengthening [118]. In this regard, aluminum composite reinforced with 5 wt.% Al–20Cu–10Mg showed an increase in modulus and hardness which further increased with an increase in the amount of reinforcement [91]. The yield strength showed a similar trend with an increase of 95% [91]. Thus, with an increase in the amount of reinforcement, more load was transferred to the reinforcement increasing the strength [91]. This was facilitated by the clean interface between the H/MEA particles and the matrix resulting in higher strength compared to conventional composites. Similarly, Lu et al. found that, in contrast to the Al–SiC composite, the

Al–SiC–CoNiFe$Al_{0.4}Ti_{0.6}Cr_{0.5}$ composite exhibited higher strength and modulus with better ductility where an average value of 712 MPa, 171 GPa, and 0.82% were obtained [66]. Investigations on the effect of solution treatment for 1–4 h in AA7075– $Co_{0.5}$NiFeCr$Ti_{0.5}$ composite showed that the 2 h heat-treated composite possesses the highest yield strength (~1.2 times) followed by the 1 and 4 h heat-treated composites [97]. The results suggested that these improvements in mechanical properties were mainly dependent on [66,97].

1. A suitable length of Al/HEA interface where increased solutionizing time, widened the interface significantly from being 100 nm in 1 h to 500 nm after 4 h.
2. Dissolution of secondary phases (Al_2CuMg) in the matrix with an increase in the solutionizing time.
3. Precipitate-free zones formed in the vicinity of the Al/HEA interface with an increase in solutionizing time.
4. Growth of HEA particles from 15.4 and 24.4 nm for 1 and 2 h solutionized MMC to 44.2 nm in 4 h solutionized MMC.

Thus, the increase in strength of H/MEA reinforced composites formed by casting can be attributed to [91] (a) refined grain size of matrix, (b) reduced inter-particle distance, (c) good interfacial bond between the reinforcement and the matrix, (d) enhanced dislocation density, and (e) higher constraint to localized matrix deformation due to presence of reinforcement and particle solubility in the matrix [119].

12.4.6.2 Mechanical and Corrosion Properties of Powder Metallurgy Processed MMCs

A combination of ball milling followed by high-temperature sintering has been the most sought method to obtain high strength and better ductility in H/MEA reinforced composites. In this regard, the yield strength of the Cu matrix increased by ~160–220% when reinforced with 10–20% of AlCoCrFeNi particles by powder metallurgy route [98]. The HEA showed a remarkable strengthening effect and the yield strength of the composite was close to value calculated using Voigt model given as [120,121]

$$\sigma_c = V_m \sigma_m + V_r \sigma_r \tag{12.10}$$

where σ is the yield stress, V is the volume fraction, and the subscripts c, m, and r refer to the composite, the matrix, and the reinforcement, respectively. The different strengthening effect of HEAs is thought to arise from their different plasticity [98]. Along with high strength, HEA alloys have good plasticity (compression strain of AlCoCrFeNi is ~24%) [122]. Hence, part of the total strain during plastic deformation can be accommodated by the HEA particle itself [98]. Again, pre-heating before hot extrusion of the Cu-10 vol.% HEA composite significantly improved its mechanical properties [99]. For a pre-heating time of 6 h, the yield strength of the composite increased from 160 to 317 MPa with an increase in elongation of 11.7% [99]. This may be attributed to the decreased hardness of the reinforcement with a ductile FCC phase [99]. However, a higher pre-heating time of 8 h reduced the strengthening effect of reinforcements with subsequent improvement in the plasticity of the composite [99].

Similarly, at a sintering temperature of 540 °C, 5 vol.% of AlCoCrFeNi reinforced in an aluminum matrix generated a yield strength and plastic strain of 96 MPa and 36% [100]. Further increase in sintering temperature up to 580 °C improved the yield strength up to 137 MPa [100]. Comparable work by Wang et al. observed that a reinforcement mass fraction of 40% of the AlCoCrFeNi alloy increased the yield strength of the aluminum composite by ~2 times [102]. Such high strength was mainly attributed to the nanocrystalline grains of α(Al) and nano-scaled Al_3CoCrFeNi particles with fine secondary Al_2Cu and Fe-rich phases [102]. In another work, a fine-grained aluminum matrix composite reinforced with 7.5 vol.% nanocrystalline $CoNiFeCrAl_{0.6}Ti_{0.4}$ exhibited a yield strength of 419 MPa, which is 112 MPa higher than the aluminum matrix [101]. The yield strength of the composite increased to ~500 MPa with an increase to 30 vol.% of the reinforcement [101]. Noticeably, the accumulation of dislocations rendered a higher strain hardening rate of the composites than in the AA2024 alloy matrix [101]. Similarly, in $Al_{0.6}CoCrFeNi$ reinforced aluminum composite, the yield strength was found to be as high as ~3120 MPa with low plasticity [108]. Similarly, in CoCrFeMnNi reinforced aluminum composite, the hardness improved by ~64% [109]. In the case of microwave sintered Al–$FeCoNi_{1.5}CrCu$ composite, the matrix and the reinforcement were observed to have outstanding interface wettability and interface compatibility [103]. This resulted in maximum yield strength and hardness of 227.4 MPa, and 93.1 HV, respectively, after 40 min of sintering [103]. However, with increased sintering time, the HEA particles agglomerated [103]. Similarly, Al–CrMnFeNiCu also demonstrated an improvement in hardness with an increase in reinforcement fraction and sintering time/temperature with the hardness at 20 wt.% HEA increasing to approximately two times at 600 °C [113]. The improved strength of such composites is attributed to stable crack propagation, increase of fracture energy, and significantly low sintering defects in the matrix [108].

Another strategy carried out by Tun et al. to develop lightweight MMCs was by reinforcing 2.5–7.5 wt.% AlMgLiCuZn particles in a magnesium matrix [64]. The reinforcement addition led to an improved hardness of ~47, 51, and 54% in Mg-2.5, 5, and 7.5 wt.% HEA reinforced composites, respectively [64]. The results suggest that the strain hardening effect from the particles enhances with the increasing presence of reinforcement particles in the magnesium matrix [64]. Compression testing indicated a high yield and ultimate strength (~1.7 times) with a simultaneous increase in ductility when compared to unreinforced metal [64]. This was mainly due to an efficient load transfer, Orowan, and Hall–Petch grain boundary strengthening mechanisms [64]. Similarly, in the CrMnFeCoNi alloy reinforced NiAl composite, the compressive strength improved with an increase in reinforcement fraction [110]. The composite containing 10 wt.% HEA exhibited significant room temperature compressive properties, with an ultimate compressive strength of 2692 MPa and a compressive strain of 34.2%, respectively [110]. Similarly, in TiB_2–TiNiFeCoCrAl composites, a high hardness of ~1784–2376 HV and fracture toughness of 8.7–12.8 MPa was observed for 5–20 wt.% HEA [111]. The high hardness and fracture toughness was mainly attributed to the fine grain size and the toughening mechanisms, including crack deflection due to the high plasticity of the FCC HEA, along with the absence of brittle intermetallic phases [111]. Similarly, in spark plasma sintered W–CoCrFeMnNi composite, fine grains along with the Cr–Mn-rich oxide phase resulted in improved compressive strength (~2041 MPa) as compared to microwave (~1962 MPa) and conventional (~1758 MPa) sintering [115]. Likewise, in the Al–MoTaNbVW composite, the hardness values increased with an increase in reinforcement, indicating the reinforcing action of the hard and strong HEA phase [65]. However, increased reinforcement of the H/MEA particles are also limited as higher amounts of reinforcement increase strain at the boundaries resulting in initiation of crack and subsequent failure. This was highlighted when 20% CrMnFeNiCu particles showed lower strength than the 10% counterpart when reinforced in aluminum matrix [113].

In aluminum MMC, the corrosion resistance of the composites in 3.5 wt.% NaCl solution with low HEA particulate (1–3 vol.% HEA) appeared to be similar to aluminum [65]. However, an increase in reinforcement up to 5 vol.% increased the corrosion current density, which was attributed to the increase in cathodic area in the composite [65]. The surface appeared to be degraded by pit formation where the interface underwent preferential dissolution due to the higher nobility of the reinforcing phase compared to the matrix [65]. The galvanic effect between aluminum and the HEA particles was further increased due to the high amount of oxygen at the interface [65]. Again, the potential difference between the dendrites and inter-dendrites resulted in uniform and inter-granular corrosion [65]. However, spark plasma sintered Al–CuZrNiAlTiW composite showed higher pitting resistance (in seawater solution) as compared to pure aluminum [114]. The pitting resistance of the composites increased with an increase in reinforcement content from 10 to 30 vol.% [114]. This was mainly attributed to the enrichment of the translation layer with nickel and tungsten [114].

12.4.6.3 Effect of Additive Manufacturing on the Strength of MMCs

Additively manufactured composites showed significantly higher tensile and compressive strength compared to conventional wrought-processed alloys [116]. Additionally, such manufacturing processes offered a better combination of strength and ductility as compared to conventional aluminum matrix composites [116]. Thus, the incorporation of 12 vol.% CoCrFeNi reinforcement into the aluminum matrix by FSP increased the tensile yield by ~1.4 times while the compressive yield strength increased by ~1.8 times as compared to FSPed aluminum matrix [116]. Similarly, in the FSPed Al–FeCoNiCrAl composite, the hardness values improved from ~78 to ~95 HV, which further increased to ~158 HV with multiple passes due to improved particle distribution [117]. Further, submerged FCP of the composite showed 25.1% higher yield stress and 31.9% higher ultimate tensile strength in comparison to the base metal while maintaining an acceptable ductility of 18.9% [8]. When compared with conventional aluminum composites with ceramic particles, it is observed that in conventional MMCs, severe deterioration in ductility is inevitable [8]. For example, Al-10 wt.% SiC composites produced by powder metallurgy exhibited a UTS value of 134 MPa and ductility of 8.6% [123]. A UTS value of 283 MPa and ductility of ~2% were observed in Al-3 vol.% nano TiB_2 composites prepared by casting [26]. Figure 12.12 presents the yield strength of the various H/MEA reinforced composites and compares it with the conventional aluminum-based composites [37,124,125].

12.5 SUMMARY

To adhere to standards for industrial applications, it is crucial to tailor alloy designs with suitable compositions and mechanical and corrosion properties. This chapter offers an in-depth exploration of Al–TiB_2 composites and enhancing their microstructural and mechanical properties through

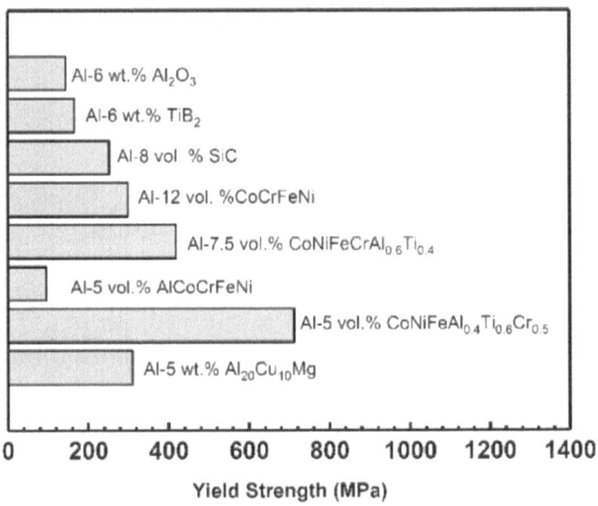

FIGURE 12.12 Comparison of the yield strength of various H/MEA reinforced composites with conventional Al-based composites. (Data from [16,31,32,66,91,100,101,116].)

the integration of H/MEAs reinforcements and suggests various paths for future investigations. Following noteworthy points are highlighted:

1. The Al–TiB_2 composites have enormous potential for use in automobile and aerospace applications. However, proper reinforcement distribution, brittle phase reduction, and strength improvement at low TiB_2 contents remain some of the key challenges in their use.
2. H/MEAs offer superior component distribution, resulting in unique properties. They serve as optimal reinforcements for PMMCs due to continuous advancements across sectors. Offering excellent high-temperature stability, strength, and corrosion resistance, they minimize interfacial reactions for enhanced overall performance.
3. H/MEA reinforced composites are fabricated through methods like casting, powder metallurgy, and FCP. The choice of method depends on application and economic needs. Such composites offer significant microstructural improvements across various fabrication methods, enhancing both microstructural and mechanical properties. These enhancements often resemble a shell-type structure, akin to an ID layer.
4. High dislocation density, unique to H/MEA reinforced composites, enhances strength and toughness by impeding dislocation motion at grain boundaries, leading to grain refinement. Improved mechanical properties stem from more uniform reinforcement dispersion, facilitated by higher dislocation density. Such composites have shown enhanced yield strength and corrosion resistance due to factors like transition layer formation between matrix and reinforcement, diverse reinforcement methods, and varying heat treatment conditions.

12.6 FUTURE SCOPE

Significant opportunities for research and advancement exist in the realm of high-entropy composites. Key areas for future investigation include the following:

i. New fabrication routes to facilitate mass production of the composites need exploration.
ii. With the development of increasingly sustainable materials, it is imperative to assess the impact of high-entropy composites in terms of circular economy.
iii. Homologation for H/MEA reinforced composites enables more efficient comparison and evaluation with conventional materials, thereby increasing their adoption in product development.

NOTE

1 Hamaker constant: It is a parameter for defining the strength of van der Waals interaction of substances and is determinable by using different methods out of which wetting is one of the possible methods [Ref: Q. Shen, New insight on critical Hamaker constant of solid materials, *Mater. Res. Bull.* 133 (2021) 111082–111086].

REFERENCES

[1] Y. Zhang, N. Ma, H. Wang, Y. Le, X. Li, Damping capacity of in situ TiB_2 particulates reinforced aluminum composites with Ti addition, *Mater. Des.* 28 (2007) 628–632.

[2] S. Jayalakshmi, S. Gupta, S. Sankaranarayanan, S. Sahu, M. Gupta, Structural and mechanical properties of $Ni_{60}Nb_{40}$ amorphous alloy particle reinforced Al-based composites produced by microwave-assisted rapid sintering, *Mater. Sci. Eng. A* 581 (2013) 119–127.

[3] B.G. Narasimha, V.M. Krishna, A.M. Xavior, A review on processing of particulate metal matrix composites and its properties, *Int. J. Appl. Eng. Res.* 8 (2013) 647–666.

[4] M. Wang, D. Chen, Z. Chen, Y. Wu, F. Wang, N. Ma, H. Wang, Mechanical properties of in-situ TiB_2/A356 composites, *Mater. Sci. Eng. A* 590 (2014) 246–254.

[5] C.S. Kim, I. Sohn, M. Nezafati, J.B. Ferguson, B.F. Schultz, Z. Bajestani-Gohari, P.K. Rohatgi, K. Cho, Prediction models for the yield strength of particle-reinforced unimodal pure magnesium (Mg) metal matrix nanocomposites (MMNCs), *J. Mater. Sci.* 48 (2013) 4191–4204.

[6] S.Z. Lu, A. Hellawell, Modification of Al-Si alloys: Microstructure, thermal analysis, and mechanisms, *JOM* 47 (1995) 38–40.

[7] A. Sanaty-Zadeh, Comparison between current models for the strength of particulate-reinforced metal matrix nanocomposites with emphasis on consideration of Hall–Petch effect, *Mater. Sci. Eng. A* 531 (2012) 112–118.

[8] X. Yang, P. Dong, Z. Yan, B. Cheng, X. Zhai, H. Chen, H. Zhang, W. Wang, AlCoCrFeNi high-entropy alloy particle reinforced 5083Al matrix composites with fine grain structure fabricated by submerged friction stir processing, *J. Alloys Compd.* 836 (2020) 155411–155423.

[9] W. Chen, Z. Li, T. Lu, T. He, R. Li, B. Li, B. Wan, Z. Fu, S. Scudino, Effect of ball milling on microstructure and mechanical properties of 6061Al matrix composites reinforced with high-entropy alloy particles, *Mater. Sci. Eng. A* 762 (2019) 138116–138127.

[10] A. Mandal, R. Maiti, M. Chakraborty, B.S. Murty, Effect of TiB_2 particles on aging response of Al-4Cu alloy, *Mater. Sci. Eng. A* 386 (2004) 296–300.

[11] G.M. Janowski, B.J. Pletka, The effect of particle size and volume fraction on the aging behavior of a liquid-phase sintered SiC/aluminum composite, *Metall. Mater. Trans. A* 26 (1995) 3027–3035.

[12] C.S. Pande, K.P. Cooper, Nanomechanics of Hall–Petch relationship in nanocrystalline materials, *Prog. Mater. Sci.* 54 (2009) 689–706.

[13] L. Jiang, H. Yang, J.K. Yee, X. Mo, T. Topping, E.J. Lavernia, J.M. Schoenung, Toughening of aluminum matrix nanocomposites via spatial arrays of boron carbide spherical nanoparticles, *Acta Mater.* 103 (2016) 128–140.

[14] F. Khodabakhshi, A.P. Gerlich, A. Simchi, A.H. Kokabi, Cryogenic friction-stir processing of ultrafine-grained Al-Mg-TiO_2 nanocomposites, *Mater. Sci. Eng. A* 620 (2015) 471–482.

[15] T. Das, P.R. Munroe, S. Bandyopadhyay, The effect of Al_2O_3 particulates on the precipitation behaviour of 6061 aluminum-matrix composites, *J. Mater. Sci.* 31 (1996) 5351–5361.

[16] J.M. Papazian, P.N. Adler, Tensile properties of short fiber-reinforced SiC/Al composites: Part I. Effects of matrix precipitates, *Metall. Trans. A* 21 (1990) 401–410.

[17] S. Kumar, M. Chakraborty, V. Subramanya Sarma, B.S. Murty, Tensile and wear behaviour of in situ Al-7Si/TiB_2 particulate composites, *Wear* 265 (2008) 134–142.

[18] A. Mandal, M. Chakraborty, B.S. Murty, Ageing behaviour of A356 alloy reinforced with in-situ formed TiB_2 particles, *Mater. Sci. Eng. A* 489 (2008) 220–226.

[19] L.H. Hihara, R.M. Latanision, Galvanic corrosion of aluminum-matrix composites, *Corrosion* 48 (1992) 1–33.

[20] L.H. Hihara, R.M. Latanision, Corrosion of metal matrix composites, *Int. Mater. Rev.* 39 (1994) 245–264.

[21] E.-S.M. Sherif, A.A. Almajid, F.H. Latif, Effects of graphite on the corrosion behavior of aluminum-graphite composite in sodium chloride solutions, *Int. J. Electrochem. Sci.* 6 (2011) 1085–1099.

[22] A. Pardo, M.C. Merino, S. Merino, F. Viejo, M. Carboneras, R. Arrabal, Influence of reinforcement proportion and matrix composition on pitting corrosion behaviour of cast aluminum matrix composites (A3xx.x/SiCp), *Corros. Sci.* 47 (2005) 1750–1764.

[23] A.S. Verma, N.M. Suri, Corrosion behavior of aluminum base particulate metal matrix composites: A review, *Mater. Today Proc.* 2 (2015) 2840–2851.

[24] G. S. Frankel, Corrosion: Fundamentals, testing, and protection, *ASM Int.* 13A (2003).

[25] D.M. Aylor, P.J. Moran, Effect of reinforcement on the pitting behavior of aluminum-base metal matrix composites, *Proc. Electrochem. Soc.* 84 (1984) 584–595.

[26] M. Karbalaei Akbari, H.R. Baharvandi, K. Shirvanimoghaddam, Tensile and fracture behavior of nano/micro TiB_2 particle reinforced casting A356 aluminum alloy composites, *Mater. Des.* 66 (2015) 150–161.

[27] R.G. Munro, Material properties of titanium diboride, *J. Res. Natl. Inst. Stand. Technol.* 105 (2000) 709–720.

[28] V. Milman, M.C. Warren, Elastic properties of TiB_2 and MgB_2, *J. Phys. Condens. Matter.* 13 (2001) 5585–5595.

[29] M. Emamy, M. Mahta, J. Rasizadeh, Formation of TiB_2 particles during dissolution of $TiAl_3$ in Al-TiB_2 metal matrix composite using an in situ technique, *Compos. Sci. Technol.* 66 (2006) 1063–1066.
[30] H. Abdizadeh, H.R. Baharvandi, K.S. Moghaddam, Comparing the effect of processing temperature on microstructure and mechanical behavior of ($ZrSiO_4$ or TiB_2)/aluminum composites, *Mater. Sci. Eng. A* 498 (2008) 53–58.
[31] Y. Pazhouhanfar, B. Eghbali, Microstructural characterization and mechanical properties of TiB_2 reinforced Al6061 matrix composites produced using stir casting process, *Mater. Sci. Eng. A* 710 (2018) 172–180.
[32] V. Bharath, M. Nagaral, V. Auradi, S.A. Kori, Preparation of 6061Al-Al_2O_3 MMC's by stir casting and evaluation of mechanical and wear properties, *Procedia Mater. Sci.* 6 (2014) 1658–1667.
[33] J.J. Moses, I. Dinaharan, S.J. Sekhar, Characterization of silicon carbide particulate reinforced AA6061 aluminum alloy composites produced via stir casting, *Procedia Mater. Sci.* 5 (2014) 106–112.
[34] Y.M. Youssef, R.J. Dashwood, P.D. Lee, Effect of clustering on particle pushing and solidification behaviour in TiB_2 reinforced aluminum PMMCs, *Compos. Part A* 36 (2005) 747–763.
[35] H.B.M. Rajan, I. Dinaharan, S. Ramabalan, E.T. Akinlabi, Influence of friction stir processing on microstructure and properties of AA7075/TiB_2 in situ composite, *J. Alloys Compd.* 657 (2016) 250–260.
[36] D. Yadav, R. Bauri, Friction stir processing of Al-TiB_2 in situ composite: Effect on particle distribution, microstructure and properties, *J. Mater. Eng. Perform.* 24 (2015) 1116–1124.
[37] K.L. Tee, L. Lu, M.O. Lai, In situ stir cast Al-TiB_2 composite: processing and mechanical properties, *Mater. Sci. Tech.* 17 (2001) 201–206.
[38] F. Chen, Z. Chen, F. Mao, T. Wang, Z. Cao, TiB_2 reinforced aluminum based in situ composites fabricated by stir casting, *Mater. Sci. Eng. A* 625 (2015) 357–368.
[39] L. Lü, M.O. Lai, Y. Su, H.L. Teo, C.F. Feng, In situ TiB_2 reinforced Al alloy composites, *Scr. Mater.* 45 (2001) 1017–1023.
[40] R. Sharma, A.K. Singh, A. Arora, S. Pati, Effect of friction stir processing on corrosion of Al–TiB_2 based composite in 3.5 wt.% sodium chloride solution, *Trans. Nonferrous Met. Soc. China* 29 (2019) 1383–1392.
[41] S. Riva, A. Tudball, S. Mehraban, N.P. Lavery, S. Brown, K.V. Yusenko, A novel high-entropy alloy-based composite material, *J. Alloys Compd.* 730 (2018) 544–551.
[42] A.L. Greer, A.M. Bunn, A. Tronche, P. V. Evans, D.J. Bristow, Modelling of inoculation of metallic melts: application to grain refinement of aluminum by Al–Ti–B, *Acta Mater.* 48 (2000) 2823–2835.
[43] A.L. Greer, Overview: Application of heterogeneous nucleation in grain-refining of metals, *J. Chem. Phys.* 145 (2016) 211704–211718.
[44] S. Carlsson, *Grain refinement of aluminum by titanium diboride particles: The importance of nucleation, growth restriction, and cooling rate*, Master's thesis KTH Material Science and Engineering (2019).
[45] N. Iqbal, N.H. Van Dijk, S.E. Offerman, M.P. Moret, L. Katgerman, G.J. Kearley, Real-time observation of grain nucleation and growth during solidification of aluminum alloys, *Acta Mater.* 53 (2005) 2875–2880.
[46] M.A. Easton, D.H. Stjohn, A model of grain refinement incorporating alloy constitution and potency of heterogeneous nucleant particles, *Acta Mater.* 49 (2001) 1867–1878.
[47] N. Kumar, G. Gautam, R.K. Gautam, A. Mohan, S. Mohan, Synthesis and characterization of TiB_2 reinforced aluminum matrix composites: A Review, *J. Inst. Eng. Ser. D* 97 (2016) 233–253.
[48] N.R. Rajasekaran, V. Sampath, Effect of in-situ TiB_2 particle addition on the mechanical properties of AA 2219 Al alloy composite, *J. Miner. Mater. Charact. Eng.* 10 (2011) 527–534.
[49] Y. Han, K. Li, J. Wang, D. Shu, B. Sun, Influence of high-intensity ultrasound on grain refining performance of Al–5Ti–1B master alloy on aluminum, *Mater. Sci. Eng. A* 405 (2005) 306–312.
[50] T.C. Tszeng, The effects of particle clustering on the mechanical behavior of particle reinforced composites, *Compos. Part B* 29 (1998) 299–308.
[51] J. Xue, J. Wang, Y. Han, P. Li, B. Sun, Effects of CeO_2 additive on the microstructure and mechanical properties of in situ TiB_2/Al composite, *J. Alloys Compd.* 509 (2011) 1573–1578.

[52] A. Jha, C. Dometakis, The dispersion mechanism of TiB_2 ceramic phase in molten aluminum and its alloys, *Mater. Des.* 18 (1997) 297–301.

[53] K.R. Ravi, M. Saravanan, R.M. Pillai, A. Mandal, B.S. Murty, M. Chakraborty, B.C. Pai, Equal channel angular pressing of Al-5 wt% TiB_2 in situ composite, *J. Alloys Compd.* 459 (2008) 239–243.

[54] M.J. Tan, X. Zhang, Powder metal matrix composites: Selection and processing, *Mater. Sci. Eng. A* 244 (1998) 80–85.

[55] R.S. Mishra, Z.Y. Ma, Friction stir welding and processing, *Mater. Sci. Eng. R Rep.* 50 (2005) 1–78.

[56] J.D. Majumdar, B.R. Chandra, A.K. Nath, I. Manna, In situ dispersion of titanium boride on aluminum by laser composite surfacing for improved wear resistance, *Surf. Coatings Technol.* 201 (2006) 1236–1242.

[57] M. Trueba, S.P. Trasatti, Study of Al alloy corrosion in neutral NaCl by the pitting scan technique, *Mater. Chem. Phys.* 121 (2010) 523–533.

[58] T. Ramgopal, P.I. Gouma, G.S. Frankel, Role of grain-boundary precipitates and solute- depleted zone on the intergranular corrosion of aluminum alloy 7150, *Mater. Sci.* 58 (2002) 687–697.

[59] H. Sun, N. Ma, D. Chen, X. Li, H. Wang, Fabrication and analysis of anti-corrosion coatings on in-situ TiB_{2p} reinforced aluminum matrix composite, *Surf. Coatings Technol.* 203 (2008) 329–334.

[60] S. Poria, G. Sutradhar, P. Sahoo, Corrosion behavior of stir-cast Al–TiB_2 metal matrix composites, *Int. J. Mater. Res.* 110 (2019) 148–154.

[61] D.S. Sekhawat, M. Chakraborty, U.K. Chatterjee, Wear and corrosion behaviour of in-situ Al-TiB_2 metal matrix composites, *Mater. Sci. Forum.* 475–479 (2005) 449–452.

[62] K. Surekha, B.S. Murty, K. Prasad Rao, Comparison of corrosion behaviour of friction stir processed and laser melted AA 2219 aluminum alloy, *Mater. Des.* 32 (2011) 4502–4508.

[63] N. Abu-warda, M.D. López, B. González, E. Otero, M.D. Escalera-Rodríguez, S. Cruz, P. Rey, D. Verdera, M. V. Utrilla, Precipitation Hardening and corrosion behavior of friction stir welded A6005-TiB_2 nanocomposite, *Met. Mater. Int.* 27 (2021) 2867–2878.

[64] K.S. Tun, M. Gupta, Enhanced mechanical properties and near unity yield asymmetry in equiatomic high entropy alloy particles reinforced magnesium composites, *J. Alloys Compd.* 810 (2019) 151909–151916.

[65] E. Ananiadis, K.T. Argyris, T.E. Matikas, A.K. Sfikas, A.E. Karantzalis, Microstructure and corrosion performance of aluminium matrix composites reinforced with refractory high-entropy alloy particulates, *Appl. Sci.* 11 (2021) 1300–1311.

[66] T. Lu, S. Scudino, W. Chen, P. Wang, D. Li, M. Mao, Z. Fu, The influence of nanocrystalline $CoNiFeAl_{0.4}Ti_{0.6}Cr_{0.5}$ high-entropy alloy particles addition on microstructure and mechanical properties of SiC_p/7075Al composites, *Mater. Sci. Eng. A* 726 (2018) 126–136.

[67] J.W. Yeh, S.K. Chen, S.J. Lin, J.Y. Gan, T.S. Chin, T.T. Shun, S.Y. Chang, Nanostructured high-entropy alloys with multiple principal elements novel alloy design concepts and outcomes, *Adv. Eng. Mater.* 6 (2004) 299–303.

[68] B. Cantor, I. Chang, P. Knight, A. Vincent, Microstructural development in equiatomic multicomponent alloys, *Mater. Sci. Eng. A* 375–377 (2004) 213–218.

[69] B.Y. Zhang, Y.J. Zhou, J.P. Lin, G.L. Chen, P.K. Liaw, Solid-solution phase formation rules for multi-component Alloys, *Adv. Eng. Mater.* 10 (2008) 534–538.

[70] A. Roy, M. Ghosh, H. Gourkar, P.S. De, Microstructure, indentation and first principles study of AlCuFeMn alloy, *Materialia* 5 (2019) 100206–100216.

[71] S. Guo, C. Ng, J. Lu, C.T. Liu, Effect of valence electron concentration on stability of fcc or bcc phase in high entropy alloys, *J. Appl. Phys.* 109 (2016) 103505–103010.

[72] J. Chen, X. Zhou, W. Wang, B. Liu, Y. Lv, W. Yang, D. Xu, Y. Liu, A review on fundamental of high entropy alloys with promising high-temperature properties, *J. Alloys Compd.* 760 (2018) 15–30.

[73] B. Schuh, F. Mendez-martin, B. Völker, E.P. George, H. Clemens, R. Pippan, A. Hohenwarter, Mechanical properties, microstructure and thermal stability of a nanocrystalline CoCrFeMnNi high-entropy alloy after severe plastic deformation, *Acta Mater.* 96 (2015) 258–268.

[74] N.D. Stepanov, N.Y. Yurchenko, S. V Zherebtsov, M.A. Tikhonovsky, G.A. Salishchev, Aging behavior of the HfNbTaTiZr high entropy alloy, *Mater. Lett.* 211 (2018) 87–90.

[75] C.Y. Hsu, C.C. Juan, W.R. Wang, T.S. Sheu, J.W. Yeh, S.K. Chen, On the superior hot hardness and softening resistance of $AlCoCr_xFeMo_{0.5}Ni$ high-entropy alloys, *Mater. Sci. Eng. A* 528 (2011) 3581–3588.

[76] O.N. Senkov, G.B. Wilks, J.M. Scott, D.B. Miracle, Mechanical properties of $Nb_{25}Mo_{25}Ta_{25}W_{25}$ and $V_{20}Nb_{20}Mo_{20}Ta_{20}W_{20}$ refractory high entropy alloys, *Intermetallics* 19 (2011) 698–706.

[77] M.H. Tsai, C.W. Wang, C.W. Tsai, W.J. Shen, J.W. Yeh, J.Y. Gan, W.W. Wu, Thermal stability and performance of NbSiTaTiZr high-entropy alloy barrier for copper metallization, *J. Electrochem. Soc.* 158 (2011) H1161–H1165.

[78] M.H. Tsai, J.W. Yeh, J.Y. Gan, Diffusion barrier properties of AlMoNbSiTaTiVZr high-entropy alloy layer between copper and silicon, *Thin Solid Films.* 516 (2008) 5527–5530.

[79] K. Tsai, M. Tsai, J. Yeh, Sluggish diffusion in Co–Cr–Fe–Mn–Ni high-entropy alloys, *Acta Mater.* 61 (2013) 4887–4897.

[80] W.H. Liu, Y. Wu, J.Y. He, T.G. Nieh, Z.P. Lu, Grain growth and the Hall–Petch relationship in a high-entropy FeCrNiCoMn alloy, *Scr. Mater.* 68 (2013) 526–529.

[81] B.S. Murty, J. Yeh, S. Ranganathan, *High Entropy Alloys*, Elsevier (2019).

[82] D.B. Miracle, O.N. Senkov, A critical review of high entropy alloys and related concepts, *Acta Mater.* 122 (2017) 448–511.

[83] J. Yeh, S. Chen, J. Gan, S. Lin, T. Chin, Formation of simple crystal structures in Cu–Co–Ni–Cr–Al–Fe–Ti–V alloys with multiprincipal metallic elements, *Metall. Mater. Trans. A* 35 (2010) 2533–2536.

[84] S. Ranganathan, Alloyed pleasures: Multimetallic cocktails, *Curr. Sci.* 85 (2003) 1404–1406.

[85] T. Saito, T. Furuta, J.H. Hwang, S. Kuramoto, K. Nishino, N. Suzuki, R. Chen, A. Yamada, K. Ito, Y. Seno, T. Nonaka, H. Ikehata, N. Nagasako, C. Iwamoto, Y. Ikuhara, T. Sakuma, Multifunctional alloys obtained via a dislocation-free plastic deformation mechanism, *Science* 300 (2003) 464–467.

[86] J.W. Yeh, Recent progress in high-entropy alloys, *Ann. Chim. Sci. Des. Mater.* 31 (2006) 633–648.

[87] F.J. Wang, Y. Zhang, G.L. Chen, H.A. Davies, Cooling rate and size effect on the microstructure and mechanical properties of AlCoCrFeNi high entropy alloy, *J. Eng. Mater. Technol.* 131 (2009) 1–3.

[88] C.M. Lin, H.L. Tsai, Evolution of microstructure, hardness, and corrosion properties of high-entropy $Al_{0.5}CoCrFeNi$ alloy, *Intermetallics* 19 (2011) 288–294.

[89] Y. Wang, S. Ma, X. Chen, J. Shi, Y. Zhang, J. Qiao, Optimizing mechanical properties of AlCoCrFeNiTix high-entropy alloys by tailoring microstructures, *Acta Metall. Sin. (English Lett.)* 26 (2013) 277–284.

[90] V. Soare, D. Mitrica, I. Constantin, V. Badilita, F. Stoiciu, A.J. Popescu, V. Soare, D. Mitrica, I. Constantin, V. Badilita, F. Stoiciu, A.J. Popescu, Influence of remelting on microstructure, hardness and corrosion behaviour of AlCoCrFeNiTi high entropy alloy, *Mater. Sci. Technol.* 31 (2015) 1194–1200.

[91] K.P. Kumar, M.G. Krishna, J.B. Rao, N.R.M.R. Bhargava, Fabrication and characterization of 2024 aluminum-high entropy alloy composites, *J. Alloys Compd.* 640 (2015) 421–427.

[92] D. Poirier, R.A.L. Drew, M.L. Trudeau, R. Gauvin, Fabrication and properties of mechanically milled alumina/aluminum nanocomposites, *Mater. Sci. Eng. A* 527 (2010) 7605–7614.

[93] B.S. Ünlü, Investigation of tribological and mechanical properties Al_2O_3–SiC reinforced Al composites manufactured by casting or P/M method, *Mater. Des.* 29 (2008) 2002–2008.

[94] X.C. Tong, H.S. Fang, Al-TiC Composites in situ – Processed by ingot metallurgy and rapid solidification technology: Part I. Microstructural Evolution, *Metall. Mater. Trans. A* 29 (1998) 875–891.

[95] H. Lu, J. Hu, C. Chen, H. Sun, X. Hu, D. Yang, Characterization of Al_2O_3–Al nano-composite powder prepared by a wet chemical method, *Ceram. Int.* 31 (2005) 481–485.

[96] X. Li, Y. Yang, D. Weiss, Ultrasonic cavitation-based nanomanufacturing of bulk aluminum matrix nanocomposites, *AFS Trans.* 133 (2007) 497–501.

[97] T. Lu, T. He, W. Chen, H. Chen, Y. Liu, B. Wan, Z. Fu, S. Scudino, Effect of solution time on the microstructure, precipitation behavior and mechanical properties of $(Co_{0.5}NiFeCrTi_{0.5}+SiC)_p/7075Al$ hybrid composite, *Mater. Charact.* 170 (2020) 110702–110713.

[98] J. Chen, P. Niu, T. Wei, L. Hao, Y. Liu, X. Wang, Y. Peng, Fabrication and mechanical properties of AlCoNiCrFe high-entropy alloy particle reinforced Cu matrix composites, *J. Alloys Compd.* 649 (2015) 630–634.

[99] Y. Liu, J. Chen, Z. Li, X. Wang, P. Zhang, J. Liu, AlCoCrFeNi high entropy alloy reinforced Cu-based composite with high strength and ductility after hot extrusion, *Vacuum* 184 (2021) 109882–109889.

[100] Y. Liu, J. Chen, Z. Li, X. Wang, X. Fan, J. Liu, Formation of transition layer and its effect on mechanical properties of AlCoCrFeNi high-entropy alloy/Al composites, *J. Alloys Compd.* 780 (2019) 558–564.

[101] T. Lu, W. Chen, Z. Li, T. He, B. Li, R. Li, Z. Fu, S. Scudino, Processing and mechanical properties of fine grained Al matrix composites reinforced with a uniform dispersion of nanocrystalline high-entropy alloy particles, *J. Alloys Compd.* 801 (2019) 473–477.

[102] Z.W. Wang, Y.B. Yuan, R.X. Zheng, K. Ameyama, C.L. Ma, Microstructures and mechanical properties of extruded 2024 aluminum alloy reinforced by $FeNiCrCoAl_3$ particles, *Trans. Nonferrous Met. Soc. China* 24 (2014) 2366–2373.

[103] H. Wang, W. Ren, G. Li, H. Wen, C. Wang, J. Chen, Y. Zhao, G. Chen, X. Kai, Microstructure and properties of $FeCoNi_{1.5}CrCu$/2024Al composites prepared by microwave sintering, *Mater. Sci. Eng. A* 801 (2021) 140406–140416.

[104] M.J. Shen, X.J. Wang, C.D. Li, M.F. Zhang, X.S. Hu, M.Y. Zheng, K. Wu, Effect of bimodal size SiC particulates on microstructure and mechanical properties of AZ31B magnesium matrix composites, *Mater. Des.* 52 (2013) 1011–1017.

[105] E.A. Ball, P.B. Prangnell, Tensile-compressive yield asymmetries in high strength wrought magnesium alloys, *Scr. Metall. Mater.* 31 (1994) 111–116.

[106] X.J. Wang, X.S. Hu, K.B. Nie, K.K. Deng, K. Wu, M.Y. Zheng, Dynamic recrystallization behavior of particle reinforced Mg matrix composites fabricated by stir casting, *Mater. Sci. Eng. A* 545 (2012) 38–43.

[107] B. Inem, Dynamic recrystallization in a thermomechanically processed metal matrix composite, *Mater. Sci. Eng. A* 197 (1995) 91–95.

[108] Z. Tan, L. Wang, Y. Xue, P. Zhang, T. Cao, X. Cheng, High-entropy alloy particle reinforced Al-based amorphous alloy composite with ultrahigh strength prepared by spark plasma sintering, *Mater. Des.* 109 (2016) 219–226.

[109] Z. Yuan, W. Tian, F. Li, Q. Fu, Y. Hu, X. Wang, Microstructure and properties of high-entropy alloy reinforced aluminum matrix composites by spark plasma sintering, *J. Alloys Compd.* 806 (2019) 901–908.

[110] S.Q. Zhou, X.Y. Liu, Y. Xu, Microstructure, mechanical properties and tribological properties of NiAl Composites reinforced by CrMnFeCoNi High-entropy alloy, *Materials* 11 (2018) 1850–1862.

[111] Z. Fu, R. Koc, Processing and characterization of TiB_2-TiNiFeCrCoAl high-entropy alloy composite, *J. Am. Ceram. Soc.* 100 (2017) 2803–2813.

[112] Z. Fu, R. Koc, Ultrafine TiB_2-TiNiFeCrCoAl high-entropy alloy composite with enhanced mechanical properties, *Mater. Sci. Eng. A* 702 (2017) 184–188.

[113] R.K. Prabakaran, A.N. Sait, V. Senthilkumar, Synthesis and characterization of high entropy alloy (CrMnFeNiCu) reinforced AA6061 aluminum matrix composite, *Mech. Mech. Eng.* 21 (2017) 415–424.

[114] N. Wang, B. Wu, W. Wu, J. Li, C. Ge, Y. Dong, L. Zhang, Y. Wang, Microstructure and properties of aluminum-high entropy alloy composites fabricated by mechanical alloying and spark plasma sintering, *Mater. Today Commun.* 25 (2020) 1–8.

[115] P.V. Satyanarayana, R. Sokkalingam, P.K. Jena, K. Sivaprasad, K.G. Prashanth, Tungsten matrix composite reinforced with CoCrFeMnNi high-entropy alloy: Impact of processing routes on microstructure and mechanical properties, *Metals* 9 (2019) 1–12.

[116] G.M. Karthik, S. Panikar, G.D.J. Ram, R.S. Kottada, Additive manufacturing of an aluminum matrix composite reinforced with nanocrystalline high-entropy alloy particles, *Mater. Sci. Eng. A* 679 (2017) 193–203.

[117] J. Gao, X. Wang, S. Zhang, L. Yu, J. Zhang, Y. Shen, Producing of FeCoNiCrAl high-entropy alloy reinforced Al composites via friction stir processing technology, *Int. J. Adv. Manuf. Technol.* 110 (2020) 569–580.

[118] T. Ram Prabhu, Y. Chodancar, M. Arivarasu, N. Arivazhagan, R.K. Mishra, High entropy based composites – An overview, *Mater. Sci. Forum* 969 (2019) 98–103.

[119] N.R.M.R. Bhargava, I. Samajdar, S. Ranganathan, M.K. Surappa, Role of cold work and SiC reinforcements on the β'/β precipitation in Al-10 pct Mg alloy, *Metall. Mater. Trans. A Phys. Metall. Mater. Sci.* 29 (1998) 2835–2842.

[120] S. Scudino, G. Liu, K.G. Prashanth, B. Bartusch, K.B. Surreddi, B.S. Murty, J. Eckert, Mechanical properties of Al-based metal matrix composites reinforced with Zr-based glassy particles produced by powder metallurgy, *Acta Mater.* 57 (2009) 2029–2039.

[121] H.S. Kim, On the rule of mixtures for the hardness of particle reinforced composites, *Mater. Sci. Eng. A* 289 (2000) 30–33.

[122] S.G. Ma, Y. Zhang, Effect of Nb addition on the microstructure and properties of AlCoCrFeNi high-entropy alloy, *Mater. Sci. Eng. A* 532 (2012) 480–486.

[123] Ş. Karabulut, U. Gökmen, H. Çinici, Study on the mechanical and drilling properties of AA7039 composites reinforced with $Al_2O_3/B_4C/SiC$ particles, *Compos. Part B* 93 (2016) 43–55.

[124] L.J. Zhang, F. Qiu, J.G. Wang, H.Y. Wang, Q.C. Jiang, Microstructures and mechanical properties of the Al2014 composites reinforced with bimodal sized SiC particles, *Mater. Sci. Eng. A* 637 (2015) 70–74.

[125] H.R. Ezatpour, M. Torabi-Parizi, S.A. Sajjadi, Microstructure and mechanical properties of extruded Al/Al_2O_3 composites fabricated by stir-casting process, *Trans. Nonferrous Met. Soc. China* 23 (2013) 1262–1268.

13 Introduction to Functionally Graded Materials

R.G. Abaszade, E.A. Khanmamadova, and F.G. Aliyev

13.1 INTRODUCTION

From 1991 on, carbon nanotubes (CNTs) have attracted enormous scientific and commercial interest within a multidisciplinary, interdisciplinary basis embracing chemistry, physics, biology, medicine, and engineering. CNTs are cylinder-shaped allotropic modifications of carbon, visualized as seamless rolls of graphitic sheets, closed at the ends by which half of a molecule resembles a hollow fullerene. They have an extremely high aspect ratio and generally have a diameter within the range of 1–100 nm though it may extend to several centimeters in length. They can be mainly classified into two forms: single-walled nanotubes (SWNTs) and multi-walled nanotubes (MWNTs). MWNTs contain a number of cylinders which are placed in a layer. However, in the SWNT, it has only a simple single-layered geometry with a diameter running between 0.4 and 3 nm. Depending on the orientation of the hexagon rings along the tubular surface, CNTs can either be metallic or semiconducting [1–7]. Successful surface functionalization lies at the heart of many applications of CNTs. CNTs are cylindrical nanostructures with special electrical, mechanical, and thermal properties attributed to the small nanoscale dimensions and carbon structure. This is directly related to one of the paramount advantages of CNTs that different chemical groups can be attached to their surfaces. The functionalization of CNTs will, therefore, enable their properties to be manipulated so that they are suitable for many uses in science, engineering, the environment, and even in the biomedical field. In the field of cancer therapy, CNTs have opened up vast possibilities for drug delivery and other therapeutic applications. Their large surface area and the possibility of modification of their surface render them an ideal carrier for therapeutic molecules, including chemotherapy drugs, peptides, and genes, to their specific target sites in a body. It makes such targeted drug delivery potentially capable of enhancing the efficiency while reducing the side effects of cancer treatment. The review provides a systematic analysis regarding the published scientific works during the past decade pertaining to the synthesis, properties, and biomedical applications of CNTs, considering that problems regarding biocompatibility, possible toxicity, and regulatory considerations are still to be solved with the aim of guaranteeing the biomedical potential use of these systems for delivery. We also present some relevant clinical trials assessing CNT-based formulations and discuss the risk of bias in randomized studies under Cochrane assessment [6–8].

Global energy demand is increasing way faster than energy produced by renewable sources. For that reason, nearly all the needed energy is obtained from fossil fuels. On the other hand, permanent use of fossil fuels leads to global warming, environment pollution, and energy resource depletion, which has become one of the biggest problems we have ever had. Hydrogen is considered to be a perfect replacement for fossil fuel due to its promising ability to be used continuously as renewable clean energy in the near future. The existence of distinctive features like zero CO_2 emission, high

DOI: 10.1201/9781003664611-14

efficiency, and various use abilities make hydrogen very promising as an energy carrier. However, the future of a hydrogen-based economy is heavily handicapped by finding a cheap, safe, and practical method to store hydrogen at ambient temperature. Generally, two common methods of liquid hydrogen storage and high-pressure gas have not been able to meet this demand because of high prices and safety problems [9–17]. Nanotechnology is a potential game-changer in this respect. Nanotechnology is a niche for the emerging branches of science with several applications in industrial, medicinal, and energy uses. During the past years, adsorption-based hydrogen storage in nanostructured materials has been attracting much attention. Efforts are made to assess porous material candidates for hydrogen storage and purification. Those materials have high capacity but usually under extreme operational conditions [18-30]. Carbon-based materials have also been studied for hydrogen storage; some of the examples include activated carbons, graphene sheets, and CNTs [31–38]. Among the many nanomaterials, SWCNT (Single-Walled Carbon Nanotube) has taken much attention because of the significant benefits in surface area, light weight, pore structure, and the capability of adsorption at ambient temperature [39–46].

13.2 FUNCTIONAL MATERIALS

13.2.1 Manufacturing Methods

13.2.1.1 Arc Method

As shown in Figure 13.1, the arc discharge technique generally uses two high-pure graphite electrodes. One of them, the anode, is either pure graphite or contains metals. In the latter case, metals are mixed with the graphite powder and introduced into a hole made at the anode center. These electrodes are suddenly brought into contact and an arc is struck. The synthesis takes place at low pressure—30–130 Torr or even 500 Torr—in an atmosphere comprising inert and/or reactant gas. The inter-electrode distance is decreased until a current (of 50–150 A) passes. The temperature in the inter-electrode zone is high that carbon sublimes from the positive electrode, which is consumed. The anode position is adjusted so as to maintain a constant gap between the anode and the cathode. Plasma is formed between the electrodes. The voltage controls the spacing of the electrodes. It stabilizes plasma for long reaction times from 25 to 40 V. The reaction time varies from 30–60 s to 2–10 min. Products of different nature are formed in different parts of the reactor: (1) large amounts of rubbery soot on the walls of the reactor, (2) web-like structures between the cathode and the chamber walls, (3) grey hard deposit at the end of the cathode, and (4) spongy collaret around the cathodic deposit. The metals normally used were Fe, Ni, Co, Mo, Y either alone or in combination. Better results were obtained using bimetallic catalysts. Amorphous carbon, encapsulated metal nanoparticles, polyhedral carbon are also present in the product [47,48]. When no catalyst is used, only the soot and the deposit are formed.

The soot contains fullerenes while MWNTs together with graphite carbon nanoparticles are found in the carbon deposit. The inner diameter of the MWNTs varies from 1 to 3 nm, the outer one is in the range of 2–25 nm, the tube length does not exceed 1 μm and the tubes have closed tips. When metal catalysts are co-evaporated with carbon in the DC arc discharge, the core of the deposit contains MWNTs, metal filled MWNTs, graphitic carbon nanoparticles, metal filled graphite carbon nanoparticles, and metal nanoparticles, while the powder-like or spongy soot contains MWNTs, Functionalized Multi-Walled Carbon Nanotube (FMWNTs), and SWNTs. The SWNTs have closed tips and are free of any catalyst material; they are also either isolated or in bundles. Diameters of most the SWNTs fall within 1.1–1.4 nm, while their lengths run into several microns. The collarette is formed mainly of SWNTs (80%), isolated or in bundles; however, it forms only in the presence of certain catalysts. The physical and chemical factors affecting the arc discharge process are dependent on the concentration of carbon vapor, dispersion of the carbon vapor in inert gas, temperature within the reactor, composition of the catalyst, addition of promoters, and presence of hydrogen. These

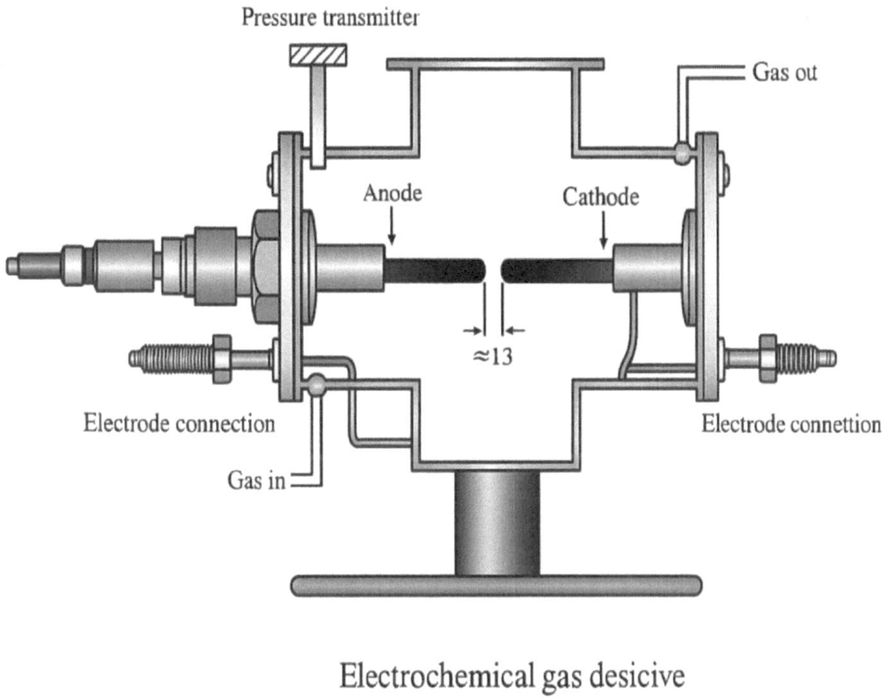

FIGURE 13.1 Schematic representation of arc discharge apparatus.

factors are going to influence them in nucleation and growth in nanotubes, inner and outer diameters in nanotubes, and type of nanotubes (SWNTs, MWNTs). In case of pure hydrogen, it was observed that the amount of the carbon nanoparticles decreases during the reaction.

13.2.1.2 Laser Ablation Method

A pulsed laser-ablation process was developed for the production of single-wall CNTs [49,50]. The method they used is described below. In 1996, a laser ablation method was developed by Smalley et al. for the mass production of SWNTs [51]. Other improvements were made and using double beam laser [52,53]. Nanotubes produced by laser ablation have higher purity (up to about 90% pure) and their structure is better graphitized than those produced in the arc process. The drawback of this method is the small carbon deposit. The laser ablation technique favors the growth of SWNTs. MWNTs are generated only when special reaction conditions are employed. Figure 13.2 represents schematic classical setup of this method. The oven laser-vaporization apparatus used was the same as the one used to produce fullerenes, metallofullerenes, and MWNTs [54]. The Laser beam, of wavelength 532 nm, was focused onto a metal–graphite composite target which was placed in a high-temperature furnace at 1200°C. The laser beam scans across the target surface under computer control to maintain a smooth face for vaporization. The soot produced by the laser vaporization was entrained by the flowing Ar gas out of the high-temperature zone and deposited onto a water-cooled copper collector positioned downstream, just outside the furnace. The targets used were composite rods mixed uniformly in a three-step process: (1) a paste made from the mixture of high-purity metal or metal-oxide with graphite powder and carbon cement at room temperature was dropped in a mold; (2) the mold was kept in a hydraulic press equipped with heating plates and baked at 130°C for 4–5 h under constant pressure; (3) the baked rod was cured for 8 h at 810°C under Ar flow. Fresh targets were heated at 1200°C under flowing Ar for 12 h. Subsequent runs with the same target proceeded after two additional hours heating at 1200°C. Metals used were Co, Cu, Nb, Ni, Pt, Co/Ni, Co/Pt,

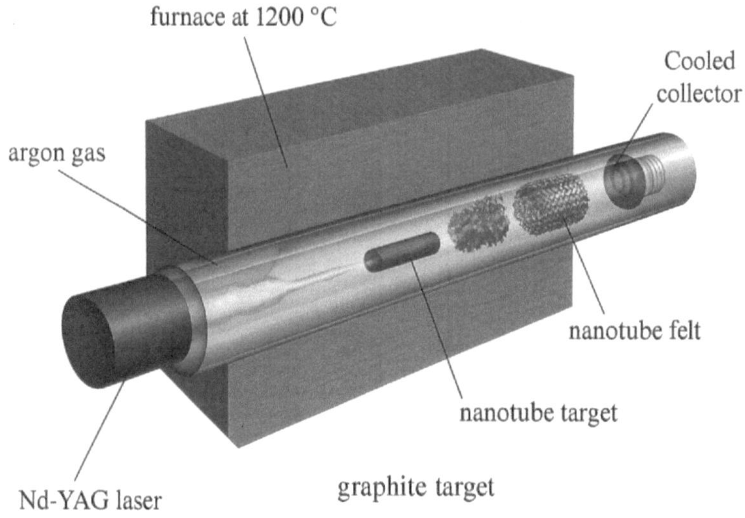

FIGURE 13.2 Schematic synthesis apparatus.

Co/Cu, Ni/Pt. Of the configurations tested, it is found that the composite ablation target consisting of 1 at% each of Ni and Co, uniformly mixed with graphite, results in the best purity and yield of SWNTs. The major impurities in the product are amorphous carbon, graphite particles, catalysts, and fullerenes. Other impurities that have been detected are silicon and various hydrocarbons, which are probably due to impurities introduced from unknown sources [50–55].

13.2.1.3 Chemical Vapor Deposition Method

While the arc discharge method is able to generate large quantities of unpurified nanotubes, a significant effort is currently invested in production processes offering more controllable routes to the nanotube synthesis. One class of processes that apparently offers the best chance for obtaining a controllable process for the selective production of nanotubes with predefined properties is chemical vapor deposition (CVD). The process of carbon filaments formation by catalytic decomposition of carbon-containing gas on metal surfaces is very ancient [56]. Nevertheless, until the success of [57], there was no evidence that this technique could be used for CNTs synthesis. Today, Catalytic Chemical Vapor Deposition (CCVD) still remains the only economically viable technique for mass CNT production and integration into various devices [58], which is proven by NANOCYL large-scale production of quite pure CNTs founded by one of the authors [59]. In principle, catalytic decomposition of hydrocarbon or carbon monoxide feedstock using supported transition metal catalysts, under either atmospheric or lowered pressures, constitutes the process of CVD. In general, the experiment is carried out with the aid of atmospheric pressure in a flow furnace. There are two types of furnace modality: one is horizontal and the other is vertical, as shown in Figure 13.3. Horizontal furnaces find the most popular application. Here, the catalyst is placed in a ceramic or quartz boat which is introduced into a quartz tube. The reaction mixture containing a source of hydrocarbon along with an inert gas is passed over the catalyst bed at temperatures ranging from 500°C to 1100°C. The system is cooled down to room temperature. The vertical furnace configuration is normally used for the continuous mass production of carbon fibers/nanotubes. The resultant filaments grow during flight and are collected at the bottom of the chamber where the catalyst and carbon source are injected at the top of the furnace. To that effect, ultrafine metal catalyst particles are either introduced in the reactor directly or formed in-situ using precursors such as metallocene's. A modification of the vertical furnace is the fluidized bed reactor. The supported

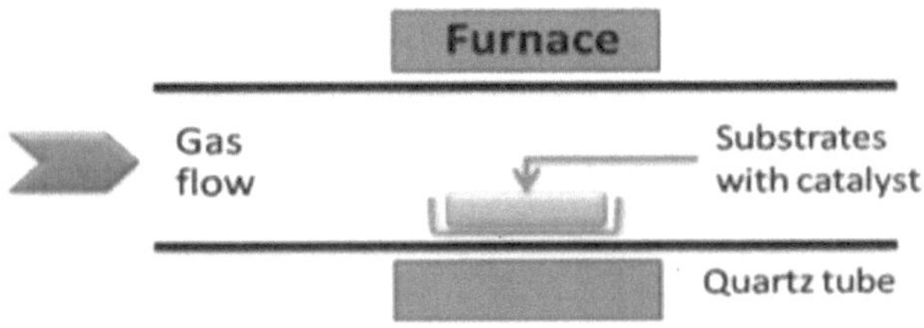

FIGURE 13.3 Schematic demonstration of CVD method.

catalysts are usually placed in the center of the furnace and an upward flow of carbon feedstock gases is used. During fluidization, the supported catalysts remain much longer in the furnace than in vertical floating technique [55]. The general mechanism of nanotube growth in the CVD process involves catalytic dissociation of the hydrocarbon molecule by the transition metal and saturation of carbon atoms in the metal nanoparticle. The precipitation of carbon from the metal particle leads to the formation of tubular carbon solids in sp^2 structure. Indeed, the CNTs' characteristics produced depend on working conditions that include temperature and operation pressure, type/volume/concentration of hydrocarbon, nature/size/pretreatment of metallic catalyst, nature of the support, and reaction time. These active particles on the surface of the catalyst may thus be varied to control the nanotube diameter. Certainly, the length of such tubes depends on the reaction time; even 60 mm long tubes can be produced [58,59].

13.2.1.4 Flame Synthesis Method

Another approach for the preparation of CNTs is by the flame synthesis method. In this technique, hydrocarbon flames are utilized. These flames assist in the nucleation and growth of CNTs. Gases such as CO, CH_4, C_2H_2, C_2H_4, and C_2H_6, existing in the post-flame region, are good sources of carbon. The reaction is exothermic and the chemical energy released in the form of heat in the flame supports endothermic carbon deposition reactions. Catalysts are also essential to provide reaction sites for the deposition of solid black carbon. In many respects, growth of the CNTs occurs in a similar way as for the CVD process. If provided with a proper catalyst and the flame and reaction conditions are proper, a large quantity of the CNTs can be obtained commercially [60].

13.2.1.5 Vapor-Phase Growth

This is a comparatively new method and a variant of CVD. The key difference is that the CNTs grow directly from the reaction gas and catalytic metal in the chamber without a substrate. There are two furnaces inside the reaction chamber. The catalyst used here is ferrocene. The vaporization of catalytic carbon at a comparatively low temperature is kept constant in the first furnace. Here, fine catalytic particles are formed and, upon reaching the second furnace, the decomposed carbons are absorbed in this catalyst through diffusion, where they turn into CNTs.

13.3 RECENT TRENDS IN THE SYNTHESIS OF CNTS

MWNTs were also prepared using a recent method of nebulized spray pyrolysis. In this method, the nebulized spray is the key factor and is obtained by the use of a special ultrasonic atomizer. MWNTs have been produced with this technique in aligned bundles with quite uniform diameters. Using an ultrasonic nebulizer, ethanol is sprayed together with ferrocene (catalyst) and ethanol (as a solvent and carbon source) into a tubular furnace with a fixed temperature of 800°C under an argon flow of 1 l/min. Ethanol will be used as a solvent as well as a source of carbon because it is non-polluting, cheap, produces harmless byproducts, such as CO; it is also easily handled. High growth of MWNTs can be realized on a surface. The advantage of using a nebulized spray is the ease of scaling into an industrial-scale process, as the reactants are fed into the furnace continuously [61,62].

13.3.1 Functionalization of Materials

13.3.1.1 Covalent Functionalization

Based on the number of graphene walls, CNTs can be classified as single-walled or multi-walled. The latter could be divided further into multiple concentric tubes rolled in a spiral manner itself, while the former one could take armchair, zigzag, or chiral configurations.

In this kind of functionalization, the desired groups are permanently attached onto the sidewall or tips of the CNTs in an irreversible way. Various chemical groups, such as carboxylic, p-aminobenzoic acid, fluorine, added dichloro-carbon groups, are attached to the surface or at the ends of the CNTs [85–88]. The advantages of the chemical functionalization method are that it links covalently with polymeric materials and the dispersion in different solvents is good. The main disadvantage of the chemical function is the formation of defects on CNTs. A schematic representation of the most common types of carbon nanotube functionalization is shown (Figure 13.4).

13.3.1.2 Non-Covalent Functionalization

This type of functionalization is based on Van der Waal's forces. Beside these forces, hydrophobic or Pi–Pi interactions have also been reported. Non-covalent functionalization is important and interesting because it does not damage or modify the CNTs structure to a great extent. In solution beside Van der Waal's forces some other forces are also involved, which render the efficiency of this type of functionalization. Some typical examples of such functionalization are the use of surfactant, especially with SWNTs, CNT wrapping, and non-covalent protein interactions [63,64]. The non-covalent bonding preserves the sp^2 hybridization and π-conjugated structure of the CNTs and also its optical and electronic properties. The chemical interactions are weaker, such as CH–π, electrostatic, hydrogen bonding, ionic bonds, π–π, and Van der Waals, and may be reversible, all of which would

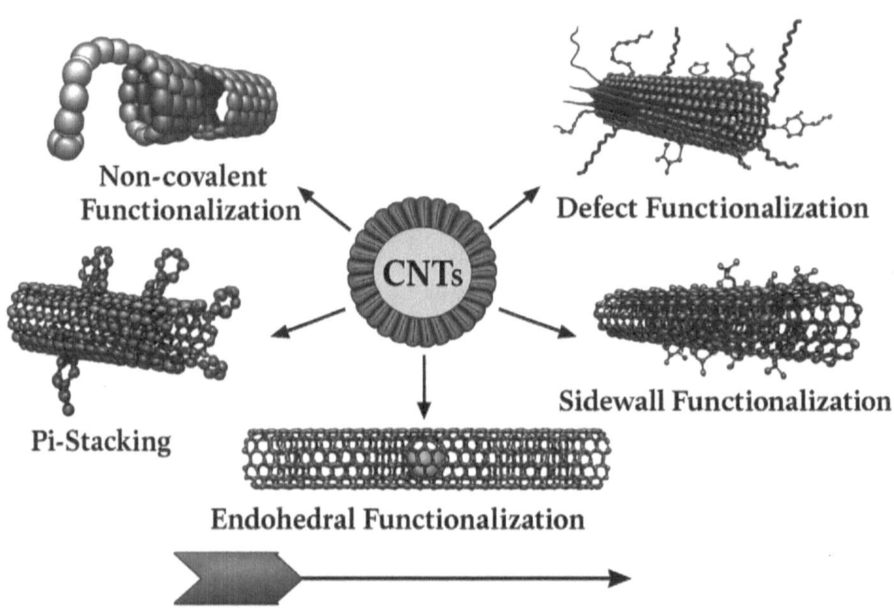

FIGURE 13.4 Most common types of functionalization. CNT, carbon nanotube. (Reprinted from Hirsch [84] with permission from Willey-VCH Verlag GmbH.)

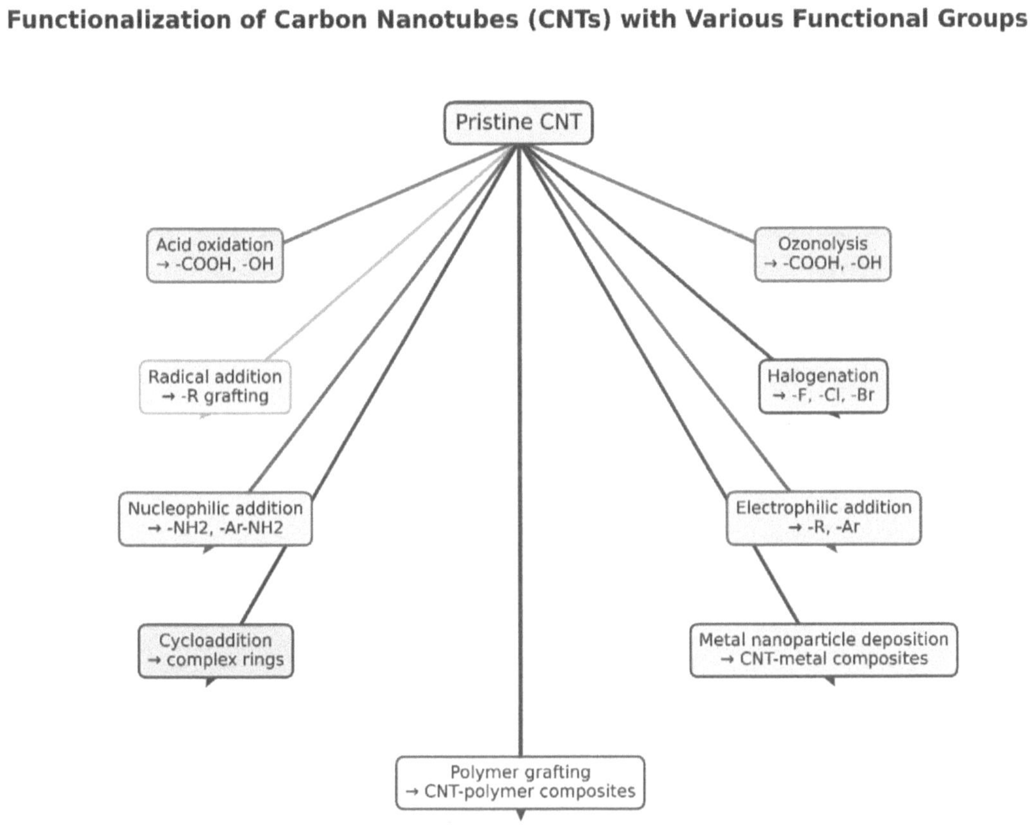

FIGURE 13.5 Functionalization of carbon nanotubes.

help in maintaining organic and inorganic molecules adsorbed on the CNT surface. Normally, in a non-covalent functionalization process, polymer chains or surfactants surround the structure of the CNT (Figures 13.5 and 13.6).

13.3.2 Doping with Metals

An amount of 0.005 g of Gd_2S_3 and 0.95 g of functionalized CNTs were mixed together in a Petri dish and put into a 150-ml flask. An amount of 50 ml of distilled water was added to this solation. The mixture was stirred for 6 h at 750°C in a magnetic stirrer equipped with a reflux condenser and a thermometer. New compound was filtered, separated, and dried in vacuum [65–69]. In this chapter, results of first-principles simulations are presented for structural, electronic, and magnetic properties of Al, Fe, Mn, and Ti atom-doped CNTs, interacting with vitamin C molecules based on density functional theory. The charge transfers are obtained from the vitamins into the tubes for adsorption and substitutional doping cases. It estimated the highest binding energies of about 1.20 and 3.26 eV for the Al substitutional and Ti-adsorbed cases of the vitamin C molecule. Indeed, depending on the kind of doping, one may change the spin polarizations and the conductance character of the systems, which could be relevant to improve molecule adsorption in carbon nanostructures.

Concerns are rising over the long-lived radioactive isotope iodine-129, associated with nuclear fuel cycle processes, because of its inability to efficiently separate it from other nuclear wastes, posing projected long-term risks to human health and the environment. In this work, we have investigated the reaction of iodide in alkaline solutions with silver-doped CNTs as a means of addressing

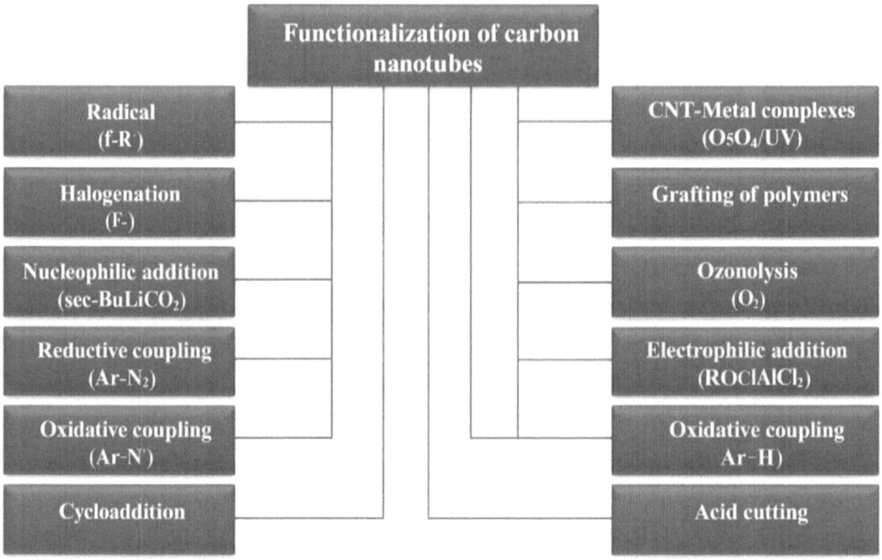

FIGURE 13.6 Schematic diagrams functionnalization of carbon nanotubes.

concerns associated with this radioisotope. In this research, an in-situ CNT doping method in which silver is added to the CNT during the CVD process, and a chemical filling process in which silver is added to the CNT by solution contact followed by reduction under heat, is utilized for synthesizing Ag-CNTs. On the other hand, the iodide adsorption behavior of the Ag-CNTs tended to be significantly controlled by the content of silver in the CNTs.

13.4 APPLICATIONS OF FUNCTIONAL MATERIALS

13.4.1 Applications of Carbon Nanotubes

Nanotechnology is the latest and one of the most developed technologies that offers a lot of advantages and benefits for new materials with improved properties. Nanotechnology has an application in various fields, such as nanomedicine, energy, the environment, and in sensors [68]. CNTs were initially functionalized. Obtained by the arc discharge method, 0.5 g of high-purity CNTs were mixed with 250 ml of 8 M sulfuric acid in a 500 mm beaker. Sulfuric acid was mixed continuously with CNTs in a magnetic mixer equipped with a heater [69,70].

13.4.2 CNTs as Fillers

Among the fields that have been developed in nanotechnology, one of the biggest involves using CNTs as fillers in different materials to form nanocomposites. Almost all researchers dealing with nanocomposites have attempted to use CNTs as a filler. The basic idea behind incorporating CNTs into different polymeric and other materials is generally the improvement of their different properties. This improves the mechanical, electrical, and thermal properties to near the ideal range. Garcia-Gutierrez et al. [71] prepared melt-processed injected molded polybutylene terephthalate, Polybutylene terephthalate/single-walled carbon nanotubes (PBT/SWCNTs) nanocomposites, and presented how SWNTs affect the polymer orientation during shearing and template the crystallization behavior of PBT. In [72] reported that the tensile modulus and yield strength increased with the addition of SWNT loading in a SWNTs/polyimide nanocomposite. They showed that with good dispersion, the mechanical properties of polyimide are enhanced, and [73] fabricated poly (adipic acid-hexamethylene diamine) (PA66) and F-MWNTs. In the fabrication process, amino MWNTs were used and the dispersion of A-MWNTs was increased in formic acid (HCOOH) after PA66 functionalization. The PA66 was grafted onto the surface of A-MWNTs. Conclusions were made that the chain length of the PA decreased with the addition of A-MWNTs, the thermal decomposition temperature of the PA-MWNTs was higher than that of pure PA66 composite, and with the addition of A-MWNTs, the storage modulus improved.

13.4.3 In Electronic Devices as Field-Emission Sources

CNTs can act as field-emission sources in electric devices. This is realized by applying a potential between the surface of CNTs and the anode. The curvature in the CNTs available in the form of pentagons [74,75] or the presence of oxidized tips [76,77] can easily emit electrons from their tips. This makes CNTs useful for the fabrication of various electronic devices, such as flat-panel displays, and intense light sources that can be used as bright lamps or X-ray sources. CNTs themselves have been reported to be good emitters, but a nanocomposite of CNTs is a good electron-emission surface with the added advantage of being vacuum-stable. There are many advantages in using CNTs as electron emitters (Figure 13.7). These include long lifetimes of the components, stable field emission over prolonged time periods, low emission threshold potentials, the absence of need for an ultra-high vacuum, and high current densities. Large current densities as large as 4 A/cm^2 have reportedly been achieved [77].

13.4.4 Batteries (Lithium Ions Batteries)

Lithium (Li) is an element that could be very useful due to its lowest electronegativity and since electrons are easily donated from Li. Therefore, it is the best candidate to make lightweight and efficient batteries. However, in spite of the above advantages, the high reactivity of Li reduces the possibility of applicability because this metal loses its efficiency. This problem can be overcome by combining an application of CNTs with Li by intercalating Li ions within CNTs. This will enable Li^+ ions to migrate from the graphitic anode to the cathode. A separating medium is required between the anode and the cathode, normally polyolefin. Theoretically, storage capacity is expected to be 372 mAh/g when Li intercalates with CNTs. In these batteries, charge and discharge phenomena are controlled by Li^+ intercalation and de intercalation rates [78–80].

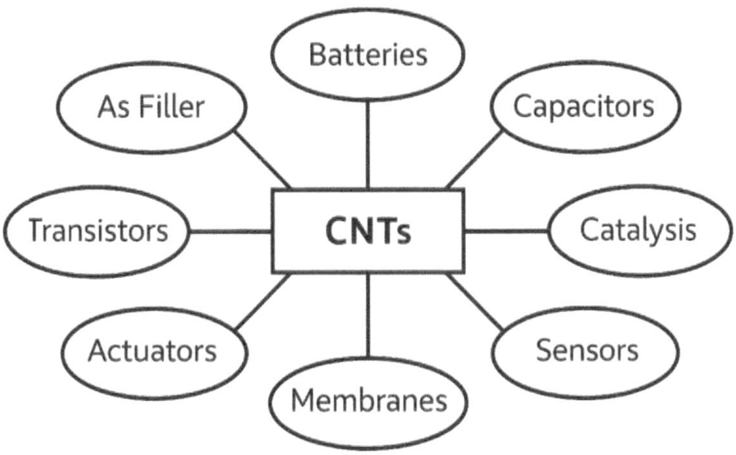

FIGURE 13.7 Applications of CNTs.

13.4.5 CNTs in Supercapacitors and Actuators

Due to the fact that CNTs have a large surface area and high electrical conductivity, their electrochemical devices would be quite good. It was first demonstrated by Niu et al. [83] and Ma et al. [84] that very high specific capacitances can be attained in individual cells with sheet electrodes of pyrolytically grown MWNTs in devices containing 38 wt% of H_2SO_4 as the electrolyte. The cells could achieve power densities that went above 8,000 W/kg. Recently, similar research was conducted by Frackowiak and Beguin [82] and Ma et al. [84], which revealed that MWNT–polypyrrole composites can attain specific capacitances of 163 F/g. CNT supercapacitors find their application where high power and higher storage capacities are required in devices. At energy densities of 7 Wh/kg, power densities of up to 20 kW/kg are reached [81–85].

13.4.6 Gas and Hydrogen Storage

The hollow cylindrical nature of CNTs enables them to act as very efficient containers for gases and metals. The core of the CNTs has been found to take up a lot of vital chemical species, such as metals, metal carbides, and oxides, by: (1) chemical treatments; (2) arc discharge methods; (3) solid-state reactions; and (4) electrochemical techniques. However, the major problems associated with these methods are based on the fact that CNTs cannot store gaseous materials. It has been recently reported that H_2 and Ar can be stored within SWNTs and MWNTs, respectively [86–88]. Terrones et al. [89] showed experimentally that gaseous nitrogen can be inculcated within MWNTs by a single-step method.

13.4.7 Electronic Devices Using CNTs

CNTs become very important nanomaterials that can be used in electronic devices, aiming to improve the properties of the devices. Tans et al. [90] fabricated a three-terminal switch-able device based on SWNTs. In the study, the SWNT molecule was semiconducting and connected to metal nanoelectrodes. The low capacitance level performance was excellent. The major problem associated with nanotubes is the difficulty in manipulating them; thus, another method was devised for controlling the length and the electronic properties of individual nanotubes by Scanning Tunneling Microscope (STM) nanostructuring.

13.4.8 Sensors

Sensors are another type of detecting devices which are being used extensively in various fields nowadays. The efficiency of biosensors and molecular sensors can be enhanced by attaching CNTs onto them. Wong et al. used the chemical force microscopy techniques to demonstrate the possibility of sensing functional chemical groups attached onto the ends of CNTs. Thus, it is possible to construct various types of sensors containing nanotube composite pellets, which are very sensitive to gases and which can be used to monitor leaks in chemical plants. Collins et al. [91–93] reported after working in a similar field that SWNTs are extremely sensitive to air and vacuum conditions by noting large variations in the electrical resistance levels of their SWNT samples. They further added that MWNTs can act as effective sensors for NH_3, H_2O, CO_2, and CO. In this chapter, we have used computational screening to study systematically the application of transition metal-doped CNTs for chemical gas sensing. It computes the binding energy and conductance variation upon adsorption on a metal atom occupying a vacancy in a (6,6) CNT for a set of target molecules relevant to CO, NH_3, H_2S, and the main constituents of air, like N_2, O_2, and H_2O. Using these descriptors, we are able to highlight the most promising dopant candidates for the detection of a given target molecule. Knowing the fractional coverage of the metal sites in thermal equilibrium with air, we can estimate the change in the nanotube resistance per doping site as a function of target molecule concentration assuming charge transport in the diffusive regime. Our analysis singles out Ni-doped nanotubes as promising candidates for CO sensors working under ordinary atmospheric conditions.

13.4.9 Scanning Probe Tips

Scanning probe tips are very important because using these substrates, images can be achieved at better resolutions. If a scanning probe with an attached MWNT is used instead of a regular probe, a very good image resolution is obtained compared to previous results. Currently, Seiko Instruments and Daiken Chemical Company are making an effort for the commercialization of these probes. Such chemically modified nanotube tips can be used as receptors for recognition of specific chemical and/or biological groups [91–97].

13.4.10 Medical Applications of CNTs

The enhanced profits accrued with the medical technology associated with gene therapy, cancer treatments, and innovative new answers for life-threatening diseases have made the science of nanomedicine the fastest-growing field. The unique properties and characteristics of CNTs facilitate scientists in developing new areas in nanomedicine. SWNTs and MWNTs have already shown their potential to act as a safer and effective alternative to the previous drug delivery methods. They can diffuse across membranes and transport therapeutic drugs, vaccines, and nucleic acids deep into the cell to their substrate targets. They act as ideal, nontoxic vehicles in which, in some cases, the solubility of the drug increases, making them more efficient and safer. On the whole, the recent investigations on CNTs have opened a very bright future for them in medicine.

13.4.11 Functionalized Carbon Nanostructures for Flexible Electronics

FES is a young, developing field of science and engineering. This technology becomes very promising due to the flexibility, which is specific, novel, and highly demanded in the existing electronic industry, offering a broad variety of flexible applications in many sectors. Recently, research and focus have been directed to carbon nanostructures and their functionalized derivatives for use as building blocks in the production of FES products. In this respect, optimum selection from CNTs, graphene, and CNFs may differ. For instance, CNTs exhibit high electrical conductivity and mechanical flexibility; thus, they are very suitable for interconnects and transistors in flexible electronic systems. Graphene is used in most cases in transparent conductive electrodes and sensors due to its high carrier mobility and transparency. The high mechanical strength of CNFs renders them appropriate for use as reinforcement in flexible substrates. The optimum quantity of required FCN will be decided by the balance required between conductivity, flexibility, and durability for the specific application in hand, considering other factors such as ease of production and cost involved. An overview of the application of FCN in flexible electronics is presented in detail in this chapter.

13.4.12 2D Materials for Quantum Information Science and Spintronics

New materials and processing technologies that allow ultrahigh integration of solid-state electronic switching devices have propelled the evolution of digital computers from large machines to a portable system. Along this conventional scaling pathway, the constituent nanomaterials—SiO_2 gate dielectrics, poly-Si floating gates, and Co–Cr–Pt ferromagnetic alloys—increasingly possess properties dominated by quantum physics as this pathway has approached atomic-scale dimensions. Parallel to that the quantum information science has been evolving as the new alternative to transistor technology and foresees to bring new paradigms in computation, communication, and sensing. More specifically, it is in the field of two-dimensional (2D) materials that the convergence of quantum material properties and prototype quantum devices can be best expounded. In this review, we describe the quantum properties and prospects for the use of 2D materials as solid-state platforms for quantum-dot qubits, single-photon emitters, superconducting qubits, and topological quantum computing elements. By focusing on the interplay between quantum physics and materials science, we identify key opportunities and challenges for the use of 2D materials in quantum information science. The discovery of 2D magnetism within atomically thin structures obtained from layered magnetic crystals opened a new realm to explore magnetic heterostructures. This opens a new frontier in the study of unique physical properties and exquisite phenomena at the nanometer and molecular/atomic scales. Physically engineering the 2D interfaces and interlayer interactions can give rise to extraordinary exchange dynamics, thereby extending high-performance and high-density magnetic memory applications to future neuromorphic and quantum computing. This review highlights recent advances in magnetic 2D materials, explains the mechanisms of 2D interfaces, and points out the development of 2D devices for spintronics and quantum information processing. We consider in more detail 2D magnetic heterostructures with topological properties, which promise a robust and low-error information system. Finally, we would like to draw some conclusions on the prospects of 2D heterostructures for future electronics, regarding the challenges and opportunities arising from physics, material synthesis, and technological perspectives [98–108].

13.5 CONCLUSION

In brief, CNTs represent one of the most gigantic strides taken in materials science and are deployable in very expansive applications against most divergent scientific and industrial themes. Among others, uniqueness in their cylindrical structure, high aspect ratio, and having an electronic range spanning from metallic to semiconducting properties make them very valuable and useful. Functionalization of CNTs offers a wide range of possible uses in electronics, storage, medicine, and environmental management. Some examples of areas being developed with functionalized CNTs are targeted drug delivery systems and improved hydrogen storage. Some of the challenges in these otherwise very promising applications include biocompatibility and toxicity, followed by cost-effective storage solutions. Flexible electronics systems have come to denote one of the fastest-growing areas in science and engineering today. It is fueled by an ever-growing demand for varied and innovative electronic solutions. Carbon nanostructures and their functionalized derivatives, such as CNTs, graphene, and CNFs, represent a recent focus in FES products as central core components. Each material has different benefits: CNTs possess very good electrical conductivity and mechanical flexibility, so they are appropriate for interconnects and transistors; with high carrier mobility and transparency, graphene becomes very appropriate for transparent conductive electrodes and sensors; and due to their mechanical strength, CNFs provide the required reinforcement in flexible substrates. The optimal functional carbon nanostructure will be chosen on the basis of balanced factors: conductivity, flexibility, durability, ease of production, and cost according to specific FES applications.

The discovery of 2D materials has revolutionized quantum information science and spintronics by providing previously unachieved integration and performance in electronic devices. The power of quantum physics and materials science comes together primarily through 2D materials to drive forward the advancement in quantum-dot qubits, single-photon emitters, superconducting qubits, and topological quantum computing elements. The discovery of 2D magnetism in layered magnetic crystals further expands such possibilities for magnetic heterostructures that have made high-density magnetic memory applications and innovations in neuromorphic and quantum computing possible. The review shows the recent progress of magnetic 2D materials, 2D interfaces mechanisms, and the actual development in 2D devices for spintronics and quantum information processing. Thus, challenges but also opportunities are brought to the future for 2D heterostructures in further evolving electronic and quantum systems. CNTs, like many other nanomaterials, hold a huge future potential for research in technological and scientific advancement with respect to energy, medicine, and environment management. Only through further research, by conquering the existing limitations and optimization of these materials, can their maximum benefits be exploited to derive their full potential.

REFERENCES

[1] S. Iijima, Helical microtubules of graphitic carbon, *Nature* 354 (1991) 56–58.

[2] M. Meyyappan, *Carbon Nanotubes: Science and Applications*, CRC Press, Boca Raton, FL, 2004.

[3] N. Hamada, S. Sawada and A. Oshiyama, New one-dimensional conductors: graphitic microtubules, *Phys. Rev. Lett.* 68 (1992) 1579–1581.

[4] J.W. Mintmire, B.I. Dunlap and C.T. White, Are fullerene tubules metallic? *Phys. Rev. Lett.* 68 (1992) 631–634.

[5] R. Saito, M. Fujita, G. Dresselhaus and M.S. Dresselhaus, Electronic structure of graphene tubules based on C_{60}, *Phys. Rev. B: Condens. Matter Mater. Phys.* 46 (1992) 1804–1811.

[6] F.S. Lu, L.R. Gu, M.J. Meziani, X. Wang, P.G. Luo, L.M. Veca, L. Cao and Y.P. Sun, Advances in bioapplications of carbon nanotubes, *Adv. Mater.* 21 (2009) 139–152.

[7] R.G. Abaszade, A.G. Mammadov, E.A. Khanmamedova, F.G. Aliyev, V.O. Kotsyubynsky, Gür E., B.D. Soltabayev, T.O. Margitych, M.O. Stetsenko, A. Singh, S. Arya, Photoconductivity of functionalized darbon nanotubes, *Digest J. Nanomater. Biostruct.* 19 (2) (2024) 837–843. https://doi.org/10.15251/DJNB.2024.192.837

[8] L.R.M. de Andrade, L.N. Andrade, J.O. Bahú, V.O. Cárdenas Concha, A.T. Machado, D.S. Pires, R. Santos, T.F.M. Cardoso, J.C. Cardoso, R.L.C. Albuquerque-Junior, P. Severino, E.B. Souto, Biomedical applications of carbon nanotubes: A systematic review of data and clinical trials, *J. Drug Deliv. Sci. Technol.* https://doi.org/10.1016/j.jddst.2024.105932

[9] D. Parra, L. Valverde, F.J. Pino, M.K. Patel, A review on the role, cost and value of hydrogen energy systems for deep decarbonization, *Renew, Sustain. Energy Rev.* 101 (2019) 279–294, https://doi.org/10.1016/j.rser.2018.11.010. Elsevier Ltd.

[10] J.O. Abe, A.P.I. Popoola, E. Ajenifuja, O.M. Popoola, Hydrogen energy, economy and storage: Review and recommendation, *Int. J. Hydrogen Energy* 44 (29) (2019) 15072–15086, https://doi.org/10.1016/j.ijhydene.2019.04.068. Elsevier Ltd.

[11] S. Sharma, S. Basu, N.P. Shetti, T.M. Aminabhavi, Waste-to-energy nexus for circular economy and environmental protection: Recent trends in hydrogen energy, *Sci. Total Environ.* 713 (2020), https://doi.org/10.1016/j.scitotenv.2020.136633

[12] Y. Zhou, R. Li, Z. Lv, J. Liu, H. Zhou, C. Xu, Green hydrogen: A promising way to the carbon-free society, *Chin. J. Chem. Eng.* 43 (2022) 2–13, https://doi.org/10.1016/j.cjche.2022.02.001. Materials China.

[13] S. Hu, Y. Yong, Z. Zhao, R. Gao, Q. Zhou, Y. Kuang, C_7N_6 monolayer as high capacity and reversible hydrogen storage media: A DFT study, *Int. J. Hydrogen Energy* 46 (42) (2021) 21994–22003, https://doi.org/10.1016/j.ijhydene.2021.04.053

[14] S. Feng, et al., Biohydrogen production, storage, and delivery: A comprehensive overview of current strategies and limitations, *Chem. Eng. J.* 471 (2023), https://doi.org/10.1016/j.cej.2023.144669

[15] F. Emir, F.V. Bekun, Energy intensity, carbon emissions, renewable energy, and economic growth nexus: New insights from Romania, *Energy Environ.* 30 (3) (2019) 427–443, https://doi.org/10.1177/0958305X18793108

[16] A. Züttel, Hydrogen storage methods, *Naturwissenschaften* 91 (4) (2004) 157–172, https://doi.org/10.1007/s00114-004-0516-x

[17] S. Niaz, T. Manzoor, A.H. Pandith, Hydrogen storage: Materials, methods and perspectives, *Renew. Sustain. Energy Rev.* 50 (2015) 457–469, https://doi.org/10.1016/j.rser.2015.05.011

[18] M.H. Ahmadi, et al., Renewable energy harvesting with the application of nanotechnology: A review, *Int. J. Energy Res.* 43 (4) (2019) 1387–1410, https://doi.org/10.1002/er.4282

[19] M.A. Shah, B.M. Pirzada, G. Price, A.L. Shibiru, A. Qurashi, Applications of nanotechnology in smart textile industry: A critical review, *J. Adv. Res.* 38 (2022) 55–75, https://doi.org/10.1016/j.jare.2022.01.008

[20] M. Zain, et al., "Applications of nanotechnology in biological systems and medicine", in Nanotechnology for hematology, blood transfusion, and artificial blood, *Elsevier* (2021) 215–235, https://doi.org/10.1016/B978-0-12-823971-1.00019-2

[21] N. Dey, et al., Nanotechnology-assisted production of value-added biopotent energy-yielding products from lignocellulosic biomass refinery – A review, *Bioresour. Technol.* 344 (2022), https://doi.org/10.1016/j.biortech.2021.126171

[22] L.H. Yao, W.Q. Cao, J.C. Shu, M.S. Cao, X. Di Sun, Tailoring adsorption for tunable lithium ion storage and devices, *Chem. Eng. J.* 413 (2021), https://doi.org/10.1016/j.cej.2020.127428

[23] A. Ahmed, et al., Exceptional hydrogen storage achieved by screening nearly half a million metal-organic frameworks, *Nat. Commun.* 10 (1) (2019), https://doi.org/10.1038/s41467-019-09365-w

[24] P. Ramirez-Vidal, et al., A step forward in understanding the hydrogen adsorption and compression on activated carbons, *ACS Appl. Mater. Interfaces* 13 (10) (2021) 12562–12574, https://doi.org/10.1021/acsami.0c22192

[25] S. Cao, B. Li, R. Zhu, H. Pang, Design and synthesis of covalent organic frameworks towards energy and environment fields, *Chem. Eng. J.* 355 (2019) 602–623, https://doi.org/10.1016/j.cej.2018.08.184. Elsevier B.V.

[26] M. Hirscher, et al., Materials for hydrogen-based energy storage – Past, recent progress and future outlook, *J. Alloy. Compd.* 827 (2020), https://doi.org/10.1016/j.jallcom.2019.153548

[27] R. Kumar, A. Khan, A.M. Asiri, H. Dzudzevic-Cancar, Preparation methods of hydrogen storage materials and nanomaterials, *Nanomater. Hydrogen Storage Appl.* (2021) 1–16, https://doi.org/10.1016/b978-0-12-819476-8.00002-5.

[28] Y. Fu, et al., Catalytic effect of bamboo-like carbon nanotubes loaded with NiFe nanoparticles on hydrogen storage properties of MgH_2, *Chem. Eng. J.* 458 (2023), https://doi.org/10.1016/j.cej.2023.141337

[29] S.H. Sang, H. Furukawa, O.M. Yaghi, W.A. Goddard, Covalent organic frameworks as exceptional hydrogen storage materials, *J. Am. Chem. Soc.* 130 (35) (2008) 11580–11581, https://doi.org/10.1021/ja803247y

[30] K. Suresh, D. Aulakh, J. Purewal, D.J. Siegel, M. Veenstra, A.J. Matzger, Optimizing hydrogen storage in MOFs through engineering of crystal morphology and control of crystal size, *J. Am. Chem. Soc.* 143 (28) (2021) 10727–10734, https://doi.org/10.1021/jacs.1c04926

[31] M. Erdos, et al., In silico screening of zeolites for high-pressure hydrogen drying, *ACS Appl. Mater. Interfaces* 13 (7) (2021) 8383–8394, https://doi.org/10.1021/acsami.0c20892.

[32] N.P. Stadie, J.J. Vajo, R.W. Cumberland, A.A. Wilson, C.C. Ahn, B. Fultz, Zeolitetemplated carbon materials for high-pressure hydrogen storage, *Langmuir* 28 (26) (2012) 10057–10063, https://doi.org/10.1021/la302050m

[33] Q. Hu, Y. Lu, G.P. Meisner, Preparation of nanoporous carbon particles and their cryogenic hydrogen storage capacities, *J. Phys. Chem. C* 112 (5) (2008) 1516–1523, https://doi.org/10.1021/jp076409t

[34] C. da Wu, T.H. Fang, J.Y. Lo, Effects of pressure, temperature, and geometric structure of pillared graphene on hydrogen storage capacity, *Int. J. Hydrogen Energy* 37 (19) (2012) 14211–14216, https://doi.org/10.1016/j.ijhydene.2012.07.040

[35] Z. Jin, et al., Nano-engineered spacing in graphene sheets for hydrogen storage, *Chem. Mater.* 23 (4) (2011) 923–925, https://doi.org/10.1021/cm1025188

[36] X. Peng, J.M. Vicent-Luna, Q. Jin, Water-gas shift reaction to capture carbon dioxide and separate hydrogen on single-walled carbon nanotubes, *ACS Appl. Mater. Interfaces* 13 (9) (2021) 11026–11038, https://doi.org/10.1021/acsami.1c00145

[37] V. Jim´enez, A. Ramírez-Lucas, P. S´anchez, J.L. Valverde, A. Romero, Hydrogen storage in different carbon materials: Influence of the porosity development by chemical activation, *Appl. Surf. Sci.* 258 (7) (2012) 2498–2509, https://doi.org/10.1016/j.apsusc.2011.10.080

[38] M. Mohan, V.K. Sharma, E.A. Kumar, V. Gayathri, Hydrogen storage in carbon materials—A review, *Energy Storage* 1 (2) (2019) e35.

[39] E. Boateng, A. Chen, Recent advances in nanomaterial-based solid-state hydrogen storage, *Mater Today Adv.* 6 (Jun. 2020), https://doi.org/10.1016/j.mtadv.2019.100022

[40] K. Kajiwara, H. Sugime, S. Noda, N. Hanada, Fast and stable hydrogen storage in the porous composite of MgH_2 with Nb_2O_5 catalyst and carbon nanotube, *J. Alloy. Compd.* 893 (2022), https://doi.org/10.1016/j.jallcom.2021.162206

[41] S. Ullah Rather, Preparation, characterization and hydrogen storage studies of carbon nanotubes and their composites: A review, *Int. J. Hydrogen Energy* 45 (7) (2020) 4653–4672, https://doi.org/10.1016/j.ijhydene.2019.12.055. Elsevier Ltd.

[42] M. Liu, et al., Novel 1D carbon nanotubes uniformly wrapped nanoscale MgH2 for efficient hydrogen storage cycling performances with extreme high gravimetric and volumetric capacities, *Nano Energy* 61 (2019) 540–549, https://doi.org/10.1016/j.nanoen.2019.04.094

[43] J. Yuan, et al., Ni-doped carbon nanotube-$Mg(BH_4)_2$ composites for hydrogen storage, *ACS Appl. Nano Mater.* 4 (2) (2021) 1604–1612, https://doi.org/10.1021/acsanm.0c02738

[44] J.F.N. Dethan, V. Swamy, Mechanical and thermal properties of carbon nanotubes and boron nitride nanotubes for fuel cells and hydrogen storage applications: A comparative review of molecular dynamics studies, *Int. J. Hydrogen Energy* 47 (59) (2022) 24916–24944, https://doi.org/10.1016/j.ijhydene.2022.05.240

[45] S.I. Hwang, et al., Metal-organic frameworks on palladium nanoparticle functionalized carbon nanotubes for monitoring hydrogen storage, *ACS Appl. Nano Mater.* (2022), https://doi.org/10.1021/acsanm.2c00998

[46] S. Rezaie, D.M.J. Smeulders, A. Luna-Triguero, Enhanced hydrogen storage in gold-doped carbon nanotubes: A first-principles study, *Chem. Eng. J.* 476 (2023) Article 146525, https://doi.org/10.1016/j.cej.2023.146525

[47] A. Szabó, C. Perri, A. Csató, G. Giordano, D. Vuono, J.B. Nagy, Synthesis methods of carbon nanotubes and related materials, *Materials* 3 (2010) 3092–3140.

[48] A. Fonseca, J. B.Nagy, Carbon nanotubes formation in the arc discharge process. In *Carbon Filaments and Nanotubes: Common Origins, Differing Applications*, Kluwer Academic Publishers, Dordrecht, The Netherlands, 2001, Volume 372, pp. 75–84.

[49] J. Journet, W.K. Maser, P. Bernier, A. Loiseau, M.L. de la Chapelle, S. Lefrant, P. Deniard, R. Lee, J.E Fisher, Large-scale production of single-walled carbon nanotubes by the electric-arc technique, *Nature* 388 (1997) 756–758.

[50] T. Guo, P. Nikolaev, A.G. Rinzler, D. Tomanek, D.T. Colbert, R.E. Smalley, Self-assembly of tubular fullerenes, *J. Phys. Chem.* 99 (1995) 10694–10697.

[51] J.H. Hafner, M.J. Bronikowski, B.R. Azamian, P. Nikolaev, A.G. Rinzler, D.T. Colbert, K.A. Smith, R.E. Smalley, Catalytic growth of single-wall carbon nanotubes from metal particles, *Chem. Phys. Lett.* 296 (1998) 195–202.

[52] A. Thess, R. Lee, P. Nikolaev, H. Dai, P. Petit, J. Robert, C. Xu, Y.H. Lee, S.G. Kim, A.G. Rinzler, D.T. Colbert, G.E. Scuseria, D. Tománek, J.E. Fischer, R.E. Smalley, Crystalline ropes of metallic carbon nanotubes, *Science* 273 (1996) 483–487.

[53] A.M. Rao, E. Richter, S. Bandow, B. Chase, P.C. Eklund, K.A. Williams, S. Fang, K.R. Subbaswamy, M. Menon, A. Thess, R.E. Smalley, G. Dresselhaus, M.S. Dresselhaus, Diameter-selective Raman scattering from vibrational modes in carbon nanotubes, *Science* 275 (1997) 187–191.

[54] T. Guo, P. Nikolaev, A. Thess, D.T. Colbert, R.E. Smalley, Catalytic growth of single-walled nanotubes by laser vaporization, *Chem. Phys. Lett.* 243 (1995) 49–45

[55] K.B.K. Teo, C. Singh, M. Chhowalla, W.I. Milne, Catalytic synthesis of carbon nanotubes and nanofibers. In *Encyclopedia of Nanoscience and Nanotechnology*, American Scientific Publisher, Valencia, CA, (2003), 665–668.

[56] C. Journet, P. Bernier, Production of carbon nanotubes, *Appl. Phys. A* 67 (1998) 1–9.

[57] K. Wang, F. Wang, Q. Jiang, P. Zhu, K. Leu, R. Zhang, Controlled synthesis, properties, and applications of ultralong carbon nanotubes, *Nanoscale Adv.*, 6, (2024), 4504–4521https://doi.org/10.1039/D4NA00437J

[58] C.N.R. Rao, A. Govindaraj, G. Gundiah, S.R.C. Vivekchand, Nanotubes and nanowires, *Chem. Eng. Sci.* 59 (2004) 4665, http://dx.doi.org/10.1016/j.ces.2004.07.067

[59] G.B. Novais, M.A. Dias, A.A.M. Santana, T.C. Batista, M.N. Marques, C.R. Melo, R.L.C. Albuquerque, M.Z. Gomes, P. Severino, E.B. Souto, J.C. Cardoso, Isoflavones-functionalized single-walled and multi-walled carbon nanotubes: Synthesis and characterization of new nanoarchitetonics for biomedical uses, *J. Mol. Struct.* 1294 (2023) 136351, https://doi.org/10.1016/j.molstruc.2023.136351

[60] L. Lavagna, R. Nisticò, S. Musso, M. Pavese, Functionalization as a way to enhance dispersion of carbon nanotubes in matrices: A review, *Mater. Today Chem.* 20 (2021) 100477.

[61] V. Negri, J. Pacheco-Torres, D. Calle, P. López-Larrubia, Carbon nanotubes in biomedicine, *Top. Curr. Chem.* 378 (2020) 177–217.
[62] Hirsch A. Functionalization of single-walled carbon nanotubes, *Angew. Chem. Int. Ed.* 41 (2002) 1853.
[63] A.M. Díez-Pascual, Chemical functionalization of carbon nanotubes with polymers: A brief overview, *Macromol.* 1 (2021) 64–83.
[64] M.N. Norizan, M.H. Moklis, S.Z.N. Demon, N.A. Halim, A. Samsuri, I.S. Mohamad, V.F. Knight, N. Abdullah, Carbon nanotubes: Functionalisation and their application in chemical sensors, *RSC Adv.* 10 (2020) 43704–43732.
[65] R.G. Abaszade, O.A. Kapush, A.M. Nabiev, Properties of carbon nanotubes doped with gadolinium, *J. Optoelectron. Biomed. Mater.* 12 (3) (2020) 61–65.
[66] A.R. Cadore, I. Zanella, V.M. de Menezed, J. Rossato, R. Mota, S.B. Fagan, Metal-doped carbon nanotubes interacting with vitamin C, *Phys. Chem. Chem. Phys.* 14 (2012) 16737–16744.
[67] A.L. Pishko, S.M. Serkiz, K.E. Zeigler, A.M. Rao, Removal and sequestration of iodide from alkaline solutions using silverdoped carbon nanotubes, *J. South Carolina Acad. Sci.* 9 (1) (2011) 37.
[68] K.S. Ibrahim, Carbon nanotubes–properties and applications: A review, *Carbon Lett.* 14 (3) (2013), 131–144.
[69] R. Abaszade, O. Kapush, S. Mamedova, A. Nabiyev, S. Melikova, S. Budzulyak, (2020). Gadolinium doping influence on the properties of carbon nanotubes, *Phys. Chem. Solid State* 21 (3) 404–408, https://doi.org/10.15330/pcss.21.3.404-408
[70] Y.T. Kim, T. Mitani, *Appl.Phys.Lett.* 89 (2006) 033107, https://doi.org/10.1063/1.2370876
[71] Garcia-Gutierrez M.C., Nogales A., Rueda D.R., Domingo C., Garcia-Ramos J.V., Broza G., Roslaniec Z., Schulte K., Davies R.J., Ezquerra T.A. Templating of crystallization and shear-induced self-assembly of single-wall carbon nanotubes in a polymer-nanocomposite, *Polymer* 47 (2006) 341, http://dx.doi.org/10.1016/j.polymer.2005.11.018
[72] E.J. Siochi, D.C. Working, C. Park, P.T. Lillehei, J.H. Rouse, C.C. Topping, A.R. Bhattacharyya, S Kumar, Melt processing of SWCNTpolyimide nanocomposite fibers, *Composites B* 35 (2004) 439, http://dx.doi.org/10.1016/j.compositesb.2003.09.007
[73] X.X. Zhang, Q.J. Meng, X.C. Wang, S.H. Bai, Poly (adipic acid-hexamethylene diamine)-functionalized multi-walled carbon nanotube nanocomposites, *J. Mater. Sci.* 46 (2011) 923, http://dx.doi.org/10.1007/s10853-010-4836-2
[74] A. De Vita, J.C. Charlier, X. Blase, R. Car, Electronic structure at carbon nanotube tips, *Appl. Phys. A* 68 (1999) 283, http://dx.doi.org/10.1007/s003390050889
[75] J.M. Bonard, T. Stockli, F. Maier, W.A. de Heer, A. Chatelain, J.P. Salvetat, L. Forro, Field-emission-induced luminescence from carbon nanotubes, *Phys. Rev. Lett.* 81 (1998) 1441, http://dx.doi.org/10.1103/PhysRevLett.81.1441
[76] Y. Saito, S. Uemura, K. Hamaguchi, Cathode ray tube lighting elements with carbon nanotube field emitters, *Jpn. J. Appl. Phys.* 37 (1998) L346, http://dx.doi.org/10.1143/JJAP.37.L346
[77] D. Rotman, The nanotube computer, *MIT Technol. Rev.* 36 (2002) 105,
[78] H. Xia, Y. Wang, J. Lin, L. Lu, Hydrothermal synthesis of MnO_2/CNT nanocomposite with a CNT core/porous MnO_2 sheath hierarchy architecture for supercapacitors, *Nanoscale Res. Lett.* 7 (2012) 33, http://dx.doi.org/10.1186/1556-276X-7-33
[79] K. Evanoff, J. Benson, M. Schauer, I. Kovalenko, D. Lashmore, W.J. Ready, G. Yushin, Ultra-strong silicon-coated carbon nanotube nonwoven fabric as a multifunctional lithium-ion battery anode, *ACS Nano* 6 (2012) 9837, http://dx.doi.org/10.1021/nn303393p
[80] N.S. Lee, D.S. Chung, I.T. Han, J.H. Kang, Y.S. Choi, H.Y. Kim, S.H. Park, Y.W. Jin, W.K. Yi, M.J. Yun, J.E. Jung, C.J. Lee, J.H. You, S.H. Jo, C.G. Lee, J.M. Kim, Application of carbon nanotubes to field emission displays, *Diamond Relat. Mater.* 10 (2001) 265, http://dx.doi.org/10.1016/S0925-9635(00)00478-7
[81] A. Ahmad, M.M. Kholoud, E. Abou, A.A. Reda, A.W. Abdulrahman, Carbon nanotubes, science and technology part (I) structure, synthesis and characterization, *Arabian J. Chem.* 5 (1) (2012), http://dx.doi.org/10.1016/j.arabjc.2010.08.022
[82] E. Frackowiak, F. Beguin, Electrochemical storage of energy in carbon nanotubes and nanostructured carbons, *Carbon* 40 (2002) 1775, http://dx.doi.org/10.1016/S0008-6223(02)00045-3

[83] C. Niu, E.K. Sichel, R. Hoch, D. Moy, H. Tennent, High power electrochemical capacitors based on carbon nanotube electrodes, *Appl. Phys. Lett.* 70 (1997) 1480, http://dx.doi.org/10.1063/1.118568
[84] R.Z. Ma, J. Liang, B.Q. Wei, B. Zhang, C.L. Xu, D.H. Wu, Processing and performance of electric double-layer capacitors with blocktype carbon nanotube electrodes, *Bull. Chem. Soc. Jpn.* 72 (1999) 2563.
[85] F. Banhart, N. Grobert, M. Terrones, J.C. Charlier, P.M. Ajayan, Metal atoms in carbon nanotubes and related nanoparticles, *Int. J. Mod. Phys. B* 15 (2001) 4037, http://dx.doi.org/10.1142/S0217979201007944
[86] Park S.J., Lee S.Y. Hydrogen storage behaviors of carbon nanotubes/metal-organic frameworks-5 hybrid composites, *Carbon Lett.* 10 (2009) 19, http://dx.doi.org/10.5714/CL.2009.10.1.019
[87] C. Liu, Y.Y. Fan, M. Liu, H.T. Cong, H.M. Cheng, M.S. Dresselhaus, Hydrogen storage in single-walled carbon nanotubes at room temperature, *Science*, 286 (1999), 1127, http://dx.doi.org/10.1126/science.286.5442.1127
[88] G.E. Gadd, M. Blackford, S. Moricca, N. Webb, P.J. Evans, A.M. Smith, G. Jacobsen, S. Leung, A. Day, Q. Hua, The world's smallest gas cylinders? *Science* 277 (1997) 933, http://dx.doi.org/10.1126/science.277.5328.933
[89] M. Terrones, R. Kamalakaran, T. Seeger, M. Ruhle, Novel nanoscale gas containers: Encapsulation of N_2 in CN_x nanotubes, *Chem. Commun.* 23 (2000) 2335, http://dx.doi.org/10.1039/B008253H
[90] S.J. Tans, A.R.M. Verschueren, C. Dekker, Room-temperature transistor based on a single carbon nanotube, *Nature* 393 (1998) 49, http://dx.doi.org/10.1038/29954
[91] L.C. Venema, J.W.G. Wildoer, H.L.J.T. Tuinstra, C. Dekker, A.G. Rinzler, R.E. Smalley, Length control of individual carbon nanotubes by nanostructuring with a scanning tunneling microscope, *Appl. Phys. Lett.* 71 (1997) 2629, http://dx.doi.org/10.1063/1.120161
[92] S.S. Wong, E. Joselevich, A.T. Woolley, C.L. Cheung, C.M. Lieber, Covalently functionalized nanotubes as nanometre-sized probes in chemistry and biology, *Nature* 394 (1998) 52, http://dx.doi.org/10.1038/27873
[93] P.G. Collins, K. Bradley, M. Ishigami, A. Zettl, Extreme oxygen sensitivity of electronic properties of carbon nanotubes, *Science* 287 (2000) 1801, http://dx.doi.org/10.1126/science.287.5459.1801
[94] J.M. García-Lastra, D.J. Mowbray, K.S. Thygesen, A. Rubio, K.W. Jacobsen, Computational design of chemical nanosensors: Metal doped carbon nanotubes, *Phys. Rev. B* 81 (2010) 245429, https://doi.org/10.1103/PhysRevB.81.245429
[95] H. Dai, J.H. Hafner, A.G. Rinzler, D.T. Colbert, R.E.Smalley, Nanotubes as nanoprobes in scanning probe microscopy, *Nature* 384 (1996) 147, http://dx.doi.org/10.1038/384147a0
[96] S.J. Park, S.Y. Lee, Hydrogen storage behaviors of carbon nanotubes/metal-organic frameworks-5 hybrid composites, *Carbon Lett.* 10 (2009) 19, http://dx.doi.org/10.5714/CL.2009.10.1.019
[97] D. Akiladevi, S. Basak, Carbon nanotubes (CNTs) production, characterization and its applications, *Int. J. Adv. Pharm. Sci.* 1 (2010) 187.10.5138/ijaps.2010.0976.1055.01024
[98] R. Moradi, N. Pour Khalili, E. Khanmamadova, R. Abaszade, Functionalized carbon nanostructures for flexible electronics, In *Handbook of Functionalized Carbon Nanostructures*, Springer, Cham, (2024), 2581–2614.
[99] E.A. Khanmamadova, R.G.Abaszade, Diodes made from carbon nanotubes, *Ecoenergetics* 1 (2024) 48–52.
[100] R.G. Abaszade, O.A. Kapush, S.A. Mamedova, A.M. Nabiyev, S.Z. Melikova, S.I. Budzulyak, Gadolinium doping influence on the properties of carbon nanotubes, *Phys. Chem. Solid State* 21 (3) (2020) 404–408, https://doi.org/10.15330/pcss.21.3.404-408
[101] R.G. Abaszade, M.B. Babanli, V.O. Kotsyubynsky, A.G. Mammadov, E. Gür, O.A. Kapush, M.O. Stetsenko, R.I. Zapukhlyak, Influence of gadolinium doping on structural properties of carbon nanotube, *Phys. Chem. Solid State* 24 (1) (2023) 53–158, https://doi.org/10.15330/pcss.24.1.153-158
[102] A.G. Mammadov, R.G. Abaszade, M.B. Babanli, V.O. Kotsyubynsky, E. Gur, B.D. Soltabayev, T.O. Margitych, M.O. Stetsenko, Photoconductivity of gadolinium-doped carbon nanotubes, *Int. J. Tech. Phys. Prob. Eng.* 15 (3) (2023) 53–58, www.iotpe.com/IJTPE/IJTPE-2023/IJTPE-Issue56-Vol15-No3-Sep2023/7-IJTPE-Issue56-Vol15-No3-Sep2023-pp53-58.pdf
[103] A.G. Mammadov, R.G. Abaszade, V.O. Kotsyubynsky, E.Y. Gur, I.Y. Bayramov, E.A. Khanmamadova, O.A. Kapush, Photoconductivity of carbon nanotubes, *Tech. Phys. Prob. Eng.* 14 (3) (2022) 155–160,

www.iotpe.com/IJTPE/IJTPE-2022/IJTPE-Issue52-Vol14-No3-Sep2022/21-IJTPE-Issue52-Vol14-No3-Sep2022-pp155-160.pdf

[104] R.G. Abaszade, E.M. Aliyev, M.B. Babanli, V.O. Kotsyubynsky, R.I. Zapukhlyak, A.G. Mamedov, H.F. Budak, A.E. Kasapoglu, E. Gur, T.O. Margitych, M.O. Stetsenko, Investigation of thermal properties of carbon nanotubes and carboxyl group – Functionalized carbon nanotubes, *Phys. Chem. Solid State* 24 (3) (2023) 530–535, https://doi.org/10.15330/pcss.24.3.530-535

[105] R.G. Abaszade, E.M. Aliyev, A.G. Mammadov, E.A. Khanmamadova, A.A. Guliyev, F.G. Aliyev, R.I. Zapukhlyak, H.F. Budak, A.E. Kasapoglu, T.O. Margitych, A. Singh, S. Arya, E. Gür, M.O. Stetsenko, Investigation of thermal properties of gadolinium doped carbon nanotubes, *Phys. Chem. Solid State* 25 (1) (2024) 142–147, https://doi.org/10.15330/pcss.25.1.142-147

[106] R.G. Abaszade, A.G. Mammadov, E.A. Khanmamedova, F.G. Aliyev, V.O. Kotsyubynsky, E. Gür, B.D. Soltabayev, M.O. Margitych, M.O. Stetsenko A. Singh, S. Arya, Photoconductivity of functionalized carbon nanotubes, *Dig. J. Nanomater.Biostruct.* 19 (2) (2024) 837–843, https://doi.org/10.15251/DJNB.2024.192.837

[107] X. Liu, M.C. Hersam, 2D materials for quantum information science, *Nat. Rev. Mater.* 4 (2019) 669–684.

[108] Z. Bingyu, L. Pengcheng, T. Roozbeh, X.-L.F. Philip, W. Yingying, 2D Magnetic heterostructures: Spintronics and quantum future, *npj Spintronics* 2 (2024) Article number: 6, https://doi.org/10.1038/s44306-024-00011-w

14 Functionally Graded Materials in Medical Applications (Implants, Drug Delivery, Tissue Engineering)

I.A. Asgarli, E.A. Khanmamadova, T.O. Margitych, M.O. Stetsenko, and R.G. Abaszade

14.1 INTRODUCTION

The current scientific and technological progress shows interest in the synthesis of new-generation materials. These materials appear to attract special attention due to the unique properties they possess in relation to the various branches of science where they are applied. Modern approaches allow applying in the most diverse fields of medicine, starting from implant technologies, due to its unique properties. These applications are designed to be quite secure, long-lasting, and apply modern approaches. Implant-associated infection (IAI) is a problem confronted worldwide with significant morbidity, mortality, and a burden of high healthcare cost [1,2]. The rate of IAI with some devices is as high as 30%. It varies from a low of A$19,688 for debridement, antibiotics, and implant retention to a high of A$23,805 for excision arthroplasty, A$26,722 for one-stage revision, and A$44,744 for two-stage revision, as per an Australian study on 114 patients who underwent a total of 178 surgical procedures to treat IAI. IAI is a result of the ingress of bacteria into the body via the implantation site, which then adhere to the device and form biofilm [3–5]. These biofilms show high resistance to antibiotics compared to what is presented in planktonic cells; an infection is then difficult to clear. The number of such antibiotic-resistant infections is increasing every day, and the trend is going to continue. Consequently, there is a great demand for developing new coatings with advanced properties to avoid the risk of IAI. Graphene-based coatings provided an application in wound healing. It has been revealed that graphene coatings could act to inhibit the growth of a wide range of bacteria [6–9]. Graphene oxide (GO) sheets were deposited generally by electrophoretic deposition on polyethylene oxide (PEO)-pretreated Ti substrates. The GO improved antibacterial activity of the PEO-Ti samples by reducing approximately 80% and 100% of the *Escherichia coli* and *Staphylococcus aureus* viability, respectively. In studies taken further, Spark Plasma Sintering (SPS) for incorporation of graphene in a magnesium alloy implant significantly depressed *S. aureus* and *E. coli* viability up to five times [10–13]. Graphene-modified surfaces also improved cytocompatibility with human mesenchymal stromal cells (hMSCs) and osteogenic properties [14,15].

These shortcomings take the form of sudden metabolism and elimination of drugs from the body before acting on their target sites, poor water solubility of some drugs, non-specific targeting, low retention efficacy on dead regions, and toxic effects on normal cells. Recent development in the field of nanotechnology and nanomaterials has opened up a bright avenue in biomedicine: improved biodistribution of the drug, reduced adverse effects on healthy cells, specific selectivity, and higher local therapeutic absorption with the aid of newly proposed drug delivery systems. In the nanobiomedicine

DOI: 10.1201/9781003664611-15

domain, a huge quantity of nanomaterials has been explored by researchers during the last decades, including polymeric nanoparticles, micelles, liposomes, mesoporous inorganic nanoparticles, metal oxides, noble materials, and carbon nanomaterials.

Bone is one of the most critical tissues in the body because it forms a framework for both mechanical support and organ protection and skeletal continuity. Different factors can lead to damage to the integrity of the bone, including trauma, surgery, tumors, and osteoporosis. In most cases, bone has big potential for regeneration and self-repair. Full regeneration, however, is not possible in some cases and requires further stimulation. Biomaterials are one of the potential substitutes that can replace bone transplants in bone tissue engineering. Some of the biomaterials used significantly in bone and hard tissue engineering include synthetic hydroxyl apatite, tri-calcium phosphate, other bioceramics, polymeric scaffolds, and metallic implants [2–4]. Nanotechnology has made nanomedical research clinical with the help of new nanodevices and nanosystems dependent on functional nanomaterial design and perfect integration. Graphene family derivatives are of interest among the many synthetic nanobiomaterials for biomedical applications. Graphene oxide is a hydrophilic version of graphene sheet with hybridized sp^2 carbon atoms and has shown a very bright promise as a biomedical application [7]. The reduced form of graphene oxide, with minimal toxicity, is biocompatible and has reaction sites which could be utilized to stimulate cellular activities. It increased the osteogenic capacity of the bone tissue [5–13]. Interest has lately been placed on the orthopedical therapeutic properties of strontium, Sr. Sr can accumulate in human hard tissues and displace calcium from the apatite phase of bone mineral. Sr is also associated with an increase in bone compressive strength, whereas its shortage is associated with negative consequences in hard tissues. Sr ions have been found to stimulate bone formation and inhibit bone resorption in vitro and in vivo studies, which made them a potential agent for the treatment of osteoporosis. Properties of Sr have made it an ingredient of choice in the synthesis of bioactive glasses and bio ceramics because of their several advantages. Sr has been used for bone remodeling applications in view of its structural and physico-chemical similarity to calcium ions, Ca^{2+}. For example, incorporation of Sr into calcium phosphates and titanium implants was studied to enhance bone-forming properties of such materials. Due to the large surface area of graphene nanosheets, researchers have recently synthesized SrNPs on the surface of graphene by co-reduction of GO and Sr. In this case, hydrazine as a reducing agent was used to reduce, despite SrNPs on the surface of rGO reported by Kumar et al. Qi et al. recently synthesized Sr-decorated rGO nanosheets using hydrazine incorporated into the poly(L-lactide) (PLLA) for fabrication of 3D scaffold. While Sr-rGO has far surpassed the osteogenesis induction of the rGO and pure PLLA scaffolds, these studies have recently raised major concerns: the high capability of the Sr-rGO nanosheets in bone tissue engineering, enhancing graphene strength and Sr osteogenesis properties using hydrazine in the synthesis process. First, water solubility of rGO nanosheets has been significantly reduced, which caused problems during the fabrication process of the graphene-based composite. Second, hydrazine was a toxic reagent that could affect the health of users during the experiment. Indeed, many efforts have been made to introduce a green and safe reducer to overcome the limitation of hydrazine. For example, protein-like bovine serum albumin (BSA) and ascorbic acid are safe reducers for GO. Primary damage to the spinal cord from extrinsic, either direct or indirect, factors will usually be followed by significant sensory, motor, and autonomic dysfunction below the level of the injury. According to the World Health Organization, between 250,000 and 500,000 new cases of spinal injuries occur each year. The estimated annual incidence in Canada is 1,785; the total annual economic burden of citation indexing service (SCI) in Canada is \$2.67 billion [14–23]. Though the advances in clinical management have brought in place minimal morbidity and resulted in enhanced patient outcomes, effective repair schemes that result in the actual recovery of SCI remain few. Before the development of new restore approaches for a spinal cord, knowledge of the drastic alterations that follow injury in the surrounding microenvironment is a requisite. In the microenvironment of SCI, dynamic changes in the number of cells, phenotypes, and spatiotemporal distribution occur. Primary injury includes compression, splitting, twisting, or shearing of the spinal cord due to a direct or indirect external mechanical force, resulting in axonal breakage, nerve tissue destruction, and neuronal death. Primary injuries such as this induce a sequela of secondary pathophysiological events. This damage to the blood–spinal cord barrier allows invasion by inflammatory cells into the site, which in turn secrete various inflammatory factors. The various Reactive Oxygen Species (ROS) are produced by the phagocytic inflammatory cells, leading to cellular necrosis and apoptosis. Later, at the site of the lesion, glial and fibrous scars form. These secondary injuries impede neurodegeneration and lengthen the duration of the injury, which leads to hindrance in tissue regeneration and functional recovery [24–26].

They are the physical sensing platforms that detect the surroundings, monitor and communicate with acquired physical data in terms of pressure, shear, strain, torsion, temperature, and humidity, which are the building blocks of a huge number of advanced applications, such as wearable consumer electronics, soft robotics, smart medical prosthetics and electronic skins, and real-time healthcare monitoring [27–38]. With increased demand for such applications, there is a corresponding increase in the requirements and standards for the design and proper deployment of these sensors. As a result, in recent years, there has been the sensing of a specific group of physical sensors—a light-weight, flexible, and wearable physical sensor, characterized by highly deformable and conformable natures, long-term stability, increased sensitivity, and high optical transparency [39–42]. Cancer is an extremely fatal disease and caused close to 9.6 million deaths globally in 2018, according to the reports of the World Health Organization.

Cancer detection at an initial stage, as well as its timely treatment, is quite necessary in view of improving the survival rate of the cancer patients [43]. Early detection of cancer guarantees better response to an effective treatment and possibly higher survival, less morbidity, and affordable treatment. Thus, the development of sensitive, specific, fast, and economical methods is of prime interest for early diagnosis in cancer detection [44]. Whereas these include mammography, Prostate-Specific Antigen (PSA) level in blood, Papanicolaou test, occult blood analysis, endoscopy, CT scan, X-ray, ultrasound imaging, MRI, and others, the tests mentioned above are all considered unspecified inferences for the detection of some kinds of cancer in very initial stages. More so, some of the above-noted diagnostic tests normally take relatively long periods and are costly, while all of them are not universally available. Thereby, there lies the highest demand to develop such specific and reliable techniques that can be suitable for the timely detection of various cancer biomarkers [45]. The infection and the development of diseases generally are involved with the production or the change in the level of some of the biomolecules that remain present in the human body, and through the suitable disease biomarkers, the detection, as well as the monitoring of the disease, can be achieved. Cancer biomarkers can diagnose the biological state of cancer progression and help in understanding the cellular processes of cancer cells. Samples from urine, blood, saliva, tears, or sweat can be collected very easily, and it is a less invasive collection of samples. It is to be collected at any time for monitoring the disease. Therefore, this makes it possible to develop effective techniques for the rapid diagnosis and treatment of cancer [46]. The detection of biomarkers in body fluids is, however, a challenging task because of its low concentration and complex biological assays. During this period, nanotechnology arrived and, with applications in almost all the branches of science, and offered a new way into the potential use of diverse nanomaterials for cancer theranosis [47–50].

However, most medical implants activate undesired Foreign Body Reactions (FBRs) upon implantation, and this generally leads to fibrosis, after which irreparable damage of the device occurs [51,52]. Upon contacting body fluids, certain proteins such as albumin, fibronectin, fibrinogen, and complements directly absorb to the surface of implants. This protein deposition will trigger the innate immune response, which subsequently causes acute inflammation. One of the earliest recruited white cells to the site of implantation is macrophages. At this location, they can polarize into pro-inflammatory M_1 and alternative M_2 phenotypes. This continuous stimulation at the implant site leads eventually to the development of chronic inflammation following some days of acute inflammation. In contrast, during chronic inflammation, the macrophages polarize to an alternatively activated M_2 phenotype that secretes profibrotic factors like platelet-derived growth factor and transforming growth factor-β1. These factors are reported to directly induce the trans-differentiation of fibroblasts into myofibroblasts. The latter secrete large quantities of extracellular matrixes (ECM), mainly type I collagen, leading to the formation of a dense, fibrous capsule around the implant. It has been reported that macrophages, by their regulation of the ECM environment, play an important role in tissue regeneration. However, extreme polarization of both M_1 and M_2 phenotypes results in fibrosis.

Traditional implant materials—stainless steels, titanium alloys, and cobalt–chromium alloys, in particular—have good resistance to corrosion, wear, and fatigue, apart from their excellent load-bearing capacity. However, these alloys are known to build up several adverse effects, among them stress shielding [53–58] and inflammation of the local tissues by possibly released cytotoxic ions [59]; in many cases, an implant removal surgery is necessary. Therefore, temporary implants that "biocorrode" have gained significant interest in biomedical applications [60–64]. Basically, two classes of materials have been widely considered for biodegradable implant applications: polymers and metals, which have inherent ability to corrode away within the human body, abrogating the second surgery for implant removal [65–67]; however, the corrosion/degradation products should be harmless. Among them, metallic materials, especially magnesium alloys, are of more interest in their mechanical properties. However, the fast corrosion rate of magnesium-based implants in body fluid is the main bottleneck that restricts their further application [68–83].

14.2 GRAPHENE-BASED MATERIALS

One of the simple methods of activation has been applied to obtain porous carbon material from different materials as example from walnut shells. In our previous research, an investigation on the effect of activation time at 400°C in an atmosphere with limited air access on structural, morphological, and electrochemical properties of the obtained porous carbon material has been conducted. The thermal activation time variation at 400°C shows transformation of structural, morphological, and electrochemical performance of porous carbon material derived from walnut shells. Structural and morphological characterization of pristine and activated carbon samples has been confirmed by Small-Angle X-ray Scattering (SAXS) and low-temperature adsorption porosimetry, Scanning Electron Microscopy (SEM), and Raman spectroscopy. It is therefore deduced from the results that increasing the duration of activation at constant temperature has the effect of reducing both the thickness and lateral dimensions of the graphite crystallite. It has been shown that thermal activation allows for an approximate doubling of the specific Brunauer–Emmett–Teller (SBET) surface area of the initial carbon material and contributes to the development of its mesoporous structure with a relative content of mesopores of about 75–78% and an average pore diameter of about 5 nm. The fractal dimension of the obtained carbon materials was estimated by the Frenkel–Halsey–Hill method in the area of low- and high-relative pressure. This, therefore, got the prepared porous carbon materials in the application of the preparation of electrochemical capacitor electrodes. These cells were electrochemically characterized in 6M KOH aqueous electrolyte by cyclic voltammetry, galvanostatic charge/discharge, and impedance spectroscopy. It was noted that the material activated at 400°C has a specific capacity of about 110–130 F/g at currents in the range of 4–100 mA/g during discharging [84–94, 194–203].

The carbon material was inserted into a ceramic vessel and then introduced into the muffle furnace for further activation. Activation of this carbon material took place at the temperature of 400°C; the activation times of 1 hour, 2 hours, and 3 hours were denoted by Carbon Coated Alumina / Canonical Correlation Analysis (CCA1), CCA2, and CCA3, respectively. Figure 14.1 shows the process scheme of obtaining PCM from walnut shells [95–110].

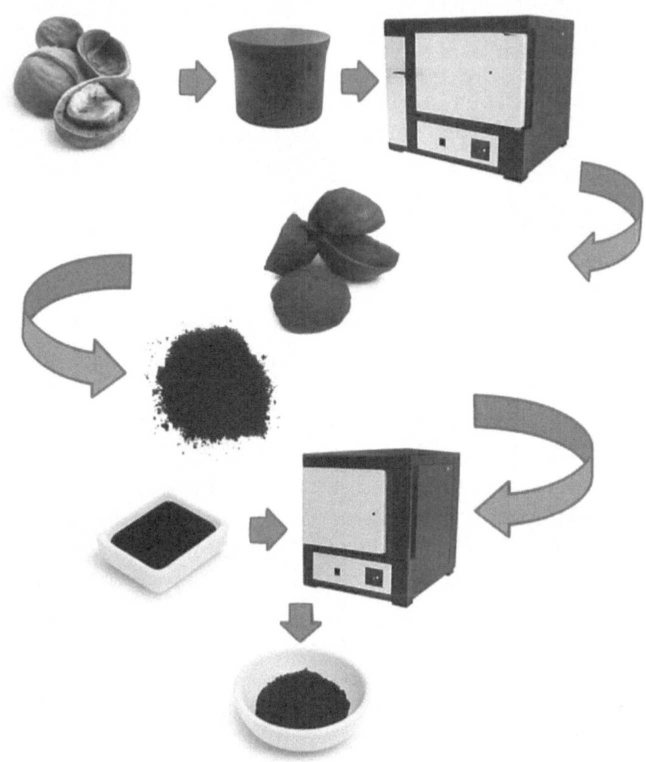

FIGURE 14.1 A schematic illustration of obtaining PCM from walnut shells.

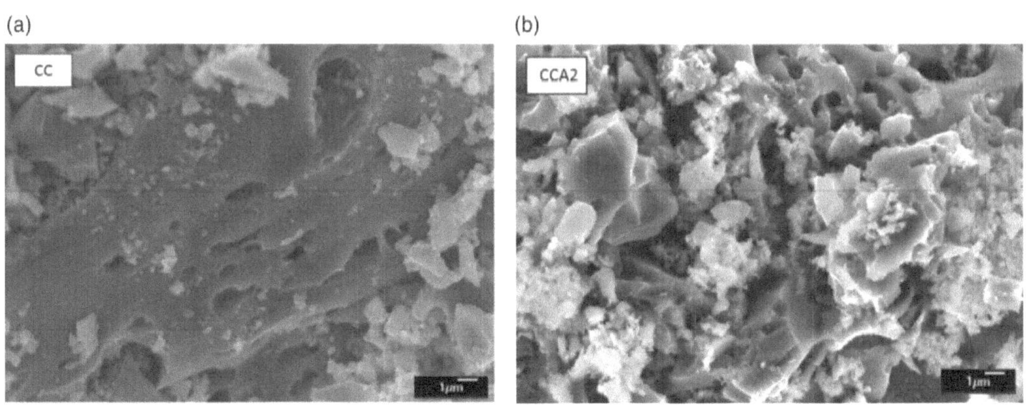

FIGURE 14.2 SEM image of carbonized (a) and activated at 400 °C for 2 hours (b) carbon material.

Figure 14.2 represents the SEM images of the surface morphology of the investigated PCM. In the thermally activated material, carbon particles seemed to become more porous and rougher, and small fragments of carbon particles can be clearly noted on their surface. From an optical point of view, one may postulate that the pores in the carbon particles are like a root-like structure and undergo expansion and branching due to thermal activation [95–110].

14.3 MEDICAL APPLICATIONS OF GRAPHENE-BASED MATERIALS

14.3.1 In Implant Technology

Infections at the site of implantation are serious concerns to health and health expenditure. In this work, we deposited graphene onto a medical-grade cobalt–chromium alloy surface using radiofrequency plasma-enhanced chemical vapor deposition with *Origanum vulgare* as the precursor material. Deposition of Gr on CoCr was confirmed by Raman spectroscopy and X-ray Photoelectron Spectroscopy (XPS) and SEM. Biocompatibility and antibacterial properties of CoCr-Gr were investigated. CoCr-Gr presented biocompatibility, adhesion, and spreading of RAW 267.4 macrophage cells. CoCr-Gr inhibited growth against *S. aureus* and *Pseudomonas aeruginosa* and prevented attachment of *P. aeruginosa*. The results indicate that CoCr-Gr can be a potential antibacterial coating material for implantable devices. Gr was deposited on CoCr substrates by RF-PECVD using *O. vulgare* as Gr precursor. Raman spectroscopy and X-ray photoelectron spectroscopy confirmed the presence of Gr in CoCr-Gr sample with a high content of defects, which was multilayered. Plasma deposition of Gr on CoCr surfaces did not influence the hydrophilicity of samples. SEM microscopies detected Gr in the form of nanostructures, showing various nanometric shapes and sizes. The biocompatibility results indicated that there was no difference in viability, adhesion, and morphology between CoCr and CoCr-Gr RAW macrophages after 48 hours of incubation. The bactericidal and antifouling effect of CoCr-Gr against *P. aeruginosa* was significantly higher in comparison with that of the control. CoCr-Gr also exhibited a bactericidal effect against *S. aureus*. The synthesis of Gr by RF-PECVD on several substrates, including a renewable volatile carbon source such as *O. vulgare*, allows versatility and possibly customization of implant design and production. The high biocompatibility exerted by Gr coatings, with antibacterial and antifouling properties in general, makes them very promising coating materials for medical implants. Further research is therefore required to understand the potential of Gr coatings within implant design, production, and use, and their safety and efficacy in the clinical setting (Figure 14.3).

14.3.2 Application of Graphene for Dental Implants

Innovations based on materials have led to many advanced treatments and technological marvels to cure a different disease in the last few decades. The use of new metallic alloys such as stainless steel and nitinol (NiTi) is often common in biomedical implants owing to their better mechanical properties [111–115]. The integrating biological system and tooth implant underwent fibro-osseous

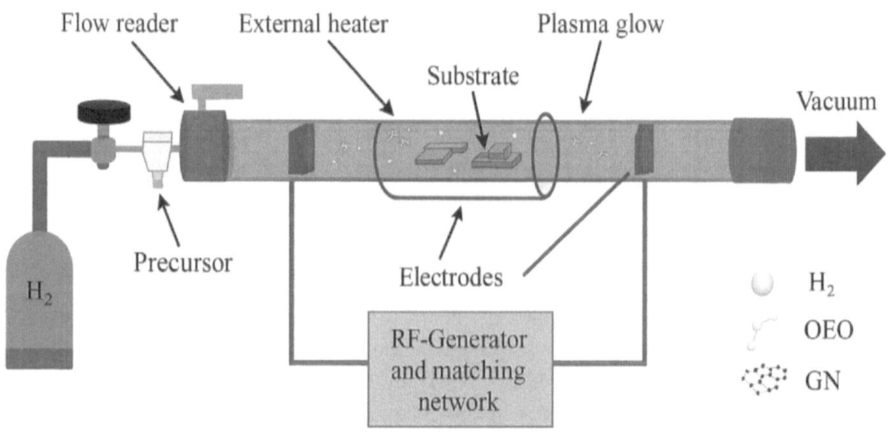

FIGURE 14.3 Representation scheme of the RF-PECVD system utilized to fabricate Gr on CoCr substrate.

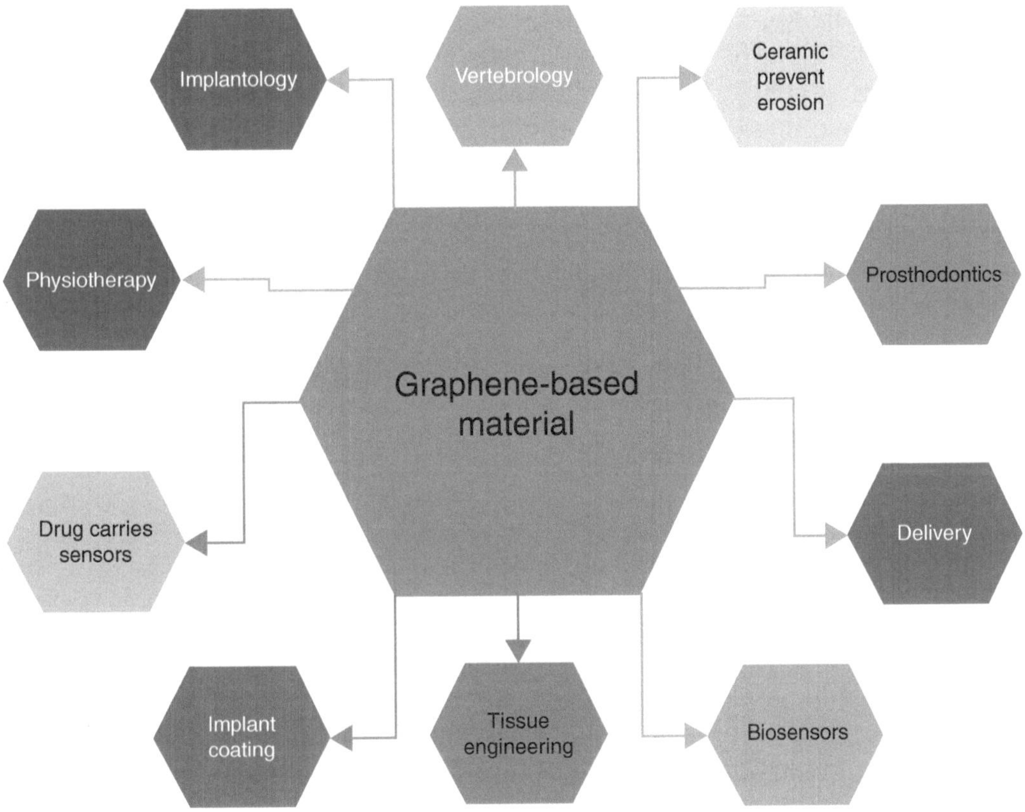

FIGURE 14.4 Applications of graphene nanoparticles and their derivatives in the field of dentistry.

fusion following dental insertion. For osseointegration to occur at the complex tissue interface, the implant material needs to exhibit osteogenic features. In contrast, at the soft tissue interface, the maintenance of tight epithelial sealing is necessary because modifications to titanium implant-related factors, such as topography, hydrophilicity, surface roughness, topography, and shape, have been demonstrated to recently affect the rates and/or quality of osseointegration in studies involving nanotechnology [116,117]. Dentistry, referred to as oral medicine, deals with identifying the irregularities of dental structures at an early stage. Recently, researchers have conducted a somewhat exhaustive study on graphene-based nanoparticles for dentistry. In the domain of dentistry, graphene has the following major features, outlined in Figure 14.4.

14.3.3 Applications in Tissue Engineering

Tissue engineering repairs and replaces multiple imperfections caused by tumors, injuries, illnesses, and other conditions. Basically, it is acknowledged that scaffolds offer a surface on which the various stem cells adhere, proliferate, and differentiate during tissue creation. Many studies have demonstrated the possibility of using graphene-based materials to construct or protect tissue engineering scaffolds. These different studies show the promotion of osteoblasts in various stem cell types by graphene. The hybrid sheet consisting of graphene and HAp showed outstanding biomimetic weathering. In addition to high surface area, mechanical strength, electrical conductivity, and other mentioned advantages make graphene a very promising substance for promoting cell adhesion, proliferation, and differentiation. Graphene-based scaffolds have already shown success in aiding the regeneration of bone, neural, cardiac, and skin tissues. Further investigation is therefore required to shed full light on the long-term biocompatibility and safety of graphene-based materials in tissue engineering applications. With the development in this area, graphene is going to have huge potential to solve many medical problems and improve the quality of life for patients in tissue engineering.

14.3.4 Osseointegration and Applications on Implant Surfaces

Since the invention of ceramic and titanium implants, their osseointegration has remained a challenge. Currently, researchers are studying surface treatments to enhance antimicrobial properties and tissue-implant interaction in a bid to improve the success rate. This interface is critical in various processes, such as inflammation, cell recruitment, protein adsorption, and biofilm formation. Such application of GO-coated titanium implants demonstrated a variety of positive effects on cell proliferation, Alkaline Phosphatase (ALP) activity, enhanced gene expression involved in osteogenesis, promotion of protein expression associated with bone formation markers Bone Sialoprotein (BSP), Runx2, and Organic Carbon Nanostructures (OCN). It was noticed that with the increasing thickness of the graphene oxide layer, there is enhancement of ALP-positive areas and increase in the mineralization of the extracellular matrix [117–123]. However, the three-dimensional (3D) morphology was not provided by those initial graphene-based coatings, which is a key factor necessary for the osseointegration process. The first 3D porous coatings thus were developed by Qiu's group using GO/rGO onto pure titanium plates. These products demonstrated very high osteoinduction capacity with good biocompatibility. In their observations, Li et al. commented that coating titanium with GO increased new bone formation with less gap between the implants and surrounding peri-implant bone tissue [124–126].

14.3.5 Anti-Fibrotic Coating Graphene Oxide-Based Hybrid Interface for Metal Implants

In tissue engineering, such FBRs could result following the insertion of implants used for medical reasons; this is a big challenge. Materials mainly used in implants today are metals that are non-organic; as such, since it lacks biocompatibility and lack of immune regulations often lead to fibrosis after long periods of implantation. In [127] they showed a highly biocompatible hybrid interface based on graphene oxide and collagen I, wherein topological nanostructures act to effectively inhibit the differentiation of fibroblasts into myofibroblasts. By changing the GO concentration, modulation of the structure and roughness of this coating interface could easily be made at the nanoscale level and would effectively induce polarization of macrophages into the M1 state without producing excessive pro-inflammatory factors. Compared to nanomaterials or the extracellular matrix alone, this hybrid bio-interface exhibited better mechanical strength and displayed physical structures with high inflammation as an anti-fibrotic interface. GO-COL has shown great potential in applicability within medical implants and cell–material interface, proven by many inorganic materials like glass, titanium, and nitinol [127].

14.3.6 Graphene–Calcium Carbonate Coating for Biodegradable Implant Properties

Biodegradable implants are of interest in regenerative orthopedic procedures, but they could be exposed to too-fast corrosion in the human-body environment. This thus calls for the synthesis of a suitable coating that might improve the corrosion resistance of such implants without prejudicing their mechanical integrity. The thin coating of reduced graphene oxide–calcium carbonate composite was applied on an AZ91 magnesium alloy, representing a biodegradable Mg implant material. Such a coating is comprehensive in terms of analytical and in vitro electrochemical characterization, which may improve corrosion resistance and mechanical integrity; therefore, it has huge potential related to applications.

14.3.7 Toward Smart Biomaterials for Bone Research

The demand for such biomaterials, able to promote either the repair, replacement, or restoration of hard and soft tissues, increases with aging populations. Traditionally, smart biomaterials are defined as those that respond to stimuli. However, considering the rapid development in this field, it is time to change how the smartness of biomaterials is conceptualized. In [128,129] the authors redefine the term "Smart Biomaterial" and address recent advances and applications of smart biomaterials for hard tissue restoration and regeneration. To clarify the use of the term, "smart biomaterials" were proposed four degrees of smartness according to the level of the interaction of the biomaterials with the bioenvironment and the biological/cellular responses they elicit: these are defined as inert, active, responsive, and autonomous. Possibility of uses of smart biomaterials for hard tissues, mainly based on the materials' responses—both external and internal stimuli—and their use as immunomodulatory biomaterials were analyzed in [129]. We finish by addressing the limitations and challenges to translation from bench to bedside that is needed for the creation of clinically relevant uses for these technologies [129].

14.3.8 Immunomodulatory Materials Used in Bone Regeneration

Multiple mechanisms secure interactions between bone tissue and the immune system. Information about these interactions at a detailed level will provide a scientific basis for the development of a new group of smart biomaterials in bone regeneration: immunomodulatory materials. The latter can manipulate and control the host immune system to generate key regenerative effects, such as the initiation of tissue repair, inflammation reduction, or osteoblast differentiation. Applications of such immune-modulatory smart materials can help to overcome complications in bone healing and bone-related diseases, enabling the development of more efficient and safer therapeutics and more efficient regenerative engineering approaches. Immunomodulatory biomaterials are relatively recent newcomers to the family of smart biomaterials for bone regeneration. Any biomaterial possessing the capability to interact with and manipulate host immune response, leading to the induction of hard tissue repair and/or bone regeneration, either locally or systemically, could be called an immunomodulatory biomaterial for this review.

14.3.9 From Basic Research to Bedside

Many challenges have often been translated into avenues of biomedical investigations and finally to clinical applications because these interactions among the major stakeholders involved—academia, regulatory authorities, and industry—are not well coordinated. These three stakeholders are major players in the quite elaborate process of bringing a scientific discovery from the laboratory bench to the patient's bedside. While the academic community is likely to assume responsibilities for the initial discovery and early development of new technologies and therapies, the lack of regulatory input and guidance may result in failure to meet the standards of safety and efficacy that must be fulfilled for such an application in the clinic. Furthermore, the participation of the industrial sector in increasing the scale of production, organizing extensive clinical trials, and ultimately offering new treatments to the general public becomes indispensable.

All too often, however, there is almost a break in continuity between these sectors, thus delaying the process and creating a logjam in the development pipeline. One of the reasons this happens is that academia is usually poorly placed in terms of resources or expertise to navigate the complex regulatory environment, and industry may be driven by other priorities—marketability and profit margins—which sometimes conflict with securing the best clinical outcomes. On the other side, the regulatory agencies face pressure to strike a balance between the need for rigorous safety and the need for rapid, innovative, life-saving treatments to reach the marketplace.

Addressing such issues will raise communications and collaboration among these players and fill the dwindling pipelines of translational research. With such well-orchestrated interactions, streamlining the translation process could be possible, hence fast and safe conversions of scientific discoveries of biomedical breakthroughs into practical clinical applications. Not only in this way will the development of new therapies be accelerated, but also patients will benefit more quickly and fully from the latest innovations in biomedical science.

14.3.10 New Materials for Bone Tissue

Material design in bone tissue engineering involves understanding the composition and structure of native bone tissue, along with the proper selection of biomimetic natural or tunable synthetic materials—specifically, polymers, bio ceramics, metals, and composites. Such scalable fabrication technologies—such as 3D printing and electric-field-assisted techniques—can then be used to process these biomaterials into forms appropriate for bone-tissue engineering. In this review, we outline considerations of material design for bone tissue engineering applications in both disease modeling and treatment of injuries and diseases in humans. We present the pathway to material design, starting from the strategy in implementation to material and fabrication method selection to, at the very end, evaluation. We further discuss some of the unmet needs in the development of ideal materials for bone tissue regeneration, current challenges, and emerging strategies in this field. Successful design of these materials for bone tissue engineering is contingent on the clear understanding of the defect site, the end user's needs, and the functional deficits the materials will address. Bone defects at both orthopedic and craniofacial locations are very serious but differ appreciably in their detriments to a patient's quality of life. For example, such requirements for each specific clinical application must be defined at the outset of the materials' design process.

14.3.11 Graphene-Doped Strontium as a Powerful Complex for Bone Tissue Engineering

Graphene and its derivatives have huge potential in tissue engineering applications due to their super mechanical properties, electrical conductivity, and antibacterial activity. Considering some other characteristics of graphene, such as high surface area and ready-to-use functionalization according to the high oxygen-containing groups in the graphene oxide family, some needs in bone tissue engineering could be covered. In [130–134] researchers synthesized and decorated strontium nanoparticles on graphene oxide during the reduction process using a green and novel method. Strontium NPs were synthesized and decorated on the surface of rGO simultaneously without using hydrazine or chemical linkers by means of BSA. The results of UV–Vis, FTIR, and Raman spectroscopy demonstrated that BSA could reduce graphene oxide and decorate SrNPs on the surface of rGO. The present study demonstrated that in situ synthesized SrNPs were nearly 25–30 nm in diameter by Field Emission Scanning Electron Microscopy (FESEM) and Transmission Electron Microscopy (TEM). The cell viability for MC3T3-E1 cells treated with SrNPs-rGO, in this regard, was interestingly significantly higher than BSA-rGO and GO at a constant concentration. The alkaline phosphatase activity of the nanosheets was also studied, and the results obtained revealed that Sr-BSA-rGO enhanced ALP activity against GO and BSA-rGO. The research revealed that using proteins and other biomolecules as green and facile agent for decoration of smart nanoparticles on the surface of nanosheets would be promising, and it would assist researchers to replace harsh and toxic hydrazine-like materials with bio-friendly methods. These results indicated that Sr-BSA-rGO had excellent capability in terms of regeneration of bone tissue and could be used as an osteogenesis booster in implants.

14.3.12 As a Cell Layer in Bone and Cartilage Regeneration

Bone defects result from trauma, tumor resection, infection, and congenital deformities. Together with defects in articular cartilage and those in the cartilage–subchondral bone complex resulting from trauma and degenerative diseases, they remain significant challenges for clinicians. Novel strategies that make use of cell sheet technology in enhancing bone and cartilage regeneration are under development. The cell sheet technology showed huge clinical potential in regenerative medicine because of its effective preservation of cell–cell connections and the extracellular matrix, and its scaffold-free nature. In the [130–134] review will introduce some of the widely used cell sheet preparation systems, including traditional approaches, recent improvements, and their advantages and shortcomings. In this regard, recent advances in the application of cell sheet technology for the regeneration of bone or cartilage defects and bone–cartilage complex defects are reviewed. The major challenges and future research directions for the application of cell sheet technology in bone and cartilage regeneration will also be discussed [135]. Results with the application of cell sheets alone have been quite encouraging in ectopic sites or bone defects. After osteogenic induction, the rat Bone Marrow-derived Mesenchymal Stem Cells (BMSCs) were prepared into cell sheets by means of a mechanical system and then subcutaneously transplanted into the rat thigh. New bone formation was promoted without scaffolds [136]. Much progress in the preparation and application of cell sheets has been made during recent decades. With great efforts from numerous researchers, cell sheet technology is becoming a very promising strategy for the repair and regeneration of bone and cartilage defects. Some obstacles remain in applying cell sheet technology. For instance, difficulty in constructing tissue or organ-like tissue with complicated structure and morphology. There are cases of necrosis of cells in multilayered cell sheets. Moreover, the composition of cells and ECM in the cell sheets is very different from that of native cartilage [137].

14.3.13 Application of Chitosan–Graphene Oxide 3D Scaffolds in Bone Regeneration

One of the challenging factors when reconstructing critical sized bone defects is that the self-regenerative capacity of the human skeleton is extremely low. The study was carried out focusing on bone tissue regeneration with 3D chitosan scaffolds enriched with graphene oxide. This combination was tested in a critical-sized mouse calvarial defect model for its efficacy in promoting osteogenic differentiation and bone repair.

Histopathological and histomorphometry analyses, together with SEM on the implants, have shown that there was a significantly higher amount of new bone formation in defects filled with Chitosan (CHT)/GO compared to CHT alone, with a p-value less than 0.001. Moreover, in the presence of GO, it also synergistically advanced the osteogenic potential of CHT scaffolds, as evidenced by the increased ALP activity in both in vitro and in vivo studies. It was further supported by the upregulation of BMP and Runx2 in the first four weeks after implantation, which suggested that GO acts to enhance the osteoprogenitor cell differentiation process.

Furthermore, this osteogenesis-promoting activity of GO was shown to have continued later through the increased expression of osteopontin at week 8 and osteocalcin at week 18. These findings suggest that CHT/GO biomaterials could be a very promising strategy for the reconstruction of large bone defects without using exogenous living cells or growth factors. This work offers insights into the great potential of GO-enriched scaffolds for the advancement of bone tissue engineering and clinical bone repair solutions [138].

14.4 APPLICATIONS IN SPINAL CORD, INFLAMMATION MANAGEMENT AND DRUG DELIVERY

Since SCIs are among one of the most refractory neurological diseases, permanent neurological dysfunction generally occurs in patients. Unfortunately, given the complex pathological mechanism and unique microenvironment resulting from injury, to date, there is no way to totally repair the injured spinal cord. In recent years, with the rapid development of tissue engineering technology, a combination of biomaterials and medicine would give a new idea for treating SCI. Herein, we will summarize representative natural, synthetic, nano-, and hybrid biomaterials and their applications in treating spinal cord injury. In this paper, some state-of-the-art fabrication techniques in tissue engineering are presented. Authors have tried to give several new insights on the use of biomaterial-based therapeutic strategies in attempts to reduce the secondary damage and to promote repair. Finally, some clinical studies of biomaterials are presented. In [139] the review provides a reference and new insights into further exploring spinal cord regeneration strategies in the future. After SCI, the activated immune cells infiltrate into the lesion site and release inflammatory factors and ROS. These continuous immune and inflammatory cascades result in secondary damage. Therefore, delivering anti-inflammatory treatments to modulate the SCI immune microenvironment is of help [140]. Three physiologically relevant ROS include superoxide, hydrogen peroxide, and hydroxyl radicals, whose production is tightly controlled with their elimination. Following SCI, the excess ROS released into the extracellular microenvironment disrupts this very equilibrium. Reducing ROS is often a by-product of reducing inflammation; however, it deserves equal attention, as inflammation is not the only source of ROS. One of the countermeasures to reduce ROS is the intake of antioxidant molecules; however, there exist certain limitations to them such as easy degradation, poor water solubility, and low bioavailability. Fortunately, the antioxidant activity of a wide array of natural and synthetic compounds has been previously scrupulously investigated. This eventually gave way to new therapeutic strategies. The Blood-Spinal Cord Barrier (BSCB) represents a blood–brain barrier-like structure existing in spinal cord tissue. Under normal conditions, Tight Junctions (TJs) tightly regulate the diffusion of both molecules and cells across the BSCB, allowing for the CNS to be considered essentially as an isolated system. As a result of SCI, regardless of trauma method used, there will be a structural breakage and raised permeability in the BSCB with an influx of immune cells and inflammatory factors into the stroma. For example, reducing astrocyte activation or infiltration may limit glial scar formation, which applies more to acute SCI. These biomaterials work through their inherent properties. In this case, alginate/chitosan/genipin hydrogel composites lowered astrocyte reactivity in vitro while sequestering extracellular Ca^{2+} to prevent Ca^{2+}-induced secondary injury. Specifically, recently developed aligned hydrogel microfiber scaffolds constructed with photo crosslinked gelatin methacryloyl and electrospinning technology significantly reduced astrocyte activity in vivo, thus inhibiting glial scar formation [141–146].

14.4.1 In Applications to Enhance Neurorestoration

In addition, neurorestoration based on the management of secondary inflammatory cascades should also be enhanced, since this is of necessity for the achievement of the functional recovery that is the ultimate goal of injured spinal cord repair. The next sections detail three biomaterial-based strategies to enhance neural repair. With the continuous progress in cell culture techniques, cell transplantation presents a new-generation treatment for SCI not only in replacing or protecting damaged tissue but also in promoting axonal regrowth through both physical and trophic support. Inflammation is of central importance in the response to danger signals that emanate from damage to our body and in restoring homeostasis. Indeed, dysregulated inflammatory responses underline a broad span of diseases, ranging from cancer to sepsis and autoimmunity [147]. Biomaterials can potentiate the potency of anti-inflammatory drug therapies developed against dysregulated inflammation by both improving the bioavailability of drugs and reducing side effects. In this review, we outline key elements and stages of the inflammatory environment before discussing strategies for the design of biomaterials to deliver different anti-inflammatory therapeutics. In that regard, biomaterials can be designed to scavenge danger signals such as reactive oxygen and nitrogen species and cell-free DNA at the early stages of inflammation. Materials can also be designed in order to prevent adhesive interactions of leukocytes and endothelial cells that lead to inflammatory responses. Another way nanoscale platforms work is by delivering anti-inflammatory agents to sites of inflammation. We end with a discussion of the challenges and opportunities for innovations in biomaterials to address inflammation [145].

14.4.2 Non-Covalent Functionalization of Carbon Nanotubes for Tissue Engineering

It is another hot topic of recent research and requires special requirements. It depends on the development of scaffolds allowing tissue formation with specific features. Carbon nanotubes–collagen composite attracts much attention from this point of view for researchers. However, Carbon Nanotubes (CNTs) suffer from low water dispersibility, hampering their utilization. Hence, we intend to functionalize CNTs non-covalently by using an appropriate hydrophilic linker derivatized from polyethylene glycol (PEG) terminated with either a hydroxyl or carboxyl group in order to disperse them in water. Functionalization of the CNTs was confirmed successfully by TEM, absorption spectroscopy, Thermogravimetric Analysis (TGA), and zeta potential analysis. Engineered Cardiac Tissues / Electrical Conductivity Tests (ECTs) generated using 3T3 cells were prepared with different concentrations of the functionalized CNTs. These tissues show significantly enhanced electrical conductivity at a concentration of 0.025%, but cell viability is reduced by about 10–20%. All ECTs containing f-CNTs demonstrate significant reduction in tissue fibrosis and matrix porosity against the control tissues. All these constructs developed in this study are therefore quite promising for further in vivo studies as an engineered tissue. Effective non-covalent functionalization of CNTs was achieved, which provided stable dispersion of these macromolecules in water. Collagen-based ECTs containing 3T3 cells showed a significant increase in electrical conductivity due to dispersed f-CNTs within the collagen matrix and superior performance of the f-MWCNTs compared with the f-SWCNTs. These concentrations of f-CNTs showed only an acceptable level of toxicity. All together, the results from the constructs developed in this work could form a basis for further optimization of excitable engineered tissues for various in vivo applications [146,197–203].

14.4.3 Applications in Drug Transportation

Graphene, in general, is viewed as a wonder material that sweeps many different areas of scientific studies, ranging from materials science to electronics, condensed physics, quantum physics, and finally, energy systems. Since 2004, it has been subjected to extensive studies to understand its physical and chemical properties. Having unique features, it rapidly became a potential candidate for nano-bio researchers to study its applications in biomedical fields. In the past decade, tremendous efforts have been devoted to exploring biomedical applications for graphene and related materials, especially in the fields of smart drug and gene delivery and cancer therapy. With many successful integrations of graphene-based materials into the biomedical field in mind, we will attempt in this work to summarize the very latest developments regarding graphene applications in biomedicine. The state-of-the-art advances of graphene-based materials in drug delivery applications, targeted drug/gene delivery, delivery of antitumor drugs, controlled and stimuli-responsive drug release, photodynamic therapy applications, and optical imaging and theranostics were reviewed in this paper. We investigate the future trends and succeeding challenges in this topic to provide an outlook for further research.

14.4.4 Graphene in Smart Gene Delivery Systems

The smart gene delivery (SGD) systems are sophisticated techniques of gene delivery that involve transferring foreign DNA into a cell, for instance, gene therapy or genetic research. They are classified based on the mode by which the gene is introduced into the cytosol:

- **Chemical Methods:** These are the methods involving gene delivery through organic complexes. Common examples include lipid-based carriers, cationic polymers, and calcium phosphate. These methods have wide applications because of their efficiency and ease of application.
- **Biological Methods:** It makes use of viral or bacterial vectors in delivering genes. This method involves infecting target cells with specific genes. The process uses the natural process of viruses and bacteria for gene transfer.
- **Mechanical Methods:** These are physical techniques used in the direct introduction of naked nucleic acids into cells. Adopted techniques for direct transfer include electroporation, microinjection, and biolistics, more popularly known as gene guns.

Each of these methods has associated advantages and limitations, and system choice largely depends on specific requirements of the gene therapy or research application [147–151].

14.4.5 Applications in Optical Imaging and Theranostics Systems

The American Cancer Society has estimated that one in every four deaths will be caused by cancer; thus, this is a highly serious disease. One of the main complications of cancer treatment is that it tends to disseminate in the blood, which consequently develops metastasis. Early detection in the process is very important for management and treatment of this disease.

The use of nanomaterials in imaging applications has been receiving loads of efforts in the recent past due to their promising role in changing diagnosis and treatment of cancer. One of the most pioneering developments in this regard is nanotechnology-aided theranostics. It is an integrated approach that should bind diagnostic imaging with therapeutic interventions, thus ensuring the right drug for the right patient at the right time and the use of noninvasive but selective cancer therapies.

Noticeably under development are Photothermal Therapy (PTT) and Photodynamic Therapy (PDT), together with their imaging techniques for the diagnosis of diseases. These methods have received significant attention because, unlike traditional drugs and diagnostic dyes, they have very distinct advantages in terms of safety: they are associated with minimal toxicity and side effects, which greatly sets them apart from the conventional anticancer treatment means and increases their potential to make a difference in effective and friendly cancer care.

14.4.6 Graphene Angiogenesis and Anti-Angiogenesis

Besides their high capability in drug/gene delivery, some studies have focused on the angiogenesis/anti-angiogenesis properties of GO nanoplatforms. In the process of angiogenesis, new vessels form in the primary vessel channels to supply the essential nutrition and oxygen for rapid proliferative cells and take away the useless biochemical products. According to many publications, it participates in wound healing and the treatment of cardiovascular diseases, as well as in tumor growth [148–150]. Stimuli-responsive drug delivery has been paid much attention during the past decades. It provides spatial and temporal-controlled release in response to different triggers, thus making highly efficient drug delivery with minimized drug side effects possible. In this regard, graphene-based nanomaterials have been widely investigated, which proved very promising in smart drug delivery due to stimuli-responsive behavior and high loading capacity against an extended range of drug molecules. These features are attributed to large surface areas, mechanical and chemical stability, and exceptional optical, electrical, and thermal properties. It is this great and infinite functionalization potential that also enables them to be integrated into several types of polymers, macromolecules, or other nanoparticles to create novel nanocarriers with greatly enhanced biocompatibility and trigger sensitive properties. Thus, numerous studies have been dedicated to graphene modification and functionalization. The present review covers graphene derivatives and different sorts of graphene-based nanomaterials applied for drug delivery, explaining related significant developments of their functionalization and modification. Besides, we will talk about their potential and advancement in intelligent drug release in response to various types of stimuli, either endogenous (pH, redox conditions, and reactive oxygen species) or exogenous (temperature, near-infrared radiation, and electric field).

14.5 IN THE DIAGNOSIS OF CARDIOVASCULAR DISEASES AND CANCER CELLS

With gigantic surface areas and a multivalent structure, two-dimensional (2D) GO will exhibit remarkable binding with single-stranded DNA/RNA through hydrophobic and π-stacking interactions. This aptamer binds with very high affinity and specificity due to its small size, exceptional chemical stability, and low immunogenicity. The synthesis of GO–aptamer conjugate materials, which involves integrating aptamers with GO, is a method that can help researchers offer better alternatives to traditional antibody-based approaches for cancer diagnostics and therapy. Moreover, GO's advanced fluorescence quenching properties open up ways to develop efficient fluorescence-sensing platforms. This paper reviews the recent progress in the development of GO–aptamer conjugate materials and their application for the detection of major cancer biomarkers [151,152].

14.5.1 Aptamers as an Alternative to Antibodies

Aptamers have recently been developed as an ultra-powerful replacement for antibodies, mostly in the area of molecular recognition. Aptamers resemble antibodies in many aspects: high specificity and affinity for target molecules, and the ability to bind in a manner usually compared to that by which antibodies attach to antigens. This makes aptamers quite useful in diagnostics, therapeutics, and biosensing applications. It is very stable and versatile under different conditions, one of the main advantages of aptamers in comparison with conventional antibodies. While antibody based devices are useful, they mostly require strict control of such environmental parameters as a precise temperature range, relative humidity, and ionic strength of the buffer solution for proper functioning. Such requirements limit the use of antibodies in field conditions, where such parameters are hard to control. In contrast, aptamers are more rugged and can retain their binding properties even under very challenging environmental conditions. This makes them especially suitable for applications in portable devices and in-field diagnostics, where the environment cannot be precisely controlled. Aptamers can easily be engineered for resistance against temperature, humidity, and ionic strength changes, greatly expanding their use beyond the laboratory environment.

Moreover, the aptamers are chemically synthesized and therefore have a number of advantages over antibodies. First, the chemical synthesis of aptamers enables levels of reproducibility and consistency of production that are often lacking in the case of antibodies produced through biological systems. This will then allow easy modification of the aptamers to improve stability, binding affinity, or other properties, which in fact confers flexibility in the design of molecular recognition tools on researchers [154–156].

Another characteristic is that aptamers can be selected against a wide array of targets, including proteins and small molecules, and even whole cells or viruses. In vitro selection methods, SELEX, for the systematic evolution of ligands by exponential enrichment, ultimately lead to the isolation of high-affinity aptamers for nearly any chosen target. This makes them one of the most promising alternatives to antibodies in many applications because it is coupled with stability and ease of production.

Aptamers also have some distinct advantages in therapeutic applications. They are, by and large, nonimmunogenic—that is, they do not provoke an immune response to the extent that other molecules do when used in vivo. This is a very important consideration for therapeutic agents, since an immune response can result in a reduction of treatment efficacy and in the instigation of side effects. Further, aptamers can easily be conjugated to other molecules or therapeutic agents, enabling drug targeting and increasing the therapeutic potential.

14.5.2 Application of Graphene Oxide Materials for Cancer Diagnosis

This has led to an increased number of reports on biomarkers, therefore necessitating rapid and robust methods for cancer diagnosis. GO has a very large surface area, high conductivity for surface functionalization, and efficient fluorescence quenching, which are of prime importance for application in biological sensors and the detection of biomolecules. In this respect, the GO–aptamer conjugate materials have been hybridized as probes for cancer diagnosis, in which its unique fluorescence quenching properties are utilized for the detection of various cancer biomarkers. The following sections will further probe into the GO–aptamer conjugate materials in terms of their effectiveness in detecting main types of cancer biomarkers, which include PSA, mucin 1, CCRF-CEM, and circulating tumor cells (CTCs). Acute lymphoblastic leukemia is one of the most common childhood malignancies. It is a disorder that emanates from the uncontrollable production and accumulation of lymphoblasts, a variety of cancerous and immature white blood cells. These lymphoblasts proliferate rapidly in the bone marrow and cause damage and death by inhibiting the normal process of replication of other blood cells and through infiltration into other organs. Due to the short ALL pathogenesis time and the average survival period of patients with no treatment of three months, an early diagnosis of ALL at a mixed reality stage of the disease may improve the chance of survival. Some possible reasons for the low cure rate may be the drug resistance, poor tolerance, and less effective treatment.

14.5.3 In Diagnostics of Circulating Tumor Cells

CTCs are circulated via the bloodstream and thereby to various tissues of the body after being normally shed from metastatic tumors. That can lead to metastasis and is a major way of cancer dissemination that causes many deaths in cases of cancer [157]. Well, it is inarguably a fact that the transfer of findings on CTC research from the laboratory to clinical practice is slow, and the unduly vague benefit of using CTCs in decision making for the treatment is widely accepted. An increasing knowledge base of CTCs and related materials can give further insights, and future studies should be related to using proteomics and genomics to evaluate CTCs. Besides effective CTC detection, a growing need is the development of highly efficient methods for isolation and purification of CTCs for molecular and functional analyses. Functionalized GO exhibits the selective biosensing property coupled with the therapeutic capability, and an aptamer-conjugated theragnostic GO-based assay has been applied to cancer diagnosis using the detection of CTCs from blood. Apart from the most common cancer biomarkers above-mentioned (PSA, MUC1, CCRF-CEM, and CTCs), very few other cancer cell lines are reported, some of which have been briefed herein. A dual-aptamer-functionalized GO complex uses two kinds of aptamers, Sgc8c and ATP, and GO to detect Molt-4 cells, a T-cell line of ALL. The aptamer Sgc8c binds with high sensitivity and selectivity to protein tyrosine kinase-7, a major biomarker for T cells ALL. A dual-aptamer-functionalized GO complex showed significant fluorescence emission for target cells, while it does not result in internalization into normal cells (U266 cells). In another study, a RuOMO label-free optical sensing platform was developed by using GO and a TB-specific pair of aptamers, which captures TB as a biomarker of pulmonary metastases [158–160].

14.5.4 Properties of GO-Based Samples in Cancer Diagnosis

The fluorescence quenching capability of graphene, which is induced through the process of fluorescence resonance energy transfer, is utilized in graphene-based fluorescence biosensors. Graphene-based materials, G, GO, and rGO, have been used rather extensively as the key material component in biosensors for the detection of cancer biomarkers. Control over graphene property through reduction or functionalization-mediated pathway is also found to be of much use. The large number of carboxyl groups present and the strong adsorption of certain biomolecules by graphene help in the functionalization of GO. GO interacts with ssDNA concurrently and can quench fluorescence efficiently as they are strong acceptors of Förster Resonance Energy Transfer (FRET) because of their broad absorption in the visible region. Fluorescence quenching by GO is very effective compared to other carbon nanomaterials. According to the known facts, energy transfer efficiency depends inversely on the sixth power of the donor–acceptor distance; therefore, FRET will be very sensitive to even a small change in this distance. Efficient quenching of dye fluorescence occurs when the dye lies close to the surface of GO. Further, GO showed enhanced electron transfer ability besides high conductivity, unique pliability, and biocompatibility. These properties are useful for the application of GO in biosensing techniques for cancer detection, and various fluorescent biosensors have been reported for the detection of cancerous cells and cancer markers [161–167].

14.5.5 Application of Nanotherapy

The foremost source of mortality across the globe, cardiovascular disease, proves a challenging condition for researchers, clinicians, and patients alike. Cardiovascular disease includes a wide collection of diseases that fills the heart, the vasculature, and the blood that runs through and interconnects them. Constraints toward therapies and diagnostics have generated an enormous interest in the progress toward nanomedicine in Chemical Vapor Deposition (CVD). Focused on the improvement of patient outcomes with transformative therapies, imaging agents, and ex vivo diagnostics, the therapeutic and diagnostic landscapes in CVD have been profound. Clinical applications fundamentally drive the development of CVD nano systems and include the following: (1) improved function of either a cardiac or heart-related biomaterial alone or combined with the functional advantages of nanomaterials, such as mechanical, immunological, and electrical; (2) the vasculature, where systemically injected nanotherapeutics image nano-diagnostics, nano-enabled biomaterials, or tissue-nanoengineered solutions; and (3) enhancing a sensitivity and/or specificity improvement when used with ex vivo diagnostic devices for patient samples. Whereas immunotherapy has become an onco-nanotechnology cornerstone over the past decade, CVD immunotherapy and immunoimaging are the new kids on the block and promise to play significant roles in the CVD armamentarium over the next decade. The range of nanomaterials that are part of current CVD-related trials and the myriad promising preclinical approaches are indicative that nanomedicine abounds at a moment when it is poised to have a huge effect on companion animals with CVD. Here we revisit these recent advances, underlining with comments key clinical opportunities in this rapidly emerging field of CVD nanomedicine. Nanotechnology is very much more than just the progressive reduction in size of current technology; it is an enabler of new possibilities across the spectrum of medicine. Such miniaturization has driven advancements in the imaging of anatomic structures and functional biological states in health and disease. Innovations exist at the present time that can target active moieties at any cellular and even molecular level, whereby directed therapy will be feasible using some specific, smart therapeutics, materials, and devices, such that the interactions with the individual are programmed rather than only tolerated. Thus, it provides the opportunity to specifically and intelligently modulate immune response and reactivity of implants, both supported by stunning progress in basic immunological knowledge and with the opportunity for real-time, iterative feedback from implants and injections of the local environmental state. A future integration of developments in other fields, such as artificial intelligence (AI) and systems biology, with nanotechnology, may make the whole spectrum of health, from its monitoring and disease detection to programs interlinked in therapy and in vivo prognosis with response-to-therapy imaging, possible. Nowhere is this more evident, important, and impactful than in cardiovascular medicine, in which the future will allow for the first time physiologic rather than pharmacologic therapeutics and patient-specific and environmentally specific interventions and will reveal a new appreciation for biology on the cellular-to-subcellular scale [168]. Intravenously injected vascular nanotherapies are generally intended to target sites of atherosclerosis, thrombosis, or intimal hyperplasia; these vascular diseases form an interconnected arc from atherosclerotic plaque development to plaque-related thrombosis to aftermath blockage, that is restenosis, which may be caused by thrombosis or intimal hyperplasia. Nanotherapies to treat these interrelated yet distinct CVDs can be generalized into two categories: nanomaterials that encapsulate presently used therapeutic molecules to increase payload delivery and/or to restrict delivery to specific cell subsets or tissues (targeted delivery) in order to decrease adverse effects, those that enable completely unprecedented CVD therapies, for example, which includes encapsulating drugs that could not be delivered otherwise, or that transduce intrinsic or applied energy in order to physically modulate tissues. For all CVD therapies, minimizing exposure to off-target sites is critical. Nanotherapies have been designed rationally to these ends and often achieve minimized side effects as well as optimized therapeutic efficacy [169]. Major classes of non-immune-mediated nanotherapies include those for the modulation of lipid and glycolytic metabolism and those for the prevention of plaque neoangiogenesis. For instance, liposomal nanoparticles have been designed to target altered lipid metabolism through small interfering RNA that silences the apolipoprotein B and proprotein convertase subtilisin–kexin type 9 (PCSK9) expression, thus reducing low-density lipoprotein levels in small and large animal models. Polysaccharide nanoparticles activated macrophage metabolism and normalized vasculature by delivering a glycolysis inhibitor, which in turn decreased atherosclerotic plaque inflammation in murine models [170–173]. Other non-immune-mediated approaches involve polymeric nanodelivery of the biochemical machinery to drive antioxidant responses [174].

14.5.6 In Applications to the Myocardium and Electrophysiology

Heart muscle is a functional syncytium of cardiomyocytes that generates and conducts electrical impulses to contract and produce force. In function, cardiomyocytes need to be healthy and viable and in their proper 3D structure. Given the complexity, specialized nature of cells, and precise 3D spatiotemporal structure critical to its function, it is not surprising that this myocardium suffers from a wide variety of disorders and dysfunctions, ranging from diverse cardiomyopathies to ischemia. Myocardial therapies usually consist of intracardiac or IV injections or implants whose properties may be endowed or substantially improved by nanomaterials [175].

14.5.7 In Cardiovascular Nano-Immunotherapy

It has been very clear that the immune system is strongly implicated in most vascular diseases, and in many cases, immunity really drives disease pathophysiology, thus forming a very compelling clinical target. While cancer immunotherapy rapidly matures, that of cardiovascular disease remains at an early stage. Indeed, the Canakinumab Anti-Inflammatory Thrombosis Outcomes Study trial was the first proof of concept for the clinical efficacy of cardiovascular disease immunotherapy. However, despite successes, there are key risks: the systemic targeting of innate immunity can induce profound effects on organism defenses, risking infection and sepsis. Nanomaterials are exciting in large part for their ability to minimize such systemic effects through targeting only the cell subsets or tissues of interest. Cell selectivity can thus restrict injected drug loads and decrease adverse effects, all while increasing the drug's efficacy [174–177].

14.6 APPLICATIONS IN WEARABLE PHYSICAL, PHYSIOLOGICAL, BIOCHEMICAL SENSORS AND BIOELECTRONICS IN HEALTHCARE

There are now large numbers of flexible and wearable sensing technologies in the pipeline, which can run a whole gamut of physical and physiological measurements. In the past couple of years, the rapid pace of development and deployment of such sensors has testified to the increasing importance and potential utility of this singular class of sensing platforms in applications such as wearable consumer electronics, soft robotics, medical prosthetics, electronic skin, and health monitoring. In this review, we present the current state-of-the-art overview of the emerging flexible and wearable sensing platforms for healthcare and biomedical applications. First, we introduce the selection of flexible and stretchable materials, and the fabrication of sensors based on these materials. Sensing platforms based on different solid-state, and liquid-state physical types are compared, and working mechanisms of the mechanical deforming based on such sensing principles are examined. We then highlight several exciting flexible and wearable physical sensing applications in a variety of new healthcare and biomedical applications, especially artificial electronic skins, health monitoring, and assessment of physiology, and therapy and drug delivery. Finally, this review concludes with the presentation of some views on the challenges and opportunities that lie ahead for this field [177].

14.6.1 Sensor Materials

Basically, flexible/wearable sensors are expected to provide sensing that is both accurate and reliable without marring natural movement and comfort. An important characteristic defining flexible sensors is, therefore, their skin-like stretchability and conformability. Such flexible materials are made of flexible thermoplastic polymers like polyethylene terephthalate, polycarbonate (PC), and polyurethane (PU), which have excellent optical transparency, can be easily processed, and show great deformability. Besides these materials, another category of flexible substrates that has attracted tremendous interest recently is soft silicone elastomers, particularly Polydimethylsiloxane (PDMS) and trademarked silicone rubbers such as EcoFlex, DragonSkin, and Silbione. This family of flexible elastomers provides high deformability and conformability on a variety of surfaces with different textures and geometries; hence, they could be potential candidates to serve as one of the basic components in stretchable and wearable sensing devices. Moreover, flexible silicone elastomers are usually inert to chemical reaction and biocompatible, making them superb for implantable flexible sensors [178–185].

14.6.2 Applications in Artificial Electronic Skins

Biological skin-based sensory receptors collect and transmit rich streams of physical variables from an external environment to the mechano- and thermoreceptors. Despite major strides made in understanding mechano- and thermo-sensations, replicating these unique sensory capabilities in artificial skins and prosthetics is mostly an elusive task. Consequently, prostheses and artificial skins are worn not as functional substitutes for amputated natural limbs but rather for supplementary movement aid [186–188] or for cosmetic utility. Another growing use of soft, flexible, and wearable physical sensing platforms other than exciting applications like artificial e-skins or physiological monitoring and assessment systems is for therapeutic and drug delivery purposes [178–181]. One of the emerging technological goals in a wide spectrum of applications related to healthcare and medicine is developing flexible and wearable physical sensing devices. In these sorts of applications, the sensing platforms should design and tailor their sensing parameters—particularly sensitivity and range of detection—to lie in the ranges required for differentiating different types of body movement and/or extracting other physiological information.

14.6.3 Biochemical Composition of Sensors

Athletes are continually seeking new technologies and therapies to gain a competitive edge to maximize their health and performance. Athletes have gravitated toward the use of wearable sensors to monitor their training and recovery. Wearable technologies currently utilized by sports teams monitor both the internal and external workload of athletes. However, there remains an unmet medical need by the sports community to gain further insight into the internal workload of the athlete to tailor recovery protocols to each athlete. The ability to monitor biomarkers from saliva or sweat in a noninvasive and continuous manner remain the next technological gap for sports medical personnel to tailor hydration and recovery protocols per the athlete. The emergence of flexible and stretchable electronics coupled with the ability to quantify biochemical analytes and physiological parameters have enabled the detection of key markers indicative of performance and stress, as reviewed in this paper [189]. Electrochemical wearable sensors have received considerable attention recently because of their potential to monitor a wide array of biomarkers in a continuous and noninvasive manner [190]. Most existing wearable devices utilized in the sporting community currently focus on monitoring physical or physiological parameters (e.g., motion, HR, respiration rate, RR). However, such devices do not permit team trainers and physicians to quantify the biochemical profile of an athlete in a real-time manner with the goal of alleviating soft-tissue injuries, dehydration, or cramping. Proper hydration is key to success as under drinking can lead to hypohydration and over-drinking can lead to hyponatremia (low-serum sodium concentration) [191].

14.6.4 Sweat Sensor Technology for Sports

Microfluidic platforms make possible the collection of sweat from the skin surface by interfacing with eccrine sweat glands at various body locations, with the ability to scale the size and form factor of the device to the application of interest. The ability to extract sweat in this manner for laboratory analysis is of importance, but it relies on post-collection analysis (Figure 14.5).

Clinical studies with wearable sensors incorporated in sports medicine can allow access to a lot of data sets to train analytical models that can accurately, efficiently, and precisely predict athlete injuries based on workload profiles. This translates the acquired sensor data into actionable protocols for sports-medical personnel. One could imagine a platform that, with such advanced measurements, would interpret physiological parameters and biomarkers to give key medical personnel real-time information on how the athlete is performing and to help advise the coach and team doctor on the recovery program required for that individual. Open-source solutions, such as the American Heart Association Precision Medicine Platform, powered by Amazon Web Services (AWS), which allow researchers to share the data from clinical trials will be crucial to the advancement of the translation utility of this field for sports medicine.

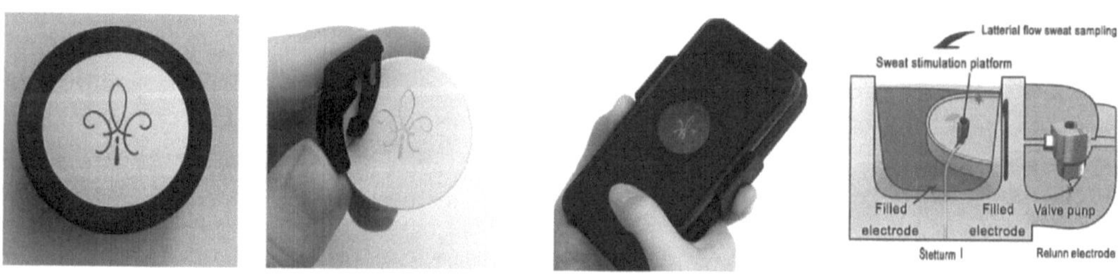

FIGURE 14.5 Wearable sensors to monitor the biochemical status of the athlete by detecting biomarkers from eccrine sweat.

The final goal of wearable sensors for sports medicine is to develop a multimodal, nonintrusive device that enables noninvasive, continuous measurement with combinatorial capabilities of both physiological parameters and biomarkers. Such existing technological roadblocks, coming both from the device side and from the systems side, will enable the translation of this technology to dramatically help practitioners in sports medicine and treatment trainers to properly and effectively monitor programs and adjust treatment plans to provide optimal performance while minimizing the risk of injury.

14.6.5 Graphene-Based Materials Used for Biotechnological and Biomedical Applications

The bulkiness and energy greed of radiofrequency wireless components, usually necessary for bioelectronic systems interfacing with the human body, limit the overall system performance. Artificial 2D materials with subwavelength structure, known as metasurfaces, can be designed to control electromagnetic fields around the human body and could help solve these limitations of current bioelectronic interfaces. The current state of development on metasurfaces for bioelectronics and their ability to create a wide set of contemporary and evolving healthcare technologies was analyzed in [193–197]. The manipulation of the electromotive force (EMF) near a human body with the help of metasurfaces is discussed, followed by applications in microwave imaging, magnetic resonance imaging, biosensors, body networks, and wireless power transfer. The authors [192–203] considered further developments in materials science and AI as well to better metasurface properties for bioelectronics. Because of their flexibility and compactness, the metasurfaces could provide functionalities that conventional, bulky radio-wave components cannot in biomedical applications like medical imaging and wireless power transfer. More specifically, metasurfaces offer optimum control of the electromagnetic waves both inside and around the body for connected personal health systems. Recent developments in medical applications of such structures are expected soon. In this integration process of bioelectronics with the human body and the devices for wireless health monitoring along with therapy, there have been recent breakthroughs. However, medical applications will be explored in due course of time upon testing.

Graphene is an extremely versatile, carbon-based 2D material that is utilized in several research areas that have progressive characteristics, such as properties like high surface area, transport, thermal, optical, elasticity, and strength. Graphene-based composite materials are considered as potentially smart candidates for various biomedical application areas and often used within medical applications. The applications of graphene composite cover electronics, implants, devices, and other biomedical-related applications. In addition to that, in comparison to graphene, its derivatives like graphene oxide and to an extent, reduced graphene oxide, are very easily exfoliated from chemically oxidized graphite and are, therefore, the best candidates for biomedical applications, having exhibited outstanding properties like surface functionalizability, aqueous processability, amphiphilicity, fluorescence quenching. In this proposed review article, some of the latest researches on biomedical applications of graphene materials, in particular, graphene and its derivatives with polymer/nanoparticles, were selectively assessed in drug/gene delivery, tissue engineering, bioimaging, biosensing, and magnetic resonance imaging areas. The effects of graphene materials exposure in biotechnological/biomedical applications have been summarized together with their effects on human health and the environment. The present review article is envisaged to contribute toward existing knowledge and to create new ideas to enhance the development of secure and effective biotechnological/biomedical devices based on graphene materials [195]. Results regarding graphene and its polymer and metal oxide derivatives for biomedical and biotechnological applications are reviewed here in view of the amazing features of graphene-based materials. In addition, a number of bio-applications are pointed out with respect to tissue engineering, biosensing, bioimaging, drug delivery, and gene delivery (Figure 14.6).

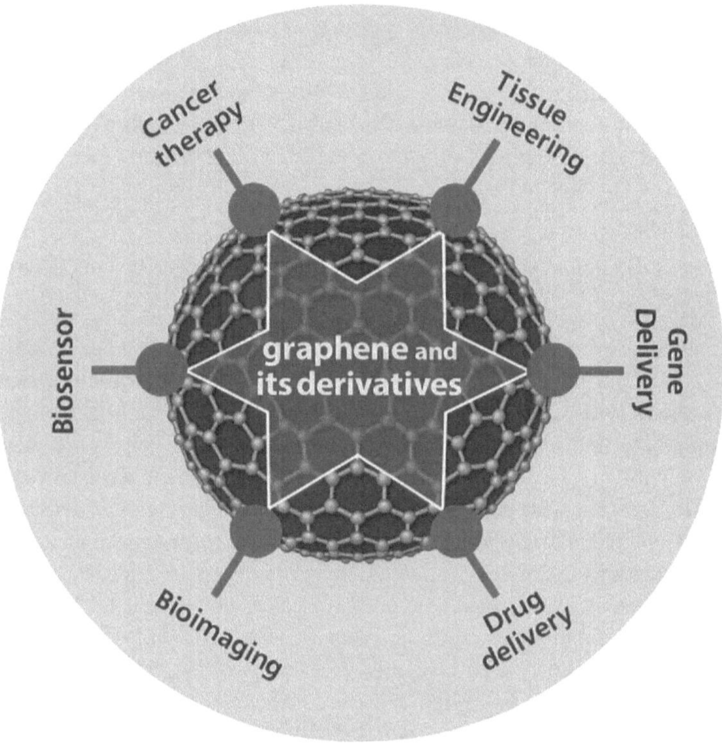

FIGURE 14.6 Schematic diagram of graphene-based materials and applications.

14.7 CONCLUSION

Functionally graded materials (FGMs) have garnered significant interest in medical applications, including implants, drug delivery systems, and tissue engineering. These materials, with their progressive variation in composition and structure, offer tailored properties that closely mimic the gradients found in natural biological tissues. FGMs enhance biocompatibility, reduce stress shielding, and improve the integration between implants and surrounding tissues. They also offer controlled release profiles in drug delivery, leading to increased therapeutic efficacy and reduced side effects. In tissue engineering, FGMs facilitate the development of scaffolds that support cell growth and differentiation, closely resembling natural tissue environments. Their versatility and adaptability address many contemporary challenges in biomedical technology, holding great promise for better patient outcomes. In the context of physiotherapy and rehabilitation, FGMs play a crucial role in shortening recovery periods following surgery, particularly for musculoskeletal conditions. Traditional implants often pose limitations that can prolong recovery or hinder complete rehabilitation. This has spurred research into graphene-based materials for implant development. Graphene enhances the durability and longevity of implants by improving their production process. Coating layers applied to implants are critical for enhancing both functionality and resilience. Additionally, techniques like galvanoplasty and galvanostrategy further strengthen implant surfaces, contributing to more effective and lasting recovery. This approach not only reduces recovery time but also optimizes surgical outcomes, representing a significant advancement in postoperative care.

REFERENCES

[1] Z. Saleem, B. Godman, M.A. Hassali, F.K. Hashmi, F. Azhar, I.U. Rehman, Point prevalence surveys of health-care-associated infections: A systematic review, *Pathog. Glob. Health* 113 (4) (2019) 191–205.

[2] J. Sahoo, S. Sarkhel, N. Mukherjee, A. Jaiswal, Nanomaterial-based antimicrobial coating for biomedical implants: New age solution for biofilm-associated infections, *ACS Omega* 7 (50) (2022) 45962–45980.

[3] A. Trampuz, W. Zimmerli, Diagnosis and treatment of infections associated with fracture-fixation devices, *Injury* 37 (2) (2006) S59–S66.

[4] K.M.D. Merollini, R.W. Crawford, N. Graves, Surgical treatment approaches and reimbursement costs of surgical site infections post hip arthroplasty in Australia: A retrospective analysis, *BMC Health Serv. Res.* 13 (1) (2013) 91.

[5] R. Bright, A. Hayles, J. Wood, N. Ninan, D. Palms, R.M. Visalakshan, A. Burzava, T. Brown, D. Barker, K. Vasilev, Bio-inspired nanostructured Ti-6Al-4V alloy: The role of two alkaline etchants and the hydrothermal processing duration on antibacterial activity, *Nanomaterials* 12 (7) (2022) 1140.

[6] A.F.P. Carreiro, J.A. Delben, S. Guedes, E.J.D. Silveira, M.N. Janal, C.E. Vergani, S. Pushalkar, S. Duarte, Low-temperature plasma on peri-implant–related biofilm and gingival tissue, *J. Periodontol.* 90 (5) (2019) 507–515 1970.

[7] T.-F.C. Mah, G.A. O'Toole, Mechanisms of biofilm resistance to antimicrobial agents, *Trends Microbiol.* 9 (1) (2001) 34–39.

[8] A.G. Emelianova, N.V. Petrova, C. Fremez, M. Fontanié, S.A. Tarasov, O.I. Epstein, Therapeutic potential of highly diluted antibodies in antibiotic-resistant infection, *Eur. J. Pharm. Sci.* 173 (2022) 106161.

[9] A. Hayles, J. Hasan, R. Bright, D. Palms, T. Brown, D. Barker, K. Vasilev, Hydrothermally etched titanium: A review on a promising mechano-bactericidal surface for implant applications, *Mater. Today Chem.* 22 (2021) 100622.

[10] S. Beyranvand, Z. Pourghobadi, S. Sattari, K. Soleymani, I. Donskyi, M. Gharabaghi, W.E.S. Unger, G. Farjanikish, H. Nayebzadeh, M. Adeli, Boronic acid functionalized graphene platforms for diabetic wound healing, *Carbon N. Y.* 158 (2020) 327–336.

[11] M. Shahnawaz Khan, H.N. Abdelhamid, H F Wu, Near infrared (NIR) laser mediated surface activation of graphene oxide nanoflakes for efficient antibacterial, antifungal and wound healing treatment, *Colloids Surf. B* 127 (2015) 281–291.

[12] S. Pandit, Z. Cao, V.R.S.S. Mokkapati, E. Celauro, A. Yurgens, M. Lovmar, F. Westerlund, J. Sun, I. Mijakovic, Vertically aligned graphene coating is bactericidal and prevents the formation of bacterial biofilms, *Adv. Mater. Interfaces* 5 (7) (2018) 1701331.

[13] S.Y. Bhong, N. More, M. Choppadandi, G. Kapusetti, Review on carbon nanomaterials as typical candidates for orthopaedic coatings, *SN Appl. Sci.* 1 (1) (2018) 76.

[14] A. Mazinani, M.J. Nine, R. Chiesa, G. Candiani, P. Tarsini, T.T. Tung, D. Losic, Graphene oxide (GO) decorated on multi-structured porous titania fabricated by plasma electrolytic oxidation (PEO) for enhanced antibacterial performance, *Mater. Des.* 200 (2021) 109443.

[15] N. Safari, N. Golafshan, M. Kharaziha, M.R. Toroghinejad, L. Utomo, J. Malda, M. Castilho, Stable and antibacterial magnesium–graphene nanocomposite-based implants for bone repair, *ACS Biomater. Sci. Eng.* 6 (11) (2020) 6253–6262.

[16] J. Liu, L. Cui, D. Losic, Graphene and graphene oxide as new nanocarriers for drug delivery applications, *Acta Biomater.* 9 (12) (2013) 9243–9257, https://doi.org/10.1016/j.actbio.2013.08.016

[17] Z. Li, J. Fan, C. Tong, et al., A smart drug-delivery nanosystem based on carboxylated graphene quantum dots for tumor-targeted chemotherapy, *Nanomedicine* 14 (15) (2019) 2011–2025, https://doi.org/10.2217/nnm-2018-0378

[18] K.S. Novoselov, A. Geim, The rise of graphene, *Nat. Mater.* 6 (3) (2007) 183–191, https://doi.org/10.1038/nmat1849

[19] A.M. Pinto, I.C. Goncalves, F.D Magalhães, Graphene-based materials biocompatibility: A review, *Colloids Surf. B Biointerfaces* 111 (2013) 188–202, https://doi.org/10.1016/j.colsurfb.2013.05.022

[20] P.S. Zangabad, S. Mirkiani, S. Shahsavari, et al., Stimulus-responsive liposomes as smart nanoplatforms for drug delivery applications, *Nanotechnol. Rev.* 7 (1) (2018) 95–122.

[21] F. Farjadian, A. Ghasemi, O. Gohari, A. Roointan, M. Karimi, M.R. Hamblin, Nanopharmaceuticals and nanomedicines currently on the market: Challenges and opportunities, *Nanomedicine* 14 (1) (2019) 93–126, https://doi.org/10.2217/nnm-2018-0120

[22] F. Farjadian, A. Roointan, S. Mohammadi-Samani, M. Hosseini, Mesoporous silica nanoparticles: Synthesis, pharmaceutical applications, biodistribution, and biosafety assessment, *Chem. Eng. J.* 359 (2019) 684–705, https://doi.org/10.1016/j.cej.2018.11.156

[23] E. Entezar-Almahdi, S. Mohammadi-Samani, L. Tayebi, F. Farjadian, Recent advances in designing 5-fluorouracil delivery systems: A stepping stone in the safe treatment of colorectal cancer, *Int. J. Nanomed.* 15 (2020) 5445, https://doi.org/10.2147/IJN.S257700

[24] M. Karimi, A. Ghasemi, S. Mirkiani, S.M.M. Basri, M.R. Hamblin, *Carbon Nanotubes in Drug and Gene Delivery*. San Rafael, CA: Morgan & Claypool Publishers; 2017.

[25] C. Qi, et al., A sericin/graphene oxide composite scaffold as a biomimetic extracellular matrix for structural and functional repair of calvarial bone, *Teranostics* 10 (2020) 741–756.

[26] M. Dang, L. Saunders, X. Niu, Y. Fan, P.X. Ma, Biomimetic delivery of signals for bone tissue engineering, *Bone Res.* 6 (1) (2018).

[27] E. Askari, et al., Reduced graphene oxide-grafted bovine serum albumin/bredigite nanocomposites with high mechanical properties and excellent osteogenic bioactivity for bone tissue engineering, *Bio-Design Manuf.* (2021).https://doi.org/10.1007/s42242-020-00113-4

[28] E. Askari, et al., A hybrid approach for in-situ synthesis of bioceramic nanocomposites to adjust the physicochemical and biological characteristics, *J. Mater. Res. Technol.* 14 (2021) 464–474.

[29] E. Askari, S.M. Naghib, A. Seyfoori, A. Maleki, M. Rahmanian, Ultrasonic-assisted synthesis and in vitro biological assessments of a novel herceptin-stabilized graphene using three dimensional cell spheroid, *Ultrason. Sonochem.* 58 (2019) 104615.

[30] E. Askari, S.M. Naghib, A novel approach to facile synthesis and biosensing of the protein-regulated graphene, *Int. J. Electrochem. Sci.* 13 (2018) 886–897.

[31] J. Ruan, et al., Enhanced physiochemical and mechanical performance of chitosan-grafted graphene oxide for superior osteoinductivity, *Adv. Funct. Mater.* 26 (2016) 1085–1097.

[32] S. Kumar, S. Raj, K. Sarkar, K. Chatterjee, Engineering a multi-biofunctional composite using poly(ethylenimine) decorated graphene oxide for bone tissue regeneration, *Nanoscale* 8 (2016) 6820–6836.

[33] C.S.D. Cabral, S.P. Miguel, D. de Melo-Diogo, R.O. Louro, I.J Correia, In situ green reduced graphene oxide functionalized 3D printed scaffolds for bone tissue regeneration, *Carbon N. Y.* 146 (2019) 513–523.

[34] Y. Chen, et al., Developing a strontium-releasing graphene oxide-/collagen-based organic-inorganic nanobiocomposite for large bone defect regeneration via MAPK signaling pathway, *ACS Appl. Mater. Interfaces* 11 (2019) 15986–15997.

[35] L. Maïmoun, et al., Strontium ranelate improves implant osseointegration, *Bone* (2010), https://doi.org/10.1016/j.bone.2010.01.379

[36] F.C. Yogui, et al., Strontium ranelate promotes increased peri-implant bone formation in ovariectomized rats, *Res. Soc. Dev.* (2020), https://doi.org/10.33448/rsd-v9i10.9092

[37] U. Patel, et al., In vitro cellular testing of strontium/calcium substituted phosphate glass discs and microspheres shows potential for bone regeneration, *J. Tissue Eng. Regen. Med.* (2019), https://doi.org/10.1002/term.2796

[38] J.W. Park, Y.J. Kim, J.H. Jang, Enhanced osteoblast response to hydrophilic strontium and/or phosphate ions-incorporated titanium oxide surfaces, *Clin. Oral Implants Res.* (2010), https://doi.org/10.1111/j.1600-0501.2009.01863.x

[39] F. Qi, et al., A co-dispersed nanosystem of strontium-anchored reduced graphene oxide to enhance the bioactivity and mechanical property of polymer scaffolds, *Mater. Chem. Front.* (2021), https://doi.org/10.1039/d0qm00958j

[40] S. Kumar, K. Chatterjee, Strontium eluting graphene hybrid nanoparticles augment osteogenesis in a 3D tissue scaffold, *Nanoscale* 7 (2015) 2023–2033.

[41] Y. Lei, Z. Tang, R. Liao, B. Guo, Hydrolysable tannin as environmentally friendly reducer and stabilizer for graphene oxide, *Green Chem.* (2011), https://doi.org/10.1039/c1gc15081b

[42] J. Zhang, et al., Reduction of graphene oxide via L-ascorbic acid, *Chem. Commun. (Camb)* 46 (2010) 1112–1114.
[43] S. Khoee, A. Sadeghi, An NIR-triggered drug release and highly efficient photodynamic therapy from PCL/PNIPAm/porphyrin modified graphene oxide nanoparticles with the Janus morphology, *RSC Adv.* (2019), https://doi.org/10.1039/c9ra06058h
[44] J.W. McDonald, C. Sadowsky, Spinal-cord injury, *Lancet* 359 (2002) 417–425.
[45] G. Courtine, M.V. Sofroniew, Spinal cord repair: Advances in biology and technology, *Nat. Med.* 25 (2019) 898–908.
[46] M.G. Fehlings, et al., A clinical practice guideline for the management of acute spinal cord injury: Introduction, rationale, and scope, *Glob. Spine J.* 7 (2017) 84s–94s.
[47] J.P. Li, L.M. He, W.T Wu, Pathological changes and repair strategies for spinal cord injury (in Chinese), *Sci. Sin. Vitae* 51 (2021).
[48] A. Ulndreaj, J.C. Chio, C.S. Ahuja, M.G. Fehlings, Modulating the immune response in spinal cord injury, *Expert Rev. Neurother.* 16 (2016) 1127–1129.
[49] C.S. Ahuja, et al., Traumatic spinal cord injury-repair and regeneration, *Neurosurgery* 80 (2017) S9–s22.
[50] M.A. Anwar, T.S. Al Shehabi, A.H. Eid, Inflammogenesis of secondary spinal cord injury, *Front. Cell. Neurosci.* 10 (2016) 98.
[51] S. Xu, Y. Zhang, L. Jia, et al., Soft microfluidic assemblies of sensors, circuits, and radios for the skin, *Science* 344 (2014) 70–74.
[52] S. Gong, W. Schwalb, Y. Wang, et al., A wearable and highly sensitive pressure sensor with ultrathin gold nanowires, *Nat. Commun.* 5 (2014) 3132.
[53] F. Ilievski, A.D. Mazzeo, R.F. Shepherd, et al., Soft robotics for chemists, *Angew. Chem. Int. Ed.* 50 (2011) 1890–1895.
[54] R.F. Shepherd, F. Ilievski, W. Choi, et al., Multigait soft robot, *Proc. Natl. Acad. Sci. U.S.A.* 108 (2011) 20400–20403.
[55] S.A. Morin, R.F. Shepherd, S.W. Kwok, et al., Camouflage and display for soft machines, *Science* 337 (2012) 828–832.
[56] J. Park, Y. Lee, J. Hong, et al., Tactile-direction-sensitive and stretchable electronic skins based on human-skin-inspired interlocked microstructures, *ACS Nano* (2014)
[57] J. Kim, M. Lee, H.J. Shim, et al., Stretchable silicon nanoribbon electronics for skin prosthesis, *Nat. Commun.* 5 (2014) 5747.
[58] M. Vosgueritchian, J.B.H. Tok, Z. Bao, Stretchable LEDs: Light-emitting electronic skin, *Nat. Photonics* 7 (2013) 769–771.
[59] I.E. Araci, B. Su, S.R. Quake, et al., An implantable microfluidic device for self-monitoring of intraocular pressure, *Nat. Med.* 20 (2014) 1074–1078.
[60] M. Kaltenbrunner, T. Sekitani, J. Reeder, et al., An ultra-lightweight design for imperceptible plastic electronics, *Nature* 499 (2013) 458–463.
[61] J.A. Rogers, T. Someya, Y. Huang, Materials and mechanics for stretchable electronics, *Science* 327 (2010) 1603–1607.
[62] B.Y. Ahn, E.B. Duoss, M.J. Motala, et al., Omnidirectional printing of flexible, stretchable, and spanning silver microelectrodes, *Science* 323 (2009) 1590–1593.
[63] T. Sekitani, Y. Noguchi, K. Hata, et al., A rubberlike stretchable active matrix using elastic conductors, *Science* 321 (2008) 1468–1472.
[64] Y. Kim, J. Zhu, B. Yeom, et al., Stretchable nanoparticle conductors with self-organized conductive pathways, *Nature* 500 (2013) 59–63.
[65] R. Etzioni, et al., The case for early detection, *Nat. Rev. Cancer* 3 (2003) 243–252.
[66] Y. Zhou, et al., Diagnosis of cancer as an emergency: A critical review of current evidence, *Nat. Rev. Clin. Oncol.* 14 ((2017)) 45–56.
[67] J.D. Cohen, et al., Detection and localization of surgically resectable cancers with a multi-analyte blood test, *Science* 359 (2018) 926–930.
[68] M.D. Kalinich, A. Haber, Cancer detection: Seeking signals in blood, *Science* 359 (2018) 866–867.
[69] S. Sengupta, R. Sasisekharan, Exploiting nanotechnology to target cancer, *Br. J. Cancer* 96 (2007) 1315–1319.

[70] V.S. Madamsetty, M.K. Paul, A. Mukherjee, S. Mukherjee, Functionalization of nanomaterials and their application in melanoma cancer theranostics, *ACS Biomater. Sci. Eng.* 61 (2020) 167–181.
[71] W. Song, A.C. Anselmo, L. Huang, Nanotechnology intervention of the microbiome for cancer therapy, *Nat. Nanotechnol.* 14 (2019) 1093–1103.
[72] J. Chen, et al., Advances in nanomaterials for photodynamic therapy applications: Status and challenges, *Biomaterials* 237 (2020) 119827.
[73] M. Kastellorizios, N. Tipnis, D.J. Burgess, Foreign body reaction to subcutaneous implants, *Immune Responses to Biosurfaces*, 2015, pp. 93–108.
[74] A. Carnicer-Lombarte, S.-T. Chen, G.G. Malliaras, D.G. Barone, Foreign body reaction to implanted biomaterials and its impact in nerve neuroprosthetics, *Front. Bioeng. Biotechnol.* (2021) 271.
[75] A. Boddupalli, L. Zhu, K.M. Bratlie, Methods for implant acceptance and wound healing: Material selection and implant location modulate macrophage and fibroblast phenotypes, *Adv. Health. Mater.* 5 (20) (2016) 2575–2594.
[76] J.M. Anderson, A. Rodriguez, D.T. Chang, Foreign body reaction to biomaterials, *Semin. Immunol.* 20 (2) (2008) 86–100.
[77] P. Pakshir, B. Hinz, The big five in fibrosis: Macrophages, myofibroblasts, matrix, mechanics, and miscommunication, *Matrix Biol.* 68–69 (2018) 81–93.
[78] C.E. Witherel, D. Abebayehu, T.H. Barker, K.L. Spiller, Macrophage and fibroblast interactions in biomaterial-mediated fibrosis, *Adv. Health. Mater.* 8 (4) (2019) 1801451.
[79] S. Mukherjee, S. Darzi, K. Paul, J.A. Werkmeister, C.E. Gargett, Mesenchymal stem cell-based bioengineered constructs: Foreign body response, cross-talk with macrophages and impact of biomaterial design strategies for pelvic floor disorders, *Interface Focus* 9 (4) (2019) 15.
[80] B.N. Brown, S.F. Badylak, Expanded applications, shifting paradigms and an improved understanding of host-biomaterial interactions, *Acta Biomater.* 9 (2) (2013) 4948–4955.
[81] C. Li, C. Guo, V. Fitzpatrick, A. Ibrahim, M.J. Zwierstra, P. Hanna, A. Lechtig, A. Nazarian, S.J. Lin, D.L. Kaplan, Design of biodegradable, implantable devices towards clinical translation, *Nat. Rev. Mater.* 5 (1) (2019) 61–81
[82] F. Witte, The history of biodegradable magnesium implants: a review, *Acta Biomater.* 6 (5) (2011) 1680–92.
[83] J.Y. Wang, et al., Titanium, chromium and cobalt ions modulate the release of bone-associated cytokines by human monocytes/macrophages in vitro, *Biomaterials* 17 (23) (1996) 2233–40.
[84] P. Erne, M. Schier, T.J. Resink, The road to bioabsorbable stents: reaching clinical reality?, *Cardiovasc. Intervent. Radiol.* 29 (1) (2006) 11–6. DOI: 10.1007/s00270-004-0341-9
[85] B. Zberg, P.J. Uggowitzer, J.F. Loffler, MgZnCa glasses without clinically observable hydrogen evolution for biodegradable implants, *Nat. Mater.* 8 (11) (2009) 887–91.
[86] R. Kumar, P. Katyal, Bioresorbable magnesium-based alloys containing strontium doped hydroxyapatite: A review, *Mater. Today: Proc.* 56 (2022) 2443–50.
[87] D. Bairagi, S. Mandal, A comprehensive review on biocompatible Mg-based alloys as temporary orthopaedic implants: Current status, challenges, and future prospects, *J. Magnes. Alloys* 10 (3) (2022) 627–69.
[88] S. Kumar, et al., Assessment of factors influencing bio-corrosion of magnesium based alloy implants: a review, *Mater. Today: Proc.* 56 (2022) 2680–89.
[89] M.P. Staiger, et al., Magnesium and its alloys as orthopedic biomaterials: a review, *Biomaterials* 27 (9) (2006) 1728–34.
[90] J. Walker, et al., Magnesium-based temporary implants: potential, current status, and future prospects, *J. Biomed. Mater. Res. Part B: Appl. Biomater.* 102 (6) (2014) 1316–31.
[91] M. Kasaeian-Naeini, M. Sedighi, R. Hashemi, In vitro biocorrosion of Ti–6Al–4V implant alloy by a mouse macrophage cell line, *J. Magnes. Alloys* 10 (4) (2022) 938–55.
[92] N.T. Kirkland, N. Birbilis, M.P. Staiger, Magnesium and its alloys as orthopedic biomaterials: a review, *Acta Biomater.* 8 (3) (2012) 925–36.
[93] Y. Xin, T. Hu, P.K. Chu, Magnesium and its alloys as orthopedic biomaterials: a review, *Acta Biomater.* 7 (4) (2011) 1452–59.
[94] N. Ivanichok, P. Kolkovskyi, O. Ivanichok, V. Kotsyubynsky, V. Boychuk, B. Rachiy, M. Bembenek, Ł. Warguła, R. Abaszade, L. Ropyak, Effect of thermal activation on the structure and electrochemical

properties of carbon material obtained from walnut shells, *Materials* 17 (2024) 2514, https://doi.org/10.3390/ma17112514
[95] R.G. Abaszade, Synthesis and analysis of flakes graphene oxide, *J. Optoelectron. Biomed. Mater.* 14 (3) (2022) 107–114, https://doi.org/10.15251/JOBM.2022.143.107
[96] S.R.Figarova, E.M.Aliyev, R.G.Abaszade, V.R.Figarov, Negative thermal expansion of sulphur-doped graphene oxide, *Adv. Mater. Res.* 1175 (2023) 55–62, https://doi.org/10.4028/p-rppn12
[97] N.A. Guliyeva, R.G. Abaszade, E.A. Khanmammadova, E.M. Azizov, Synthesis and analysis of nanostructured graphene oxide, *J. Optoelectron. Biomed. Mater.* 15 (1) (January–March 2023) 23–30
[98] E.M. Aliyev, M.M. Khan, A.M. Nabiyev, R.M. Alosmanov, I.A. Bunyad-zadeh, S. Shishatskiy, V. Filiz, Covalently modified graphene oxide and polymer of intrinsic microporosity (PIM-1) in mixed matrix thin-film composite membranes, *Nanoscale Res. Lett.* 13 (359) (2018), https://doi.org/10.1186/s11671-018-2771-3
[99] S. Figarova, E. Aliyev, R. Abaszade, V. Figarov, Negative thermal expansion of sulphur-doped graphene oxide, *Adv. Mater. Res.* 1175 (2023) 55–62.
[100] R.G. Abaszade, A.G. Mammadov, V.O. Kotsyubynsky, E.Y. Gur, I.Y. Bayramov, E.A. Khanmamadova, O.A. Kapush, Modeling of voltage-ampere characteristic structures on the basis of graphene oxide/sulfur compounds, *Tech. Phys. Prob. Eng.* 14 (2) (2022) 302–306.
[101] R.G. Abaszade, A.G. Mammadova, E.A. Khanmammadova, İ.Y. Bayramov, R.A. Namazov, K.M. Popal, S.Z. Melikova, R.C. Qasımov, M.A. Bayramov, N.İ. Babayeva, Electron paramagnetic resonance study of gadoliniumum doped graphene oxide, *J. Ovonic Res.* 19 (2) (2023) 259–263.
[102] B. Sharma, A. Singh, A. Sharma, A. Dubey, V. Gupta, R.G. Abaszade, A.K. Sundramoorthy, N. Sharma, S. Arya, Synthesis and characterization of zinc selenide/graphene oxide (ZnSe/GO) nanocomposites for electrochemical detection of cadmium ions, *Appl. Phys. A* 130 (297) (2024), https://doi.org/10.1007/s00339-024-07472-0
[103] V.M. Boychuk, V.O. Kotsyubynsky, K.V. Bandura, I.P. Yaremiy, S.V. Fedorchenko, Reduced graphene oxide obtained by hummers and marcano-tour methods: Comparison of electrical properties, *J. Nanosci. Nanotechnol.* 19 (11) (2019) 7320–7329, https://doi.org/10.1166/jnn.2019.16712
[104] V. Boychuk, R. Zapukhlyak, M. Hodlevsky, V. Kotsyubynsky, L. Turovska, Surfactant-assisted hydrothermal synthesis of $NiFe_2O_4$/reduced graphene oxide composites, *Mater. Today: Proc.* 62 (2022) 5705–5711.
[105] V.M. Boychuk, R. Zapukhlyak, R. Abaszade, V. Kotsyubynsky, M. Hodlevsky, B. Rachiy, S. Fedorchenko, Solution combustion synthesized $NiFe_2O_4$/reduced graphene oxide composite nanomaterials: Morphology and electrical conductivity, *Phys. Chem. Solid State* 23 (4) (2022) 815–824, https://doi.org/10.15330/pcss.23.4.815-824
[106] P.H. Phong, H.-S. Chuang, D.T. Thuong, N.N. Sang, N.T.H. Lien, N.T. Nghia, N.D. Toan, L.M Thanh, Graphene oxide-decorated hydrogel inverse opal photonic crystal improving colorimetric and fluorescent responses for rapid detection of lipocalin-1, *Photonics Nanostruct. – Fundam. Appl.* 58 (2024) 101237, https://doi.org/10.1016/j.photonics.2024.101237
[107] S.A. Baskakov, Y.V. Baskakova, E.N. Kabachkov, M.V. Zhidkov, A.V. Alperovich, S.S. Krasnikova, D.A. Chernyaev, Y.M. Shulga, G.L. Gutsev, Hydrophobization of reduced graphene oxide aerogel using soy wax to improve sorption properties, *Materials* 17 (11) (2024) 2538, https://doi.org/10.3390/ma17112538
[108] H. Liu, J. Li, S. Chen, J. Cao, B. Wei, J. Liu, Y. Zhang, Graphene oxide for nonvolatile memory application by using electrophoretic technique, *Mater. Today Commun.* 25 (2020) 101537, https://doi.org/10.1016/j.mtcomm.2020.101537
[109] H.Y. Jeong, J.Y. Kim, J.W. Kim, J.O. Hwang, J.E. Kim, J.Y. Lee, T.H. Yoon, B.J. Cho, S.O. Kim, R.S. Ruoff, et al., Graphene oxide thin films for flexible nonvolatile memory applications, *Nano Lett.* 10 (2010) 4381–4386.
[110] R. Moradi, N. Pour Khalili, E. Khanmamadova, R. Abaszade, Functionalized Carbon Nanostructures for Flexible Electronics. *Handbook of Functionalized Carbon Nanostructures*. Springer, Cham; 2024, 2581–2614, https://doi.org/10.1007/978-3-031-14955-9_71-1
[111] J. Romo-Ricoa, R. Bright, S.M. Krishna, K. Vasilev, J. Golledg, M.V. Jacob, Antimicrobial graphene-based coatings for biomedical implant applications, *Carbon Trends* 12 (2023) 100282.

[112] M.H. Mobarak, N. Hossain, A. Hossain, J.J. Mim, F. Khan, M.T. Rayhan, M.A. Islam, M.A. Chowdhury, Advances of graphene nanoparticles in dental implant applications – A review, *Appl. Surf. Sci. Adv.* 18 (2023) 100470, https://doi.org/10.1016/j.apsadv.2023.100470

[113] R.K. Roy, K.R. Lee, Biomedical applications of diamond-like carbon coatings: A review, *J. Biomed. Mater. Res. B Appl. Biomater.* 83 (1) (2007) 72–84.

[114] R. Rasouli, A. Barhoum, H., Uludag, A review of nanostructured surfaces and materials for dental implants: Surface coating, patterning and functionalization for improved performance, *Biomater. Sci.* 6 (6) (2018) 1312–1338.

[115] X. Li, X. Liang, Y. Wang, D. Wang, M. Teng, H. Xu, B. Zhao, L. Han, Graphene-based nanomaterials for dental applications: Principles, current advances, and future outlook, *Front. BioengBiotechnol.* 10 (2022) 804201, https://doi.org/10.3389/fbioe.2022.804201. Mar 10PMID: 35360406; PMCID: PMC8961302.

[116] Z. Fan, J. Wang, Z. Wang, H. Ran, Y. Li, L. Niu, S Yang, One-pot synthesis of graphene/hydroxyapatite nanorod composite for tissue engineering, *Carbon* 66 (2014) 407–416.

[117] Z. Fan, J. Wang, Z. Wang, H. Ran, Y. Li, L. Niu, S Yang, One-pot synthesis of graphene/hydroxyapatite nanorod composite for tissue engineering, *Carbon* 66 (2014) 407–416.

[118] C. Chu, J. Deng, Y. Hou, L. Xiang, Y. Wu, Y. Qu, Y. Man, Application of PEG and EGCG modified collagen-base membrane to promote osteoblasts proliferation, *Mater. Sci. Eng.* C 76 (2017) 31–36.

[119] R.G. Bai, K. Muthoosamy, S. Manickam, A. Hilal-Alnaqbi, Graphene-based 3D scaffolds in tissue engineering: Fabrication, applications, and future scope in liver tissue engineering, *Int. J. Nanomed.* 14 (2019) 5753–5783, https://doi.org/10.2147/ijn.s192779

[120] E.A. Abou Neel, W. Chrzanowski, V.M. Salih, H. Kim, J.C Knowles, Tissue engineering in dentistry, *J. Dent.* 42 (8) (2014) 915–928, https://doi.org/10.1016/j.jdent.2014.05.008

[121] S. Dinescu, M. Ionita, A.M. Pandele, B. Galateanu, H. Iovu, A. Ardelean, M. Costache, A. Hermenean, In vitro cytocompatibility evaluation of chitosan/graphene oxide 3D scaffold composites designed for bone tissue engineering, *BioMed. Mater. Eng.* 24 (6) (2014) 2249–2256, https://doi.org/10.3233/bme141037

[122] A.G. Gristina, Biomaterial-centered infection: Microbial adhesion versus tissue integration, *Science* 237 (4822) (1987) 1588–1595, https://doi.org/10.1126/science.3629258

[123] Q. Zhou, P. Yang, X. Li, H. Liu, S. Ge, Bioactivity of periodontal ligament stem cells on sodium titanate coated with graphene oxide, *Sci. Rep.* 6 (1) (2016), https://doi.org/10.1038/srep19343

[124] J. Qiu, H. Geng, D. Wang, Q. Shi, H. Zhu, Y. Qiao, W. Qian, X. Liu, Layer-number dependent antibacterial and osteogenic behaviors of graphene oxide electrophoretic deposited on titanium, *ACS Appl. Mater. Interfaces* 9 (14) (2017) 12253–12263, https://doi.org/10.1021/acsami.7b00314

[125] J. Xia, N. Zhang, S. Chong, D. Li, Y. Chen, C. Sun, Three-dimensional porous graphene-like sheets synthesized from biocarbon via low-temperature graphitization for a supercapacitor, *Green Chem.* 20 (3) (2018) 694–700, https://doi.org/10.1039/c7gc03426a

[126] H. Xie, T. Cao, A. Franco-Obregon, V. Rosa, Graphene-induced osteogenic differentiation is mediated by the Integrin/FAK axis, *Int. J. Mol. Sci.* 20 (3) (2019) 574, https://doi.org/10.3390/ijms20030574

[127] C.-Y. Chen, P.-H. Tsai, Y.-H. Lin, C.-Y. Huang, J.H.Y. Chung, G.-Y. Chen, Controllable graphene oxide-based biocompatible hybrid interface as an anti-fibrotic coating for metallic implants, *Mater. Today Bio* 15 (2022) 100326, https://doi.org/10.1016/j.mtbio.2022.100326

[128] L. Choudharya, P. Chakraborty Banerjee, R.K. Singh Ramana, D.E. Loboa, C.D. Eastond, M. Majumder, F. Wittee, J.F. Löfflerf, Graphene–calcium carbonate coating to improve the degradation resistance and mechanical integrity of a biodegradable implant, *J. Magnesium Alloys* 12 (2024) 394–404, https://doi.org/10.1016/j.jma.2023.12.012

[129] C. Montoya, Y. Du, A.L. Gianforcaro, S. Orrego, M. Yang, P.I. Lelkes, On the road to smart biomaterials for bone research: Definitions, concepts, advances, and outlook, *Bone Res.* 9 (12) 2021, https://doi.org/10.1038/s41413-020-00131-z

[130] J. Lee, H. Byun, S.K. Madhurakkat Perikamana, S. Lee, H. Shin, Current advances in immunomodulatory biomaterials for bone regeneration, *Adv. Health. Mater.* 8 (2019) e1801106.

[131] S. Freedman, K. Mullane, The academic–industrial complex: Navigating the translational and cultural divide, *Drug Discov. Today* 22 (2017) 976–993.

[132] G.L. Koons, M. Diba, A.G. Mikos, Materials design for bone-tissue engineering, *Nat. Rev. Mater.* 5 (2020) 584–603.

[133] W.R. Walsh, et al., Does implantation site influence bone ingrowth into 3D-printed porous implants? *Spine J.* 19 (2019) 1885–1898.

[134] H. Akbari, E. Askari, S.M. Naghib, Z. Salehi, Bovine serum albumin-functionalized graphene-decorated strontium as a potent complex nanoparticle for bone tissue engineering, *Sci. Rep.* 12 (2022) 12336, https://doi.org/10.1038/s41598-022-16568-7

[135] Y. Lu, W. Zhang, J. Wang, G. Yang, S. Yin, T. Tang, C. Yu, X. Jiang, Recent advances in cell sheet technology for bone and cartilage regeneration: From preparation to application, *Int. J. Oral Sci.* 11 (17) (2019), https://doi.org/10.1038/s41368-019-0050-5

[136] M. Akahane, et al., Osteogenic matrix sheet-cell transplantation using osteoblastic cell sheet resulted in bone formation without scaffold at an ectopic site, *J. Tissue Eng. Regen. Med.* 2 (2008) 196–201.

[137] A. Hermenean, A. Codreanu, H. Herman, C. Balta, M. Rosu, C.V. Mihali, A. Ivan, S. Dinescu, M. Ionita, M. Costache, Chitosan-graphene oxide 3D scaffolds as promising tools for bone regeneration in critical-size mouse calvarial defects, *Sci. Rep.* 7 16641, https://doi.org/10.1038/s41598-017-16599-5

[138] K. Chen, W. Yu, G. Zheng, Z. Xu, C. Yang, Y. Wang, Z. Yue, W. Yuan, B. Hu, H. Chen, Biomaterial-based regenerative therapeutic strategies for spinal cord injury, *NPG Asia Mater.* 16 (5) (2024).

[139] T.R. Ham, N.D. Leipzig, Biomaterial strategies for limiting the impact of secondary events following spinal cord injury, *Biomed. Mater.* 13 (2018) 024105.

[140] A. Al Mamun, et al., Advances in immunotherapy for the treatment of spinal cord injury, *Immunobiology* 226 (2021) 152033.

[141] C. Martinelli, C. Pucci, M. Battaglini, A. Marino, G. Ciofani, Antioxidants and nanotechnology: Promises and limits of potentially disruptive approaches in the treatment of central nervous system diseases, *Adv. Health. Mater.* 9 (2020) e1901589.

[142] Y. Cho, R. Shi, A. Ivanisevic, R.B. Borgens, Functional silica nanoparticle-mediated neuronal membrane sealing following traumatic spinal cord injury, *J. Neurosci. Res.* 88 (2010) 1433–1444.

[143] C.A. McKay, et al., An injectable, calcium-responsive composite hydrogel for the treatment of acute spinal cord injury, *ACS Appl. Mater. Interfaces* 6 (2014) 1424–1438.

[144] C. Chen, et al., Bioinspired hydrogel electrospun fibers for spinal cord regeneration, *Adv. Funct. Mater.* 29 (2019) 1806899.

[145] Z. Tu, Y. Zhong, H. Hu, D. Shao, R. Haag, M. Schirner, J. Lee, B. Sullenger, K.W. Leong, Design of therapeutic biomaterials to control inflammation, *Materials* 7 (2022) 557.

[146] M. Assali, N. Kittana, S. Alhaj-Qasem, M. Hajjyahya, H. Abu-Rass, W Alshaer, R. Al-Buqain, Noncovalent functionalization of carbon nanotubes as a scaffold for tissue engineering, *Sci. Rep.* 12 (2022) 12062, https://doi.org/10.1038/s41598-022-16247-7

[147] M. Hoseini-Ghahfarokhi, S. Mirkiani, N. Mozaffari, M.A. Abdolahi Sadatlu, A. Ghasemi, S. Abbaspour, M. Akbarian, F. Farjadian, M. Karim, Applications of graphene and graphene oxide in smart drug/gene delivery: Is the world still flat?, *Int. J. Nanomed.* 15 (2020) 9469–9496.

[148] D.W. Pack, A.S. Hoffman, S. Pun, P.S. Stayton, Design and development of polymers for gene delivery, *Nat. Rev. Drug Discov.* 4 (7) (2005) 581, https://doi.org/10.1038/nrd1775

[149] J. Panyam, V. Labhasetwar, Biodegradable nanoparticles for drug and gene delivery to cells and tissue, *Adv. Drug Deliv. Rev.* 55 (3) (2003) 329–347, https://doi.org/10.1016/S0169-409X(02)00228-4

[150] Y. Xiang, N.N.L. Oo, J.P. Lee, Z. Li, X.J. Loh, Recent development of synthetic nonviral systems for sustained gene delivery, *Drug Discov. Today* 22 (9) (2017) 1318–1335, https://doi.org/10.1016/j.drudis.2017. 04.001

[151] P. Midoux, C. Pichon, J.J. Yaouanc, P.A. Jaffrès, Chemical vectors for gene delivery: A current review on polymers, peptides and lipids containing histidine or imidazole as nucleic acids carriers, *Br. J. Pharmacol.* 157 (2) (2009) 166–178, https://doi.org/10.1111/j.1476-5381.2009.00288.x

[152] J. Xiong, Q. Yang, J. Li, S. Zhou, Effects of MDM2 inhibitors on vascular endothelial growth factor-mediated tumor angiogenesis in human breast cancer, *Angiogenesis* 17 (1) (2014) 37–50, https://doi.org/10.1007/s10456-013-9376-3

[153] R. Khurana, M. Simons, J.F. Martin, I.C. Zachary, Role of angiogenesis in cardiovascular disease: A critical appraisal, *Circulation* 112 (12) (2005) 1813–1824, https://doi.org/10.1161/CIRCULATIONAHA.105.535294

[154] P. Carmeliet, R.K. Jain, Angiogenesis in cancer and other diseases, *Nature* 407 (6801) (2000) 249–257, https://doi.org/10.1038/35025220

[155] E. Khakpour, S. Salehi, S.M. Naghib, S. Ghorbanzadeh, W. Zhang, Graphene-based nanomaterials for stimuli-sensitive controlled delivery of therapeutic molecules, *Front. Bioeng. Biotechnol.* 11 (2023) 1129768, https://doi.org/10.3389/fbioe.2023.1129768

[156] S.S. Sekhon, P. Kaur, Y.-H. Kim, S.S. Sekhon, 2D graphene oxide–aptamer conjugate materials for cancer diagnosis, *npj 2D Mater. Appl.* 5 (2021) 21, https://doi.org/10.1038/s41699-021-00202-7

[157] L.C. Bock, L.C. Griffin, J.A. Latham, E.H. Vermaas, J.J. Toole, Selection single-stranded DNA molecules that bind and inhibit human thrombin, *Nature* 355 (1992) 564–566.

[158] S.S. Sekhon, S.H. Lee, K. Lee, J. Min, B.T. Lee, K.W. Kim, J.Y. Ahn, Y.H. Kim, Defining the copper binding aptamotif and aptamer integrated recovery platform (AIRP), *Nanoscale* 9 (2017) 2883–2894.

[159] Y. Ueno, et al., On-chip graphene oxide aptasensor for multiple protein detection, *Anal. Chim. Acta* 866 (2015) 1–9.

[160] K. Xiao, J. Liu, H. Chen, S. Zhang, J. Kong, A label-free and high-efficient GO-based aptasensor for cancer cells based on cyclic enzymatic signal amplification, *Biosens. Bioelectron.* 91 (2017) 76–81.

[161] L. Cao, et al., Visual and high-throughput detection of cancer cells using a graphene oxide-based FRET aptasensing microfluidic chip, *Lab Chip* 12 (2012) 4864–4869.

[162] Z. Shen, A. Wu, X. Chen, Current detection technologies for circulating tumor cells, *Chem. Soc. Rev.* 46 (2017) 2038–2056.

[163] B.P.V. Nellore, et al., Aptamer-conjugated theranostic hybrid graphene oxide with highly selective biosensing and combined therapy capability, *Faraday Discuss.* 175 (2015) 257–271.

[164] A. Bahreyni, R.Y. Robati, S. Hashemitabar, M. Ramezani, P. Ramezani, K. Abnous, S.M. Taghdisi, A new chemotherapy agent-free theranostic system composed of graphene oxide nano-complex and aptamers for treatment of cancer cells, *Int. J. Pharm.* 526 (2017) 391–399.

[165] J. Li, X. Hu, S. Shi, Y. Zhang, T. Yao, Three label-free thrombin aptasensors based on aptamers and $[Ru(bpy)_2(o\text{-mopip})]^2$, *J. Mater. Chem. B* 4 (2016) 1361–1367.

[166] Y. Wang, et al., Aptamer/graphene oxide nanocomplex for in situ molecular probing in living cells, *J. Am. Chem. Soc.* 132 (2010) 9274–9276.

[167] L. Feng, Y. Chen, J. Ren, X. Qu, A graphene functionalized electrochemical aptasensor for selective label-free detection of cancer cells, *Biomaterials* 32 (2011) 2930–2937.

[168] B.R. Smith, E.R. Edelman, Nanomedicines for cardiovascular disease, nature cardiovascular research, 2 (April 2023) 351–36, https://doi.org/10.1038/s44161-023-00232-y

[169] A.P. Singh, A. Biswas, A. Shukla, P. Maiti, Targeted therapy in chronic diseases using nanomaterial-based drug delivery vehicles, *Signal Transduct. Target. Ther.* (2019) 4, 33.

[170] M. Frank-Kamenetsky, et al., Therapeutic RNAi targeting PCSK9 acutely lowers plasma cholesterol in rodents and LDL cholesterol in nonhuman primates, *Proc. Natl. Acad. Sci. U.S.A.* 105 (2008) 11915–11920.

[171] Y. Kanthi, A. de la Zerda, B.R. Smith, Nanotherapeutic shots through the heart of plaque, *ACS Nano* 14 (2020) 1236–1242.

[172] M. Tadin-Strapps, et al., siRNA-induced liver ApoB knockdown lowers serum LDL-cholesterol in a mouse model with human-like serum lipids, *J. Lipid Res.* 52 (2011) 1084–1097.

[173] T.J. Beldman, et al., Nanoparticle-aided characterization of arterial endothelial architecture during atherosclerosis progression and metabolic therapy, *ACS Nano* 13 (2019) 13759–13774.

[174] K. Ashtari, H. Nazari, H. Ko , P. Tebon, M. Akhshik, M. Akbari, S.N. Alhosseini, M. Mozafari, B. Mehravi, M. Soleimani, R. Ardehali, M.E. Warkiani, S. Ahadian, A. Khademhosseini, Electrically conductive nanomaterials for cardiac tissue engineering, *Adv. Drug Deliv. Rev.* 144 (2019) 162–179.

[175] P.M. Ridker, et al., Anti inflammatory therapy with canakinumab for atherosclerotic disease, *N. Engl. J. Med.* 377 (2017) 1119–1131.

[176] R.A. Baylis, D. Gomez, Z. Mallat, G. Pasterkamp, G.K. Owens, The CANTOS trial: One important step for clinical cardiology but a giant leap for vascular biology, *Arterioscler. Thromb. Vasc. Biol.* 37 (2017) e174–e177.

[177] S.C. Jacobson, D.F. Knutti, R.T. Johnson, et al., Development of the Utah artificial arm, *IEEE Trans. Biomed. Eng.* 29 (1982) 249–269.

[178] S. Choi, J. Park, W. Hyun, et al., Stretchable heater using ligand-exchanged silver nanowire nanocomposite for wearable articular thermotherapy, *ACS Nano* 9 (2015) 6626–6633.

[179] I.R. Minev, P. Musienko, A. Hirsch, et al., Electronic dura mater for long-term multimodal neural interfaces, *Science* 347 (2015) 159–163.

[180] J. Di, S. Yao, Y. Ye, et al., Stretch-triggered drug delivery from wearable elastomer films containing therapeutic depots, *ACS Nano* 9 (2015) 9407–9415.

[181] J.Y. Lim, S.Y. Kim, Yield strain behavior of poly(ethylene terephthalate): Correlation with yield stress behavior in strain rate, temperature, and structure dependence, *Polym. J.* 36 (2004) 769–773.

[182] J.H. Waller, L. Lalande, Y. Leterrier, et al., Modelling the effect of temperature on crack onset strain of brittle coatings on polymer substrates, *Thin Solid Films* 519 (2011) 4249–4255.

[183] S. Khan, L. Lorenzelli, R.S. Dahiya, Technologies for printing sensors and electronics over large flexible substrates: A review, *IEEE Sens. J.* 15 (2015) 3164–3185.

[184] R.D. Widdle, Jr, A.K. Bajaj, P. Davies, Measurement of the Poisson's ratio of flexible polyurethane foam and its influence on a uniaxial compression model, *Int. J. Eng. Sci.* 46 (2008) 31–49.

[185] H.-Y. Mi, Z. Li, L.-S. Turng, et al., Silver nanowire/thermoplastic polyurethane elastomer nanocomposites: Thermal, mechanical, and dielectric properties, *Mater. Des.* 56 (2014) 398–404.

[186] C.H. Lee, Y. Ma, K.-I. Jang, et al., Soft core/shell packages for stretchable electronics, *Adv. Func. Mater.* 25 (2015) 3698–3704.

[187] K.-I. Jang, S.Y. Han, S. Xu, et al., Rugged and breathable forms of stretchable electronics with adherent composite substrates for transcutaneous monitoring, *Nat. Commun.* 5 (2014) 4779.

[188] K.-I. Jang, H.U. Chung, S. Xu, et al., Soft network composite materials with deterministic and bio-inspired designs, *Nat. Commun.* 6 (2015) 6566.

[189] D.R. Seshadri, R.T. Li, J.E. Voos, J.R. Rowbottom, C.M. Alfes, C.A. Zorman, C.K. Drummond, Wearable sensors for monitoring the physiological and biochemical profile of the athlete, *npj Digital Med.* 2 (2019) 72, https://doi.org/10.1038/s41746-019-0150-9

[190] A.J. Bandodkar, J. Wang, Non-invasive wearable electrochemical sensors: A review, *Trends Biotechnol.* 32 (2014) 363–371.

[191] T. Hew-Butler, et al., Statement of the third international exercise-associated hyponatremia consensus development conference, Carlsbad, California, 2015, *Clin. J. Sport Med.* 25 (2015) 303.

[192] A. Hauke, et al., Complete validation of a continuous and blood-correlated sweat biosensing device with integrated sweat stimulation, *Lab. Chip* (2018), https://doi.org/10.1039/C8LC01082J

[193] Y. Sekine, S.B. Kim, Y. Zhang, A.J. Bandodkar, S. Xu, J. Choi, M. Irie, T.R. Ray, P. Kohli, N. Kozai, T. Sugita, Y. Wu, K.H. Lee, K.-T. Lee, R. Ghaffari, J.A. Rogers, A fluorometric skin-interfaced microfluidic device and smartphone imaging module for in situ quantitative analysis of sweat chemistry, *Lab. Chip* 18 (2018) 2178–2186.

[194] Kass-Hout, T.A., Stevens, L.M., Hall, J.L. American Heart Association precision medicine platform, *Circulation* 137 (2018) 647–649.

[195] Z. Li, X. Tian, C.-W. Qiu, J.S. Ho, Metasurfaces for bioelectronics and healthcare, *Nat. Electron.* 4 (2021) 382–391, https://doi.org/10.1038/s41928-021-00589-7

[196] R. Kumar, D.P. Singh, R. Munoz, M. Amami, R.K. Singh, S. Singh, V. Kumar, Graphene-based materials for biotechnological and biomedical applications: Drug delivery, bioimaging and biosensing, *Mater. Today Chem.* 33 (2023) 101750.

[197] R.G. Abaszade, M.B. Babanli, V.O. Kotsyubynsky, A.G. Mammadov, E. Gür, O.A. Kapush, M.O. Stetsenko, R.I. Zapukhlyak, Influence of gadolinium doping on structural properties of carbon nanotube, *Phys. Chem. Solid State* 24 (1) (2023) 53–158, https://doi.org/10.15330/pcss.24.1.153-158

[198] A.G. Mammadov, R.G. Abaszade, M.B. Babanli, V.O. Kotsyubynsky, E. Gur, B.D. Soltabayev, T.O. Margitych, M.O. Stetsenko, Photoconductivity of gadolinium-doped carbon nanotubes, *Int. J. Tech. Phys. Prob. Eng.* 15 (3) (2023) 53–58.

[199] A.G. Mammadov, R.G. Abaszade, V.O. Kotsyubynsky, E.Y. Gur, I.Y. Bayramov, E.A. Khanmamadova, O.A. Kapush, Photoconductivity of carbon nanotubes, *Tech. Phys. Prob. Eng.* 14 (3) (2022) 155–160.

[200] R.G. Abaszade, E.M. Aliyev, M.B. Babanli, V.O. Kotsyubynsky, R.I. Zapukhlyak, A.G. Mamedov, H.F. Budak, A.E. Kasapoglu, E. Gur, T.O. Margitych, M.O. Stetsenko, Investigation of thermal properties of carbon nanotubes and carboxyl group – Functionalized carbon nanotubes, *Phys. Chem.Solid State* 24 (3) (2023) 530–535, https://doi.org/10.15330/pcss.24.3.530-535

[201] R.G. Abaszade, A.G. Mammadov, E.A. Khanmamadova, F.G. Aliyev, V.O. Kotsyubynsky, E. Gur, B.D. Soltabayev, T.O. Margitych, M.O. Stetsenko, A. Singh, S. Arya, Photoconductivity of functionalized carbon nanotubes, *Dig. J. Nanomater. Biostruct.* 19 (2) (2024) 837–843.
[202] R.G. Abaszade, O.A. Kapush, S.A. Mamedova, A.M. Nabiyev, S.Z. Melikova, S.I. Budzulyak, Gadolinium doping influence on the properties of carbon nanotubes, *Phys. Chem. Solid State* 21 (3) (2020) 404–408.
[203] R.G. Abaszade, O.A. Kapush, A.M. Nabiyev, Properties of carbon nanotubes doped with gadolinium, *J. Optoelectron. Biomed. Mater.* 12 (3) (2020) 61–65.

Index

L

M

N

O

P

R

S

T

V

W